T0329168

Fundamentals of Continuum Mechanics

Fundamentals of Continuum Mechanics

With Applications to Mechanical, Thermomechanical, and Smart Materials

Stephen E. Bechtel

Robert L. Lowe

AMSTERDAM • BOSTON • HEIDELBERG • LONDON
NEW YORK • OXFORD • PARIS • SAN DIEGO
SAN FRANCISCO • SINGAPORE • SYDNEY • TOKYO
Academic Press is an imprint of Elsevier

Academic Press is an imprint of Elsevier
525 B Street, Suite 1800, San Diego, CA 92101-4495, USA
225 Wyman Street, Waltham, MA 02451, USA
The Boulevard, Langford Lane, Kidlington, Oxford OX5 1GB, UK

Library of Congress Cataloging-in-Publication Data
A catalog record for this book is available from the Library of Congress

British Library Cataloguing-in-Publication Data
A catalogue record for this book is available from the British Library

For information on all Academic Press publications
visit our web site at store.elsevier.com

Printed and bound in the United States
14 15 16 17 10 9 8 7 6 5 4 3 2 1

ISBN: 978-0-12-394600-3

This book is dedicated to my wife Barbara and my father Robert.

Contents

Preface

CONTINUUM MECHANICS: THE NEW PEDAGOGY

Since my days as a graduate student at Berkeley in the early 1980s, the graduate engineering mechanics curriculum has undergone three major changes: First, in previous years, the curriculum left time for courses in linear elasticity, finite elasticity, plasticity, viscoelasticity, inviscid fluid dynamics, and viscous fluid dynamics, followed by a unifying course in continuum mechanics. Today, with the onslaught of new materials and technological advances competing for space in the curriculum, the reality is that there is no longer room for this many traditional mechanics courses. Second, back in the day, much of the teaching and learning was accomplished through homework exercises worked by the students on their own; graded and annotated by the professor; and then returned to the students. Although the most effective, this approach is impractical given today's time constraints on both the student and the professor. As such, there is a risk that important concepts may be overlooked if they are not illuminated through worked examples. Third, mathematics and applied mechanics have diverged, and this gap continues to widen. As Courant and Hilbert [1] mused in their treatise *Methods of Mathematical Physics*, and is still the case today:

> *"Since the seventeenth century, physical intuition has served as a vital source for mathematical problems and methods. Recent trends and fashions have, however, weakened the connection between mathematics and physics; mathematicians, turning away from the roots of mathematics in intuition, have concentrated on refinement and emphasized the postulational side of mathematics, and at times have overlooked the unity of their science with physics and other fields. In many cases, physicists have ceased to appreciate the attitudes of mathematicians. This rift is unquestionably a serious threat to science as a whole; the broad stream of scientific development may split into smaller and smaller rivulets and dry out."*

This textbook adjusts to each of these realities: First, the material is covered in the most time-efficient manner, that is, by first giving the unified situation (continuum mechanics), then applying it to special cases (finite elasticity, viscous fluid dynamics, and so on). Because these special cases are presented in a single textbook, the handoff between one subject and another is cleaner, and undue redundancy is avoided. Second, the majority of the problems in the textbook are presented as worked examples with full, detailed solutions. Each of these problems is designed to convey an important concept. Third, we place a strong emphasis on explicitly connecting the mathematics to the continuum physics. Indicial notation is jettisoned almost entirely in favor of the more compact and elegant direct notation, allowing us to be more fundamental in our treatment and cover much more material in a single book. Furthermore, we restrict our development of the mathematical parlance to only that which is required to rigorously present the physical concepts.

Continuum mechanics is presented here as a unifying course in the sense that it separates those concepts that are true for all materials (i.e., the fundamental laws) from those that vary from material to material (i.e., the constitutive equations). Our textbook is structured as follows: In Chapter 2, we discuss the mathematical world in which continuum mechanics lives and acquaint the reader with direct notation. Once the reader is fluent in direct notation, which is the primary goal of Chapter 2, he or she will be able to think of physical quantities such as velocity and stress as elements of a vector space rather than just in terms of their components. This insight is powerful, allowing the reader to proceed further conceptually than is possible with component notation and enabling a more transparent interplay between the mathematics and the physics. Chapter 3 covers motion and deformation, which is merely a discussion of geometry. (My apologies to those researchers who have devoted their lives to the study of geometry.) Chapter 4 develops the fundamental laws (or first principles), valid for all materials. Chapter 5 introduces the notion of constitutive equations, which describe how different materials respond to loading and deformation.

A distinguishing feature of this book is the postulation of constitutive equations for various materials in their most general form, and their subsequent simplification using restrictions imposed by the second law of thermodynamics, invariance, conservation of angular momentum, and material symmetry. This is illustrated for elastic solids in Chapter 6 and viscous fluids in Chapter 7. Chapter 8 presents both traditional and modern approaches to modeling constraints such as incompressibility and explores the role of stability in constitutive modeling. Finally, Chapter 9 discusses coupled thermo-electro-magneto-mechanical behavior and illustrates the development of field theories for smart materials such as piezoelectrics.

The material in this book is presented in the most powerful, straightforward, and understandable manner that I have arrived at in my 30 years of teaching. One thing I've noticed in my line of work is that clarity is sometimes equated with simplicity, with the attitude: 'If I understand it, it must not be that difficult.' In this case, I'll have to take that reaction in stride.

As I am now retired and no longer check my e-mail on a regular basis, I would be most appreciative if you would bring any comments, criticisms, or errors to our attention via my co-author Robert Lowe (lowe.194@osu.edu).

ACKNOWLEDGMENTS

I would first like to acknowledge the contributions and dedication of my gifted Ph.D. student Robert Lowe, without whom this project would not have been accomplished. I am indebted to the late Prof. Paul Naghdi, whose lectures at Berkeley inspired me and shaped my way of thinking as a continuum mechanician. Prof. Naghdi's teachings were influential as I developed notes for my graduate-level mechanics courses at

Ohio State, which ultimately evolved into this textbook. I am also indebted to Prof. Morton Gurtin's textbook *An Introduction to Continuum Mechanics* for its elegant presentation of tensor algebra in direct notation, which guided the development of Sections 2.2 and 2.3. Thanks are due to my former Ph.D. student and postdoctoral researcher Dr. Sushma Santapuri for her contributions to Chapter 9 and her careful review of several manuscript chapters.

I gratefully acknowledge Prof. David Bogy and Prof. Michael Carroll for guiding and inspiring me during my graduate studies at Berkeley. I thank my friends and colleagues Prof. Greg Forest, Dr. Frank Rooney, Prof. Qi Wang, and Prof. Marcelo Dapino for productive collaborations over the years, whose fruits are reflected most notably in Chapter 8 (Forest, Rooney, and Wang) and Chapter 9 (Dapino). I extend much gratitude to Joe Hayton, Chelsea Johnston, Jason Mitchell, Christina Edwards, and the other publishers, project managers, copy editors, and typesetters at Elsevier who provided excellent support and showed extraordinary patience.

This book has been a battle against the encroachment of Parkinson's disease. I'd like to thank my colleagues for their encouragement during this difficult struggle. Most of all, though, I'd like to thank my family–particularly my wife Barbara, daughters Joanna and Clara, and father Robert–for their unwavering patience and support.

Stephen E. Bechtel
Columbus, Ohio
September 2014

PART I

The Beginning

CHAPTER

What Is a Continuum?

1

Physics may be defined as attempts to gain an understanding of nature. A **model** is proposed for nature that is an idealization of nature: much of nature is left unrepresented, and some features not observed in nature may be added (e.g., the concepts of the infinite and the infinitesimal). Natural phenomena are converted into problems in the context of this model. The solutions of the mathematical problems are compared with experimental observation. The degree of agreement between a particular natural phenomenon and the solution to the corresponding problem is a measure of the relevance of the model to that particular phenomenon. A comparison of a wide range of phenomena with the corresponding problem solutions gives a measure of the relevance of the model in general.

One important task in physics is to work toward the discovery of the basic quantities that constitute all of nature, and the laws that govern their interaction. This motivation has led to the concept of the molecule, followed by the notions of atom, proton, neutron, and electron, and now all of the various types of quarks, together with quantum mechanics to govern their behavior. Such models in physics can be described as **discrete**, since they regard matter as being composed ultimately of atomic and subatomic particles. Discrete models underpin modern molecular dynamics simulation tools, which predict the evolving kinematics and kinetics of individual atoms in a deforming material.

In this book, we adopt a different approach to physics. We assume that matter is a **continuum**, infinitely divisible without modifying any of its properties. Physical quantities are regarded as continuous functions of space and time. Continuum models underpin modern finite element and computational fluid dynamics codes, which are widely used in industry and engineering practice.

The continuum and discrete approaches are eventually contradictory and irreconcilable. For instance, in a continuum model, the smallest portion of a steel specimen is still steel, with all of the properties of the original specimen. In a discrete model, as we narrow our scope of interest to smaller and smaller volumes of the steel specimen, we see that it is composed of several distinct types of atoms, such as iron and carbon, the properties of each alone much different than those of steel. We then say that these atoms consist of widely spaced subatomic particles, and so on.

A physical theory must have relevance to the observations made of natural phenomena. In general, a particular theory is in good agreement with some types of phenomena, but in error for, or even unable to handle, other types of phenomena. Failures of theories obtained via the continuum approach are well known; these failures are what motivated the development of the predecessors of the current discrete models. But there are also many successes: continuum models are quite capable of describing much of what we experience in our world. The continuum approach is therefore assured of worth and permanence.

CHAPTER

Our Mathematical Playground

2

This textbook is primarily a course in physics. The physical notions, however, must be expressed through the language of mathematics. When this mathematical language becomes cumbersome, there is a danger that the mathematics will obscure the physics, and the subject will appear to be mere symbol manipulation. It is therefore desirable to present the physics in the simplest possible mathematics, which is **direct notation.** This direct presentation of the mathematics exists independently of any coordinate system. Once the theory has been developed and presented in direct form, it may be referred to any coordinate system when applied to a particular problem. This chapter will acquaint the student with, or serve as a review of, direct notation.

In this chapter, as well as the remainder of the book, we employ for brevity the following logical notation (beyond the customary operational and ordering symbols $+$, $-, =, <, >, \leq, \geq$): $\forall$ abbreviates "for all" or "for any," $\in$ abbreviates "an element of," $\exists$ abbreviates "there exists," $\ni$ abbreviates "such that," $\subset$ or $\subseteq$ abbreviates "a subset of," $\mathbb{R}$ abbreviates "the set of real numbers," $\cup$ abbreviates "the union of," $\Rightarrow$ abbreviates "implies," $\Leftrightarrow$ abbreviates "if and only if," and $\equiv$ abbreviates "is defined as."

2.1 REAL NUMBERS AND EUCLIDEAN SPACE

The interplay of mathematics and physics in the development of continuum mechanics was as follows[1]: In their observations of the world around them, physically minded scientists encountered two types of quantities. Some quantities, such as temperature, mass, and pressure, were ordered sets (see Figure 2.1). From this concept were constructed the **real numbers.** Other quantities, such as velocity, acceleration, and force, had both a magnitude and a direction, and combined as shown in Figure 2.2. From these observations came a vector space endowed with an inner product called a **Euclidean space.**

2.1.1 PROPERTIES OF REAL NUMBERS

Physical quantities such as temperature, pressure, and mass are described by real numbers. For all scalars α, β, γ that are elements of the set of real numbers $\mathbb{R}$ (or, in simplified notation, $\forall\, \alpha, \beta, \gamma \in \mathbb{R}$), the following properties hold:

[1] This might not be exactly how it happened, but it sure sounds good!

FIGURE 2.1

The real number line, illustrating an ordered set of masses $0 < m_1 < m_2$.

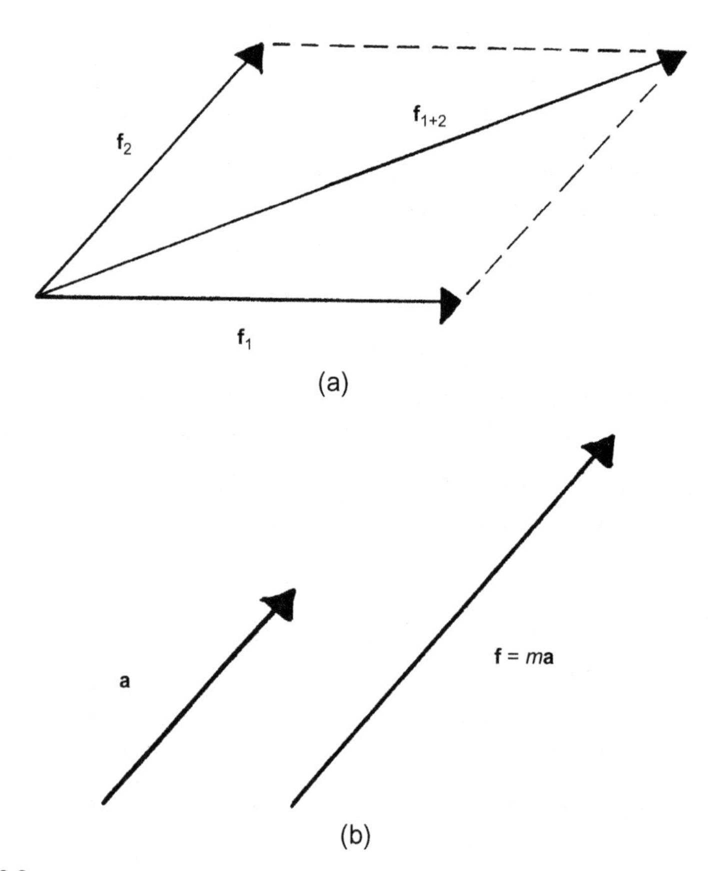

FIGURE 2.2

Physical observations of the interaction between kinematic and kinetic quantities in mechanics. (a) The combination of two forces $\mathbf{f}_1$ and $\mathbf{f}_2$, yielding the resultant force $\mathbf{f}_{1+2}$ (concept of vector addition). (b) The product of mass m with acceleration **a**, yielding force **f** (concept of scalar multiplication of a vector).

closure of addition,	$\alpha + \beta \in \mathbb{R};$
commutativity of addition,	$\alpha + \beta = \beta + \alpha;$
associativity of addition,	$\alpha + (\beta + \gamma) = (\alpha + \beta) + \gamma;$
existence of an additive identity,	$\exists 0 \ni \alpha + 0 = \alpha;$
existence of an additive inverse,	$\exists (-\alpha) \ni \alpha + (-\alpha) = 0;$
closure of multiplication,	$\alpha \beta \in \mathbb{R};$ (2.1)
commutativity of multiplication,	$\alpha \beta = \beta \alpha;$
associativity of multiplication,	$(\alpha \beta) \gamma = \alpha (\beta \gamma);$
existence of a multiplicative identity,	$1 \alpha = \alpha;$
existence of a multiplicative inverse (or reciprocal),	$\exists \frac{1}{\alpha} \ni \alpha \frac{1}{\alpha} = 1, \quad \alpha \neq 0;$
zero product,	$0 \alpha = 0;$
distributivity of multiplication over addition,	$(\alpha + \beta) \gamma = \alpha \gamma + \beta \gamma.$

Real numbers are an ordered set, so any pair of scalars α and β that are elements of the set of real numbers $\mathbb{R}$ satisfy one and only one of

$$\alpha < \beta, \qquad \alpha = \beta, \qquad \alpha > \beta. \tag{2.2}$$

2.1.2 PROPERTIES OF EUCLIDEAN SPACE

In this section, we arrive at Euclidean space by progressing from vector spaces, to metric spaces, to normed spaces, and finally to inner product spaces. The vector space (whose elements are called vectors) postulates the *algebraic* concepts of vector addition, scalar multiplication, and the zero element (or origin) of the space. The metric space (whose elements are called points) postulates *topological* concepts such as the distance between two points. The normed space (a vector space endowed with a norm) postulates the concept of the length of a vector. Finally, the inner product space (a vector space endowed with an inner product) postulates the concept of an angle between two vectors. *Ultimately, we illustrate that every inner product space is also a vector space, a metric space, and a normed space, and is hence endowed with all of their separate properties* (refer to Figure 2.3). An n-dimensional inner product space, where n is a positive integer, is known as a **Euclidean space** $\mathcal{E}^n$.

Vector space X. The elements of vector space X are called **vectors**. For all vectors $\mathbf{u}$, $\mathbf{v}$, $\mathbf{w}$ in vector space X, and for all scalars α, β that are elements of the set of real numbers $\mathbb{R}$ (or, in simplified notation, $\forall\, \mathbf{u}, \mathbf{v}, \mathbf{w} \in X$ and $\forall\, \alpha, \beta \in \mathbb{R}$), the following properties hold:

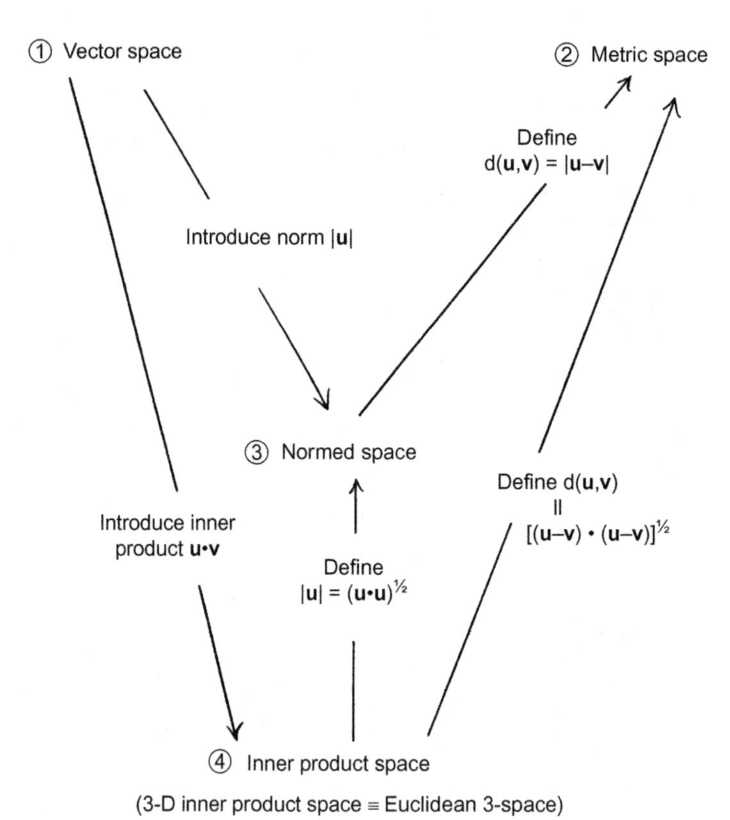

FIGURE 2.3

A schematic illustrating the interplay between the properties of vector, metric, normed, and inner product spaces.

$$
\begin{array}{ll}
\text{closure of vector addition,} & \mathbf{u}+\mathbf{v}\in X; \\
\text{commutativity of vector addition,} & \mathbf{u}+\mathbf{v}=\mathbf{v}+\mathbf{u}; \\
\text{associativity of vector addition,} & \mathbf{u}+(\mathbf{v}+\mathbf{w})=(\mathbf{u}+\mathbf{v})+\mathbf{w}; \\
\text{existence of an additive identity,} & \exists\,\mathbf{0}\ni \mathbf{u}+\mathbf{0}=\mathbf{u}; \\
\text{existence of an additive inverse,} & \exists\,(-\mathbf{u})\ni \mathbf{u}+(-\mathbf{u})=\mathbf{0}; \\
\text{closure of scalar multiplication,} & \alpha\,\mathbf{u}\in X; \\
\text{associativity of scalar multiplication,} & \alpha(\beta\mathbf{u})=(\alpha\beta)\mathbf{u}; \\
\text{existence of a multiplicative identity,} & 1\mathbf{u}=\mathbf{u}; \\
\text{distributivity of scalar multiplication over scalar addition,} & (\alpha+\beta)\mathbf{u}=\alpha\mathbf{u}+\beta\mathbf{u}; \\
\text{distributivity of scalar multiplication over vector addition,} & \alpha(\mathbf{u}+\mathbf{v})=\alpha\mathbf{u}+\alpha\mathbf{v}.
\end{array}
\tag{2.3}
$$

For a vector space we have the *algebraic* concepts of linear combination, independence, dependence, span, linear manifold, basis, and dimension.

Metric space X. The elements of metric space X are called **points**. The real-valued function $d(\mathbf{u}, \mathbf{v})$ is called the **metric** of X; it accepts points $\mathbf{u}$ and $\mathbf{v}$ as inputs, and provides the real-valued **distance** between points $\mathbf{u}$ and $\mathbf{v}$ as output. The metric $d(\mathbf{u}, \mathbf{v})$ is defined such that the following properties hold $\forall\, \mathbf{u}, \mathbf{v}, \mathbf{w} \in X$:

$$\mathbf{u} \neq \mathbf{v} \Rightarrow d(\mathbf{u}, \mathbf{v}) > 0, \quad d(\mathbf{u}, \mathbf{u}) = 0, \quad d(\mathbf{u}, \mathbf{v}) = d(\mathbf{v}, \mathbf{u}), \quad d(\mathbf{u}, \mathbf{w}) \leq d(\mathbf{u}, \mathbf{v}) + d(\mathbf{v}, \mathbf{w}). \tag{2.4}$$

For a metric space we have the *topological* concepts of open sets, closed sets, continuity, convergence, completeness, compactness, connectedness, and boundedness. Note that we can have a vector space without the notion of a metric, and a metric space without the notions of scalar multiplication or a zero element (i.e., an origin).

Normed space X. The normed space X is a vector space in which there exists a real-valued function $|\mathbf{u}|$ known as the **norm** of the vector $\mathbf{u}$; the norm accepts vector $\mathbf{u}$ as input, and provides the real-valued **length** of $\mathbf{u}$ as output. The norm is defined such that the following properties hold $\forall\, \mathbf{u}, \mathbf{v} \in X$ and $\forall\, \alpha \in \mathbb{R}$:

$$\mathbf{u} \neq \mathbf{0} \Rightarrow |\mathbf{u}| > 0, \qquad |\mathbf{0}| = 0, \qquad |\alpha \mathbf{u}| = |\alpha|\,|\mathbf{u}|, \qquad |\mathbf{u} + \mathbf{v}| \leq |\mathbf{u}| + |\mathbf{v}|. \tag{2.5}$$

By definition, every normed space is a vector space (properties (2.5) of a normed space are defined on a vector space). In addition, we can show (refer to Problem 2.1 and Figure 2.3) that every normed space is also a metric space if we define the metric $d(\mathbf{u}, \mathbf{v}) \equiv |\mathbf{u} - \mathbf{v}|$, called the *natural* metric generated by the norm. Hence, the elements of a normed space can be referred to as either vectors or points. That is, we may think of $\mathbf{x}$ as a point in space, or a vector from the zero element (origin) to the point. The length $|\mathbf{x}|$ of the vector $\mathbf{x}$ is the distance between point $\mathbf{x}$ and the origin. The distance $d(\mathbf{u}, \mathbf{v})$ between two points $\mathbf{u}, \mathbf{v}$ is the length $|\mathbf{u} - \mathbf{v}|$ of the difference $\mathbf{u} - \mathbf{v}$ of the vectors $\mathbf{u}, \mathbf{v}$.

Inner product space X. The inner product space X is a vector space in which there exists a real-valued function $\mathbf{u} \cdot \mathbf{v}$ known as the **inner product** of vectors $\mathbf{u}$ and $\mathbf{v}$; the inner product accepts vectors $\mathbf{u}$ and $\mathbf{v}$ as inputs, and provides a real-valued quantity related to the **angle** between $\mathbf{u}$ and $\mathbf{v}$ as output. (Look ahead to Eq. (2.54).) The inner product is defined such that the following properties hold $\forall\, \mathbf{u}, \mathbf{v}, \mathbf{w} \in X$ and $\forall\, \alpha \in \mathbb{R}$:

$$\mathbf{u} \cdot \mathbf{v} = \mathbf{v} \cdot \mathbf{u}, \quad (\alpha \mathbf{u}) \cdot \mathbf{v} = \alpha(\mathbf{u} \cdot \mathbf{v}), \quad (\mathbf{u} + \mathbf{v}) \cdot \mathbf{w} = \mathbf{u} \cdot \mathbf{w} + \mathbf{v} \cdot \mathbf{w}, \quad \mathbf{u} \neq \mathbf{0} \Rightarrow \mathbf{u} \cdot \mathbf{u} > \mathbf{0}. \tag{2.6}$$

By definition, every inner product space is a vector space (properties (2.6) of an inner product space are defined on a vector space). It can be shown (refer to Problem 2.3 and Figure 2.3) that every inner product space is also a normed space and a metric space if $|\mathbf{u}| = (\mathbf{u} \cdot \mathbf{u})^{1/2}$ and $d(\mathbf{u}, \mathbf{v}) = [(\mathbf{u} - \mathbf{v}) \cdot (\mathbf{u} - \mathbf{v})]^{1/2}$. These are called the *natural* norm and *natural* metric, respectively, generated by the inner product. An n-dimensional inner product space, where n is a positive integer, is known as a Euclidean space $\mathcal{E}^n$. We hereafter specialize to three-dimensional Euclidean space $\mathcal{E}^3$.

All results obtained in this treatise follow rigorously from the postulated properties (2.1)–(2.6).

PROBLEM 2.1

Prove that every normed space is also a metric space if the metric is defined $d(\mathbf{u},\mathbf{v}) = |\mathbf{u}-\mathbf{v}|$.

Solution

To accomplish this, we must show that properties (2.5) of a normed space, together with the particular definition $d(\mathbf{u},\mathbf{v}) = |\mathbf{u}-\mathbf{v}|$ of the metric (called the natural metric generated by the norm), satisfy properties (2.4) of a metric space:

(i) Show that property $(2.4)_1$ is satisfied:

$$\begin{aligned} \mathbf{u} \neq \mathbf{v} \;&\Rightarrow\; \mathbf{u} + (-\mathbf{v}) \neq \mathbf{v} + (-\mathbf{v}) \\ &\Rightarrow\; \mathbf{u} - \mathbf{v} \neq \mathbf{0} && \text{(property } (2.3)_5) \\ &\Rightarrow\; |\mathbf{u}-\mathbf{v}| > 0 && \text{(property } (2.5)_1) \\ &\Rightarrow\; d(\mathbf{u},\mathbf{v}) > 0 && \text{(definition of metric).} \end{aligned}$$

(ii) Show that property $(2.4)_2$ is satisfied:

$$\begin{aligned} d(\mathbf{u},\mathbf{u}) &= |\mathbf{u}-\mathbf{u}| && \text{(definition of metric)} \\ &= |\mathbf{0}| && \text{(property } (2.3)_5) \\ &= 0 && \text{(property } (2.5)_2). \end{aligned}$$

(iii) Show that property $(2.4)_3$ is satisfied:

$$\begin{aligned} d(\mathbf{u},\mathbf{v}) &= |\mathbf{u}-\mathbf{v}| && \text{(definition of metric)} \\ &= |(-1)(\mathbf{v}-\mathbf{u})| && \text{(property } (2.3)_{10}) \\ &= |-1||\mathbf{v}-\mathbf{u}| && \text{(property } (2.5)_3) \\ &= |\mathbf{v}-\mathbf{u}| \\ &= d(\mathbf{v},\mathbf{u}) && \text{(definition of metric).} \end{aligned}$$

(iv) Show that property $(2.4)_4$ is satisfied:

$$\begin{aligned} d(\mathbf{u},\mathbf{w}) &= |\mathbf{u}-\mathbf{w}| && \text{(definition of metric)} \\ &= |(\mathbf{u}+\mathbf{0})-\mathbf{w}| && \text{(property } (2.3)_4) \\ &= |\mathbf{u}+[(-\mathbf{v})+\mathbf{v}]-\mathbf{w}| && \text{(property } (2.3)_5) \\ &= |(\mathbf{u}-\mathbf{v})+(\mathbf{v}-\mathbf{w})| && \text{(property } (2.3)_3) \\ &\leq \underbrace{|\mathbf{u}-\mathbf{v}|}_{d(\mathbf{u},\mathbf{v})} + \underbrace{|\mathbf{v}-\mathbf{w}|}_{d(\mathbf{v},\mathbf{w})} && \text{(property } (2.5)_4). \end{aligned}$$

PROBLEM 2.2

Prove the Cauchy-Schwarz inequality $(\mathbf{u}\cdot\mathbf{v})^2 \le (\mathbf{u}\cdot\mathbf{u})(\mathbf{v}\cdot\mathbf{v})$.

Solution

For *any* vectors **u** and **w**, the properties of normed and inner product spaces demand that

$$(\mathbf{u}-\mathbf{w})\cdot(\mathbf{u}-\mathbf{w}) \ge 0.$$

Then, once again using the properties of an inner product space, this becomes

$$\mathbf{u}\cdot\mathbf{u} \ge 2\,(\mathbf{u}\cdot\mathbf{w}) - \mathbf{w}\cdot\mathbf{w}.$$

We now set

$$\mathbf{w} = \frac{(\mathbf{u}\cdot\mathbf{v})}{(\mathbf{v}\cdot\mathbf{v})}\mathbf{v},$$

with **v** arbitrary. Then

$$\mathbf{u}\cdot\mathbf{u} \ge 2\left[\mathbf{u}\cdot\frac{(\mathbf{u}\cdot\mathbf{v})}{(\mathbf{v}\cdot\mathbf{v})}\mathbf{v}\right] - \frac{(\mathbf{u}\cdot\mathbf{v})}{(\mathbf{v}\cdot\mathbf{v})}\mathbf{v}\cdot\frac{(\mathbf{u}\cdot\mathbf{v})}{(\mathbf{v}\cdot\mathbf{v})}\mathbf{v}.$$

It follows that

$$\mathbf{u}\cdot\mathbf{u} \ge \frac{(\mathbf{u}\cdot\mathbf{v})^2}{\mathbf{v}\cdot\mathbf{v}},$$

i.e.,

$$(\mathbf{u}\cdot\mathbf{v})^2 \le (\mathbf{u}\cdot\mathbf{u})(\mathbf{v}\cdot\mathbf{v}).$$

PROBLEM 2.3

Prove that every inner product space is also a normed space if the norm is defined $|\mathbf{u}| = (\mathbf{u}\cdot\mathbf{u})^{\frac{1}{2}}$.

Solution

To accomplish this, we must show that properties (2.6) of an inner product space, along with the particular definition $|\mathbf{u}| = (\mathbf{u}\cdot\mathbf{u})^{\frac{1}{2}}$ of the norm (called the natural norm generated by the inner product), satisfy properties (2.5) of a normed space:

(i) Show that property $(2.5)_1$ is satisfied:

$$\begin{aligned} \mathbf{u}\neq\mathbf{0} \;&\Rightarrow\; \mathbf{u}\cdot\mathbf{u} > 0 && \text{(property } (2.6)_4\text{)}\\ &\Rightarrow\; (\mathbf{u}\cdot\mathbf{u})^{\frac{1}{2}} > 0 \\ &\Rightarrow\; |\mathbf{u}| > 0 && \text{(definition of norm).} \end{aligned}$$

(ii) Show that property $(2.5)_2$ is satisfied:

$$\begin{aligned} \mathbf{0}\cdot\mathbf{v} &= (0\mathbf{u})\cdot\mathbf{v} \\ &= 0\,(\mathbf{u}\cdot\mathbf{v}) && \text{(property } (2.6)_2\text{)}\\ &= 0 && \text{(property } (2.1)_{11}\text{).} \end{aligned}$$

The result $\mathbf{0} \cdot \mathbf{v} = 0$ is true for any vector $\mathbf{v}$ in $\mathcal{E}^3$. In particular, if we choose $\mathbf{v} = \mathbf{0}$, then

$$\mathbf{0} \cdot \mathbf{0} = 0 \quad \Rightarrow \quad (\mathbf{0} \cdot \mathbf{0})^{\frac{1}{2}} = 0 \quad \Rightarrow \quad |\mathbf{0}| = 0,$$

where we have used the definition of the norm.

(iii) Show that property $(2.5)_3$ is satisfied:

$$\begin{aligned} |\alpha \mathbf{u}| &= (\alpha \mathbf{u} \cdot \alpha \mathbf{u})^{\frac{1}{2}} && \text{(definition of norm)} \\ &= \left[\alpha^2 (\mathbf{u} \cdot \mathbf{u})\right]^{\frac{1}{2}} && \text{(property } (2.6)_2) \\ &= (\alpha^2)^{\frac{1}{2}} (\mathbf{u} \cdot \mathbf{u})^{\frac{1}{2}} \\ &= |\alpha||\mathbf{u}| && \text{(definition of norm).} \end{aligned}$$

(iv) Show that property $(2.5)_4$ is satisfied:

It follows from the definition of the norm and the properties (2.6) of inner product spaces that

$$|\mathbf{u} + \mathbf{v}|^2 = (\mathbf{u} + \mathbf{v}) \cdot (\mathbf{u} + \mathbf{v}) = \mathbf{u} \cdot \mathbf{u} + 2(\mathbf{u} \cdot \mathbf{v}) + \mathbf{v} \cdot \mathbf{v}.$$

The Cauchy-Schwarz inequality (refer to Problem 2.2) then implies that

$$|\mathbf{u} + \mathbf{v}|^2 \leq \mathbf{u} \cdot \mathbf{u} + 2(\mathbf{u} \cdot \mathbf{u})^{\frac{1}{2}} (\mathbf{v} \cdot \mathbf{v})^{\frac{1}{2}} + \mathbf{v} \cdot \mathbf{v}.$$

Then, from the definition of the norm, it follows that

$$|\mathbf{u} + \mathbf{v}|^2 \leq |\mathbf{u}|^2 + 2|\mathbf{u}||\mathbf{v}| + |\mathbf{v}|^2,$$

from which we have

$$|\mathbf{u} + \mathbf{v}|^2 \leq (|\mathbf{u}| + |\mathbf{v}|)^2,$$

and, finally,

$$|\mathbf{u} + \mathbf{v}| \leq |\mathbf{u}| + |\mathbf{v}|.$$

2.2 TENSOR ALGEBRA

The presentation of the conceptual material in this section follows [2].

2.2.1 SECOND-ORDER TENSORS, ZERO TENSOR, IDENTITY TENSOR

A **second-order tensor** (or **tensor**, for short) is defined *only by how it acts on an arbitrary vector* (a vector is also known as a **first-order tensor**). In particular, a second-order tensor $\mathbf{T}$ is an operation that assigns to each vector $\mathbf{v}$ in vector space $\mathcal{E}^3$ a vector $\mathbf{Tv}$ in vector space $\mathcal{E}^3$ such that

$$\mathbf{T}(\mathbf{v} + \mathbf{w}) = \mathbf{Tv} + \mathbf{Tw}, \qquad \mathbf{T}(\alpha \mathbf{v}) = \alpha(\mathbf{Tv}) \tag{2.7}$$

for any vectors $\mathbf{v}, \mathbf{w} \in \mathcal{E}^3$ and scalars $\alpha \in \mathbb{R}$. Property $(2.7)_1$ indicates that the map of the sum is the sum of the maps, and property $(2.7)_2$ indicates that the map of the product is the product of the map; thus, a second-order tensor is a *linear* map from vector space $\mathcal{E}^3$ to vector space $\mathcal{E}^3$. The set of all second-order tensors is denoted by $\mathcal{L}$.

Addition and scalar multiplication of second-order tensors are defined by

$$(\mathbf{S}+\mathbf{T})\mathbf{v} = \mathbf{S}\mathbf{v} + \mathbf{T}\mathbf{v}, \qquad (\alpha\mathbf{S})\mathbf{v} = \alpha(\mathbf{S}\mathbf{v}) \tag{2.8}$$

for any tensors $\mathbf{S}, \mathbf{T} \in \mathcal{L}$, vectors $\mathbf{v} \in \mathcal{E}^3$, and scalars $\alpha \in \mathbb{R}$. Note that the operations $\mathbf{S}+\mathbf{T}$ and $\alpha\mathbf{S}$ are defined by how they act on an arbitrary vector $\mathbf{v}$. With definition (2.7), definition (2.8), and the properties of $\mathcal{E}^3$ (refer to Section 2.1.2), it can be shown (refer to Problem 2.4) that all of the requirements (2.3) are satisfied. Hence, the set $\mathcal{L}$ of all second-order tensors is a vector space.

Recall from properties (2.3) that every vector space has a zero element. The zero element of $\mathcal{L}$ is the **zero tensor 0** that maps every $\mathbf{v} \in \mathcal{E}^3$ to the zero vector $\mathbf{0}$ of $\mathcal{E}^3$, i.e.,

$$\mathbf{0}\mathbf{v} = \mathbf{0}. \tag{2.9}$$

The **identity tensor I** maps every $\mathbf{v} \in \mathcal{E}^3$ to itself, i.e.,

$$\mathbf{I}\mathbf{v} = \mathbf{v}. \tag{2.10}$$

PROBLEM 2.4

Prove that the set $\mathcal{L}$ of all second-order tensors is a vector space.

Solution

In what follows, we show that the properties of a vector space are satisfied by any arbitrary second-order tensors **R**, **S**, and **T**. That is, the algebraic properties of second-order tensors will be *deduced* using properties (2.3), which are *postulated* to hold for vectors, along with definitions (2.7) and (2.8).[2]

Closure of tensor addition

For any vectors **v**, **w** in $\mathcal{E}^3$ and scalars α in $\mathbb{R}$,

$$\begin{aligned}
(\mathbf{T}+\mathbf{S})(\alpha\mathbf{v}+\mathbf{w}) &= \mathbf{T}(\alpha\mathbf{v}+\mathbf{w}) + \mathbf{S}(\alpha\mathbf{v}+\mathbf{w}) && \text{(definition (2.8)}_1\text{)}\\
&= \alpha(\mathbf{T}\mathbf{v}) + \mathbf{T}\mathbf{w} + \alpha(\mathbf{S}\mathbf{v}) + \mathbf{S}\mathbf{w} && \text{(definition (2.7))}\\
&= \alpha(\mathbf{T}\mathbf{v}) + \alpha(\mathbf{S}\mathbf{v}) + \mathbf{T}\mathbf{w} + \mathbf{S}\mathbf{w} && \text{(property (2.3)}_2\text{)}\\
&= \alpha(\mathbf{T}\mathbf{v} + \mathbf{S}\mathbf{v}) + \mathbf{T}\mathbf{w} + \mathbf{S}\mathbf{w} && \text{(property (2.3)}_{10}\text{)}\\
&= \alpha[(\mathbf{T}+\mathbf{S})\mathbf{v}] + (\mathbf{T}+\mathbf{S})\mathbf{w} && \text{(definition (2.8)}_1\text{).}
\end{aligned}$$

[2] The properties of a first-order tensor (vector) are *postulated* in Section 2.1.2, while those of higher-order tensors must be *deduced* from those postulated properties.

Thus,

$$(\mathbf{T}+\mathbf{S})(\alpha\mathbf{v}+\mathbf{w}) = \alpha[(\mathbf{T}+\mathbf{S})\mathbf{v}] + (\mathbf{T}+\mathbf{S})\mathbf{w}.$$

It follows from (2.7) that $\mathbf{T}+\mathbf{S}$ is a linear map from $\mathcal{E}^3$ to $\mathcal{E}^3$, i.e., $\mathbf{T}+\mathbf{S} \in \mathcal{L}$.

Commutativity of tensor addition

For any vector $\mathbf{v}$ in $\mathcal{E}^3$,

$$\begin{aligned}
(\mathbf{T}+\mathbf{S})\mathbf{v} &= \mathbf{T}\mathbf{v}+\mathbf{S}\mathbf{v} && \text{(definition (2.8)}_1\text{)}\\
&= \mathbf{S}\mathbf{v}+\mathbf{T}\mathbf{v} && \text{(property (2.3)}_2\text{)}\\
&= (\mathbf{S}+\mathbf{T})\mathbf{v} && \text{(definition (2.8)}_1\text{).}
\end{aligned}$$

Since $\mathbf{v}$ is arbitrary, we conclude that $\mathbf{T}+\mathbf{S} = \mathbf{S}+\mathbf{T}$. Note that commutativity of tensor addition was *deduced* using commutativity of vector addition (*postulated* property $(2.3)_2$).

Associativity of tensor addition

For any vector $\mathbf{v}$ in $\mathcal{E}^3$,

$$\begin{aligned}
[\mathbf{T}+(\mathbf{S}+\mathbf{R})]\mathbf{v} &= \mathbf{T}\mathbf{v}+(\mathbf{S}+\mathbf{R})\mathbf{v} && \text{(definition (2.8)}_1\text{)}\\
&= \mathbf{T}\mathbf{v}+(\mathbf{S}\mathbf{v}+\mathbf{R}\mathbf{v}) && \text{(definition (2.8)}_1\text{)}\\
&= (\mathbf{T}\mathbf{v}+\mathbf{S}\mathbf{v})+\mathbf{R}\mathbf{v} && \text{(property (2.3)}_3\text{)}\\
&= (\mathbf{T}+\mathbf{S})\mathbf{v}+\mathbf{R}\mathbf{v} && \text{(definition (2.8)}_1\text{)}\\
&= [(\mathbf{T}+\mathbf{S})+\mathbf{R}]\mathbf{v} && \text{(definition (2.8)}_1\text{).}
\end{aligned}$$

Since $\mathbf{v}$ is arbitrary, we conclude that $\mathbf{T}+(\mathbf{S}+\mathbf{R}) = (\mathbf{T}+\mathbf{S})+\mathbf{R}$. Note that associativity of tensor addition was *deduced* using associativity of vector addition (*postulated* property $(2.3)_3$).

Existence of an additive identity

The *zero tensor* $\mathbf{0}$ maps any vector $\mathbf{v}$ in $\mathcal{E}^3$ to the *zero vector* $\mathbf{0}$, i.e., $\mathbf{0}\,\mathbf{v} = \mathbf{0}$ (refer to (2.9)). We have,

$$\begin{aligned}
(\mathbf{T}+\mathbf{0})\mathbf{v} &= \mathbf{T}\mathbf{v}+\mathbf{0}\mathbf{v} && \text{(definition (2.8)}_1\text{)}\\
&= \mathbf{T}\mathbf{v}+\mathbf{0} && \text{(definition (2.9))}\\
&= \mathbf{T}\mathbf{v} && \text{(property (2.3)}_4\text{).}
\end{aligned}$$

Since $\mathbf{v}$ is arbitrary, $\mathbf{T}+\mathbf{0} = \mathbf{T}$. Note that the existence of a zero tensor was *deduced* using the existence of a zero vector (*postulated* property $(2.3)_4$).

Existence of an additive inverse

For any vector $\mathbf{v}$ in $\mathcal{E}^3$,

$$\begin{aligned}
[\mathbf{T} + (-\mathbf{T})]\,\mathbf{v} &= \mathbf{T}\mathbf{v} + (-\mathbf{T})\mathbf{v} && \text{(definition (2.8)}_1\text{)}\\
&= \mathbf{T}\mathbf{v} + [-(\mathbf{T}\mathbf{v})] && \text{(definition (2.8)}_2\text{)}\\
&= \mathbf{0} && \text{(property (2.3)}_5\text{)}\\
&= \mathbf{0}\,\mathbf{v} && \text{(definition (2.9)).}
\end{aligned}$$

Thus, $\mathbf{T} + (-\mathbf{T}) = \mathbf{0}$. Note that the existence of an additive inverse for tensors was *deduced* using the existence of an additive inverse for vectors (*postulated* property $(2.3)_5$).

Closure of scalar multiplication

For any vectors $\mathbf{v}$, $\mathbf{w}$ in $\mathcal{E}^3$ and scalars β in $\mathbb{R}$,

$$\begin{aligned}
(\alpha\mathbf{T})(\beta\mathbf{v} + \mathbf{w}) &= \alpha[\mathbf{T}(\beta\mathbf{v} + \mathbf{w})] && \text{(definition (2.8)}_2\text{)}\\
&= \alpha[\beta(\mathbf{T}\mathbf{v}) + \mathbf{T}\mathbf{w}] && \text{(definition (2.7))}\\
&= \alpha[\beta(\mathbf{T}\mathbf{v})] + \alpha(\mathbf{T}\mathbf{w}) && \text{(property (2.3)}_{10}\text{)}\\
&= (\alpha\beta)(\mathbf{T}\mathbf{v}) + (\alpha\mathbf{T})\mathbf{w} && \text{(property (2.3)}_7\text{, definition (2.8)}_2\text{)}\\
&= (\beta\alpha)(\mathbf{T}\mathbf{v}) + (\alpha\mathbf{T})\mathbf{w} && \text{(property (2.1)}_7\text{)}\\
&= \beta[\alpha(\mathbf{T}\mathbf{v})] + (\alpha\mathbf{T})\mathbf{w} && \text{(property (2.3)}_7\text{)}\\
&= \beta[(\alpha\mathbf{T})\mathbf{v}] + (\alpha\mathbf{T})\mathbf{w} && \text{(definition (2.8)}_2\text{).}
\end{aligned}$$

Thus,

$$(\alpha\mathbf{T})(\beta\mathbf{v} + \mathbf{w}) = \beta[(\alpha\mathbf{T})\mathbf{v}] + (\alpha\mathbf{T})\mathbf{w}.$$

It follows from (2.7) that $\alpha\mathbf{T}$ is a linear map from $\mathcal{E}^3$ to $\mathcal{E}^3$, i.e., $\alpha\mathbf{T} \in \mathcal{L}$.

Associativity of scalar multiplication

For any vector $\mathbf{v}$ in $\mathcal{E}^3$,

$$\begin{aligned}
[\alpha(\beta\mathbf{T})]\mathbf{v} &= \alpha\,[(\beta\mathbf{T})\mathbf{v}] && \text{(definition (2.8)}_2\text{)}\\
&= \alpha\,[\beta(\mathbf{T}\mathbf{v})] && \text{(definition (2.8)}_2\text{)}\\
&= (\alpha\beta)(\mathbf{T}\mathbf{v}) && \text{(property (2.3)}_7\text{)}\\
&= [(\alpha\beta)\mathbf{T}]\,\mathbf{v} && \text{(definition (2.8)}_2\text{).}
\end{aligned}$$

Therefore, since $\mathbf{v}$ is arbitrary, $\alpha(\beta\mathbf{T}) = (\alpha\beta)\mathbf{T}$. Note that associativity of scalar multiplication of a tensor was *deduced* using associativity of scalar multiplication of a vector (*postulated* property $(2.3)_7$).

Existence of a multiplicative identity

For any vector $\mathbf{v}$ in $\mathcal{E}^3$,

$$(1\mathbf{T})\mathbf{v} = 1(\mathbf{T}\mathbf{v}) \qquad \text{(definition (2.8)}_2\text{)}$$

$$= \mathbf{T}\mathbf{v} \qquad \text{(property (2.3)}_8\text{)}.$$

Thus, $1\mathbf{T} = \mathbf{T}$. Note that the existence of a multiplicative identity for tensors was *deduced* using the existence of a multiplicative identity for vectors (*postulated* property $(2.3)_8$).

Distributivity of scalar multiplication over scalar addition

For any vector $\mathbf{v}$ in $\mathcal{E}^3$ and scalars α, β in $\mathbb{R}$,

$$[(\alpha + \beta)\mathbf{T}]\mathbf{v} = (\alpha + \beta)(\mathbf{T}\mathbf{v}) \qquad \text{(definition (2.8)}_2\text{)}$$

$$= \alpha(\mathbf{T}\mathbf{v}) + \beta(\mathbf{T}\mathbf{v}) \qquad \text{(property (2.3)}_9\text{)}$$

$$= (\alpha\mathbf{T})\mathbf{v} + (\beta\mathbf{T})\mathbf{v} \qquad \text{(definition (2.8)}_2\text{)}$$

$$= (\alpha\mathbf{T} + \beta\mathbf{T})\mathbf{v} \qquad \text{(definition (2.8)}_1\text{)}.$$

Thus, $(\alpha + \beta)\mathbf{T} = \alpha\mathbf{T} + \beta\mathbf{T}$. Note that this property for tensors was *deduced* using the analogous property $(2.3)_9$ for vectors.

Distributivity of scalar multiplication over tensor addition

For any vector $\mathbf{v}$ in $\mathcal{E}^3$ and scalar α in $\mathbb{R}$,

$$[\alpha(\mathbf{T} + \mathbf{S})]\mathbf{v} = \alpha[(\mathbf{T} + \mathbf{S})\mathbf{v}] \qquad \text{(definition (2.8)}_2\text{)}$$

$$= \alpha(\mathbf{T}\mathbf{v} + \mathbf{S}\mathbf{v}) \qquad \text{(definition (2.8)}_1\text{)}$$

$$= \alpha(\mathbf{T}\mathbf{v}) + \alpha(\mathbf{S}\mathbf{v}) \qquad \text{(property (2.3)}_{10}\text{)}$$

$$= (\alpha\mathbf{T})\mathbf{v} + (\alpha\mathbf{S})\mathbf{v} \qquad \text{(definition (2.8)}_2\text{)}$$

$$= (\alpha\mathbf{T} + \alpha\mathbf{S})\mathbf{v} \qquad \text{(definition (2.8)}_1\text{)}.$$

Since $\mathbf{v}$ is arbitrary, we have $\alpha(\mathbf{T} + \mathbf{S}) = \alpha\mathbf{T} + \alpha\mathbf{S}$. Note that this property for tensors was *deduced* using the analogous property $(2.3)_{10}$ for vectors.

2.2.2 PRODUCT, TRANSPOSE, SYMMETRY

The **product of two tensors** is defined by

$$(\mathbf{S}\mathbf{T})\mathbf{v} = \mathbf{S}(\mathbf{T}\mathbf{v}) \tag{2.11}$$

for any vector $\mathbf{v} \in \mathcal{E}^3$. Note that this product is defined such that $\mathbf{v}$ is first mapped by $\mathbf{T}$ to $\mathbf{T}\mathbf{v}$, then $\mathbf{T}\mathbf{v}$ is mapped by $\mathbf{S}$ to $\mathbf{S}(\mathbf{T}\mathbf{v})$; see Figure 2.4. It can be shown (refer to Problem 2.5) that with definition (2.11) the product of two tensors is itself a tensor. Tensor multiplication is not *commutative*, since in general $\mathbf{S}\mathbf{T} \neq \mathbf{T}\mathbf{S}$, although it is *associative*. It can be verified (refer to Problem 2.6) that tensor multiplication distributes over tensor addition as

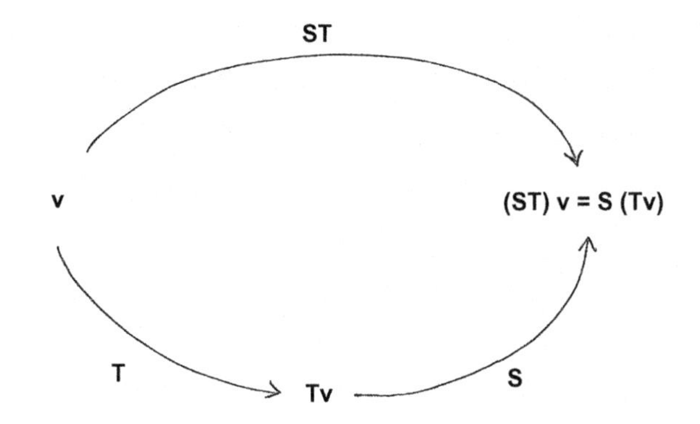

FIGURE 2.4

Schematic illustrating how the product **ST** operates on an arbitrary vector **v**.

$$(\mathbf{S}+\mathbf{T})\,\mathbf{R} = \mathbf{S}\,\mathbf{R} + \mathbf{T}\,\mathbf{R}. \tag{2.12}$$

The **transpose** $\mathbf{S}^{\mathrm{T}}$ of **S** is defined by

$$\mathbf{S}\mathbf{u}\cdot\mathbf{v} = \mathbf{u}\cdot\mathbf{S}^{\mathrm{T}}\mathbf{v} \tag{2.13}$$

for any vectors $\mathbf{u}, \mathbf{v} \in \mathcal{E}^3$. It can be shown (refer to Problem 2.7) that the transpose is unique. We have, for instance (refer to Problems 2.8–2.11),

$$(\mathbf{S}+\mathbf{T})^{\mathrm{T}} = \mathbf{S}^{\mathrm{T}} + \mathbf{T}^{\mathrm{T}}, \qquad (\mathbf{S}\mathbf{R})^{\mathrm{T}} = \mathbf{R}^{\mathrm{T}}\mathbf{S}^{\mathrm{T}}, \qquad (\mathbf{S}^{\mathrm{T}})^{\mathrm{T}} = \mathbf{S}, \qquad \mathbf{I}^{\mathrm{T}} = \mathbf{I}. \tag{2.14}$$

A tensor for which

$$\mathbf{S}^{\mathrm{T}} = \mathbf{S} \tag{2.15}$$

is called a **symmetric tensor**, and a tensor for which

$$\mathbf{S}^{\mathrm{T}} = -\mathbf{S} \tag{2.16}$$

is called a **skew tensor**. It can be shown (refer to Problem 2.12) that every tensor **S** can be additively decomposed into a symmetric part **D** and a skew part **W**, i.e.,

$$\mathbf{S} = \mathbf{D} + \mathbf{W}, \tag{2.17}$$

where

$$\mathbf{D} = \frac{1}{2}(\mathbf{S}+\mathbf{S}^{\mathrm{T}}), \qquad \mathbf{W} = \frac{1}{2}(\mathbf{S}-\mathbf{S}^{\mathrm{T}}). \tag{2.18}$$

PROBLEM 2.5

Prove that the product of two tensors is itself a tensor.

Solution

For the product **ST** of tensors **S** and **T** to itself be a tensor, it must satisfy definition (2.7), i.e.,

$$(\mathbf{ST})(\alpha\mathbf{v}+\mathbf{w}) = \alpha[(\mathbf{ST})\mathbf{v}] + (\mathbf{ST})\mathbf{w}$$

for any tensors $\mathbf{S}, \mathbf{T}$ in $\mathcal{L}$, vectors $\mathbf{v}, \mathbf{w}$ in $\mathcal{E}^3$, and scalars α in $\mathbb{R}$.

$$\begin{aligned}
(\mathbf{ST})(\alpha\mathbf{v}+\mathbf{w}) &= \mathbf{S}[\mathbf{T}(\alpha\mathbf{v}+\mathbf{w})] && \text{(definition (2.11))}\\
&= \mathbf{S}[\alpha(\mathbf{Tv})+\mathbf{Tw}] && \text{(definition (2.7))}\\
&= \alpha[\mathbf{S}(\mathbf{Tv})]+\mathbf{S}(\mathbf{Tw}) && \text{(definition (2.7))}\\
&= \alpha[(\mathbf{ST})\mathbf{v}]+(\mathbf{ST})\mathbf{w} && \text{(definition (2.11)).}
\end{aligned}$$

PROBLEM 2.6

Prove in direct notation that $(\mathbf{S}+\mathbf{T})\,\mathbf{R}=\mathbf{S}\,\mathbf{R}+\mathbf{T}\,\mathbf{R}$.

Solution

For any vector $\mathbf{v}$ in $\mathcal{E}^3$,

$$\begin{aligned}
[(\mathbf{S}+\mathbf{T})\,\mathbf{R}]\mathbf{v} &= (\mathbf{S}+\mathbf{T})(\mathbf{Rv}) && \text{(definition (2.11))}\\
&= \mathbf{S}(\mathbf{Rv})+\mathbf{T}(\mathbf{Rv}) && \text{(definition } (2.8)_1)\\
&= (\mathbf{SR})\mathbf{v}+(\mathbf{TR})\mathbf{v} && \text{(definition (2.11))}\\
&= (\mathbf{SR}+\mathbf{TR})\mathbf{v} && \text{(definition } (2.8)_1).
\end{aligned}$$

Since $\mathbf{v}$ is arbitrary, $(\mathbf{S}+\mathbf{T})\mathbf{R}=\mathbf{SR}+\mathbf{TR}$, i.e., tensor multiplication is distributive over tensor addition.

PROBLEM 2.7

Show that the transpose is unique.

Solution

The structure of this uniqueness proof follows the customary approach: we assume at the outset that two elements satisfy a particular mathematical statement, then systematically demonstrate that these two elements are identical.

Suppose there exists a tensor $\mathbf{R}$ in $\mathcal{L}$ such that

$$\mathbf{Su}\cdot\mathbf{v}=\mathbf{u}\cdot\mathbf{S}^{\mathrm{T}}\mathbf{v}=\mathbf{u}\cdot\mathbf{Rv}$$

for any vectors $\mathbf{u}, \mathbf{v}$ in $\mathcal{E}^3$. It follows that

$$\mathbf{u}\cdot\mathbf{S}^{\mathrm{T}}\mathbf{v}-\mathbf{u}\cdot\mathbf{Rv}=0.$$

Then,

$$\mathbf{u}\cdot(\mathbf{S}^{\mathrm{T}}\mathbf{v}-\mathbf{Rv})=0 \quad\Rightarrow\quad \mathbf{u}\cdot[(\mathbf{S}^{\mathrm{T}}-\mathbf{R})\mathbf{v}]=0,$$

where we have used property $(2.6)_3$ and definition $(2.8)_1$, respectively. Since $\mathbf{u}$ is arbitrary,

$$(\mathbf{S}^{\mathrm{T}} - \mathbf{R})\mathbf{v} = \mathbf{0},$$

and since $\mathbf{v}$ is arbitrary,

$$\mathbf{S}^{\mathrm{T}} - \mathbf{R} = \mathbf{0},$$

so $\mathbf{R} = \mathbf{S}^{\mathrm{T}}$, i.e., the transpose is unique.

PROBLEM 2.8

In direct notation, prove that $(\mathbf{S} + \mathbf{T})^{\mathrm{T}} = \mathbf{S}^{\mathrm{T}} + \mathbf{T}^{\mathrm{T}}$.

Solution

Since $\mathbf{S} + \mathbf{T}$ is a tensor (refer to Problem 2.4, closure), we have by definition (2.13)

$$(\mathbf{S} + \mathbf{T})\mathbf{u} \cdot \mathbf{v} = \mathbf{u} \cdot (\mathbf{S} + \mathbf{T})^{\mathrm{T}}\mathbf{v}$$

for any vectors $\mathbf{u}$ and $\mathbf{v}$ in $\mathcal{E}^3$. Working with the left-hand side, we obtain

$$\begin{aligned}
(\mathbf{S} + \mathbf{T})\mathbf{u} \cdot \mathbf{v} &= (\mathbf{S}\mathbf{u} + \mathbf{T}\mathbf{u}) \cdot \mathbf{v} && \text{(definition } (2.8)_1\text{)} \\
&= \mathbf{S}\mathbf{u} \cdot \mathbf{v} + \mathbf{T}\mathbf{u} \cdot \mathbf{v} && \text{(property } (2.6)_3\text{)} \\
&= \mathbf{u} \cdot \mathbf{S}^{\mathrm{T}}\mathbf{v} + \mathbf{u} \cdot \mathbf{T}^{\mathrm{T}}\mathbf{v} && \text{(definition (2.13))} \\
&= \mathbf{u} \cdot (\mathbf{S}^{\mathrm{T}}\mathbf{v} + \mathbf{T}^{\mathrm{T}}\mathbf{v}) && \text{(property } (2.6)_3\text{)} \\
&= \mathbf{u} \cdot (\mathbf{S}^{\mathrm{T}} + \mathbf{T}^{\mathrm{T}})\mathbf{v} && \text{(definition } (2.8)_1\text{)}.
\end{aligned}$$

Using this result in the original expression, we obtain

$$\mathbf{u} \cdot (\mathbf{S}^{\mathrm{T}} + \mathbf{T}^{\mathrm{T}})\,\mathbf{v} = \mathbf{u} \cdot (\mathbf{S} + \mathbf{T})^{\mathrm{T}}\mathbf{v}$$

for any vectors $\mathbf{u}$ and $\mathbf{v}$ in $\mathcal{E}^3$. Since $\mathbf{u}$ and $\mathbf{v}$ are arbitrary, $(\mathbf{S} + \mathbf{T})^{\mathrm{T}} = \mathbf{S}^{\mathrm{T}} + \mathbf{T}^{\mathrm{T}}$, i.e., the transpose of the sum is the sum of the transposes.

PROBLEM 2.9

In direct notation, prove that $(\mathbf{S}\mathbf{R})^{\mathrm{T}} = \mathbf{R}^{\mathrm{T}}\,\mathbf{S}^{\mathrm{T}}$.

Solution

Since $\mathbf{ST}$ is a tensor (refer to Problem 2.5), we have by definition (2.13)

$$(\mathbf{S}\mathbf{R})\mathbf{u} \cdot \mathbf{v} = \mathbf{u} \cdot (\mathbf{S}\mathbf{R})^{\mathrm{T}}\mathbf{v}$$

for any vectors $\mathbf{u}$ and $\mathbf{v}$ in $\mathcal{E}^3$. Working with the left-hand side, we obtain

$$\begin{aligned}
(\mathbf{SR})\mathbf{u}\cdot\mathbf{v} &= \mathbf{S}(\mathbf{Ru})\cdot\mathbf{v} && \text{(definition (2.11))}\\
&= \mathbf{Ru}\cdot\mathbf{S}^{\mathrm{T}}\mathbf{v} && \text{(definition (2.13))}\\
&= \mathbf{u}\cdot\mathbf{R}^{\mathrm{T}}(\mathbf{S}^{\mathrm{T}}\mathbf{v}) && \text{(definition (2.13))}\\
&= \mathbf{u}\cdot(\mathbf{R}^{\mathrm{T}}\mathbf{S}^{\mathrm{T}})\mathbf{v} && \text{(definition (2.11))}.
\end{aligned}$$

Using this result in the original expression, we obtain

$$\mathbf{u}\cdot(\mathbf{R}^{\mathrm{T}}\mathbf{S}^{\mathrm{T}})\,\mathbf{v} = \mathbf{u}\cdot(\mathbf{SR})^{\mathrm{T}}\mathbf{v}$$

for any vectors $\mathbf{u}$ and $\mathbf{v}$ in $\mathcal{E}^3$. Since $\mathbf{u}$ and $\mathbf{v}$ are arbitrary, $(\mathbf{SR})^{\mathrm{T}} = \mathbf{R}^{\mathrm{T}}\,\mathbf{S}^{\mathrm{T}}$, i.e., the transpose of the product is the product of the transposes.

PROBLEM 2.10

In direct notation, prove that $(\mathbf{S}^{\mathrm{T}})^{\mathrm{T}} = \mathbf{S}$.

Solution

For any vectors $\mathbf{u}$ and $\mathbf{v}$ in $\mathcal{E}^3$,

$$\mathbf{Su}\cdot\mathbf{v} = \mathbf{u}\cdot\mathbf{S}^{\mathrm{T}}\mathbf{v} = \mathbf{S}^{\mathrm{T}}\mathbf{v}\cdot\mathbf{u} = \mathbf{v}\cdot(\mathbf{S}^{\mathrm{T}})^{\mathrm{T}}\mathbf{u} = (\mathbf{S}^{\mathrm{T}})^{\mathrm{T}}\mathbf{u}\cdot\mathbf{v}.$$

Since vectors $\mathbf{u}$ and $\mathbf{v}$ are arbitrary, it follows that $(\mathbf{S}^{\mathrm{T}})^{\mathrm{T}} = \mathbf{S}$.

PROBLEM 2.11

In direct notation, prove that $\mathbf{I}^{\mathrm{T}} = \mathbf{I}$.

Solution

It follows from definition (2.13) that for any vectors $\mathbf{u}$ and $\mathbf{v}$ in $\mathcal{E}^3$

$$\mathbf{Iu}\cdot\mathbf{v} = \mathbf{u}\cdot\mathbf{I}^{\mathrm{T}}\mathbf{v}.$$

Working with the left-hand side, we obtain

$$\mathbf{Iu}\cdot\mathbf{v} = \mathbf{u}\cdot\mathbf{v} = \mathbf{u}\cdot\mathbf{Iv}.$$

Upon use of this result in the original expression, it follows that

$$\mathbf{u}\cdot\mathbf{Iv} - \mathbf{u}\cdot\mathbf{I}^{\mathrm{T}}\mathbf{v} = 0.$$

Successive use of properties $(2.6)_3$ and $(2.8)_1$ allows us to write

$$\mathbf{u}\cdot(\mathbf{Iv} - \mathbf{I}^{\mathrm{T}}\mathbf{v}) = 0 \quad\Rightarrow\quad \mathbf{u}\cdot[(\mathbf{I}-\mathbf{I}^{\mathrm{T}})\mathbf{v}] = 0$$

for any vectors **u** and **v** in $\mathcal{E}^3$. Since **u** is arbitrary,

$$(\mathbf{I}-\mathbf{I}^{\mathrm{T}})\,\mathbf{v}=\mathbf{0},$$

and since **v** is arbitrary,

$$\mathbf{I}-\mathbf{I}^{\mathrm{T}}=\mathbf{0},$$

so $\mathbf{I}^{\mathrm{T}}=\mathbf{I}$.

PROBLEM 2.12

Given the decomposition $\mathbf{S}=\mathbf{D}+\mathbf{W}$, where $\mathbf{D}=\frac{1}{2}\left(\mathbf{S}+\mathbf{S}^{\mathrm{T}}\right)$ and $\mathbf{W}=\frac{1}{2}\left(\mathbf{S}-\mathbf{S}^{\mathrm{T}}\right)$, prove in direct notation that **D** is symmetric and **W** is skew.

Solution

To prove that **D** is symmetric and **W** is skew, we must show that $\mathbf{D}^{\mathrm{T}}=\mathbf{D}$ and $\mathbf{W}^{\mathrm{T}}=-\mathbf{W}$:

$$\begin{aligned}
\mathbf{D}^{\mathrm{T}} &= \frac{1}{2}(\mathbf{S}+\mathbf{S}^{\mathrm{T}})^{\mathrm{T}} && \text{(definition of } \mathbf{D}\text{)}\\
&= \frac{1}{2}\left[\mathbf{S}^{\mathrm{T}}+(\mathbf{S}^{\mathrm{T}})^{\mathrm{T}}\right] && \text{(result } (2.14)_1\text{)}\\
&= \frac{1}{2}(\mathbf{S}^{\mathrm{T}}+\mathbf{S}) && \text{(result } (2.14)_3\text{)}\\
&= \frac{1}{2}(\mathbf{S}+\mathbf{S}^{\mathrm{T}}) && \text{(commutativity of tensor addition)}\\
&= \mathbf{D} && \text{(definition of } \mathbf{D}\text{);}
\end{aligned}$$

$$\begin{aligned}
\mathbf{W}^{\mathrm{T}} &= \frac{1}{2}(\mathbf{S}-\mathbf{S}^{\mathrm{T}})^{\mathrm{T}} && \text{(definition of } \mathbf{W}\text{)}\\
&= \frac{1}{2}[\mathbf{S}^{\mathrm{T}}-(\mathbf{S}^{\mathrm{T}})^{\mathrm{T}}] && \text{(result } (2.14)_1\text{)}\\
&= \frac{1}{2}(\mathbf{S}^{\mathrm{T}}-\mathbf{S}) && \text{(result } (2.14)_3\text{)}\\
&= -\frac{1}{2}(\mathbf{S}-\mathbf{S}^{\mathrm{T}}) && \text{(commutativity of tensor addition)}\\
&= -\mathbf{W} && \text{(definition of } \mathbf{W}\text{).}
\end{aligned}$$

2.2.3 DYADIC PRODUCT

The **dyadic product** (or **tensor product**) $\mathbf{a} \otimes \mathbf{b}$ accepts two vectors $\mathbf{a}, \mathbf{b} \in \mathcal{E}^3$ as inputs and provides as output a second-order tensor that maps each $\mathbf{v} \in \mathcal{E}^3$ to the vector $(\mathbf{b} \cdot \mathbf{v})\, \mathbf{a}$. That is,

$$(\mathbf{a} \otimes \mathbf{b})\, \mathbf{v} = (\mathbf{b} \cdot \mathbf{v})\, \mathbf{a}, \tag{2.19}$$

where $(\mathbf{b} \cdot \mathbf{v})\, \mathbf{a}$ is the projection of $\mathbf{b}$ onto $\mathbf{v}$ in the direction of $\mathbf{a}$. We have, for instance (refer to Problems 2.14–2.17),

$$(\mathbf{a} \otimes \mathbf{b})^{\mathrm{T}} = \mathbf{b} \otimes \mathbf{a}, \qquad \mathbf{S}\,(\mathbf{a} \otimes \mathbf{b}) = (\mathbf{S}\,\mathbf{a}) \otimes \mathbf{b},$$

$$(\mathbf{a} \otimes \mathbf{b})\, \mathbf{S} = \mathbf{a} \otimes \left(\mathbf{S}^{\mathrm{T}}\, \mathbf{b}\right), \qquad (\mathbf{a} \otimes \mathbf{b})\,(\mathbf{c} \otimes \mathbf{d}) = (\mathbf{b} \cdot \mathbf{c})\,(\mathbf{a} \otimes \mathbf{d}). \tag{2.20}$$

PROBLEM 2.13
Verify that $\mathbf{a} \otimes \mathbf{b}$ is a tensor.

Solution
To prove that $\mathbf{a} \otimes \mathbf{b}$ is a tensor, we must demonstrate that it satisfies definition (2.7), i.e.,

$$(\mathbf{a} \otimes \mathbf{b})(\alpha \mathbf{v} + \mathbf{w}) = \alpha[(\mathbf{a} \otimes \mathbf{b})\mathbf{v}] + (\mathbf{a} \otimes \mathbf{b})\mathbf{w}$$

for all vectors $\mathbf{a}$, $\mathbf{b}$, $\mathbf{v}$, $\mathbf{w}$ in $\mathcal{E}^3$ and scalars α in $\mathbb{R}$:

$$\begin{aligned}
(\mathbf{a} \otimes \mathbf{b})(\alpha \mathbf{v} + \mathbf{w}) &= [\mathbf{b} \cdot (\alpha \mathbf{v} + \mathbf{w})]\mathbf{a} && \text{(definition (2.19))} \\
&= [\mathbf{b} \cdot (\alpha \mathbf{v}) + \mathbf{b} \cdot \mathbf{w})]\mathbf{a} && \text{(property (2.6)}_3\text{)} \\
&= [\alpha(\mathbf{b} \cdot \mathbf{v}) + \mathbf{b} \cdot \mathbf{w})]\mathbf{a} && \text{(property (2.6)}_2\text{)} \\
&= [\alpha(\mathbf{b} \cdot \mathbf{v})]\,\mathbf{a} + (\mathbf{b} \cdot \mathbf{w})\mathbf{a} && \text{(property (2.3)}_9\text{)} \\
&= \alpha[(\mathbf{b} \cdot \mathbf{v})\mathbf{a}] + (\mathbf{b} \cdot \mathbf{w})\mathbf{a} && \text{(property (2.3)}_7\text{)} \\
&= \alpha[(\mathbf{a} \otimes \mathbf{b})\mathbf{v}] + (\mathbf{a} \otimes \mathbf{b})\mathbf{w} && \text{(definition (2.19)).}
\end{aligned}$$

PROBLEM 2.14
In direct notation, prove that $(\mathbf{a} \otimes \mathbf{b})^{\mathrm{T}} = \mathbf{b} \otimes \mathbf{a}$.

Solution
Since $\mathbf{a} \otimes \mathbf{b}$ is a tensor (refer to Problem 2.13), it follows from definition (2.13) that

$$(\mathbf{a} \otimes \mathbf{b})\mathbf{u} \cdot \mathbf{v} = \mathbf{u} \cdot (\mathbf{a} \otimes \mathbf{b})^{\mathrm{T}}\mathbf{v}$$

for all vectors $\mathbf{u}$, $\mathbf{v}$ in $\mathcal{E}^3$. Working with the left-hand side, we obtain

$$\begin{aligned}
(\mathbf{a}\otimes\mathbf{b})\mathbf{u}\cdot\mathbf{v} &= [(\mathbf{b}\cdot\mathbf{u})\mathbf{a}]\cdot\mathbf{v} && \text{(definition (2.19))}\\
&= (\mathbf{b}\cdot\mathbf{u})(\mathbf{a}\cdot\mathbf{v}) && \text{(property } (2.6)_2)\\
&= (\mathbf{a}\cdot\mathbf{v})(\mathbf{b}\cdot\mathbf{u}) && \text{(property } (2.1)_7)\\
&= [(\mathbf{a}\cdot\mathbf{v})\mathbf{b}]\cdot\mathbf{u} && \text{(property } (2.6)_2)\\
&= \mathbf{u}\cdot[(\mathbf{a}\cdot\mathbf{v})\mathbf{b}] && \text{(property } (2.6)_1)\\
&= \mathbf{u}\cdot(\mathbf{b}\otimes\mathbf{a})\mathbf{v} && \text{(definition (2.19)).}
\end{aligned}$$

Using this result in the original statement, we obtain

$$\mathbf{u}\cdot(\mathbf{b}\otimes\mathbf{a})\mathbf{v} = \mathbf{u}\cdot(\mathbf{a}\otimes\mathbf{b})^{\mathrm{T}}\mathbf{v}.$$

Since $\mathbf{u}$ and $\mathbf{v}$ are arbitrary, it follows that $(\mathbf{a}\otimes\mathbf{b})^{\mathrm{T}} = \mathbf{b}\otimes\mathbf{a}$.

PROBLEM 2.15

In direct notation, prove that $\mathbf{S}\,(\mathbf{a}\otimes\mathbf{b}) = (\mathbf{S}\,\mathbf{a})\otimes\mathbf{b}$.

Solution

For any vector $\mathbf{v}$ in $\mathcal{E}^3$,

$$[\mathbf{S}(\mathbf{a}\otimes\mathbf{b})]\mathbf{v} = \mathbf{S}[(\mathbf{a}\otimes\mathbf{b})\mathbf{v}] = \mathbf{S}[(\mathbf{b}\cdot\mathbf{v})\mathbf{a}] = (\mathbf{b}\cdot\mathbf{v})(\mathbf{S}\mathbf{a}) = [(\mathbf{S}\mathbf{a})\otimes\mathbf{b}]\mathbf{v}.$$

Since $\mathbf{v}$ is arbitrary, it follows that $\mathbf{S}(\mathbf{a}\otimes\mathbf{b}) = (\mathbf{S}\mathbf{a})\otimes\mathbf{b}$.

PROBLEM 2.16

In direct notation, prove that $(\mathbf{a}\otimes\mathbf{b})\,\mathbf{S} = \mathbf{a}\otimes\left(\mathbf{S}^{\mathrm{T}}\,\mathbf{b}\right)$.

Solution

For any vector $\mathbf{v}$ in $\mathcal{E}^3$,

$$\begin{aligned}
[(\mathbf{a}\otimes\mathbf{b})\mathbf{S}]\mathbf{v} &= (\mathbf{a}\otimes\mathbf{b})(\mathbf{S}\mathbf{v}) && \text{(definition (2.11))}\\
&= (\mathbf{b}\cdot\mathbf{S}\mathbf{v})\mathbf{a} && \text{(definition (2.19))}\\
&= [\mathbf{b}\cdot(\mathbf{S}^{\mathrm{T}})^{\mathrm{T}}\mathbf{v}]\mathbf{a} && \text{(result } (2.14)_3)\\
&= (\mathbf{S}^{\mathrm{T}}\mathbf{b}\cdot\mathbf{v})\mathbf{a} && \text{(definition (2.13))}
\end{aligned}$$

$$= [\mathbf{a} \otimes (\mathbf{S}^{\mathrm{T}}\mathbf{b})]\mathbf{v} \qquad \text{(definition (2.19))}.$$

Since $\mathbf{v}$ is arbitrary, $(\mathbf{a} \otimes \mathbf{b})\mathbf{S} = \mathbf{a} \otimes (\mathbf{S}^{\mathrm{T}}\mathbf{b})$.

PROBLEM 2.17

In direct notation, prove that $(\mathbf{a} \otimes \mathbf{b})\,(\mathbf{c} \otimes \mathbf{d}) = (\mathbf{b} \cdot \mathbf{c})\,(\mathbf{a} \otimes \mathbf{d})$.

Solution

For any vector $\mathbf{v}$ in $\mathcal{E}^3$,

$$\begin{aligned}
[(\mathbf{a} \otimes \mathbf{b})(\mathbf{c} \otimes \mathbf{d})]\mathbf{v} &= (\mathbf{a} \otimes \mathbf{b})[(\mathbf{c} \otimes \mathbf{d})\mathbf{v}] && \text{(definition (2.11))} \\
&= (\mathbf{a} \otimes \mathbf{b})[(\mathbf{d} \cdot \mathbf{v})\mathbf{c}] && \text{(definition (2.19))} \\
&= (\mathbf{d} \cdot \mathbf{v})[(\mathbf{a} \otimes \mathbf{b})\mathbf{c}] && \text{(definition (2.7)}_2\text{)} \\
&= (\mathbf{d} \cdot \mathbf{v})[(\mathbf{b} \cdot \mathbf{c})\mathbf{a}] && \text{(definition (2.19))} \\
&= [(\mathbf{d} \cdot \mathbf{v})(\mathbf{b} \cdot \mathbf{c})]\mathbf{a} && \text{(property (2.3)}_7\text{)} \\
&= [(\mathbf{b} \cdot \mathbf{c})(\mathbf{d} \cdot \mathbf{v})]\mathbf{a} && \text{(property (2.1)}_7\text{)} \\
&= (\mathbf{b} \cdot \mathbf{c})[(\mathbf{d} \cdot \mathbf{v})\mathbf{a}] && \text{(property (2.3)}_7\text{)} \\
&= (\mathbf{b} \cdot \mathbf{c})\,[(\mathbf{a} \otimes \mathbf{d})\mathbf{v}] && \text{(definition (2.19))} \\
&= [(\mathbf{b} \cdot \mathbf{c})(\mathbf{a} \otimes \mathbf{d})]\,\mathbf{v} && \text{(definition (2.8)}_2\text{)}.
\end{aligned}$$

Since $\mathbf{v}$ is arbitrary, it follows that $(\mathbf{a} \otimes \mathbf{b})\,(\mathbf{c} \otimes \mathbf{d}) = (\mathbf{b} \cdot \mathbf{c})\,(\mathbf{a} \otimes \mathbf{d})$.

2.2.4 CARTESIAN COMPONENTS, INDICIAL NOTATION, SUMMATION CONVENTION

Every Euclidean space $\mathcal{E}^n$ has an **orthonormal basis**. A **basis** for $\mathcal{E}^n$ is a *linearly independent* set of vectors that *spans* $\mathcal{E}^n$. Orthonormality implies that each pair of vectors in the set is orthogonal (or perpendicular) to each other, and every vector in the set has unit norm. Three-dimensional Euclidean space $\mathcal{E}^3$ has a fixed orthonormal basis $\{\mathbf{e}_1, \mathbf{e}_2, \mathbf{e}_3\}$.[3] Any vector $\mathbf{u}$ in $\mathcal{E}^3$ can be expressed as a linear combination of the basis vectors $\mathbf{e}_1$, $\mathbf{e}_2$, and $\mathbf{e}_3$:

$$\mathbf{u} = u_1\,\mathbf{e}_1 + u_2\,\mathbf{e}_2 + u_3\,\mathbf{e}_3, \tag{2.21}$$

[3]Note that $\mathcal{E}^3$ has infinitely many other fixed orthonormal bases which are merely rotations or inversions of $\{\mathbf{e}_1, \mathbf{e}_2, \mathbf{e}_3\}$. It also has infinitely many other bases which are not orthogonal or fixed, which we discuss later in this chapter.

where

$$u_1 = \mathbf{u} \cdot \mathbf{e}_1, \qquad u_2 = \mathbf{u} \cdot \mathbf{e}_2, \qquad u_3 = \mathbf{u} \cdot \mathbf{e}_3 \tag{2.22}$$

are called the **Cartesian components** of the vector $\mathbf{u}$ (i.e., the projections of $\mathbf{u}$ along $\mathbf{e}_1$, $\mathbf{e}_2$, and $\mathbf{e}_3$). Note that since $\{\mathbf{e}_1, \mathbf{e}_2, \mathbf{e}_3\}$ is an orthonormal basis,

$$\mathbf{e}_1 \cdot \mathbf{e}_1 = 1, \qquad \mathbf{e}_1 \cdot \mathbf{e}_2 = 0, \qquad \mathbf{e}_1 \cdot \mathbf{e}_3 = 0, \qquad \mathbf{e}_2 \cdot \mathbf{e}_1 = 0, \qquad \mathbf{e}_2 \cdot \mathbf{e}_2 = 1,$$

$$\mathbf{e}_2 \cdot \mathbf{e}_3 = 0, \qquad \mathbf{e}_3 \cdot \mathbf{e}_1 = 0, \qquad \mathbf{e}_3 \cdot \mathbf{e}_2 = 0, \qquad \mathbf{e}_3 \cdot \mathbf{e}_3 = 1. \tag{2.23}$$

We now present two conventions that introduce no additional physics, but save writing. First, **indicial notation** can be used to write the three equations (2.22) as the single expression

$$u_i = \mathbf{u} \cdot \mathbf{e}_i, \tag{2.24}$$

where the subscript (or index) i takes the values 1, 2, 3. Similarly, the nine equations (2.23) can be written as the single expression

$$\mathbf{e}_i \cdot \mathbf{e}_j = \delta_{ij}, \tag{2.25}$$

where the subscripts i and j take the values 1, 2, 3, and

$$\delta_{ij} = \begin{cases} 1 & \text{if } i = j, \\ 0 & \text{if } i \neq j \end{cases} \tag{2.26}$$

is the **Kronecker delta**. The second tool to save writing is the **summation convention**

$$\sum_{i=1}^{3} u_i \, \mathbf{e}_i = u_i \, \mathbf{e}_i, \tag{2.27}$$

where repeated subscripts imply summation over the values 1, 2, 3. This allows us to write

$$\mathbf{u} = u_1 \, \mathbf{e}_1 + u_2 \, \mathbf{e}_2 + u_3 \, \mathbf{e}_3 = \sum_{i=1}^{3} u_i \, \mathbf{e}_i = u_i \, \mathbf{e}_i. \tag{2.28}$$

The **Cartesian components** S_{ij} of a second-order tensor $\mathbf{S}$ are defined by

$$S_{ij} = \mathbf{e}_i \cdot \mathbf{S} \, \mathbf{e}_j, \tag{2.29}$$

so S_{ij} is the projection of the map $\mathbf{S}\mathbf{e}_j$ onto $\mathbf{e}_i$. It can be verified (refer to Problem 2.19) that

$$\delta_{ii} = 3, \qquad \delta_{ij} \, v_j = v_i, \qquad \delta_{ik} \, T_{kj} = T_{ij}, \qquad \delta_{ij} \, T_{ij} = T_{ii}. \tag{2.30}$$

Additionally (refer to Problem 2.20),

$$\mathbf{u} \cdot \mathbf{v} = u_i \, v_i. \tag{2.31}$$

We can also show (refer to Problem 2.21) that the Cartesian component form of the vector equation $\mathbf{v} = \mathbf{S}\mathbf{u}$ is, in indicial notation,

$$v_i = S_{ij}\, u_j. \tag{2.32}$$

Recall that the set $\mathcal{L}$ of all second-order tensors is a vector space. It can be shown (refer to Problem 2.22) that a basis for $\mathcal{L}$ is $\{\mathbf{e}_i \otimes \mathbf{e}_j\}$. It follows, then, that $\mathcal{L}$ is a nine-dimensional vector space.

Also, it can be shown (refer to Problems 2.23–2.27) that

$$\mathbf{S} = S_{ij}\,(\mathbf{e}_i \otimes \mathbf{e}_j), \quad (\mathbf{I})_{ij} = \delta_{ij}, \quad (\mathbf{S}^{\mathrm{T}})_{ij} = S_{ji}, \quad (\mathbf{ST})_{ij} = S_{ik}T_{kj}, \quad (\mathbf{a} \otimes \mathbf{b})_{ij} = a_i b_j. \tag{2.33}$$

The **matrix** $[\mathbf{A}]$ of the tensor $\mathbf{A}$ is defined by

$$[\mathbf{A}] = \begin{bmatrix} A_{11} & A_{12} & A_{13} \\ A_{21} & A_{22} & A_{23} \\ A_{31} & A_{32} & A_{33} \end{bmatrix}. \tag{2.34}$$

Refer to Problem 2.28. Then, using matrix algebra, we can show that the matrix of the product $\mathbf{AB}$ is the product of the matrices $[\mathbf{A}]$ and $[\mathbf{B}]$, i.e.,

$$[\mathbf{AB}] = [\mathbf{A}][\mathbf{B}]. \tag{2.35}$$

Similarly, it can be shown that the matrix of the sum $\mathbf{A} + \mathbf{B}$ is the sum of the matrices $[\mathbf{A}]$ and $[\mathbf{B}]$, i.e.,

$$[\mathbf{A} + \mathbf{B}] = [\mathbf{A}] + [\mathbf{B}]. \tag{2.36}$$

Also,

$$[\mathbf{A}^{\mathrm{T}}] = [\mathbf{A}]^{\mathrm{T}}. \tag{2.37}$$

PROBLEM 2.18

For each of the following expressions, identify how many equations the expression represents, and then fully expand each expression for a subscript range of 3:

(a) $b_i = A_{ij}\, x_j$.
(b) $\alpha = A_{ij}\, x_i\, x_j$.
(c) $\beta = A_{ii}\, B_{jj}$.
(d) $\gamma = A_{ij}\, B_{ij}$.
(e) $C_{ij} = A_{ik} B_{kj}$.

Solution

(a) For each value of i, we sum over the repeated subscript j. Thus, this expression represents three scalar equations, each a sum of three products:

$$(i=1) \qquad b_1 = \sum_{j=1}^{3} A_{1j}x_j = A_{11}x_1 + A_{12}x_2 + A_{13}x_3,$$

$$(i=2) \qquad b_2 = \sum_{j=1}^{3} A_{2j}x_j = A_{21}x_1 + A_{22}x_2 + A_{23}x_3,$$

$$(i=3) \qquad b_3 = \sum_{j=1}^{3} A_{3j}x_j = A_{31}x_1 + A_{32}x_2 + A_{33}x_3.$$

(b) We sum over the repeated subscripts i and j. Thus, this expression represents one scalar equation with nine terms (a sum of nine products):

$$\begin{aligned}\alpha &= \sum_{j=1}^{3}\sum_{i=1}^{3} A_{ij}x_ix_j \\ &= \sum_{j=1}^{3}(A_{1j}x_1x_j + A_{2j}x_2x_j + A_{3j}x_3x_j) \\ &= A_{11}x_1x_1 + A_{12}x_1x_2 + A_{13}x_1x_3 + A_{21}x_2x_1 + A_{22}x_2x_2 \\ &\quad + A_{23}x_2x_3 + A_{31}x_3x_1 + A_{32}x_3x_2 + A_{33}x_3x_3.\end{aligned}$$

(c) We sum over the repeated subscripts i and j. Thus, this expression represents one scalar equation with nine terms:

$$\begin{aligned}\beta &= \sum_{j=1}^{3}\sum_{i=1}^{3} A_{ii}B_{jj} \\ &= (A_{11} + A_{22} + A_{33})\sum_{j=1}^{3} B_{jj} \\ &= (A_{11} + A_{22} + A_{33})(B_{11} + B_{22} + B_{33}) \\ &= A_{11}B_{11} + A_{11}B_{22} + A_{11}B_{33} + A_{22}B_{11} + A_{22}B_{22} \\ &\quad + A_{22}B_{33} + A_{33}B_{11} + A_{33}B_{22} + A_{33}B_{33}.\end{aligned}$$

(d) We sum over the repeated subscripts i and j. Thus, this expression represents one scalar equation with nine terms:

$$\gamma = \sum_{j=1}^{3}\sum_{i=1}^{3} A_{ij}B_{ij}$$

$$= \sum_{j=1}^{3} (A_{1j}B_{1j} + A_{2j}B_{2j} + A_{3j}B_{3j})$$

$$= A_{11}B_{11} + A_{12}B_{12} + A_{13}B_{13} + A_{21}B_{21} + A_{22}B_{22}$$

$$+ A_{23}B_{23} + A_{31}B_{31} + A_{32}B_{32} + A_{33}B_{33}.$$

(e) For each value of i and j, we sum over the repeated subscript k. Thus, this expression represents nine scalar equations, each a sum of three products:

$$(i = j = 1) \qquad C_{11} = \sum_{k=1}^{3} A_{1k}B_{k1} = A_{11}B_{11} + A_{12}B_{21} + A_{13}B_{31},$$

$$(i = 1, j = 2) \qquad C_{12} = \sum_{k=1}^{3} A_{1k}B_{k2} = A_{11}B_{12} + A_{12}B_{22} + A_{13}B_{32},$$

$$(i = 1, j = 3) \qquad C_{13} = \sum_{k=1}^{3} A_{1k}B_{k3} = A_{11}B_{13} + A_{12}B_{23} + A_{13}B_{33},$$

$$(i = 2, j = 1) \qquad C_{21} = \sum_{k=1}^{3} A_{2k}B_{k1} = A_{21}B_{11} + A_{22}B_{21} + A_{23}B_{31},$$

$$(i = j = 2) \qquad C_{22} = \sum_{k=1}^{3} A_{2k}B_{k2} = A_{21}B_{12} + A_{22}B_{22} + A_{23}B_{32},$$

$$(i = 2, j = 3) \qquad C_{23} = \sum_{k=1}^{3} A_{2k}B_{k3} = A_{21}B_{13} + A_{22}B_{23} + A_{23}B_{33},$$

$$(i = 3, j = 1) \qquad C_{31} = \sum_{k=1}^{3} A_{3k}B_{k1} = A_{31}B_{11} + A_{32}B_{21} + A_{33}B_{31},$$

$$(i = 3, j = 2) \qquad C_{32} = \sum_{k=1}^{3} A_{3k}B_{k2} = A_{31}B_{12} + A_{32}B_{22} + A_{33}B_{32},$$

$$(i = j = 3) \qquad C_{33} = \sum_{k=1}^{3} A_{3k}B_{k3} = A_{31}B_{13} + A_{32}B_{23} + A_{33}B_{33}.$$

PROBLEM 2.19

Verify the following results:

(a) $\delta_{ii} = 3$.

(b) $\delta_{ij} v_j = v_i$.

(c) $\delta_{ik} T_{kj} = T_{ij}$.
(d) $\delta_{ij} T_{ij} = T_{ii}$.

Solution

(a) Summing over the repeated subscript i, we obtain

$$\delta_{ii} = \sum_{i=1}^{3} \delta_{ii} = \delta_{11} + \delta_{22} + \delta_{33} = 1 + 1 + 1 = 3.$$

(b) Summing over the repeated subscript j for each value of i, we obtain

$$(i = 1) \qquad \delta_{1j} v_j = \sum_{j=1}^{3} \delta_{1j} v_j = \delta_{11} v_1 + \delta_{12} v_2 + \delta_{13} v_3 = v_1,$$

$$(i = 2) \qquad \delta_{2j} v_j = \sum_{j=1}^{3} \delta_{2j} v_j = \delta_{21} v_1 + \delta_{22} v_2 + \delta_{23} v_3 = v_2,$$

$$(i = 3) \qquad \delta_{3j} v_j = \sum_{j=1}^{3} \delta_{3j} v_j = \delta_{31} v_1 + \delta_{32} v_2 + \delta_{33} v_3 = v_3,$$

where we have used $\delta_{11} = \delta_{22} = \delta_{33} = 1$, with all other permutations vanishing.

(c) For each value of i and j, we sum over the repeated subscript k:

$$(i = j = 1) \qquad \delta_{1k} T_{k1} = \sum_{k=1}^{3} \delta_{1k} T_{k1} = \delta_{11} T_{11} + \delta_{12} T_{21} + \delta_{13} T_{31} = T_{11},$$

$$(i = 1, j = 2) \qquad \delta_{1k} T_{k2} = \sum_{k=1}^{3} \delta_{1k} T_{k2} = \delta_{11} T_{12} + \delta_{12} T_{22} + \delta_{13} T_{32} = T_{12},$$

$$(i = 1, j = 3) \qquad \delta_{1k} T_{k3} = \sum_{k=1}^{3} \delta_{1k} T_{k3} = \delta_{11} T_{13} + \delta_{12} T_{23} + \delta_{13} T_{33} = T_{13},$$

$$(i = 2, j = 1) \qquad \delta_{2k} T_{k1} = \sum_{k=1}^{3} \delta_{2k} T_{k1} = \delta_{21} T_{11} + \delta_{22} T_{21} + \delta_{23} T_{31} = T_{21},$$

$$(i = j = 2) \qquad \delta_{2k} T_{k2} = \sum_{k=1}^{3} \delta_{2k} T_{k2} = \delta_{21} T_{12} + \delta_{22} T_{22} + \delta_{23} T_{32} = T_{22},$$

$$(i = 2, j = 3) \qquad \delta_{2k} T_{k3} = \sum_{k=1}^{3} \delta_{2k} T_{k3} = \delta_{21} T_{13} + \delta_{22} T_{23} + \delta_{23} T_{33} = T_{23},$$

$$(i = 3, j = 1) \qquad \delta_{3k} T_{k1} = \sum_{k=1}^{3} \delta_{3k} T_{k1} = \delta_{31} T_{11} + \delta_{32} T_{21} + \delta_{33} T_{31} = T_{31},$$

$$(i=3, j=2) \qquad \delta_{3k}T_{k2} = \sum_{k=1}^{3} \delta_{3k}T_{k2} = \delta_{31}T_{12} + \delta_{32}T_{22} + \delta_{33}T_{32} = T_{32},$$

$$(i=j=3) \qquad \delta_{3k}T_{k3} = \sum_{k=1}^{3} \delta_{3k}T_{k3} = \delta_{31}T_{13} + \delta_{32}T_{23} + \delta_{33}T_{33} = T_{33},$$

where we have used $\delta_{11} = \delta_{22} = \delta_{33} = 1$, with all other permutations vanishing.

(d) Summing over the repeated subscripts i and j, we obtain

$$\begin{aligned}
\delta_{ij}T_{ij} &= \sum_{j=1}^{3}\sum_{i=1}^{3} \delta_{ij}T_{ij} \\
&= \sum_{j=1}^{3} (\delta_{1j}T_{1j} + \delta_{2j}T_{2j} + \delta_{3j}T_{3j}) \\
&= \delta_{11}T_{11} + \delta_{12}T_{12} + \delta_{13}T_{13} + \delta_{21}T_{21} + \delta_{22}T_{22} + \delta_{23}T_{23} \\
&\quad + \delta_{31}T_{31} + \delta_{32}T_{32} + \delta_{33}T_{33} \\
&= T_{11} + T_{22} + T_{33} \\
&= \sum_{i=1}^{3} T_{ii} \\
&= T_{ii}.
\end{aligned}$$

PROBLEM 2.20

Prove that $\mathbf{u} \cdot \mathbf{v} = u_i v_i$.

Solution I (by direct expansion)

$$\begin{aligned}
\mathbf{u} \cdot \mathbf{v} &= (u_1\mathbf{e}_1 + u_2\mathbf{e}_2 + u_3\mathbf{e}_3) \cdot (v_1\mathbf{e}_1 + v_2\mathbf{e}_2 + v_3\mathbf{e}_3) && \text{(representation (2.21))} \\
&= (u_1\mathbf{e}_1) \cdot (v_1\mathbf{e}_1) + (u_1\mathbf{e}_1) \cdot (v_2\mathbf{e}_2) + (u_1\mathbf{e}_1) \cdot (v_3\mathbf{e}_3) \\
&\quad + (u_2\mathbf{e}_2) \cdot (v_1\mathbf{e}_1) + (u_2\mathbf{e}_2) \cdot (v_2\mathbf{e}_2) + (u_2\mathbf{e}_2) \cdot (v_3\mathbf{e}_3) && \text{(property (2.6)}_3\text{)} \\
&\quad + (u_3\mathbf{e}_3) \cdot (v_1\mathbf{e}_1) + (u_3\mathbf{e}_3) \cdot (v_2\mathbf{e}_2) + (u_3\mathbf{e}_3) \cdot (v_3\mathbf{e}_3) \\
&= u_1v_1(\mathbf{e}_1 \cdot \mathbf{e}_1) + u_1v_2(\mathbf{e}_1 \cdot \mathbf{e}_2) + u_1v_3(\mathbf{e}_1 \cdot \mathbf{e}_3) \\
&\quad + u_2v_1(\mathbf{e}_2 \cdot \mathbf{e}_1) + u_2v_2(\mathbf{e}_2 \cdot \mathbf{e}_2) + u_2v_3(\mathbf{e}_2 \cdot \mathbf{e}_3) && \text{(property (2.6)}_2\text{)} \\
&\quad + u_3v_1(\mathbf{e}_3 \cdot \mathbf{e}_1) + u_3v_2(\mathbf{e}_3 \cdot \mathbf{e}_2) + u_3v_3(\mathbf{e}_3 \cdot \mathbf{e}_3) \\
&= u_1v_1 + u_2v_2 + u_3v_3 && \text{(property (2.23))} \\
&= u_i v_i && \text{(convention (2.27)).}
\end{aligned}$$

Solution II (using the summation convention)

$$\begin{aligned}
\mathbf{u}\cdot\mathbf{v} &= u_i\,\mathbf{e}_i\cdot v_j\,\mathbf{e}_j && \text{(representation (2.27))}\\
&= u_i\,v_j\,(\mathbf{e}_i\cdot\mathbf{e}_j) && \text{(property (2.6)}_2\text{)}\\
&= u_i\,v_j\,\delta_{ij} && \text{(property (2.25))}\\
&= u_i\,v_i && \text{(result (2.30)}_2\text{).}
\end{aligned}$$

PROBLEM 2.21

Prove that the Cartesian component form of the vector equation $\mathbf{v}=\mathbf{S}\mathbf{u}$ is, in indicial notation, $v_i = S_{ij}\,u_j$.

Solution

Following (2.28), we write vectors $\mathbf{v}$ and $\mathbf{Su}$ as linear combinations of the basis vectors $\mathbf{e}_1$, $\mathbf{e}_2$, and $\mathbf{e}_3$, so $\mathbf{v}=\mathbf{Su}$ implies

$$(v_i - (\mathbf{Su})_i)\,\mathbf{e}_i = \mathbf{0}.$$

Linear independence of the basis vectors $\{\mathbf{e}_i\}$ demands that their coefficients must vanish, i.e., $v_i - (\mathbf{Su})_i = 0$. Thus,

$$v_i = (\mathbf{Su})_i = (\mathbf{S}\,u_j\,\mathbf{e}_j)_i = u_j\,(\mathbf{S}\mathbf{e}_j)_i = u_j\,[\mathbf{e}_i\cdot(\mathbf{S}\mathbf{e}_j)] = S_{ij}\,u_j.$$

PROBLEM 2.22

Show that the set $\mathcal{L}$ of all second-order tensors is a nine-dimensional vector space.

Solution

Recall that in Problem 2.4, we proved that the set $\mathcal{L}$ of all second-order tensors is a vector space. To show that the **dimension** of $\mathcal{L}$ is nine, we must demonstrate that its **basis** contains nine elements. This is accomplished by showing that $\{\mathbf{e}_i\otimes\mathbf{e}_j\}$ is a basis for $\mathcal{L}$, i.e., $\{\mathbf{e}_i\otimes\mathbf{e}_j\}$ **spans** $\mathcal{L}$ *and* is **linearly independent**.

To show that $\{\mathbf{e}_i\otimes\mathbf{e}_j\}$ spans $\mathcal{L}$, we demonstrate that any tensor $\mathbf{T}$ in $\mathcal{L}$ can be written as a linear combination of $\{\mathbf{e}_i\otimes\mathbf{e}_j\}$:

$$\mathbf{Tu} = (\mathbf{Tu})_i\,\mathbf{e}_i = T_{ij}\,u_j\,\mathbf{e}_i = T_{ij}\,(\mathbf{e}_j\cdot\mathbf{u})\,\mathbf{e}_i = T_{ij}\,(\mathbf{e}_i\otimes\mathbf{e}_j)\mathbf{u}.$$

Since $\mathbf{u}$ is arbitrary, it follows that $\mathbf{T} = T_{ij}\,(\mathbf{e}_i\otimes\mathbf{e}_j)$. Thus, $\{\mathbf{e}_i\otimes\mathbf{e}_j\}$ spans $\mathcal{L}$.

The spanning set $\{\mathbf{e}_i\otimes\mathbf{e}_j\}$ for $\mathcal{L}$ is linearly independent if and only if $T_{ij}\,(\mathbf{e}_i\otimes\mathbf{e}_j)=0$ implies that $T_{ij}=0$. We proceed as follows:

$$\begin{aligned}
T_{ij}\,(\mathbf{e}_i\otimes\mathbf{e}_j) = 0 \quad &\Rightarrow \quad (T_{ij}\,\mathbf{e}_i\otimes\mathbf{e}_j)\,\mathbf{v} = 0\,\mathbf{v}\\
&\Rightarrow \quad T_{ij}\,(\mathbf{e}_j\cdot\mathbf{v})\,\mathbf{e}_i = \mathbf{0}\\
&\Rightarrow \quad T_{ij}\,v_j\,\mathbf{e}_i = \mathbf{0}.
\end{aligned}$$

This holds for any $\mathbf{v}$ in $\mathcal{E}^3$. Since $\{\mathbf{e}_i\}$ is a basis for $\mathcal{E}^3$, its elements are linearly independent. This demands that the coefficients $T_{ij}\, v_j$ must vanish, i.e., $T_{ij}\, v_j = 0$. Expanding this, we obtain

$$
\begin{aligned}
&(i = 1) \qquad T_{11}v_1 + T_{12}v_2 + T_{13}v_3 = 0,\\
&(i = 2) \qquad T_{21}v_1 + T_{22}v_2 + T_{23}v_3 = 0,\\
&(i = 3) \qquad T_{31}v_1 + T_{32}v_2 + T_{33}v_3 = 0.
\end{aligned}
$$

We exploit the fact that $\mathbf{v}$ is arbitrary and choose $v_1 = 1$, $v_2 = v_3 = 0$, so $T_{11} = T_{21} = T_{31} = 0$. Similar arguments can be used to show that $T_{12} = T_{22} = T_{32} = 0$ and $T_{13} = T_{23} = T_{33} = 0$. Thus, $T_{ij}\,(\mathbf{e}_i \otimes \mathbf{e}_j) = 0$ implies that $T_{ij} = 0$, and the set $\{\mathbf{e}_i \otimes \mathbf{e}_j\}$ is linearly independent.

$\{\mathbf{e}_i \otimes \mathbf{e}_j\}$ is a spanning set for $\mathcal{L}$ and the set is linearly independent, it is a basis for $\mathcal{L}$. The basis $\{\mathbf{e}_i \otimes \mathbf{e}_j\}$ has nine elements:

$$
\mathbf{e}_1 \otimes \mathbf{e}_1,\ \mathbf{e}_1 \otimes \mathbf{e}_2,\ \mathbf{e}_1 \otimes \mathbf{e}_3,\ \mathbf{e}_2 \otimes \mathbf{e}_1,\ \mathbf{e}_2 \otimes \mathbf{e}_2,\ \mathbf{e}_2 \otimes \mathbf{e}_3,\ \mathbf{e}_3 \otimes \mathbf{e}_1,\ \mathbf{e}_3 \otimes \mathbf{e}_2,\ \mathbf{e}_3 \otimes \mathbf{e}_3.
$$

Thus, the dimension of $\mathcal{L}$ is nine, and $\mathcal{L}$ is a nine-dimensional vector space.

PROBLEM 2.23

Prove that the Cartesian component form of $\mathbf{S}$ is, in indicial notation, $S_{ij}\,\mathbf{e}_i \otimes \mathbf{e}_j$.

Solution

For any arbitrary vector $\mathbf{u}$ in $\mathcal{E}^3$,

$$
\begin{aligned}
\mathbf{S}\mathbf{u} &= (\mathbf{S}\mathbf{u})_i\,\mathbf{e}_i && \text{(representation (2.28))}\\
&= S_{ij}\,u_j\,\mathbf{e}_i && \text{(result (2.32))}\\
&= S_{ij}\,(\mathbf{e}_j \cdot \mathbf{u})\,\mathbf{e}_i && \text{(definition (2.24))}\\
&= S_{ij}\,(\mathbf{e}_i \otimes \mathbf{e}_j)\mathbf{u} && \text{(definition (2.19)).}
\end{aligned}
$$

Since $\mathbf{u}$ is arbitrary, we have $\mathbf{S} = S_{ij}\,\mathbf{e}_i \otimes \mathbf{e}_j$. Expanding this, we obtain

$$
\begin{aligned}
\mathbf{S} = S_{ij}\,\mathbf{e}_i \otimes \mathbf{e}_j = {} & S_{11}\,\mathbf{e}_1 \otimes \mathbf{e}_1 + S_{12}\,\mathbf{e}_1 \otimes \mathbf{e}_2 + S_{13}\,\mathbf{e}_1 \otimes \mathbf{e}_3 + S_{21}\,\mathbf{e}_2 \otimes \mathbf{e}_1 + S_{22}\,\mathbf{e}_2 \otimes \mathbf{e}_2\\
& + S_{23}\,\mathbf{e}_2 \otimes \mathbf{e}_3 + S_{31}\,\mathbf{e}_3 \otimes \mathbf{e}_1 + S_{32}\,\mathbf{e}_3 \otimes \mathbf{e}_2 + S_{33}\,\mathbf{e}_3 \otimes \mathbf{e}_3,
\end{aligned}
$$

or, in matrix form,

$$
[\mathbf{S}] = \begin{bmatrix} S_{11} & S_{12} & S_{13} \\ S_{21} & S_{22} & S_{23} \\ S_{31} & S_{32} & S_{33} \end{bmatrix}.
$$

PROBLEM 2.24

Prove that the Cartesian components of the identity tensor are $(\mathbf{I})_{ij} = \delta_{ij}$ so $\mathbf{I} = \mathbf{e}_i \otimes \mathbf{e}_i$.

Solution

Using definition (2.29), we have

$$(\mathbf{I})_{ij} = \mathbf{e}_i \cdot \mathbf{I}\,\mathbf{e}_j = \mathbf{e}_i \cdot \mathbf{e}_j = \delta_{ij}.$$

From result $(2.33)_1$ we have $\mathbf{I} = (\mathbf{I})_{ij}\,\mathbf{e}_i \otimes \mathbf{e}_j$. It follows that

$$\mathbf{I} = \delta_{ij}\,\mathbf{e}_i \otimes \mathbf{e}_j = \mathbf{e}_i \otimes \mathbf{e}_i,$$

where we have used result $(2.30)_4$. As an alternative proof, for any vector $\mathbf{v}$ in $\mathcal{E}^3$,

$$(\mathbf{e}_i \otimes \mathbf{e}_i)\mathbf{v} = (\mathbf{e}_i \cdot \mathbf{v})\mathbf{e}_i = \mathbf{v}_i\,\mathbf{e}_i = \mathbf{v} = \mathbf{I}\mathbf{v}.$$

Since $\mathbf{v}$ is arbitrary, $\mathbf{e}_i \otimes \mathbf{e}_i = \mathbf{I}$. Expanding this expression, we obtain

$$\mathbf{I} = \mathbf{e}_i \otimes \mathbf{e}_i = \mathbf{e}_1 \otimes \mathbf{e}_1 + \mathbf{e}_2 \otimes \mathbf{e}_2 + \mathbf{e}_3 \otimes \mathbf{e}_3,$$

or, in matrix form,

$$[\mathbf{I}] = \begin{bmatrix} 1 & 0 & 0 \\ 0 & 1 & 0 \\ 0 & 0 & 1 \end{bmatrix}.$$

PROBLEM 2.25

Prove that the Cartesian components of $\mathbf{S}^{\mathrm{T}}$ are $(\mathbf{S}^{\mathrm{T}})_{ij} = S_{ji}$ so $\mathbf{S}^{\mathrm{T}} = S_{ji}\,\mathbf{e}_i \otimes \mathbf{e}_j$.

Solution

$$\begin{aligned} (\mathbf{S}^{\mathrm{T}})_{ij} &= \mathbf{e}_i \cdot \mathbf{S}^{\mathrm{T}}\mathbf{e}_j && \text{(definition (2.29))} \\ &= \mathbf{S}\,\mathbf{e}_i \cdot \mathbf{e}_j && \text{(definition (2.13))} \\ &= \mathbf{e}_j \cdot \mathbf{S}\,\mathbf{e}_i && \text{(property } (2.6)_1) \\ &= S_{ji} && \text{(definition (2.29)).} \end{aligned}$$

Since $\mathbf{S}^{\mathrm{T}} = (\mathbf{S}^{\mathrm{T}})_{ij}\,\mathbf{e}_i \otimes \mathbf{e}_j$, we have $\mathbf{S}^{\mathrm{T}} = S_{ji}\,\mathbf{e}_i \otimes \mathbf{e}_j$. Expanding this, we obtain

$$\begin{aligned} \mathbf{S}^{\mathrm{T}} = S_{ji}\,\mathbf{e}_i \otimes \mathbf{e}_j &= S_{11}\,\mathbf{e}_1 \otimes \mathbf{e}_1 + S_{21}\,\mathbf{e}_1 \otimes \mathbf{e}_2 + S_{31}\,\mathbf{e}_1 \otimes \mathbf{e}_3 + S_{12}\,\mathbf{e}_2 \otimes \mathbf{e}_1 + S_{22}\,\mathbf{e}_2 \otimes \mathbf{e}_2 \\ &\quad + S_{32}\,\mathbf{e}_2 \otimes \mathbf{e}_3 + S_{13}\,\mathbf{e}_3 \otimes \mathbf{e}_1 + S_{23}\,\mathbf{e}_3 \otimes \mathbf{e}_2 + S_{33}\,\mathbf{e}_3 \otimes \mathbf{e}_3, \end{aligned}$$

or, in matrix form,

$$[\mathbf{S}^{\mathrm{T}}] = \begin{bmatrix} S_{11} & S_{21} & S_{31} \\ S_{12} & S_{22} & S_{32} \\ S_{13} & S_{23} & S_{33} \end{bmatrix}.$$

PROBLEM 2.26

Prove that the Cartesian components of the product of two tensors **S** and **T** are $(\mathbf{ST})_{ij} = S_{ik} T_{kj}$.

Solution

$$
\begin{aligned}
\mathbf{ST} &= (S_{ij}\, \mathbf{e}_i \otimes \mathbf{e}_j)(T_{kl}\, \mathbf{e}_k \otimes \mathbf{e}_l) && \text{(result (2.33)}_1\text{)} \\
&= S_{ij}\, T_{kl}\, (\mathbf{e}_i \otimes \mathbf{e}_j)(\mathbf{e}_k \otimes \mathbf{e}_l) \\
&= S_{ij}\, T_{kl}\, (\mathbf{e}_j \cdot \mathbf{e}_k)(\mathbf{e}_i \otimes \mathbf{e}_l) && \text{(result (2.20)}_4\text{)} \\
&= S_{ij}\, T_{kl}\, \delta_{jk}\, \mathbf{e}_i \otimes \mathbf{e}_l && \text{(property (2.25))} \\
&= S_{ij}\, T_{jl}\, \mathbf{e}_i \otimes \mathbf{e}_l && \text{(result (2.30)}_3\text{)} \\
&= \underbrace{S_{ik}\, T_{kj}}_{(\mathbf{ST})_{ij}}\, \mathbf{e}_i \otimes \mathbf{e}_j && \text{(change of repeated indices).}
\end{aligned}
$$

Hence, $(\mathbf{ST})_{ij} = S_{ik}\, T_{kj}$.

PROBLEM 2.27

Prove that the Cartesian components of the dyadic product $\mathbf{a} \otimes \mathbf{b}$ are $(\mathbf{a} \otimes \mathbf{b})_{ij} = a_i\, b_j$.

Solution

$$(\mathbf{a} \otimes \mathbf{b})_{ij} = \mathbf{e}_i \cdot (\mathbf{a} \otimes \mathbf{b})\mathbf{e}_j = \mathbf{e}_i \cdot [(\mathbf{b} \cdot \mathbf{e}_j)\mathbf{a}] = \mathbf{e}_i \cdot (b_j\, \mathbf{a}) = b_j\, (\mathbf{e}_i \cdot \mathbf{a}) = a_i\, b_j.$$

PROBLEM 2.28

Starting with the vector equation $\mathbf{v} = \mathbf{Su}$ in direct notation, write its corresponding Cartesian component forms (i.e., indicial, expanded, and matrix).

Solution

Recall from Problem 2.21 that the Cartesian component form of the vector equation $\mathbf{v} = \mathbf{Su}$ is, in indicial notation,

$$v_i = S_{ij}\, u_j.$$

For each value of i, we sum over the repeated index j, yielding three scalar equations:

$$
\begin{aligned}
v_1 &= S_{11} u_1 + S_{12}\, u_2 + S_{13}\, u_3, \\
v_2 &= S_{21} u_1 + S_{22}\, u_2 + S_{23}\, u_3,
\end{aligned}
$$

$$v_3 = S_{31} u_1 + S_{32} u_2 + S_{33} u_3.$$

This expanded Cartesian component form can then be expressed in matrix form as

$$\begin{bmatrix} v_1 \\ v_2 \\ v_3 \end{bmatrix} = \begin{bmatrix} S_{11} & S_{12} & S_{13} \\ S_{21} & S_{22} & S_{23} \\ S_{31} & S_{32} & S_{33} \end{bmatrix} \begin{bmatrix} u_1 \\ u_2 \\ u_3 \end{bmatrix}.$$

2.2.5 TRACE, SCALAR PRODUCT, DETERMINANT

The **trace** is an operation that accepts a tensor $\mathbf{T}$ as input and provides a real number $\operatorname{tr}\mathbf{T}$ as output. It is defined such that

$$\operatorname{tr}(\mathbf{S}+\mathbf{T}) = \operatorname{tr}\mathbf{S} + \operatorname{tr}\mathbf{T}, \qquad \operatorname{tr}(\alpha\mathbf{T}) = \alpha \operatorname{tr}\mathbf{T}, \qquad \operatorname{tr}(\mathbf{a}\otimes\mathbf{b}) = \mathbf{a}\cdot\mathbf{b} \tag{2.38}$$

for any tensors $\mathbf{S}, \mathbf{T} \in \mathcal{L}$, vectors $\mathbf{a}, \mathbf{b} \in \mathcal{E}^3$, and scalars $\alpha \in \mathbb{R}$. It can be shown (refer to Problem 2.29) that the component form of $\operatorname{tr}\mathbf{T}$ is

$$\operatorname{tr}\mathbf{T} = T_{ii}. \tag{2.39}$$

Also, we have (refer to Problems 2.30–2.32)

$$\operatorname{tr}\mathbf{I} = 3, \qquad \operatorname{tr}\mathbf{T}^{\mathrm{T}} = \operatorname{tr}\mathbf{T}, \qquad \operatorname{tr}(\mathbf{TS}) = \operatorname{tr}(\mathbf{ST}). \tag{2.40}$$

The **scalar product** of two tensors is defined by

$$\mathbf{S}\cdot\mathbf{T} = \operatorname{tr}(\mathbf{S}\mathbf{T}^{\mathrm{T}}). \tag{2.41}$$

As such, it accepts two tensors as inputs and provides a real number (scalar) as output. In component form, we have (refer to Problem 2.33)

$$\mathbf{S}\cdot\mathbf{T} = S_{ij} T_{ij}. \tag{2.42}$$

It can be shown (refer to Problem 2.34) that (2.41) satisfies (2.6), so it qualifies as an inner product; hence, the set $\mathcal{L}$ of all second-order tensors is an inner product space. Since $\mathcal{L}$ is nine-dimensional and an inner product space, it follows that $\mathcal{L} = \mathcal{E}^9$.

Using (2.41), we can demonstrate that

$$\mathbf{S}\cdot(\mathbf{a}\otimes\mathbf{b}) = \mathbf{a}\cdot\mathbf{S}\mathbf{b}. \tag{2.43}$$

It can be verified (refer to Problem 2.35) that if $\mathbf{D}$ is symmetric and $\mathbf{W}$ is skew, then

$$\mathbf{D}\cdot\mathbf{W} = \mathbf{0}. \tag{2.44}$$

Also, if $\mathbf{W}$ is skew, then

$$\operatorname{tr}\mathbf{W} = 0. \tag{2.45}$$

The **determinant** of the tensor $\mathbf{S}$ is defined to be the determinant of the corresponding matrix $[\mathbf{S}]$, i.e.,

$$\det\mathbf{S} = \det[\mathbf{S}]. \tag{2.46}$$

We can then use matrix algebra to show that

$$\det(\mathbf{ST}) = (\det \mathbf{S})(\det \mathbf{T}), \qquad \det \mathbf{S}^{\mathrm{T}} = \det \mathbf{S}. \tag{2.47}$$

PROBLEM 2.29

Prove that $\operatorname{tr} \mathbf{T} = T_{ii}$.

Solution

$$\begin{aligned}
\operatorname{tr} \mathbf{T} &= \operatorname{tr}(T_{ij}\, \mathbf{e}_i \otimes \mathbf{e}_j) && \text{(result (2.33)}_1\text{)}\\
&= T_{ij} \operatorname{tr}(\mathbf{e}_i \otimes \mathbf{e}_j) && \text{(property (2.38)}_2\text{)}\\
&= T_{ij}\,(\mathbf{e}_i \cdot \mathbf{e}_j) && \text{(definition (2.38)}_3\text{)}\\
&= T_{ij}\,\delta_{ij} && \text{(property (2.25))}\\
&= T_{ii} && \text{(result (2.30)}_4\text{).}
\end{aligned}$$

Expanding this result, we obtain

$$\operatorname{tr} \mathbf{T} = T_{ii} = T_{11} + T_{22} + T_{33}.$$

PROBLEM 2.30

Prove that $\operatorname{tr} \mathbf{I} = 3$.

Solution

$$\operatorname{tr} \mathbf{I} = \operatorname{tr}(\mathbf{e}_i \otimes \mathbf{e}_i) = \mathbf{e}_i \cdot \mathbf{e}_i = \delta_{ii} = \delta_{11} + \delta_{22} + \delta_{33} = 1 + 1 + 1 = 3.$$

PROBLEM 2.31

Prove that $\operatorname{tr} \mathbf{T}^{\mathrm{T}} = \operatorname{tr} \mathbf{T}$.

Solution

$$\operatorname{tr} \mathbf{T}^{\mathrm{T}} = \operatorname{tr}(T_{ji}\, \mathbf{e}_i \otimes \mathbf{e}_j) = T_{ji} \operatorname{tr}(\mathbf{e}_i \otimes \mathbf{e}_j) = T_{ji}\,(\mathbf{e}_i \cdot \mathbf{e}_j) = T_{ji}\,\delta_{ij} = T_{ii} = \operatorname{tr} \mathbf{T}.$$

PROBLEM 2.32

Verify that $\operatorname{tr}(\mathbf{TS}) = \operatorname{tr}(\mathbf{ST})$.

Solution

$$\operatorname{tr}(\mathbf{TS}) = \operatorname{tr}[(T_{ij}\, \mathbf{e}_i \otimes \mathbf{e}_j)(S_{kl}\, \mathbf{e}_k \otimes \mathbf{e}_l)] \qquad \text{(result (2.33)}_1\text{)}$$

$$= T_{ij}\, S_{kl}\, \mathrm{tr}\,[(\mathbf{e}_i \otimes \mathbf{e}_j)(\mathbf{e}_k \otimes \mathbf{e}_l)] \qquad \text{(property } (2.38)_2)$$

$$= T_{ij}\, S_{kl}\, \mathrm{tr}\,[(\mathbf{e}_j \cdot \mathbf{e}_k)(\mathbf{e}_i \otimes \mathbf{e}_l)] \qquad \text{(result } (2.20)_4)$$

$$= T_{ij}\, S_{kl}\, \mathrm{tr}\,[\delta_{jk}\,(\mathbf{e}_i \otimes \mathbf{e}_l)] \qquad \text{(property (2.25))}$$

$$= T_{ij}\, S_{kl}\, \delta_{jk}\, \mathrm{tr}\,(\mathbf{e}_i \otimes \mathbf{e}_l) \qquad \text{(property } (2.38)_2)$$

$$= T_{ij}\, S_{jl}\,(\mathbf{e}_i \cdot \mathbf{e}_l) \qquad \text{(result } (2.30)_3\text{, definition } (2.38)_3)$$

$$= T_{ij}\, S_{jl}\, \delta_{il} \qquad \text{(property (2.25))}$$

$$= T_{ij}\, S_{ji} \qquad \text{(result } (2.30)_3).$$

Expanding this result, we obtain

$$\mathrm{tr}\,(\mathbf{TS}) = T_{ij}\, S_{ji} = T_{11}S_{11} + T_{12}S_{21} + T_{13}S_{31} + T_{21}S_{12} + T_{22}S_{22} + T_{23}S_{32} + T_{31}S_{13} + T_{32}S_{23} + T_{33}S_{33}.$$

Similarly, it can be shown that

$$\mathrm{tr}\,(\mathbf{ST}) = S_{ij}\, T_{ji} = S_{11}T_{11} + S_{12}T_{21} + S_{13}T_{31} + S_{21}T_{12} + S_{22}T_{22} + S_{23}T_{32} + S_{31}T_{13} + S_{32}T_{23} + S_{33}T_{33}.$$

With use of the commutative and associative properties of real numbers (refer to (2.1)), it follows that $\mathrm{tr}\,(\mathbf{TS}) = \mathrm{tr}\,(\mathbf{ST})$.

PROBLEM 2.33

Prove that $\mathbf{S} \cdot \mathbf{T} = S_{ij}\, T_{ij}$.

Solution

$$\begin{aligned}\mathbf{S} \cdot \mathbf{T} &= \mathrm{tr}\,(\mathbf{S}\mathbf{T}^{\mathrm{T}}) = \mathrm{tr}\,[(S_{ij}\,\mathbf{e}_i \otimes \mathbf{e}_j)(T_{lk}\,\mathbf{e}_k \otimes \mathbf{e}_l)] = S_{ij}\, T_{lk}\, \mathrm{tr}\,[(\mathbf{e}_i \otimes \mathbf{e}_j)(\mathbf{e}_k \otimes \mathbf{e}_l)] \\ &= S_{ij}\, T_{lk}\, \mathrm{tr}\,[(\mathbf{e}_j \cdot \mathbf{e}_k)(\mathbf{e}_i \otimes \mathbf{e}_l)] = S_{ij}\, T_{lk}\, \mathrm{tr}\,(\delta_{jk}\,\mathbf{e}_i \otimes \mathbf{e}_l) = S_{ij}\, T_{lk}\, \delta_{jk}\, \mathrm{tr}\,(\mathbf{e}_i \otimes \mathbf{e}_l) \\ &= S_{ij}\, T_{lj}\,(\mathbf{e}_i \cdot \mathbf{e}_l) = S_{ij}\, T_{lj}\, \delta_{il} = S_{ij}\, T_{ij}.\end{aligned}$$

PROBLEM 2.34

Show that the scalar product $\mathbf{S} \cdot \mathbf{T}$ defined in (2.41) is an inner product, so $\mathcal{L}$ is a nine-dimensional inner product space, i.e., $\mathcal{L} = \mathcal{E}^9$.

Solution

Previously, in Problems 2.4 and 2.22, we showed that the set $\mathcal{L}$ of all second-order tensors is a vector space, and that this vector space is nine-dimensional. In this problem, armed with the definition (2.41) of the scalar product of two tensors, we show that the properties (2.6) of an inner product space hold for any tensors **R**, **S**, **T** in $\mathcal{L}$, so $\mathcal{L}$ is a nine-dimensional inner product space.

(i) Show that property $(2.6)_1$ is satisfied:

$$\mathbf{S}\cdot\mathbf{T} = \mathrm{tr}\,(\mathbf{S}\mathbf{T}^{\mathrm{T}}) = \mathrm{tr}\left[(\mathbf{S}\mathbf{T}^{\mathrm{T}})^{\mathrm{T}}\right] = \mathrm{tr}\left[(\mathbf{T}^{\mathrm{T}})^{\mathrm{T}}\,\mathbf{S}^{\mathrm{T}}\right] = \mathrm{tr}\,(\mathbf{T}\,\mathbf{S}^{\mathrm{T}}) = \mathbf{T}\cdot\mathbf{S}.$$

(ii) Show that property $(2.6)_2$ is satisfied:

$$(\alpha\mathbf{S})\cdot\mathbf{T} = \mathrm{tr}\left[(\alpha\mathbf{S})\mathbf{T}^{\mathrm{T}}\right] = \mathrm{tr}\left[\alpha\,(\mathbf{S}\mathbf{T}^{\mathrm{T}})\right] = \alpha\,\mathrm{tr}\,(\mathbf{S}\mathbf{T}^{\mathrm{T}}) = \alpha\,(\mathbf{S}\cdot\mathbf{T}).$$

(iii) Show that property $(2.6)_3$ is satisfied:

$$\begin{aligned}(\mathbf{S}+\mathbf{T})\cdot\mathbf{R} &= \mathrm{tr}\,\left[(\mathbf{S}+\mathbf{T})\mathbf{R}^{\mathrm{T}}\right] = \mathrm{tr}\,(\mathbf{S}\mathbf{R}^{\mathrm{T}}+\mathbf{T}\mathbf{R}^{\mathrm{T}}) = \mathrm{tr}\,(\mathbf{S}\mathbf{R}^{\mathrm{T}}) + \mathrm{tr}\,(\mathbf{T}\mathbf{R}^{\mathrm{T}})\\ &= \mathbf{S}\cdot\mathbf{R} + \mathbf{T}\cdot\mathbf{R}.\end{aligned}$$

(iv) Show that property $(2.6)_4$ is satisfied:

$$\mathbf{T}\cdot\mathbf{T} = T_{ij}T_{ij} = T_{11}T_{11} + T_{12}T_{12} + \ldots + T_{33}T_{33}.$$

Unless *every* component of **T** is 0 (i.e., $T_{11} = T_{12} = \ldots = T_{33} = 0$), we expect $\mathbf{T}\cdot\mathbf{T} > 0$. Hence, it follows that if $\mathbf{T} \neq \mathbf{0}$, then $\mathbf{T}\cdot\mathbf{T} > 0$.

PROBLEM 2.35

In direct notation, verify that if **D** is symmetric and **W** is skew, then $\mathbf{D}\cdot\mathbf{W} = 0$.

Solution

Using the definition (2.41) of the scalar product of two tensors, the properties (2.38) of the trace, and the results (2.40), we have

$$\begin{aligned}\mathbf{D}\cdot\mathbf{W} &= \mathrm{tr}\,(\mathbf{D}\mathbf{W}^{\mathrm{T}}) = \mathrm{tr}\,(-\mathbf{D}^{\mathrm{T}}\mathbf{W}) = -\mathrm{tr}\,(\mathbf{D}^{\mathrm{T}}\mathbf{W})\\ &= -\mathrm{tr}\left[(\mathbf{D}^{\mathrm{T}}\mathbf{W})^{\mathrm{T}}\right] = -\mathrm{tr}\,(\mathbf{W}^{\mathrm{T}}\mathbf{D}) = -\mathrm{tr}\,(\mathbf{D}\mathbf{W}^{\mathrm{T}}) = -\mathbf{D}\cdot\mathbf{W}.\end{aligned}$$

Note that we have used $\mathbf{D}^{\mathrm{T}} = \mathbf{D}$ since **D** is symmetric, and $\mathbf{W}^{\mathrm{T}} = -\mathbf{W}$ since **W** is skew. Then

$$\mathbf{D}\cdot\mathbf{W} = -\mathbf{D}\cdot\mathbf{W},$$

with **D** and **W** arbitrary, implies that

$$\mathbf{D}\cdot\mathbf{W} = 0.$$

2.2.6 INVERSE, ORTHOGONALITY, POSITIVE DEFINITENESS

The **inverse** $\mathbf{S}^{-1}$ of a tensor $\mathbf{S}$ is defined by

$$\mathbf{S}\mathbf{S}^{-1} = \mathbf{S}^{-1}\mathbf{S} = \mathbf{I}. \tag{2.48}$$

It can be verified (refer to Problem 2.36) that $\mathbf{S}^{-1}$ satisfies (2.7) and is thus a tensor. Also note that $\mathbf{S}$ is invertible if and only if its determinant is nonzero. We have (refer to Problems 2.37 and 2.38)

$$\det(\mathbf{S}^{-1}) = (\det \mathbf{S})^{-1}, \qquad (\mathbf{S}\mathbf{T})^{-1} = \mathbf{T}^{-1}\mathbf{S}^{-1}, \qquad (\mathbf{S}^{-1})^{\mathrm{T}} = (\mathbf{S}^{\mathrm{T}})^{-1} \equiv \mathbf{S}^{-\mathrm{T}}. \tag{2.49}$$

A tensor $\mathbf{Q}$ is said to be **orthogonal** if

$$\mathbf{Q}\mathbf{u} \cdot \mathbf{Q}\mathbf{v} = \mathbf{u} \cdot \mathbf{v} \tag{2.50}$$

for any vectors $\mathbf{u}, \mathbf{v} \in \mathcal{E}^3$. It can be shown (refer to Problem 2.39) that if $\mathbf{Q}$ is orthogonal, then

$$\mathbf{Q}\mathbf{Q}^{\mathrm{T}} = \mathbf{Q}^{\mathrm{T}}\mathbf{Q} = \mathbf{I}, \quad \text{i.e.,} \quad \mathbf{Q}^{\mathrm{T}} = \mathbf{Q}^{-1}, \tag{2.51}$$

and vice versa. From (2.51), we can show (refer to Problem 2.40) that if $\mathbf{Q}$ is orthogonal, then

$$\det \mathbf{Q} = \pm 1. \tag{2.52}$$

We say that $\mathbf{Q}$ is **proper orthogonal** if it is orthogonal *and* $\det \mathbf{Q} = 1$.

A tensor $\mathbf{P}$ is said to be **positive definite** if, for all vectors $\mathbf{v} \neq \mathbf{0}$,

$$\mathbf{v} \cdot \mathbf{P}\mathbf{v} > 0. \tag{2.53}$$

PROBLEM 2.36

Prove that if $\mathbf{S}^{-1}$ exists, it is a tensor.

Solution

Suppose that $\mathbf{S}^{-1}$ exists, so $\mathbf{S}\mathbf{S}^{-1} = \mathbf{S}^{-1}\mathbf{S} = \mathbf{I}$. For $\mathbf{S}^{-1}$ to be a tensor, it must satisfy definition (2.7), i.e.,

$$\mathbf{S}^{-1}(\alpha\mathbf{v} + \mathbf{w}) = \alpha(\mathbf{S}^{-1}\mathbf{v}) + \mathbf{S}^{-1}\mathbf{w}.$$

$$\begin{aligned}
\mathbf{S}^{-1}(\alpha\mathbf{v} + \mathbf{w}) &= \mathbf{S}^{-1}\left[\alpha(\mathbf{I}\mathbf{v}) + \mathbf{I}\mathbf{w}\right] && \text{(definition (2.10))} \\
&= \mathbf{S}^{-1}\left\{\alpha\left[(\mathbf{S}\mathbf{S}^{-1})\mathbf{v}\right] + (\mathbf{S}\mathbf{S}^{-1})\mathbf{w}\right\} && \text{(definition (2.48))} \\
&= \mathbf{S}^{-1}\left\{\alpha\left[\mathbf{S}(\mathbf{S}^{-1}\mathbf{v})\right] + \mathbf{S}(\mathbf{S}^{-1}\mathbf{w})\right\} && \text{(definition (2.11))}
\end{aligned}$$

$$= \mathbf{S}^{-1}\left\{\mathbf{S}[\alpha\,(\mathbf{S}^{-1}\mathbf{v})] + \mathbf{S}\,(\mathbf{S}^{-1}\mathbf{w})\right\} \qquad \text{(definition (2.7)}_2\text{)}$$

$$= \mathbf{S}^{-1}\left\{\mathbf{S}\left[\alpha\,(\mathbf{S}^{-1}\mathbf{v}) + \mathbf{S}^{-1}\mathbf{w}\right]\right\} \qquad \text{(definition (2.7)}_1\text{)}$$

$$= (\mathbf{S}^{-1}\mathbf{S})\left[\alpha\,(\mathbf{S}^{-1}\mathbf{v}) + \mathbf{S}^{-1}\mathbf{w}\right] \qquad \text{(definition (2.11))}$$

$$= \mathbf{I}\left[\alpha\,(\mathbf{S}^{-1}\mathbf{v}) + \mathbf{S}^{-1}\mathbf{w}\right] \qquad \text{(definition (2.48))}$$

$$= \alpha\,(\mathbf{S}^{-1}\mathbf{v}) + \mathbf{S}^{-1}\mathbf{w} \qquad \text{(definition (2.10))}.$$

PROBLEM 2.37

Verify in direct notation that $(\mathbf{ST})^{-1} = \mathbf{T}^{-1}\mathbf{S}^{-1}$.

Solution

To show that $\mathbf{T}^{-1}\mathbf{S}^{-1}$ is the inverse of $\mathbf{ST}$, we must demonstrate that it satisfies definition (2.48), i.e.,

$$\mathbf{ST}\,(\mathbf{ST})^{-1} = (\mathbf{ST})^{-1}\,\mathbf{ST} = \mathbf{I}.$$

We have

$$\mathbf{ST}\,(\mathbf{ST})^{-1} = \mathbf{ST}\,(\mathbf{T}^{-1}\mathbf{S}^{-1}) = \mathbf{S}\,(\mathbf{T}\mathbf{T}^{-1})\,\mathbf{S}^{-1} = \mathbf{S}\,\mathbf{I}\,\mathbf{S}^{-1} = \mathbf{S}\mathbf{S}^{-1} = \mathbf{I}$$

and

$$(\mathbf{ST})^{-1}\,\mathbf{ST} = (\mathbf{T}^{-1}\mathbf{S}^{-1})\,\mathbf{ST} = \mathbf{T}^{-1}(\mathbf{S}^{-1}\mathbf{S})\,\mathbf{T} = \mathbf{T}^{-1}\,\mathbf{I}\,\mathbf{T} = \mathbf{T}^{-1}\,\mathbf{T} = \mathbf{I}.$$

PROBLEM 2.38

Prove in direct notation that $(\mathbf{S}^{-1})^{\mathrm{T}} = (\mathbf{S}^{\mathrm{T}})^{-1} \equiv \mathbf{S}^{-\mathrm{T}}$.

Solution

$$\mathbf{u}\cdot(\mathbf{S}^{-1})^{\mathrm{T}}\mathbf{v} = \mathbf{S}^{-1}\mathbf{u}\cdot\mathbf{v} = \mathbf{S}^{-1}\mathbf{u}\cdot\mathbf{I}\mathbf{v} = \mathbf{S}^{-1}\mathbf{u}\cdot\mathbf{S}^{\mathrm{T}}(\mathbf{S}^{\mathrm{T}})^{-1}\mathbf{v} = \mathbf{S}\,(\mathbf{S}^{-1}\mathbf{u})\cdot\left[(\mathbf{S}^{\mathrm{T}})^{-1}\mathbf{v}\right]$$

$$= (\mathbf{S}\mathbf{S}^{-1})\mathbf{u}\cdot(\mathbf{S}^{\mathrm{T}})^{-1}\mathbf{v} = \mathbf{I}\mathbf{u}\cdot(\mathbf{S}^{\mathrm{T}})^{-1}\mathbf{v} = \mathbf{u}\cdot(\mathbf{S}^{\mathrm{T}})^{-1}\mathbf{v}.$$

Since $\mathbf{u}$ and $\mathbf{v}$ are arbitrary, it follows that $(\mathbf{S}^{-1})^{\mathrm{T}} = (\mathbf{S}^{\mathrm{T}})^{-1} \equiv \mathbf{S}^{-\mathrm{T}}$.

PROBLEM 2.39
Show in direct notation that if $\mathbf{Q}$ is orthogonal, then $\mathbf{Q}\,\mathbf{Q}^{\mathrm{T}} = \mathbf{Q}^{\mathrm{T}}\mathbf{Q} = \mathbf{I}$.

Solution
If $\mathbf{Q}$ is orthogonal, then

$$\mathbf{Qu} \cdot \mathbf{Qv} = \mathbf{u} \cdot \mathbf{v}$$

for any vectors $\mathbf{u}$ and $\mathbf{v}$ in $\mathcal{E}^3$. The left-hand side becomes

$$\mathbf{Qu} \cdot \mathbf{Qv} = \mathbf{u} \cdot \mathbf{Q}^{\mathrm{T}}(\mathbf{Qv}) = \mathbf{u} \cdot (\mathbf{Q}^{\mathrm{T}}\mathbf{Q})\mathbf{v},$$

while the right-hand side becomes

$$\mathbf{u} \cdot \mathbf{v} = \mathbf{u} \cdot \mathbf{I}\,\mathbf{v}.$$

Using these results in the original expression, we have

$$\mathbf{u} \cdot (\mathbf{Q}^{\mathrm{T}}\mathbf{Q})\mathbf{v} = \mathbf{u} \cdot \mathbf{I}\,\mathbf{v}.$$

Since $\mathbf{u}$ and $\mathbf{v}$ are arbitrary, it follows that $\mathbf{Q}^{\mathrm{T}}\mathbf{Q} = \mathbf{I}$.

Next, we have

$$\mathbf{u} \cdot \mathbf{v} = \mathbf{Iu} \cdot \mathbf{Iv} = \mathbf{Q}^{\mathrm{T}}\mathbf{Q}\,\mathbf{u} \cdot \mathbf{Q}^{\mathrm{T}}\mathbf{Q}\,\mathbf{v} = \mathbf{Qu} \cdot \mathbf{Q}^{\mathrm{TT}}(\mathbf{Q}^{\mathrm{T}}\mathbf{Q}\,\mathbf{v}) = \mathbf{Qu} \cdot (\mathbf{QQ}^{\mathrm{T}})\mathbf{Q}\,\mathbf{v},$$

where we have used $\mathbf{Q}^{\mathrm{T}}\mathbf{Q} = \mathbf{I}$ and the definition (2.13) of the transpose of a tensor. A comparison of this result with the orthogonality condition (2.50) implies that $\mathbf{Q}\,\mathbf{Q}^{\mathrm{T}} = \mathbf{I}$.

PROBLEM 2.40
Verify in direct notation that if $\mathbf{Q}$ is orthogonal, then $\det\mathbf{Q} = \pm 1$.

Solution
Recall from Problem 2.39 that if $\mathbf{Q}$ is orthogonal, then $\mathbf{Q}^{\mathrm{T}}\mathbf{Q} = \mathbf{I}$. It follows that

$$\det(\mathbf{Q}^{\mathrm{T}}\mathbf{Q}) = \det\mathbf{I}.$$

Then, by $(2.47)_1$

$$(\det\mathbf{Q}^{\mathrm{T}})(\det\mathbf{Q}) = 1,$$

where we have used

$$\det\mathbf{I} = \det[\mathbf{I}] = \det\begin{bmatrix} 1 & 0 & 0 \\ 0 & 1 & 0 \\ 0 & 0 & 1 \end{bmatrix} = 1.$$

Then, by $(2.47)_2$,

$$(\det \mathbf{Q})^2 = 1,$$

so $\det \mathbf{Q} = \pm 1$.

2.2.7 VECTOR PRODUCT, SCALAR TRIPLE PRODUCT

The angle θ between two vectors $\mathbf{u}$ and $\mathbf{v}$ in $\mathcal{E}^3$ is defined by

$$\cos\theta = \frac{\mathbf{u}\cdot\mathbf{v}}{|\mathbf{u}|\,|\mathbf{v}|}, \tag{2.54}$$

where $|\mathbf{u}|$ is the natural norm, i.e., $|\mathbf{u}| = (\mathbf{u}\cdot\mathbf{u})^{1/2}$. Then the **vector product** in $\mathcal{E}^3$ is defined by

$$\mathbf{u}\times\mathbf{v} = |\mathbf{u}|\,|\mathbf{v}|\,\mathbf{n}\sin\theta, \qquad |\mathbf{n}| = 1, \qquad \mathbf{u}\cdot\mathbf{n} = \mathbf{v}\cdot\mathbf{n} = 0. \tag{2.55}$$

As such, the vector product accepts two vectors $\mathbf{u}$ and $\mathbf{v}$ as inputs, and provides a single vector in the direction of $\mathbf{n}$ as output. Note that $\mathbf{n}$ is the unit normal to the plane containing $\mathbf{u}$ and $\mathbf{v}$; thus, two vector products are possible in $\mathcal{E}^3$. We single out the one for which $\{\mathbf{u},\mathbf{v},\mathbf{n}\}$ form a right-handed set. We also demand that the basis $\{\mathbf{e}_1,\mathbf{e}_2,\mathbf{e}_3\}$ is right-handed, so

$$\begin{aligned}
&\mathbf{e}_1\times\mathbf{e}_2 = \mathbf{e}_3, && \mathbf{e}_2\times\mathbf{e}_3 = \mathbf{e}_1, && \mathbf{e}_3\times\mathbf{e}_1 = \mathbf{e}_2,\\
&\mathbf{e}_1\times\mathbf{e}_3 = -\mathbf{e}_2, && \mathbf{e}_2\times\mathbf{e}_1 = -\mathbf{e}_3, && \mathbf{e}_3\times\mathbf{e}_2 = -\mathbf{e}_1,\\
&\mathbf{e}_1\times\mathbf{e}_1 = \mathbf{0}, && \mathbf{e}_2\times\mathbf{e}_2 = \mathbf{0}, && \mathbf{e}_3\times\mathbf{e}_3 = \mathbf{0}.
\end{aligned} \tag{2.56}$$

Using indicial notation, we can write the nine equations in (2.56) as the single expression

$$\mathbf{e}_i\times\mathbf{e}_j = \epsilon_{ijk}\,\mathbf{e}_k, \tag{2.57}$$

where ϵ_{ijk} is the **permutation symbol**, defined such that

$$\epsilon_{ijk} = \begin{cases} 1 & \text{if } ijk = 123 = 231 = 312,\\ -1 & \text{if } ijk = 132 = 213 = 321,\\ 0 & \text{otherwise.}\end{cases} \tag{2.58}$$

Then, it can be shown (refer to Problem 2.41) that

$$\mathbf{u}\times\mathbf{v} = \epsilon_{ijk}\,u_i\,v_j\,\mathbf{e}_k, \tag{2.59}$$

i.e.,

$$\mathbf{u}\times\mathbf{v} = (u_2\,v_3 - u_3\,v_2)\,\mathbf{e}_1 + (u_3\,v_1 - u_1\,v_3)\,\mathbf{e}_2 + (u_1\,v_2 - u_2\,v_1)\,\mathbf{e}_3 = \begin{vmatrix}\mathbf{e}_1 & \mathbf{e}_2 & \mathbf{e}_3\\ u_1 & u_2 & u_3\\ v_1 & v_2 & v_3\end{vmatrix}. \tag{2.60}$$

Note that the vector product is anticommutative, i.e.,

$$\mathbf{u} \times \mathbf{v} = -\mathbf{v} \times \mathbf{u}. \tag{2.61}$$

The **scalar triple product** $[\,\mathbf{u}\,\mathbf{v}\,\mathbf{w}\,]$ of three vectors is defined by

$$[\,\mathbf{u}\,\mathbf{v}\,\mathbf{w}\,] = \mathbf{u} \cdot (\mathbf{v} \times \mathbf{w}). \tag{2.62}$$

It can be shown that

$$[\,\mathbf{u}\,\mathbf{v}\,\mathbf{w}\,] = [\,\mathbf{v}\,\mathbf{w}\,\mathbf{u}\,] = [\,\mathbf{w}\,\mathbf{u}\,\mathbf{v}\,]. \tag{2.63}$$

The absolute value of $[\,\mathbf{u}\,\mathbf{v}\,\mathbf{w}\,]$ is the volume of the parallelepiped determined by $\mathbf{u}$, $\mathbf{v}$, and $\mathbf{w}$. Also,

$$\det \mathbf{S} = \frac{[\,\mathbf{Su}\ \mathbf{Sv}\ \mathbf{Sw}\,]}{[\,\mathbf{u}\,\mathbf{v}\,\mathbf{w}\,]}. \tag{2.64}$$

Corresponding to each skew tensor $\mathbf{W} \in \mathcal{L}$ is an **axial vector** $\mathbf{w} \in \mathcal{E}^3$, i.e.,

$$\mathbf{Wv} = \mathbf{w} \times \mathbf{v} \tag{2.65}$$

for any $\mathbf{v} \in \mathcal{E}^3$. Because of this correspondence, the set of all skew tensors is a three-dimensional inner product space $\mathcal{E}^3$ (refer to Problem 2.42).

PROBLEM 2.41

Show that $\mathbf{u} \times \mathbf{v} = \epsilon_{ijk}\, u_i\, v_j\, \mathbf{e}_k$.

Solution

$$\mathbf{u} \times \mathbf{v} = u_i\, \mathbf{e}_i \times v_j\, \mathbf{e}_j = u_i\, v_j\, (\mathbf{e}_i \times \mathbf{e}_j) = \epsilon_{ijk}\, u_i\, v_j\, \mathbf{e}_k.$$

PROBLEM 2.42

Verify in Cartesian component notation that the set of all symmetric tensors is a six-dimensional inner product space $\mathcal{E}^6$, and the set of all skew tensors is a three-dimensional inner product space $\mathcal{E}^3$.

Solution

Consider an arbitrary symmetric tensor $\mathbf{D}$. The Cartesian component form of $\mathbf{D}$ is

$$\begin{aligned} \mathbf{D} = D_{ij}\, \mathbf{e}_i \otimes \mathbf{e}_j = D_{11}\, \mathbf{e}_1 \otimes \mathbf{e}_1 + D_{12}\, \mathbf{e}_1 \otimes \mathbf{e}_2 + D_{13}\, \mathbf{e}_1 \otimes \mathbf{e}_3 + D_{21}\, \mathbf{e}_2 \otimes \mathbf{e}_1 + D_{22}\, \mathbf{e}_2 \otimes \mathbf{e}_2 \\ + D_{23}\, \mathbf{e}_2 \otimes \mathbf{e}_3 + D_{31}\, \mathbf{e}_3 \otimes \mathbf{e}_1 + D_{32}\, \mathbf{e}_3 \otimes \mathbf{e}_2 + D_{33}\, \mathbf{e}_3 \otimes \mathbf{e}_3. \end{aligned}$$

But, since $\mathbf{D}$ is symmetric (i.e., $\mathbf{D}^{\mathrm{T}} = \mathbf{D}$),

$$\mathrm{D}_{ij} = \mathbf{e}_i \cdot \mathbf{D}\mathbf{e}_j = \mathbf{D}\mathbf{e}_j \cdot \mathbf{e}_i = \mathbf{e}_j \cdot \mathbf{D}^{\mathrm{T}}\mathbf{e}_i = \mathbf{e}_j \cdot \mathbf{D}\mathbf{e}_i = \mathrm{D}_{ji},$$

so

$$D_{12} = D_{21}, \qquad D_{13} = D_{31}, \qquad D_{23} = D_{32}.$$

Thus,

$$\mathbf{D} = D_{11}\mathbf{e}_1 \otimes \mathbf{e}_1 + D_{12}(\mathbf{e}_1 \otimes \mathbf{e}_2 + \mathbf{e}_2 \otimes \mathbf{e}_1) + D_{13}(\mathbf{e}_1 \otimes \mathbf{e}_3 + \mathbf{e}_3 \otimes \mathbf{e}_1) + D_{22}\,\mathbf{e}_2 \otimes \mathbf{e}_2$$
$$+ D_{23}(\mathbf{e}_2 \otimes \mathbf{e}_3 + \mathbf{e}_3 \otimes \mathbf{e}_2) + D_{33}\,\mathbf{e}_3 \otimes \mathbf{e}_3.$$

It follows that the basis of any symmetric tensor **D** has six elements, so the set of all symmetric tensors is a six-dimensional inner product space $\mathcal{E}^6$. Note that only six components $(D_{11}, D_{12}, D_{13}, D_{22}, D_{23}, D_{33})$ are required to fully specify **D**.

Similarly, since **W** is skew (i.e., $\mathbf{W}^{\mathrm{T}} = -\mathbf{W}$),

$$W_{ij} = \mathbf{e}_i \cdot \mathbf{W}\mathbf{e}_j = \mathbf{W}\mathbf{e}_j \cdot \mathbf{e}_i = \mathbf{e}_j \cdot \mathbf{W}^{\mathrm{T}}\mathbf{e}_i = -\mathbf{e}_j \cdot \mathbf{W}\mathbf{e}_i = -W_{ji},$$

so

$$W_{11} = W_{22} = W_{33} = 0, \qquad W_{12} = -W_{21}, \qquad W_{13} = -W_{31}, \qquad W_{23} = -W_{32}.$$

Thus,

$$\mathbf{W} = W_{12}(\mathbf{e}_1 \otimes \mathbf{e}_2 - \mathbf{e}_2 \otimes \mathbf{e}_1) + W_{13}(\mathbf{e}_1 \otimes \mathbf{e}_3 - \mathbf{e}_3 \otimes \mathbf{e}_1) + W_{23}(\mathbf{e}_2 \otimes \mathbf{e}_3 - \mathbf{e}_3 \otimes \mathbf{e}_2).$$

It follows that the basis of any skew tensor **W** has three elements, so the set of all skew tensors is a three-dimensional inner product space $\mathcal{E}^3$. Note that only three components (W_{12}, W_{13}, W_{23}) are required to fully specify **W**.

2.3 EIGENVALUES, EIGENVECTORS, POLAR DECOMPOSITION, INVARIANTS

The presentation of the conceptual material in this section again follows [2]. In general, a tensor **S** maps a vector **u** to a vector **Su**. The vector **Su** typically has a length and orientation different from those of **u**. However, for special unit vectors **a** called **eigenvectors,** the tensor **S** maps **a** to a vector **Sa** that is parallel to **a**, i.e.,

$$\mathbf{S}\,\mathbf{a} = \alpha\,\mathbf{a}. \tag{2.66}$$

The scalar α is referred to as an **eigenvalue** of **S** corresponding to the eigenvector **a**.

It can be shown that if **S** is positive definite, then its eigenvalues are strictly positive (refer to Problem 2.43). Further, if **S** is both symmetric and positive definite, then it can be shown that $\det \mathbf{S} > 0$, so $\mathbf{S}^{-1}$ exists (refer to Problem 2.44).

If **S** is symmetric, its eigenvectors $\{\mathbf{a}_1, \mathbf{a}_2, \mathbf{a}_3\}$ constitute an orthonormal basis for $\mathcal{E}^3$ (the corresponding eigenvalues are α_1, α_2, and α_3). Additionally, any symmetric **S** in $\mathcal{L}$ can be written with respect to a basis of its eigenvectors $\{\mathbf{a}_i \otimes \mathbf{a}_j\}$ as

$$\mathbf{S} = \alpha_1\,\mathbf{a}_1 \otimes \mathbf{a}_1 + \alpha_2\,\mathbf{a}_2 \otimes \mathbf{a}_2 + \alpha_3\,\mathbf{a}_3 \otimes \mathbf{a}_3 = \sum_{i=1}^{3} \alpha_i\,\mathbf{a}_i \otimes \mathbf{a}_i, \tag{2.67a}$$

or, in matrix form,

$$[\mathbf{S}] = \begin{bmatrix} \alpha_1 & 0 & 0 \\ 0 & \alpha_2 & 0 \\ 0 & 0 & \alpha_3 \end{bmatrix}. \tag{2.67b}$$

If **B** is a symmetric, positive-definite tensor, then there exists a unique symmetric, positive-definite tensor **V** such that

$$\mathbf{V}^2 \equiv \mathbf{V}\mathbf{V} = \mathbf{B}. \tag{2.68}$$

That is, $\mathbf{V} = \sqrt{\mathbf{B}}$.

If **F** is an invertible tensor ($\det \mathbf{F} \neq 0$), then

$$\mathbf{F} = \mathbf{R}\mathbf{U} = \mathbf{V}\mathbf{R}, \tag{2.69}$$

where **U** and **V** are symmetric, positive-definite tensors, and **R** is an orthogonal tensor. The multiplicative decomposition (2.69) is referred to as the **polar decomposition** of **F**. The tensors **U**, **V**, and **R** are unique, with

$$\mathbf{U} = \sqrt{\mathbf{F}^{\mathrm{T}}\mathbf{F}}, \qquad \mathbf{V} = \sqrt{\mathbf{F}\mathbf{F}^{\mathrm{T}}}, \qquad \mathbf{R} = \mathbf{F}\mathbf{U}^{-1}. \tag{2.70}$$

The terms in (2.70) make sense, because $\mathbf{F}^{\mathrm{T}}\mathbf{F}$ and $\mathbf{F}\mathbf{F}^{\mathrm{T}}$ can be shown to be symmetric and positive definite (refer to Problem 2.45), so we may use (2.68); also, $\det \mathbf{F} \neq 0$ implies that $\det \mathbf{U} \neq 0$, so $\mathbf{U}^{-1}$ exists. Note that if $\det \mathbf{F} > 0$, then **R** is proper orthogonal.

A nontrivial solution of the eigenvalue problem (2.66) requires

$$\det(\mathbf{S} - \alpha\mathbf{I}) = 0. \tag{2.71}$$

It can be shown that

$$\det(\mathbf{S} - \alpha\mathbf{I}) = -\alpha^3 + \alpha^2 I_1(\mathbf{S}) - \alpha\, I_2(\mathbf{S}) + I_3(\mathbf{S}), \tag{2.72}$$

so (2.71) becomes

$$-\alpha^3 + \alpha^2 I_1(\mathbf{S}) - \alpha\, I_2(\mathbf{S}) + I_3(\mathbf{S}) = 0, \tag{2.73}$$

where

$$I_1(\mathbf{S}) = \operatorname{tr}\mathbf{S}, \qquad I_2(\mathbf{S}) = \frac{1}{2}\left[(\operatorname{tr}\mathbf{S})^2 - \operatorname{tr}(\mathbf{S}^2)\right], \qquad I_3(\mathbf{S}) = \det\mathbf{S} \tag{2.74}$$

are called the **principal invariants** of **S**. Equation (2.73) is called the **characteristic equation** of **S**. According to the Cayley-Hamilton theorem, every tensor **S** satisfies its own characteristic equation, i.e.,

$$-\mathbf{S}^3 + \mathbf{S}^2 I_1(\mathbf{S}) - \mathbf{S}\, I_2(\mathbf{S}) + \mathbf{I}\, I_3(\mathbf{S}) = \mathbf{0}. \tag{2.75}$$

Note that if **S** is symmetric, then it follows from (2.67b) and (2.74) that

$$I_1(\mathbf{S}) = \alpha_1 + \alpha_2 + \alpha_3, \qquad I_2(\mathbf{S}) = \alpha_1\alpha_2 + \alpha_2\alpha_3 + \alpha_1\alpha_3, \qquad I_3(\mathbf{S}) = \alpha_1\alpha_2\alpha_3. \tag{2.76}$$

Thus, two symmetric tensors with the same eigenvalues have the same principal invariants.

PROBLEM 2.43
Prove that the eigenvalues of a positive-definite tensor are strictly positive.

Solution
Recall from (2.53) that if the tensor $\mathbf{P}$ is positive definite, then $\mathbf{v} \cdot \mathbf{P}\mathbf{v} > 0$ for any vector $\mathbf{v} \neq \mathbf{0}$. Suppose, then, that we choose $\mathbf{v} = \mathbf{a}$, where $\mathbf{a}$ is *any* eigenvector of $\mathbf{P}$. Then

$$\mathbf{a} \cdot \mathbf{P}\,\mathbf{a} > 0.$$

But, by (2.66), we have $\mathbf{P}\mathbf{a} = \alpha\mathbf{a}$, where α is the eigenvalue corresponding to $\mathbf{a}$, which implies that

$$\alpha\,(\mathbf{a} \cdot \mathbf{a}) > 0.$$

Since eigenvectors are of unit length ($\mathbf{a} \cdot \mathbf{a} = |\mathbf{a}|^2 = 1$), it follows that $\alpha > 0$, i.e., the eigenvalue α corresponding to any eigenvector $\mathbf{a}$ of tensor $\mathbf{P}$ is positive.

PROBLEM 2.44
Prove that if the tensor $\mathbf{P}$ is symmetric and positive definite, then $\det \mathbf{P} > 0$, so its inverse $\mathbf{P}^{-1}$ exists.

Solution
If $\mathbf{P}$ is a symmetric tensor, then by (2.67a) we have

$$\mathbf{P} = \alpha_1\,\mathbf{a}_1 \otimes \mathbf{a}_1 + \alpha_2\,\mathbf{a}_2 \otimes \mathbf{a}_2 + \alpha_3\,\mathbf{a}_3 \otimes \mathbf{a}_3,$$

where $\mathbf{a}_1$, $\mathbf{a}_2$, $\mathbf{a}_3$ are the eigenvectors of $\mathbf{P}$, and α_1, α_2, α_3 are the corresponding eigenvalues of $\mathbf{P}$. Hence, by (2.46),

$$\det \mathbf{P} = \det [\mathbf{P}] = \det \begin{bmatrix} \alpha_1 & 0 & 0 \\ 0 & \alpha_2 & 0 \\ 0 & 0 & \alpha_3 \end{bmatrix} = \alpha_1\,\alpha_2\,\alpha_3.$$

Recall from Problem 2.43 that if a tensor $\mathbf{P}$ is positive definite, then its eigenvalues α_1, α_2, and α_3 are all positive. Hence, $\det \mathbf{P} = \alpha_1\,\alpha_2\,\alpha_3 > 0$, and $\mathbf{P}$ is invertible.

PROBLEM 2.45
Prove that if $\mathbf{F}$ is nonsingular, then $\mathbf{F}^{\mathrm{T}}\,\mathbf{F}$ and $\mathbf{F}\,\mathbf{F}^{\mathrm{T}}$ are symmetric and positive definite.

Solution
First, we verify that $\mathbf{F}^{\mathrm{T}}\,\mathbf{F}$ and $\mathbf{F}\,\mathbf{F}^{\mathrm{T}}$ are *symmetric*. We have

$$(\mathbf{F}^{\mathrm{T}}\mathbf{F})^{\mathrm{T}} = \mathbf{F}^{\mathrm{T}}(\mathbf{F}^{\mathrm{T}})^{\mathrm{T}} = \mathbf{F}^{\mathrm{T}}\mathbf{F}, \qquad (\mathbf{F}\mathbf{F}^{\mathrm{T}})^{\mathrm{T}} = (\mathbf{F}^{\mathrm{T}})^{\mathrm{T}}\mathbf{F}^{\mathrm{T}} = \mathbf{F}\mathbf{F}^{\mathrm{T}}.$$

Thus, according to definition (2.15), $\mathbf{F}^{\mathrm{T}}\mathbf{F}$ and $\mathbf{F}\mathbf{F}^{\mathrm{T}}$ are symmetric.

Next, we verify that $\mathbf{F}^{\mathrm{T}}\mathbf{F}$ is *positive definite*. We have

$$\mathbf{v} \cdot (\mathbf{F}^{\mathrm{T}}\mathbf{F})\mathbf{v} = \mathbf{v} \cdot \mathbf{F}^{\mathrm{T}}(\mathbf{F}\mathbf{v}) = \mathbf{F}\mathbf{v} \cdot \mathbf{F}\mathbf{v}.$$

According to property $(2.6)_4$ of inner product spaces, if $\mathbf{F}\mathbf{v} \neq \mathbf{0}$, then $\mathbf{F}\mathbf{v} \cdot \mathbf{F}\mathbf{v} > 0$. Since $\mathbf{F}$ is *nonsingular,* $\det \mathbf{F} \neq 0$, i.e., $\mathbf{F}\mathbf{v} = \mathbf{0}$ if and only if $\mathbf{v} = \mathbf{0}$. Hence, it follows that

$$\mathbf{v} \cdot (\mathbf{F}^{\mathrm{T}}\mathbf{F})\mathbf{v} > 0$$

for any vector $\mathbf{v} \neq \mathbf{0}$, so, according to definition (2.53), $\mathbf{F}^{\mathrm{T}}\mathbf{F}$ is positive definite.

Lastly, we verify that $\mathbf{F}\mathbf{F}^{\mathrm{T}}$ is *positive definite*. We use result $(2.47)_2$ to argue that $\det \mathbf{F}^{\mathrm{T}} = \det \mathbf{F} \neq 0$ (i.e., if $\mathbf{F}$ is nonsingular, then so is $\mathbf{F}^{\mathrm{T}}$), so $\mathbf{F}^{\mathrm{T}}\mathbf{v} = \mathbf{0}$ if and only if $\mathbf{v} = \mathbf{0}$, which implies

$$\mathbf{v} \cdot (\mathbf{F}\mathbf{F}^{\mathrm{T}})\mathbf{v} = \mathbf{v} \cdot \mathbf{F}(\mathbf{F}^{\mathrm{T}}\mathbf{v}) = \mathbf{F}^{\mathrm{T}}\mathbf{v} \cdot \mathbf{F}^{\mathrm{T}}\mathbf{v} > 0$$

for any vector $\mathbf{v} \neq \mathbf{0}$. Thus, according to definition (2.53), $\mathbf{F}\mathbf{F}^{\mathrm{T}}$ is positive definite.

2.4 TENSORS OF ORDER THREE AND FOUR

We call linear transformations from $\mathcal{E}^3$ to $\mathcal{L}$, or from $\mathcal{L}$ to $\mathcal{E}^3$, third-order tensors. Thus, a third-order tensor $\mathbf{D}^{(3)}$ is a linear map that assigns to each vector $\mathbf{u} \in \mathcal{E}^3$ a second-order tensor $\mathbf{D}^{(3)}\mathbf{u} \in \mathcal{L}$ such that

$$\mathbf{D}^{(3)}(\mathbf{u} + \mathbf{v}) = \mathbf{D}^{(3)}\mathbf{u} + \mathbf{D}^{(3)}\mathbf{v}, \qquad \mathbf{D}^{(3)}(\alpha\mathbf{v}) = \alpha(\mathbf{D}^{(3)}\mathbf{v}), \tag{2.77}$$

or a linear map that assigns to each second-order tensor $\mathbf{T} \in \mathcal{L}$ a vector $\mathbf{D}^{(3)}\mathbf{T} \in \mathcal{E}^3$ such that

$$\mathbf{D}^{(3)}(\mathbf{S} + \mathbf{T}) = \mathbf{D}^{(3)}\mathbf{S} + \mathbf{D}^{(3)}\mathbf{T}, \qquad \mathbf{D}^{(3)}(\alpha\mathbf{T}) = \alpha(\mathbf{D}^{(3)}\mathbf{T}) \tag{2.78}$$

for any second-order tensors $\mathbf{S}, \mathbf{T} \in \mathcal{L}$, vectors $\mathbf{u}, \mathbf{v} \in \mathcal{E}^3$, and scalars $\alpha \in \mathbb{R}$. The set of all third-order tensors is denoted by $\mathcal{L}^{(3)}$.

If $\mathbf{a}$, $\mathbf{b}$, and $\mathbf{c}$ are vectors in $\mathcal{E}^3$, we define the third-order tensor $\mathbf{a} \otimes \mathbf{b} \otimes \mathbf{c}$ by

$$(\mathbf{a} \otimes \mathbf{b} \otimes \mathbf{c})\,\mathbf{v} = \mathbf{a} \otimes \mathbf{b}\,(\mathbf{c} \cdot \mathbf{v}), \qquad (\mathbf{a} \otimes \mathbf{b} \otimes \mathbf{c})\,\mathbf{v} \otimes \mathbf{w} = \mathbf{a}\,(\mathbf{c} \cdot \mathbf{v})\,(\mathbf{b} \cdot \mathbf{w}) \tag{2.79}$$

for any vectors $\mathbf{v}, \mathbf{w} \in \mathcal{E}^3$.

The Cartesian components D_{ijk} of a third-order tensor $\mathbf{D}^{(3)}$ are defined by

$$D_{ijk} = (\mathbf{e}_i \otimes \mathbf{e}_j) \cdot \left(\mathbf{D}^{(3)}\mathbf{e}_k\right), \tag{2.80}$$

and we have

$$\mathbf{D}^{(3)} = D_{ijk}\,\mathbf{e}_i \otimes \mathbf{e}_j \otimes \mathbf{e}_k. \tag{2.81}$$

$\mathbf{e}_i \otimes \mathbf{e}_j \otimes \mathbf{e}_k$ is a basis for $\mathcal{L}^{(3)} = \mathcal{E}^{27}$, a 27-dimensional inner product space.

We call linear transformations from $\mathcal{E}^3$ to $\mathcal{L}^{(3)}$, $\mathcal{L}$ to $\mathcal{L}$, or $\mathcal{L}^{(3)}$ to $\mathcal{E}^3$ fourth-order tensors. Thus, a fourth-order tensor $\mathbf{C}^{(4)}$ is a linear map that assigns to each vector $\mathbf{u} \in \mathcal{E}^3$ a third-order tensor

$$\mathbf{D}^{(3)} = \mathbf{C}^{(4)}\mathbf{u}, \tag{2.82}$$

or to each second-order tensor $\mathbf{T} \in \mathcal{L}$ a second-order tensor

$$\mathbf{S} = \mathbf{C}^{(4)}\mathbf{T}, \tag{2.83}$$

or to each third-order tensor $\mathbf{E}^{(3)} \in \mathcal{L}^{(3)}$ a vector

$$\mathbf{v} = \mathbf{C}^{(4)}\mathbf{E}^{(3)}. \tag{2.84}$$

The set of all fourth-order tensors is denoted by $\mathcal{L}^{(4)}$.

If $\mathbf{a}$, $\mathbf{b}$, $\mathbf{c}$, and $\mathbf{d}$ are vectors in $\mathcal{E}^3$, we define the fourth-order tensor $\mathbf{a} \otimes \mathbf{b} \otimes \mathbf{c} \otimes \mathbf{d}$ by

$$\begin{aligned}
(\mathbf{a} \otimes \mathbf{b} \otimes \mathbf{c} \otimes \mathbf{d})\,\mathbf{v} &= \mathbf{a} \otimes \mathbf{b} \otimes \mathbf{c}\,(\mathbf{d} \cdot \mathbf{v}), \\
(\mathbf{a} \otimes \mathbf{b} \otimes \mathbf{c} \otimes \mathbf{d})\,\mathbf{v} \otimes \mathbf{w} &= \mathbf{a} \otimes \mathbf{b}\,(\mathbf{d} \cdot \mathbf{v})\,(\mathbf{c} \cdot \mathbf{w}), \\
(\mathbf{a} \otimes \mathbf{b} \otimes \mathbf{c} \otimes \mathbf{d})\,\mathbf{v} \otimes \mathbf{w} \otimes \mathbf{z} &= \mathbf{a}\,(\mathbf{d} \cdot \mathbf{v})\,(\mathbf{c} \cdot \mathbf{w})\,(\mathbf{b} \cdot \mathbf{z})
\end{aligned} \tag{2.85}$$

for any vectors $\mathbf{v}, \mathbf{w}, \mathbf{z} \in \mathcal{E}^3$.

The Cartesian components C_{ijkl} of $\mathbf{C}^{(4)}$ are defined by

$$C_{ijkl} = \left(\mathbf{e}_i \otimes \mathbf{e}_j\right) \cdot \left(\mathbf{C}^{(4)}\,\mathbf{e}_k \otimes \mathbf{e}_l\right), \tag{2.86}$$

and we have

$$\mathbf{C}^{(4)} = C_{ijkl}\,\mathbf{e}_i \otimes \mathbf{e}_j \otimes \mathbf{e}_k \otimes \mathbf{e}_l. \tag{2.87}$$

$\mathbf{e}_i \otimes \mathbf{e}_j \otimes \mathbf{e}_k \otimes \mathbf{e}_l$ is a basis for $\mathcal{L}^{(4)} = \mathcal{E}^{81}$, an 81-dimensional inner product space.

2.5 TENSOR CALCULUS

2.5.1 PARTIAL DERIVATIVES

Let ϕ be a scalar-valued function of a tensor $\mathbf{T}$, vector $\mathbf{x}$, and scalar t, i.e.,

$$\phi = \phi(\mathbf{T}, \mathbf{x}, t). \tag{2.88}$$

We define the partial derivatives of ϕ by

$$\begin{aligned}
\frac{\partial}{\partial t}\phi\,(\mathbf{T}, \mathbf{x}, t) &= \lim_{\alpha \to 0} \frac{1}{\alpha}\left[\phi(\mathbf{T}, \mathbf{x}, t + \alpha) - \phi(\mathbf{T}, \mathbf{x}, t)\right], \\
\left[\frac{\partial}{\partial \mathbf{x}}\phi\,(\mathbf{T}, \mathbf{x}, t)\right] \cdot \mathbf{u} &= \lim_{\alpha \to 0} \frac{1}{\alpha}\left[\phi(\mathbf{T}, \mathbf{x} + \alpha\mathbf{u}, t) - \phi(\mathbf{T}, \mathbf{x}, t)\right],
\end{aligned} \tag{2.89}$$

$$\left[\frac{\partial}{\partial \mathbf{T}}\phi(\mathbf{T},\mathbf{x},t)\right]\cdot\mathbf{S}=\lim_{\alpha\to 0}\frac{1}{\alpha}\left[\phi(\mathbf{T}+\alpha\mathbf{S},\mathbf{x},t)-\phi(\mathbf{T},\mathbf{x},t)\right]$$

for all vectors $\mathbf{u}$ in $\mathcal{E}^3$ and all tensors $\mathbf{S}$ in $\mathcal{L}$. Note that α is a real number. Also note that $\partial\phi/\partial t$ is a scalar, $\partial\phi/\partial\mathbf{x}$ is a vector, and $\partial\phi/\partial\mathbf{T}$ is a tensor.

Let $\mathbf{v}$ be a vector-valued function of tensor $\mathbf{T}$, vector $\mathbf{x}$, and scalar t, i.e.,

$$\mathbf{v}=\mathbf{v}(\mathbf{T},\mathbf{x},t). \tag{2.90}$$

We define the partial derivatives of $\mathbf{v}$ by

$$\frac{\partial}{\partial t}\mathbf{v}(\mathbf{T},\mathbf{x},t)=\lim_{\alpha\to 0}\frac{1}{\alpha}\left[\mathbf{v}(\mathbf{T},\mathbf{x},t+\alpha)-\mathbf{v}(\mathbf{T},\mathbf{x},t)\right],$$

$$\left[\frac{\partial}{\partial \mathbf{x}}\mathbf{v}(\mathbf{T},\mathbf{x},t)\right]\mathbf{u}=\lim_{\alpha\to 0}\frac{1}{\alpha}\left[\mathbf{v}(\mathbf{T},\mathbf{x}+\alpha\mathbf{u},t)-\mathbf{v}(\mathbf{T},\mathbf{x},t)\right], \tag{2.91}$$

$$\left[\frac{\partial}{\partial \mathbf{T}}\mathbf{v}(\mathbf{T},\mathbf{x},t)\right]\mathbf{S}=\lim_{\alpha\to 0}\frac{1}{\alpha}\left[\mathbf{v}(\mathbf{T}+\alpha\mathbf{S},\mathbf{x},t)-\mathbf{v}(\mathbf{T},\mathbf{x},t)\right],$$

for all vectors $\mathbf{u}$ in $\mathcal{E}^3$ and all tensors $\mathbf{S}$ in $\mathcal{L}$. Note that $\partial\mathbf{v}/\partial t$ is a vector and $\partial\mathbf{v}/\partial\mathbf{x}$ is a tensor. The quantity $\partial\mathbf{v}/\partial\mathbf{T}$ maps a tensor $\mathbf{S}$ to a vector, and is called a third-order tensor (refer to Section 2.4).

Let $\mathbf{A}$ be a tensor-valued function of tensor $\mathbf{T}$, vector $\mathbf{x}$, and scalar t, i.e.,

$$\mathbf{A}=\mathbf{A}(\mathbf{T},\mathbf{x},t). \tag{2.92}$$

The partial derivatives of $\mathbf{A}$ are defined by

$$\frac{\partial}{\partial t}\mathbf{A}(\mathbf{T},\mathbf{x},t)=\lim_{\alpha\to 0}\frac{1}{\alpha}\left[\mathbf{A}(\mathbf{T},\mathbf{x},t+\alpha)-\mathbf{A}(\mathbf{T},\mathbf{x},t)\right],$$

$$\left[\frac{\partial}{\partial \mathbf{x}}\mathbf{A}(\mathbf{T},\mathbf{x},t)\right]\mathbf{u}=\lim_{\alpha\to 0}\frac{1}{\alpha}\left[\mathbf{A}(\mathbf{T},\mathbf{x}+\alpha\mathbf{u},t)-\mathbf{A}(\mathbf{T},\mathbf{x},t)\right], \tag{2.93}$$

$$\left[\frac{\partial}{\partial \mathbf{T}}\mathbf{A}(\mathbf{T},\mathbf{x},t)\right]\mathbf{S}=\lim_{\alpha\to 0}\frac{1}{\alpha}\left[\mathbf{A}(\mathbf{T}+\alpha\mathbf{S},\mathbf{x},t)-\mathbf{A}(\mathbf{T},\mathbf{x},t)\right],$$

for all vectors $\mathbf{u}$ in $\mathcal{E}^3$ and all tensors $\mathbf{S}$ in $\mathcal{L}$. Note that $\partial\mathbf{A}/\partial t$ is a tensor. The quantity $\partial\mathbf{A}/\partial\mathbf{x}$ is a third-order tensor, i.e., it maps vectors to tensors; the fourth-order tensor $\partial\mathbf{A}/\partial\mathbf{T}$ maps tensors to tensors (refer to Section 2.4).

Recall from Section 2.3 that the **principal invariants** of a tensor $\mathbf{A}$ are

$$I_1(\mathbf{A})=\operatorname{tr}\mathbf{A},\qquad I_2(\mathbf{A})=\frac{1}{2}\left[(\operatorname{tr}\mathbf{A})^2-\operatorname{tr}(\mathbf{A}^2)\right],\qquad I_3(\mathbf{A})=\det\mathbf{A}.$$

It can be shown that (refer to Problems 2.46–2.48)

$$\frac{\mathrm{d}I_1(\mathbf{A})}{\mathrm{d}\mathbf{A}}=\mathbf{I},\qquad \frac{\mathrm{d}I_2(\mathbf{A})}{\mathrm{d}\mathbf{A}}=I_1(\mathbf{A})\,\mathbf{I}-\mathbf{A}^{\mathrm{T}},\qquad \frac{\mathrm{d}I_3(\mathbf{A})}{\mathrm{d}\mathbf{A}}=I_3(\mathbf{A})\,\mathbf{A}^{-\mathrm{T}}. \tag{2.94}$$

PROBLEM 2.46

Prove in direct notation that $\dfrac{\mathrm{d}I_1(\mathbf{A})}{\mathrm{d}\mathbf{A}} = \mathbf{I}.$

Solution

Using the definition $(2.89)_3$ of the derivative of a scalar with respect to a tensor, and the properties of the trace, we have, for all tensors $\mathbf{S}$ in $\mathcal{L}$,

$$\frac{\mathrm{d}I_1(\mathbf{A})}{\mathrm{d}\mathbf{A}}\cdot\mathbf{S} = \frac{\mathrm{d}(\mathrm{tr}\,\mathbf{A})}{\mathrm{d}\mathbf{A}}\cdot\mathbf{S} = \lim_{\alpha\to 0}\frac{1}{\alpha}\left[\mathrm{tr}\,(\mathbf{A}+\alpha\mathbf{S}) - \mathrm{tr}\,\mathbf{A}\right] = \lim_{\alpha\to 0}\frac{1}{\alpha}\left[\cancel{\mathrm{tr}\,\mathbf{A}} + \mathrm{tr}\,(\alpha\mathbf{S}) - \cancel{\mathrm{tr}\,\mathbf{A}}\right]$$

$$= \lim_{\alpha\to 0}\left(\frac{\alpha\,\mathrm{tr}\,\mathbf{S}}{\alpha}\right) = \mathrm{tr}\,\mathbf{S} = \mathrm{tr}\,(\mathbf{S}\mathbf{I}) = \mathrm{tr}(\mathbf{S}\mathbf{I}^{\mathrm{T}}) = \mathbf{S}\cdot\mathbf{I} = \mathbf{I}\cdot\mathbf{S}.$$

Since $\mathbf{S}$ is arbitrary, it follows that $\dfrac{\mathrm{d}I_1(\mathbf{A})}{\mathrm{d}\mathbf{A}} = \mathbf{I}.$

PROBLEM 2.47

Prove in direct notation that $\dfrac{\mathrm{d}I_2(\mathbf{A})}{\mathrm{d}\mathbf{A}} = I_1(\mathbf{A})\,\mathbf{I} - \mathbf{A}^{\mathrm{T}}.$

Solution

Note that in what follows, we employ the notation $\mathbf{A}\mathbf{A} \equiv \mathbf{A}^2$, as is done elsewhere in this book. Using the definition $(2.89)_3$ of the derivative of a scalar with respect to a tensor, and the properties of the trace, we have, for all tensors $\mathbf{S}$ in $\mathcal{L}$,

$$\frac{\mathrm{d}I_2(\mathbf{A})}{\mathrm{d}\mathbf{A}}\cdot\mathbf{S} = \frac{\mathrm{d}}{\mathrm{d}\mathbf{A}}\left\{\frac{1}{2}\left[(\mathrm{tr}\,\mathbf{A})^2 - \mathrm{tr}\,\mathbf{A}^2\right]\right\}\cdot\mathbf{S}$$

$$= \lim_{\alpha\to 0}\frac{1}{2\alpha}\left\{\left[\mathrm{tr}\,(\mathbf{A}+\alpha\mathbf{S})\right]^2 - \mathrm{tr}\,(\mathbf{A}+\alpha\mathbf{S})^2 - (\mathrm{tr}\,\mathbf{A})^2 + \mathrm{tr}\,\mathbf{A}^2\right\}$$

$$= \lim_{\alpha\to 0}\frac{1}{2\alpha}\left\{\left[(\mathrm{tr}\,\mathbf{A} + \mathrm{tr}\,(\alpha\,\mathbf{S})\right]^2 - \mathrm{tr}\left(\mathbf{A}^2 + \alpha^2\,\mathbf{S}^2 + \alpha\,\mathbf{A}\mathbf{S} + \alpha\,\mathbf{S}\mathbf{A}\right) - (\mathrm{tr}\,\mathbf{A})^2 + \mathrm{tr}\,\mathbf{A}^2\right\}$$

$$= \lim_{\alpha\to 0}\frac{1}{2\alpha}\left[(\mathrm{tr}\,\mathbf{A} + \alpha\,\mathrm{tr}\,\mathbf{S})^2 - \cancel{\mathrm{tr}\,\mathbf{A}^2} - \alpha^2\,\mathrm{tr}\,\mathbf{S}^2 - \alpha\,\mathrm{tr}(\mathbf{A}\mathbf{S}) - \alpha\,\mathrm{tr}(\mathbf{S}\mathbf{A}) - (\mathrm{tr}\,\mathbf{A})^2 + \cancel{\mathrm{tr}\,\mathbf{A}^2}\right]$$

$$= \lim_{\alpha\to 0}\frac{1}{2\alpha}\left[\cancel{(\mathrm{tr}\,\mathbf{A})^2} + \alpha^2(\mathrm{tr}\,\mathbf{S})^2 + 2\alpha\,(\mathrm{tr}\,\mathbf{A})(\mathrm{tr}\,\mathbf{S}) - \alpha^2\,\mathrm{tr}\,\mathbf{S}^2 - 2\alpha\,\mathrm{tr}\,(\mathbf{S}\mathbf{A}) - \cancel{(\mathrm{tr}\,\mathbf{A})^2}\right]$$

$$= (\mathrm{tr}\,\mathbf{A})(\mathrm{tr}\,\mathbf{S}) - \mathrm{tr}\,(\mathbf{S}\mathbf{A})$$

$$= I_1(\mathbf{A})\,(\mathrm{tr}\,\mathbf{S}) - \mathbf{S}\cdot\mathbf{A}^{\mathrm{T}}.$$

Recall from Problem 2.46 that

$$\mathrm{tr}\,\mathbf{S} = \mathrm{tr}\,(\mathbf{S}\mathbf{I}) = \mathrm{tr}(\mathbf{S}\mathbf{I}^{\mathrm{T}}) = \mathbf{S}\cdot\mathbf{I} = \mathbf{I}\cdot\mathbf{S}.$$

Thus, it follows that

$$\frac{\mathrm{d}I_2(\mathbf{A})}{\mathrm{d}\mathbf{A}} \cdot \mathbf{S} = \left(I_1(\mathbf{A})\,\mathbf{I} - \mathbf{A}^{\mathrm{T}}\right) \cdot \mathbf{S}.$$

Since $\mathbf{S}$ is arbitrary, $\dfrac{\mathrm{d}I_2(\mathbf{A})}{\mathrm{d}\mathbf{A}} = I_1(\mathbf{A})\,\mathbf{I} - \mathbf{A}^{\mathrm{T}}$.

PROBLEM 2.48

Prove in direct notation that $\dfrac{\mathrm{d}I_3(\mathbf{A})}{\mathrm{d}\mathbf{A}} = I_3(\mathbf{A})\,\mathbf{A}^{-\mathrm{T}}$.

Solution

In this problem, we will make use of the result

$$\det(\mathbf{U}+\mathbf{I}) = 1 + \operatorname{tr}\mathbf{U} + \frac{1}{2}\left[(\operatorname{tr}\mathbf{U})^2 - \operatorname{tr}\mathbf{U}^2\right] + \det\mathbf{U} = \det(\mathbf{I}+\mathbf{U}),$$

which holds for any tensor $\mathbf{U}$ in $\mathcal{L}$ and follows from (2.72) with $\alpha = -1$. Note that $I_3(\mathbf{A}) = \det\mathbf{A}$, and $\det\mathbf{A}$ is a scalar. Using the definition $(2.89)_3$ of the derivative of a scalar with respect to a tensor, we have, for all tensors $\mathbf{S}$ in $\mathcal{L}$,

$$\begin{aligned}
\left[\frac{\mathrm{d}}{\mathrm{d}\mathbf{A}}(\det\mathbf{A})\right]\cdot\mathbf{S} &= \lim_{\alpha\to 0}\frac{1}{\alpha}\left[\det(\mathbf{A}+\alpha\,\mathbf{S}) - \det\mathbf{A}\right] \\
&= \lim_{\alpha\to 0}\frac{1}{\alpha}\left\{\det\left[(\mathbf{I}+\alpha\,\mathbf{S}\mathbf{A}^{-1})\mathbf{A}\right] - \det\mathbf{A}\right\} \\
&= \lim_{\alpha\to 0}\frac{1}{\alpha}\left\{\det\left(\mathbf{I}+\alpha\,\mathbf{S}\mathbf{A}^{-1}\right)\det\mathbf{A} - \det\mathbf{A}\right\} \\
&= \lim_{\alpha\to 0}\frac{1}{\alpha}\left\{\left\{1 + \operatorname{tr}\left(\alpha\,\mathbf{S}\mathbf{A}^{-1}\right) + \frac{1}{2}\left[\left(\operatorname{tr}\left(\alpha\,\mathbf{S}\mathbf{A}^{-1}\right)\right)^2 - \operatorname{tr}\left(\alpha\,\mathbf{S}\mathbf{A}^{-1}\right)^2\right]\right.\right. \\
&\qquad \left.\left. + \det\left(\alpha\,\mathbf{S}\mathbf{A}^{-1}\right)\right\}\det\mathbf{A} - \det\mathbf{A}\right\} \\
&= \lim_{\alpha\to 0}\frac{1}{\alpha}\left\{\left\{1 + \alpha\operatorname{tr}\left(\mathbf{S}\mathbf{A}^{-1}\right) + \frac{\alpha^2}{2}\left[\left(\operatorname{tr}\left(\mathbf{S}\mathbf{A}^{-1}\right)\right)^2 - \operatorname{tr}\left(\mathbf{S}\mathbf{A}^{-1}\right)^2\right]\right.\right. \\
&\qquad \left.\left. + \alpha^3\det\left(\mathbf{S}\mathbf{A}^{-1}\right)\right\}\det\mathbf{A} - \det\mathbf{A}\right\} \\
&= \lim_{\alpha\to 0}\frac{1}{\alpha}\left[\cancel{\det\mathbf{A}} + \alpha\operatorname{tr}(\mathbf{S}\mathbf{A}^{-1})(\det\mathbf{A}) + \mathcal{O}(\alpha^2) - \cancel{\det\mathbf{A}}\right] \\
&= (\det\mathbf{A})\operatorname{tr}(\mathbf{S}\mathbf{A}^{-1})
\end{aligned}$$

$$= (\det \mathbf{A})\, \mathbf{A}^{-\mathrm{T}} \cdot \mathbf{S}.$$

Since $\mathbf{S}$ is arbitrary and $I_3(\mathbf{A}) = \det \mathbf{A}$, it follows that $\dfrac{\mathrm{d}I_3(\mathbf{A})}{\mathrm{d}\mathbf{A}} = I_3(\mathbf{A})\, \mathbf{A}^{-\mathrm{T}}$.

2.5.2 CHAIN RULE, GRADIENT, DIVERGENCE, CURL, DIVERGENCE THEOREM

If $\mathbf{T}$ and $\mathbf{x}$ in (2.88), (2.90), and (2.92) also depend on t, then the **chain rule** implies that

$$\begin{aligned}
\frac{\mathrm{d}}{\mathrm{d}t}\phi\left(\mathbf{T}(t), \mathbf{x}(t), t\right) &= \frac{\partial \phi}{\partial \mathbf{T}} \cdot \frac{\mathrm{d}\mathbf{T}}{\mathrm{d}t} + \frac{\partial \phi}{\partial \mathbf{x}} \cdot \frac{\mathrm{d}\mathbf{x}}{\mathrm{d}t} + \frac{\partial \phi}{\partial t}, \\
\frac{\mathrm{d}}{\mathrm{d}t}\mathbf{v}\left(\mathbf{T}(t), \mathbf{x}(t), t\right) &= \frac{\partial \mathbf{v}}{\partial \mathbf{T}} \frac{\mathrm{d}\mathbf{T}}{\mathrm{d}t} + \frac{\partial \mathbf{v}}{\partial \mathbf{x}} \frac{\mathrm{d}\mathbf{x}}{\mathrm{d}t} + \frac{\partial \mathbf{v}}{\partial t}, \qquad (2.95) \\
\frac{\mathrm{d}}{\mathrm{d}t}\mathbf{A}\left(\mathbf{T}(t), \mathbf{x}(t), t\right) &= \frac{\partial \mathbf{A}}{\partial \mathbf{T}} \frac{\mathrm{d}\mathbf{T}}{\mathrm{d}t} + \frac{\partial \mathbf{A}}{\partial \mathbf{x}} \frac{\mathrm{d}\mathbf{x}}{\mathrm{d}t} + \frac{\partial \mathbf{A}}{\partial t}.
\end{aligned}$$

If the vector $\mathbf{x}$ in (2.88), (2.90), and (2.92) is a **position vector**, we define **gradients** of the scalar-valued function ϕ, vector-valued function $\mathbf{v}$, and tensor-valued function $\mathbf{A}$ by

$$\operatorname{grad}\phi = \frac{\partial}{\partial \mathbf{x}}\phi(\mathbf{T}, \mathbf{x}, t), \qquad \operatorname{grad}\mathbf{v} = \frac{\partial}{\partial \mathbf{x}}\mathbf{v}(\mathbf{T}, \mathbf{x}, t), \qquad \operatorname{grad}\mathbf{A} = \frac{\partial}{\partial \mathbf{x}}\mathbf{A}(\mathbf{T}, \mathbf{x}, t). \qquad (2.96)$$

Note that the gradients of scalars, vectors, and tensors are vectors, tensors, and third-order tensors, respectively.

The **divergence** of a vector-valued function $\mathbf{v}$ of position is defined

$$\operatorname{div}\mathbf{v} = \operatorname{tr}\left(\operatorname{grad}\mathbf{v}\right), \qquad (2.97)$$

which is a scalar. The divergence of a tensor-valued function $\mathbf{A}$ of position is defined through

$$(\operatorname{div}\mathbf{A}) \cdot \mathbf{a} = \operatorname{div}\left(\mathbf{A}^{\mathrm{T}}\mathbf{a}\right) \qquad (2.98)$$

for any vector $\mathbf{a}$. The divergence of a tensor is a vector. It can be shown (refer to Problems 2.49–2.51) that

$$\begin{aligned}
&\operatorname{grad}(\phi\,\mathbf{v}) = \phi \operatorname{grad}\mathbf{v} + \mathbf{v} \otimes \operatorname{grad}\phi, \\
&\operatorname{div}(\phi\,\mathbf{v}) = \phi \operatorname{div}\mathbf{v} + \mathbf{v} \cdot \operatorname{grad}\phi, \\
&\operatorname{grad}(\mathbf{v} \cdot \mathbf{w}) = (\operatorname{grad}\mathbf{v})^{\mathrm{T}}\,\mathbf{w} + (\operatorname{grad}\mathbf{w})^{\mathrm{T}}\,\mathbf{v}, \\
&\operatorname{div}(\mathbf{A}^{\mathrm{T}}\,\mathbf{v}) = \mathbf{A} \cdot \operatorname{grad}\mathbf{v} + \mathbf{v} \cdot \operatorname{div}\mathbf{A}, \qquad (2.99)
\end{aligned}$$

$$\text{grad}\left(\frac{1}{\phi}\right) = -\frac{1}{\phi^2}\,\text{grad}\,\phi,$$

$$\text{div}\,(\phi\,\mathbf{A}) = \phi\,\text{div}\,\mathbf{A} + \mathbf{A}\,\text{grad}\,\phi,$$

$$\text{div}\,(\mathbf{v}\otimes\mathbf{w}) = \mathbf{v}\,\text{div}\,\mathbf{w} + (\text{grad}\,\mathbf{v})\,\mathbf{w}.$$

The **curl** of a vector $\mathbf{v}$ is defined by

$$(\text{curl}\,\mathbf{v})\times\mathbf{a} = \left[\text{grad}\,\mathbf{v} - (\text{grad}\,\mathbf{v})^{\mathrm{T}}\right]\mathbf{a}. \tag{2.100}$$

We can show that the divergence of the curl of a vector $\mathbf{v}$ vanishes, i.e.,

$$\text{div}\left(\text{curl}\,\mathbf{v}\right) = 0. \tag{2.101}$$

Also,

$$\text{curl}\,(\mathbf{v}\times\mathbf{w}) = (\text{grad}\,\mathbf{v})\,\mathbf{w} - (\text{grad}\,\mathbf{w})\,\mathbf{v} + \mathbf{v}\,(\text{div}\,\mathbf{w}) - \mathbf{w}\,(\text{div}\,\mathbf{v}). \tag{2.102}$$

Note that a useful property of the gradient, divergence, and curl is distributivity over vector and tensor addition, e.g.,

$$\text{curl}\,(\mathbf{v}+\mathbf{w}) = \text{curl}\,\mathbf{v} + \text{curl}\,\mathbf{w}, \tag{2.103a}$$

$$\text{div}\,(\mathbf{A}+\mathbf{B}) = \text{div}\,\mathbf{A} + \text{div}\,\mathbf{B}. \tag{2.103b}$$

If $\mathcal{R}$ is an open region (volume) bounded by a closed surface $\partial\mathcal{R}$, and ϕ, $\mathbf{v}$, and $\mathbf{A}$ are smooth functions of position, then, according to the **divergence theorem**,

$$\int_{\mathcal{R}} \text{grad}\,\phi\,\mathrm{d}v = \int_{\partial\mathcal{R}} \phi\,\mathbf{n}\,\mathrm{d}a, \qquad \int_{\mathcal{R}} \text{div}\,\mathbf{v}\,\mathrm{d}v = \int_{\partial\mathcal{R}} \mathbf{v}\cdot\mathbf{n}\,\mathrm{d}a, \qquad \int_{\mathcal{R}} \text{div}\,\mathbf{A}\,\mathrm{d}v = \int_{\partial\mathcal{R}} \mathbf{A}\mathbf{n}\,\mathrm{d}a, \tag{2.104}$$

where $\mathrm{d}v$ is the volume element of $\mathcal{R}$, $\mathrm{d}a$ is the area element of $\partial\mathcal{R}$, and $\mathbf{n}$ is the outward unit normal on $\partial\mathcal{R}$. It follows that (refer to Problem 2.52)

$$\int_{\partial\mathcal{R}} \mathbf{v}\cdot\mathbf{A}\,\mathbf{n}\,\mathrm{d}a = \int_{\mathcal{R}} (\mathbf{A}\cdot\text{grad}\,\mathbf{v} + \mathbf{v}\cdot\text{div}\,\mathbf{A})\,\mathrm{d}v. \tag{2.105}$$

PROBLEM 2.49

Prove in direct notation that $\text{grad}\,(\phi\,\mathbf{v}) = \phi\,\text{grad}\,\mathbf{v} + \mathbf{v}\otimes\text{grad}\,\phi$.

Solution

Note that $\phi\mathbf{v}$ is a vector. Definition $(2.96)_2$ of the gradient of a vector and definition $(2.91)_2$ of the partial derivative of a vector with respect to a vector imply that

$$[\text{grad}\,(\phi\,\mathbf{v})]\,\mathbf{u} = \frac{\partial(\phi\,\mathbf{v})}{\partial\mathbf{x}}\,\mathbf{u} = \lim_{\alpha\to 0}\frac{\phi(\mathbf{x}+\alpha\mathbf{u})\,\mathbf{v}(\mathbf{x}+\alpha\mathbf{u}) - \phi(\mathbf{x})\,\mathbf{v}(\mathbf{x})}{\alpha}.$$

Then, adding and subtracting $\phi(\mathbf{x}+\alpha\mathbf{u})\,\mathbf{v}(\mathbf{x})$ to and from the numerator, we have

$$
\begin{aligned}
[\operatorname{grad}(\phi\mathbf{v})]\mathbf{u} &= \lim_{\alpha\to 0}\frac{1}{\alpha}[\phi(\mathbf{x}+\alpha\mathbf{u})\mathbf{v}(\mathbf{x}+\alpha\mathbf{u}) - \phi(\mathbf{x}+\alpha\mathbf{u})\mathbf{v}(\mathbf{x}) \\
&\qquad + \phi(\mathbf{x}+\alpha\mathbf{u})\mathbf{v}(\mathbf{x}) - \phi(\mathbf{x})\mathbf{v}(\mathbf{x})] \\
&= \lim_{\alpha\to 0}\left\{\phi(\mathbf{x}+\alpha\mathbf{u})\left[\frac{\mathbf{v}(\mathbf{x}+\alpha\mathbf{u}) - \mathbf{v}(\mathbf{x})}{\alpha}\right]\right\} + \lim_{\alpha\to 0}\left\{\left[\frac{\phi(\mathbf{x}+\alpha\mathbf{u}) - \phi(\mathbf{x})}{\alpha}\right]\mathbf{v}(\mathbf{x})\right\} \\
&= \left[\lim_{\alpha\to 0}\phi(\mathbf{x}+\alpha\mathbf{u})\right]\left[\lim_{\alpha\to 0}\frac{\mathbf{v}(\mathbf{x}+\alpha\mathbf{u}) - \mathbf{v}(\mathbf{x})}{\alpha}\right] + \left[\lim_{\alpha\to 0}\frac{\phi(\mathbf{x}+\alpha\mathbf{u}) - \phi(\mathbf{x})}{\alpha}\right]\mathbf{v}(\mathbf{x}) \\
&= \phi(\mathbf{x})\left(\frac{\partial\mathbf{v}}{\partial\mathbf{x}}\mathbf{u}\right) + \left(\frac{\partial\phi}{\partial\mathbf{x}}\cdot\mathbf{u}\right)\mathbf{v}(\mathbf{x}) \\
&= (\phi\operatorname{grad}\mathbf{v})\,\mathbf{u} + (\mathbf{v}\otimes\operatorname{grad}\phi)\,\mathbf{u} \\
&= (\phi\operatorname{grad}\mathbf{v} + \mathbf{v}\otimes\operatorname{grad}\phi)\,\mathbf{u}.
\end{aligned}
$$

Since $\mathbf{u}$ is arbitrary, it follows that $\operatorname{grad}(\phi\,\mathbf{v}) = \phi\operatorname{grad}\mathbf{v} + \mathbf{v}\otimes\operatorname{grad}\phi$.

PROBLEM 2.50

Prove in direct notation that $\operatorname{div}(\phi\,\mathbf{v}) = \phi\operatorname{div}\mathbf{v} + \mathbf{v}\cdot\operatorname{grad}\phi$.

Solution

Again, note that $\phi\mathbf{v}$ is a vector. Using the definition (2.97) of the divergence of a vector, the result $\operatorname{grad}(\phi\,\mathbf{v}) = \phi\operatorname{grad}\mathbf{v} + \mathbf{v}\otimes\operatorname{grad}\phi$ from Problem 2.49, and the properties (2.38) of the trace, we have

$$
\begin{aligned}
\operatorname{div}(\phi\,\mathbf{v}) &= \operatorname{tr}[\operatorname{grad}(\phi\,\mathbf{v})] = \operatorname{tr}(\phi\operatorname{grad}\mathbf{v} + \mathbf{v}\otimes\operatorname{grad}\phi) = \operatorname{tr}(\phi\operatorname{grad}\mathbf{v}) + \operatorname{tr}(\mathbf{v}\otimes\operatorname{grad}\phi) \\
&= \phi\operatorname{tr}(\operatorname{grad}\mathbf{v}) + \mathbf{v}\cdot\operatorname{grad}\phi = \phi\operatorname{div}\mathbf{v} + \mathbf{v}\cdot\operatorname{grad}\phi.
\end{aligned}
$$

PROBLEM 2.51

Prove in direct notation that $\operatorname{grad}(\mathbf{v}\cdot\mathbf{w}) = (\operatorname{grad}\mathbf{v})^{\mathrm{T}}\mathbf{w} + (\operatorname{grad}\mathbf{w})^{\mathrm{T}}\mathbf{v}$.

Solution

Definition $(2.96)_1$ of the gradient of a scalar and definition $(2.89)_2$ of the partial derivative of a scalar with respect to a vector imply that

$$
[\operatorname{grad}(\mathbf{v}\cdot\mathbf{w})]\cdot\mathbf{u} = \frac{\partial(\mathbf{v}\cdot\mathbf{w})}{\partial\mathbf{x}}\cdot\mathbf{u}
$$

$$= \lim_{\alpha\to 0} \frac{\mathbf{v}(\mathbf{x}+\alpha\mathbf{u})\cdot\mathbf{w}(\mathbf{x}+\alpha\mathbf{u}) - \mathbf{v}(\mathbf{x})\cdot\mathbf{w}(\mathbf{x})}{\alpha}$$

$$= \lim_{\alpha\to 0} \frac{1}{\alpha}\Big[\mathbf{v}(\mathbf{x}+\alpha\mathbf{u})\cdot\mathbf{w}(\mathbf{x}+\alpha\mathbf{u}) - \mathbf{v}(\mathbf{x})\cdot\mathbf{w}(\mathbf{x}+\alpha\mathbf{u}) + \mathbf{v}(\mathbf{x})\cdot\mathbf{w}(\mathbf{x}+\alpha\mathbf{u})$$

$$- \mathbf{v}(\mathbf{x})\cdot\mathbf{w}(\mathbf{x})\Big]$$

$$= \lim_{\alpha\to 0}\left\{\left[\frac{\mathbf{v}(\mathbf{x}+\alpha\mathbf{u}) - \mathbf{v}(\mathbf{x})}{\alpha}\right]\cdot\mathbf{w}(\mathbf{x}+\alpha\mathbf{u})\right\} + \lim_{\alpha\to 0}\left\{\mathbf{v}(\mathbf{x})\cdot\left[\frac{\mathbf{w}(\mathbf{x}+\alpha\mathbf{u}) - \mathbf{w}(\mathbf{x})}{\alpha}\right]\right\}$$

$$= \left[\lim_{\alpha\to 0}\frac{\mathbf{v}(\mathbf{x}+\alpha\mathbf{u}) - \mathbf{v}(\mathbf{x})}{\alpha}\right]\cdot\left[\lim_{\alpha\to 0}\mathbf{w}(\mathbf{x}+\alpha\mathbf{u})\right] + \mathbf{v}(\mathbf{x})\cdot\left[\lim_{\alpha\to 0}\frac{\mathbf{w}(\mathbf{x}+\alpha\mathbf{u}) - \mathbf{w}(\mathbf{x})}{\alpha}\right]$$

$$= \frac{\partial\mathbf{v}}{\partial\mathbf{x}}\mathbf{u}\cdot\mathbf{w} + \mathbf{v}\cdot\frac{\partial\mathbf{w}}{\partial\mathbf{x}}\mathbf{u}$$

$$= (\operatorname{grad}\mathbf{v})\mathbf{u}\cdot\mathbf{w} + \mathbf{v}\cdot(\operatorname{grad}\mathbf{w})\mathbf{u}$$

$$= (\operatorname{grad}\mathbf{v})^{\mathrm{T}}\mathbf{w}\cdot\mathbf{u} + (\operatorname{grad}\mathbf{w})^{\mathrm{T}}\mathbf{v}\cdot\mathbf{u}$$

$$= \left[(\operatorname{grad}\mathbf{v})^{\mathrm{T}}\mathbf{w} + (\operatorname{grad}\mathbf{w})^{\mathrm{T}}\mathbf{v}\right]\cdot\mathbf{u}.$$

Since the vector $\mathbf{u}$ is arbitrary, it follows that $\operatorname{grad}(\mathbf{v}\cdot\mathbf{w}) = (\operatorname{grad}\mathbf{v})^{\mathrm{T}}\mathbf{w} + (\operatorname{grad}\mathbf{w})^{\mathrm{T}}\mathbf{v}$.

PROBLEM 2.52

Prove in direct notation that $\int_{\partial\mathcal{R}} \mathbf{v}\cdot\mathbf{A}\,\mathbf{n}\,\mathrm{d}a = \int_{\mathcal{R}} (\mathbf{A}\cdot\operatorname{grad}\mathbf{v} + \mathbf{v}\cdot\operatorname{div}\mathbf{A})\,\mathrm{d}v$.

Solution

Recall that $\mathcal{R}$ is an open volume bounded by a closed surface $\partial\mathcal{R}$, $\mathrm{d}v$ is the volume element of $\mathcal{R}$, $\mathrm{d}a$ is the area element of $\partial\mathcal{R}$, and $\mathbf{n}$ is the outward unit normal on $\partial\mathcal{R}$. We have

$$\int_{\partial\mathcal{R}} \mathbf{v}\cdot\mathbf{A}\,\mathbf{n}\,\mathrm{d}a = \int_{\partial\mathcal{R}} \mathbf{v}\cdot(\mathbf{A}^{\mathrm{T}})^{\mathrm{T}}\mathbf{n}\,\mathrm{d}a = \int_{\partial\mathcal{R}} \mathbf{A}^{\mathrm{T}}\mathbf{v}\cdot\mathbf{n}\,\mathrm{d}a$$

$$= \int_{\mathcal{R}} \operatorname{div}(\mathbf{A}^{\mathrm{T}}\mathbf{v})\,\mathrm{d}v = \int_{\mathcal{R}} (\mathbf{A}\cdot\operatorname{grad}\mathbf{v} + \mathbf{v}\cdot\operatorname{div}\mathbf{A})\,\mathrm{d}v.$$

Note that we have used result $(2.99)_4$.

2.5.3 TENSOR CALCULUS IN CARTESIAN COMPONENT FORM

The Cartesian component form of the chain rule (2.95) is

$$\begin{aligned}
\frac{\mathrm{d}}{\mathrm{d}t}\phi\left(\mathbf{T}(t),\mathbf{x}(t),t\right) &= \frac{\partial\phi}{\partial T_{ij}}\frac{\mathrm{d}T_{ij}}{\mathrm{d}t}+\frac{\partial\phi}{\partial x_i}\frac{\mathrm{d}x_i}{\mathrm{d}t}+\frac{\partial\phi}{\partial t},\\
\frac{\mathrm{d}}{\mathrm{d}t}v_i\left(\mathbf{T}(t),\mathbf{x}(t),t\right) &= \frac{\partial v_i}{\partial T_{jk}}\frac{\mathrm{d}T_{jk}}{\mathrm{d}t}+\frac{\partial v_i}{\partial x_j}\frac{\mathrm{d}x_j}{\mathrm{d}t}+\frac{\partial v_i}{\partial t},\\
\frac{\mathrm{d}}{\mathrm{d}t}A_{ij}\left(\mathbf{T}(t),\mathbf{x}(t),t\right) &= \frac{\partial A_{ij}}{\partial T_{kl}}\frac{\mathrm{d}T_{kl}}{\mathrm{d}t}+\frac{\partial A_{ij}}{\partial x_k}\frac{\mathrm{d}x_k}{\mathrm{d}t}+\frac{\partial A_{ij}}{\partial t}.
\end{aligned} \tag{2.106}$$

It can be shown (refer to Problems 2.53–2.57) that

$$(\operatorname{grad}\phi)_i = \frac{\partial\phi}{\partial x_i} \equiv \phi_{,i}, \qquad (\operatorname{grad}\mathbf{v})_{ij} = \frac{\partial v_i}{\partial x_j} \equiv v_{i,j}, \qquad (\operatorname{grad}\mathbf{A})_{ijk} = \frac{\partial A_{ij}}{\partial x_k} \equiv A_{ij,k},$$

$$\operatorname{div}\mathbf{v} = \frac{\partial v_i}{\partial x_i} = v_{i,i}, \qquad (\operatorname{div}\mathbf{A})_i = \frac{\partial A_{ij}}{\partial x_j} = A_{ij,j}. \tag{2.107}$$

Then, it follows that the Cartesian component form of the divergence theorem is

$$\int_{\mathcal{R}} \phi_{,i}\,\mathrm{d}v = \int_{\partial\mathcal{R}} \phi\, n_i\,\mathrm{d}a, \qquad \int_{\mathcal{R}} v_{i,i}\,\mathrm{d}v = \int_{\partial\mathcal{R}} v_i\, n_i\,\mathrm{d}a, \qquad \int_{\mathcal{R}} A_{ij,j}\,\mathrm{d}v = \int_{\partial\mathcal{R}} A_{ij}\, n_j\,\mathrm{d}a. \tag{2.108}$$

The Cartesian component form of the curl of a vector $\mathbf{v}$ is

$$(\operatorname{curl}\mathbf{v})_i = \epsilon_{ijk}\, v_{k,j}. \tag{2.109}$$

PROBLEM 2.53

Prove that $(\operatorname{grad}\phi)_i = \phi_{,i}$ (i.e., the Cartesian components of $\operatorname{grad}\phi$ are $\phi_{,i}$).

Solution

Note that the gradient of a scalar ϕ is a vector. Its Cartesian components are

$$(\operatorname{grad}\phi)_i = (\operatorname{grad}\phi)\cdot\mathbf{e}_i = \frac{\partial\phi}{\partial\mathbf{x}}\cdot\mathbf{e}_i = \lim_{\alpha\to 0}\frac{1}{\alpha}\left[\phi\left(\mathbf{x}+\alpha\,\mathbf{e}_i\right)-\phi\left(\mathbf{x}\right)\right] = \frac{\partial\phi}{\partial x_i} \equiv \phi_{,i}.$$

Thus,

$$\operatorname{grad}\phi = \phi_{,i}\,\mathbf{e}_i = \phi_{,1}\,\mathbf{e}_1 + \phi_{,2}\,\mathbf{e}_2 + \phi_{,3}\,\mathbf{e}_3.$$

PROBLEM 2.54

Prove that $(\operatorname{grad}\mathbf{v})_{ij} = v_{i,j}$ (i.e., the Cartesian components of $\operatorname{grad}\mathbf{v}$ are $v_{i,j}$).

Solution

Note that the gradient of a vector $\mathbf{v}$ is a tensor. Its Cartesian components are

$$(\operatorname{grad}\mathbf{v})_{ij} = \mathbf{e}_i \cdot (\operatorname{grad}\mathbf{v})\,\mathbf{e}_j = \mathbf{e}_i \cdot \frac{\partial \mathbf{v}}{\partial \mathbf{x}}\,\mathbf{e}_j = \mathbf{e}_i \cdot \left[\lim_{\alpha \to 0} \frac{\mathbf{v}(\mathbf{x} + \alpha\,\mathbf{e}_j) - \mathbf{v}(\mathbf{x})}{\alpha}\right]$$

$$= \mathbf{e}_i \cdot \frac{\partial \mathbf{v}}{\partial x_j} = \frac{\partial}{\partial x_j}(\mathbf{e}_i \cdot \mathbf{v}) = \frac{\partial v_i}{\partial x_j} \equiv v_{i,j}.$$

Thus,

$$\operatorname{grad}\mathbf{v} = v_{i,j}\,\mathbf{e}_i \otimes \mathbf{e}_j = v_{1,1}\,\mathbf{e}_1 \otimes \mathbf{e}_1 + v_{1,2}\,\mathbf{e}_1 \otimes \mathbf{e}_2 + v_{1,3}\,\mathbf{e}_1 \otimes \mathbf{e}_3 + v_{2,1}\,\mathbf{e}_2 \otimes \mathbf{e}_1$$
$$+ v_{2,2}\,\mathbf{e}_2 \otimes \mathbf{e}_2 + v_{2,3}\,\mathbf{e}_2 \otimes \mathbf{e}_3 + v_{3,1}\,\mathbf{e}_3 \otimes \mathbf{e}_1 + v_{3,2}\,\mathbf{e}_3 \otimes \mathbf{e}_2$$
$$+ v_{3,3}\,\mathbf{e}_3 \otimes \mathbf{e}_3$$

and

$$[\operatorname{grad}\mathbf{v}] = \begin{bmatrix} v_{1,1} & v_{1,2} & v_{1,3} \\ v_{2,1} & v_{2,2} & v_{2,3} \\ v_{3,1} & v_{3,2} & v_{3,3} \end{bmatrix}.$$

PROBLEM 2.55

Verify that $(\operatorname{grad}\mathbf{A})_{ijk} = A_{ij,k}$.

Solution

Note that the gradient of a tensor $\mathbf{A}$ is a third-order tensor. Hence, according to (2.80), its Cartesian components are

$$(\operatorname{grad}\mathbf{A})_{ijk} = (\mathbf{e}_i \otimes \mathbf{e}_j) \cdot \left[(\operatorname{grad}\mathbf{A})\,\mathbf{e}_k\right].$$

Then,

$$(\operatorname{grad}\mathbf{A})_{ijk} = (\mathbf{e}_i \otimes \mathbf{e}_j) \cdot \left(\frac{\partial \mathbf{A}}{\partial \mathbf{x}}\,\mathbf{e}_k\right) \qquad \text{(definition (2.96)}_3\text{)}$$

$$= (\mathbf{e}_i \otimes \mathbf{e}_j) \cdot \left\{\lim_{\alpha \to 0} \frac{1}{\alpha}\left[\mathbf{A}\,(\mathbf{x} + \alpha\,\mathbf{e}_k) - \mathbf{A}\,(\mathbf{x})\right]\right\} \qquad \text{(definition (2.93)}_2\text{)}$$

$$= (\mathbf{e}_i \otimes \mathbf{e}_j) \cdot \frac{\partial \mathbf{A}}{\partial x_k} \qquad \text{(definition of partial derivative)}$$

$$= \frac{\partial}{\partial x_k}\left[\mathbf{A} \cdot (\mathbf{e}_i \otimes \mathbf{e}_j)\right] \qquad (\mathbf{e}_i \text{ and } \mathbf{e}_j \text{ independent of } x_k)$$

$$= \frac{\partial}{\partial x_k}\operatorname{tr}\left[\mathbf{A}\,(\mathbf{e}_i \otimes \mathbf{e}_j)^{\mathrm{T}}\right] \qquad \text{(definition (2.41))}$$

$$= \frac{\partial}{\partial x_k}\operatorname{tr}\left[\mathbf{A}\,(\mathbf{e}_j \otimes \mathbf{e}_i)\right] \qquad \text{(result (2.20)}_1\text{)}$$

$$= \frac{\partial}{\partial x_k} \operatorname{tr}\left[(\mathbf{A}\,\mathbf{e}_j) \otimes \mathbf{e}_i\right] \qquad \text{(result } (2.20)_2\text{)}$$

$$= \frac{\partial}{\partial x_k} \left(\mathbf{e}_i \cdot \mathbf{A}\,\mathbf{e}_j\right) \qquad \text{(definition } (2.38)_3\text{)}$$

$$= \frac{\partial A_{ij}}{\partial x_k} \qquad \text{(definition (2.29))}$$

$$\equiv A_{ij,k}.$$

Thus, $\operatorname{grad} \mathbf{A} = A_{ij,k}\, \mathbf{e}_i \otimes \mathbf{e}_j \otimes \mathbf{e}_k$.

PROBLEM 2.56

Show that $\operatorname{div} \mathbf{v} = v_{i,i}$.

Solution

Note that the divergence of a vector $\mathbf{v}$ is a scalar. Using definition (2.97) and the result $\operatorname{grad} \mathbf{v} = v_{i,j}\, \mathbf{e}_i \otimes \mathbf{e}_j$ from Problem 2.54, we have

$$\operatorname{div} \mathbf{v} = \operatorname{tr}(\operatorname{grad} \mathbf{v}) = \operatorname{tr}(v_{i,j}\, \mathbf{e}_i \otimes \mathbf{e}_j) = v_{i,j} \operatorname{tr}(\mathbf{e}_i \otimes \mathbf{e}_j) = v_{i,j}\,(\mathbf{e}_i \cdot \mathbf{e}_j) = v_{i,j}\,\delta_{ij} = v_{i,i}.$$

Expanding this result, we obtain

$$\operatorname{div} \mathbf{v} = v_{i,i} = v_{1,1} + v_{2,2} + v_{3,3}.$$

PROBLEM 2.57

Prove that $(\operatorname{div} \mathbf{A})_i = A_{ij,j}$.

Solution

The divergence of a tensor $\mathbf{A}$ is a vector. The ith component of this vector is

$$(\operatorname{div} \mathbf{A})_i = (\operatorname{div} \mathbf{A}) \cdot \mathbf{e}_i = \operatorname{div}(\mathbf{A}^{\mathrm{T}}\, \mathbf{e}_i),$$

where we have used definition (2.24) of the Cartesian component of a vector and definition (2.98) of the divergence of a tensor. Then

$$\operatorname{div}(\mathbf{A}^{\mathrm{T}}\, \mathbf{e}_i) = \operatorname{div}[(A_{kj}\, \mathbf{e}_j \otimes \mathbf{e}_k)\, \mathbf{e}_i] = \operatorname{div}[A_{kj}\,(\mathbf{e}_k \cdot \mathbf{e}_i)\, \mathbf{e}_j] = \operatorname{div}(A_{kj}\,\delta_{ki}\, \mathbf{e}_j) = \operatorname{div}(A_{ij}\, \mathbf{e}_j).$$

Recalling that $\operatorname{div} \mathbf{v} = \operatorname{div}(v_i\, \mathbf{e}_i) = v_{i,i}$ from Problem 2.56, we deduce by analogy that $\operatorname{div}(A_{ij}\, \mathbf{e}_j) = A_{ij,j}$. Thus,

$$(\operatorname{div} \mathbf{A})_i = A_{ij,j},$$

so

$$\text{div}\,\mathbf{A} = (\text{div}\,\mathbf{A})_i\,\mathbf{e}_i = A_{ij,j}\,\mathbf{e}_i = \left(A_{11,1} + A_{12,2} + A_{13,3}\right)\mathbf{e}_1 + \left(A_{21,1} + A_{22,2} + A_{23,3}\right)\mathbf{e}_2 + \left(A_{31,1} + A_{32,2} + A_{33,3}\right)\mathbf{e}_3.$$

2.6 CURVILINEAR COORDINATES

Recall that any basis for the three-dimensional inner product space $\mathcal{E}^3$ has three elements. Also recall that $\mathcal{E}^3$ has a *fixed orthonormal* basis $\{\mathbf{e}_i\} = \{\mathbf{e}_1, \mathbf{e}_2, \mathbf{e}_3\}$. As was demonstrated in (2.21), any vector $\mathbf{u}$ in $\mathcal{E}^3$ can be expressed in terms of this basis, i.e.,

$$\mathbf{u} = u_1\,\mathbf{e}_1 + u_2\,\mathbf{e}_2 + u_3\,\mathbf{e}_3,$$

or, compactly,

$$\mathbf{u} = u_i\,\mathbf{e}_i.$$

The scalars u_1, u_2, and u_3 are called the **Cartesian components** of vector $\mathbf{u}$.

Of course, $\mathcal{E}^3$ has infinitely many other fixed orthonormal bases, which are merely rotations or inversions of $\{\mathbf{e}_1, \mathbf{e}_2, \mathbf{e}_3\}$. It also has infinitely many other bases that are not orthogonal or fixed; components with respect to these bases are called non-Cartesian or **curvilinear components.** (Because its basis vectors $\{\mathbf{e}_1, \mathbf{e}_2, \mathbf{e}_3\}$ are fixed and do not change direction or magnitude with position, Cartesian component notation is the simplest to manipulate.)

Since $\mathcal{E}^3$ is a metric space (refer to Section 2.1.2 and Figure 2.3), the **position vector x** in $\mathcal{E}^3$ can be considered a point in space, or a vector from the origin to the point. Furthermore,

$$\mathbf{x} = x_1\,\mathbf{e}_1 + x_2\,\mathbf{e}_2 + x_3\,\mathbf{e}_3,$$

or, compactly,

$$\mathbf{x} = x_i\,\mathbf{e}_i.$$

The Cartesian components x_i of position vector $\mathbf{x}$ are called the **Cartesian coordinates** of the point. The three independent variables x_1, x_2, and x_3 therefore define the position of point $\mathbf{x}$ in $\mathcal{E}^3$.

Since $\mathbf{e}_1$, $\mathbf{e}_2$, and $\mathbf{e}_3$ are fixed, the surfaces x_i = constant, called the **coordinate surfaces,** are planes, and the intersections of two coordinate surfaces, called **coordinate curves**, are straight lines (see Figure 2.5). For example, the coordinate surface $x_1 = C =$ constant is a plane perpendicular to the unit vector $\mathbf{e}_1$, and the x_1 coordinate line given by $x_2 = D =$ constant, $x_3 = E =$ constant is a straight line in the $\mathbf{e}_1$ direction.

When solving boundary-value problems, one adopts coordinate systems (θ^1, θ^2, θ^3) in which the boundary surfaces in the physical problem (where the boundary values are specified) are given by coordinate surfaces $\theta^i =$ constant. Because the

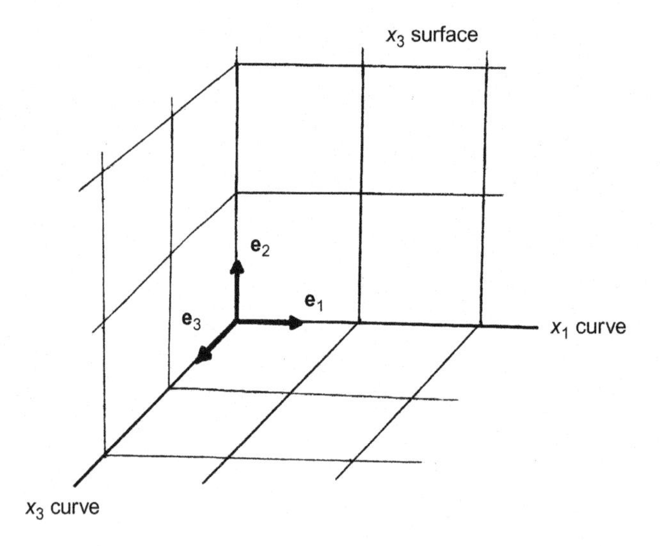

FIGURE 2.5

Cartesian coordinate curves, coordinate surfaces, and basis vectors.

coordinate surfaces are planes, Cartesian coordinates are useful for solving boundary-value problems in situations where the boundary surfaces are planes. However, in many problems of interest, the boundary surfaces are not planes, and it is therefore advantageous to adopt **curvilinear coordinates.**

Since the coordinate surfaces θ^i = constant in curvilinear coordinate systems are not planar, the coordinate curves (given by the intersection of two coordinate surfaces) are not straight lines. The basis vectors, which are tangent to the coordinate curves, are not fixed, but rather rotate in space with change of position. (Curvilinear coordinates are so called because their coordinate curves are not straight lines.) The most familiar curvilinear coordinates are **cylindrical polar coordinates** and **spherical coordinates**.

2.6.1 COVARIANT AND CONTRAVARIANT BASIS VECTORS

The **natural basis** $\{\mathbf{g}_i\} = \{\mathbf{g}_1, \mathbf{g}_2, \mathbf{g}_3\}$ of the curvilinear coordinate system (θ^1, θ^2, θ^3) is defined by

$$\mathbf{g}_i = \frac{\partial \mathbf{x}}{\partial \theta^i}, \tag{2.110}$$

where $\mathbf{x}$ is the position vector. The vectors $\mathbf{g}_i$ are also called the **covariant basis vectors** or **tangent basis vectors**. They are tangent to the coordinate curve: $\mathbf{g}_1$ is tangent to the θ^1 curve at position $\mathbf{x}$, $\mathbf{g}_2$ is tangent to the θ^2 curve at $\mathbf{x}$, and $\mathbf{g}_3$ is tangent to the θ^3 curve at $\mathbf{x}$ (see Figure 2.6).

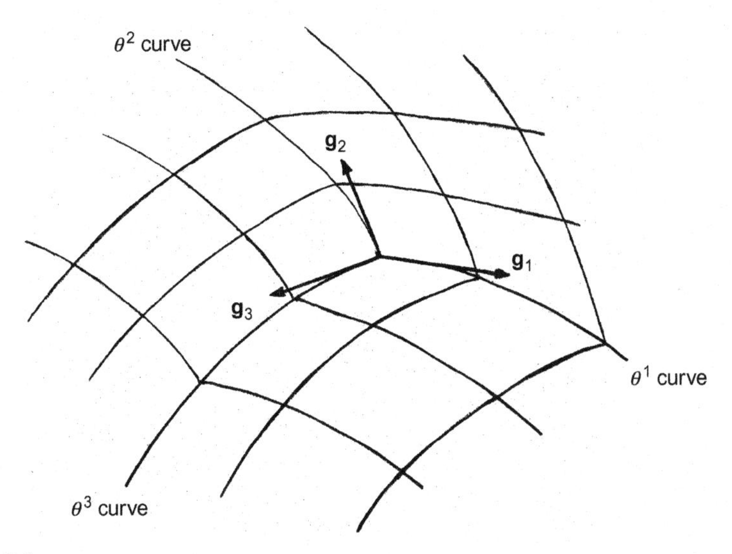

FIGURE 2.6

Curvilinear coordinate curves and natural (covariant) basis vectors.

PROBLEM 2.58

Determine the covariant basis $\{\mathbf{g}_i\} = \{\mathbf{g}_1, \mathbf{g}_2, \mathbf{g}_3\}$ of the cylindrical polar coordinate system $(\theta^1 = r, \theta^2 = \theta, \theta^3 = z)$.

Solution

The cylindrical polar coordinate system (r, θ, z) is related to the Cartesian coordinate system (x_1, x_2, x_3) through

$$x_1 = r\cos\theta, \qquad x_2 = r\sin\theta, \qquad x_3 = z, \tag{a}$$

so the position vector $\mathbf{x}$ is given by

$$\mathbf{x} = x_1\,\mathbf{e}_1 + x_2\,\mathbf{e}_2 + x_3\,\mathbf{e}_3 = r\cos\theta\,\mathbf{e}_1 + r\sin\theta\,\mathbf{e}_2 + z\,\mathbf{e}_3. \tag{b}$$

Then, from definition (2.110), the covariant basis vectors are

$$\begin{aligned} \mathbf{g}_1 &= \frac{\partial \mathbf{x}}{\partial r} = \cos\theta\,\mathbf{e}_1 + \sin\theta\,\mathbf{e}_2, \\ \mathbf{g}_2 &= \frac{\partial \mathbf{x}}{\partial \theta} = -r\sin\theta\,\mathbf{e}_1 + r\cos\theta\,\mathbf{e}_2, \\ \mathbf{g}_3 &= \frac{\partial \mathbf{x}}{\partial z} = \mathbf{e}_3. \end{aligned} \tag{c}$$

In general, the covariant basis vectors $\mathbf{g}_i$ are not fixed in either magnitude or direction, and can have dimensions. In the cylindrical polar example (c) of Problem 2.64,

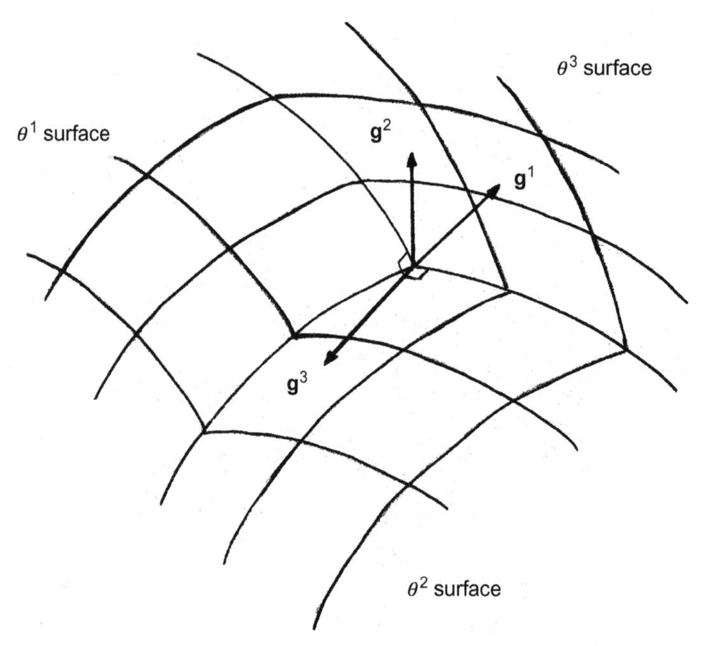

FIGURE 2.7

Curvilinear coordinate surfaces and reciprocal (contravariant) basis vectors.

the direction of basis vector $\mathbf{g}_2$ at position (r, θ, z) depends on the value of coordinate θ, and its magnitude depends on the value of coordinate r; although $\mathbf{g}_1$ and $\mathbf{g}_3$ are dimensionless unit vectors, $\mathbf{g}_2$ has dimensions of length and is not a unit vector.

The **reciprocal basis** $\{\mathbf{g}^i\} = \{\mathbf{g}^1, \mathbf{g}^2, \mathbf{g}^3\}$ is defined by

$$\mathbf{g}^i \cdot \mathbf{g}_j = \delta_j^i, \tag{2.111}$$

where δ_j^i is the Kronecker delta (refer to (2.26)), so $\mathbf{g}^1$ is perpendicular to both $\mathbf{g}_2$ and $\mathbf{g}_3$, and the inner product of $\mathbf{g}^1$ with $\mathbf{g}_1$ is 1, etc. The basis vectors $\mathbf{g}^i$ are also called the **contravariant basis vectors**. They are normal to the coordinate surfaces: $\mathbf{g}^1$ is perpendicular to the θ^1 surface at position $\mathbf{x}$, $\mathbf{g}^2$ is perpendicular to the θ^2 surface at $\mathbf{x}$, and $\mathbf{g}^3$ is perpendicular to the θ^3 surface at $\mathbf{x}$ (see Figure 2.7).

PROBLEM 2.59

Determine the contravariant basis $\{\mathbf{g}^i\} = \{\mathbf{g}^1, \mathbf{g}^2, \mathbf{g}^3\}$ of the cylindrical polar coordinate system.

Solution

For the cylindrical polar coordinate system, result (c) in Problem 2.64 and definition (2.111) imply that

$$\mathbf{g}^1 = \cos\theta\,\mathbf{e}_1 + \sin\theta\,\mathbf{e}_2 = \mathbf{g}_1,$$

$$\mathbf{g}^2 = -\frac{1}{r}\sin\theta\,\mathbf{e}_1 + \frac{1}{r}\cos\theta\,\mathbf{e}_2 = \frac{1}{r^2}\mathbf{g}_2, \qquad \text{(a)}$$

$$\mathbf{g}^3 = \mathbf{e}_3 = \mathbf{g}_3.$$

It can be seen from Figures 2.6 and 2.7 that, in general, the two bases $\mathbf{g}_1$, $\mathbf{g}_2$, $\mathbf{g}_3$ and $\mathbf{g}^1$, $\mathbf{g}^2$, $\mathbf{g}^3$ are along different directions. It is only when the coordinate system is *orthogonal*, as in the cylindrical polar example, that $\mathbf{g}_i$ and $\mathbf{g}^i$ are along the same directions; even then, however, the magnitudes and dimensions of $\mathbf{g}_i$ and $\mathbf{g}^i$ will differ (compare $\mathbf{g}_2$ in (c) of Problem 2.64 with $\mathbf{g}^2$ in (a) of Problem 2.65). It is only for a Cartesian coordinate system that the definitions (2.110) and (2.111) produce the same bases:

$$\mathbf{g}_i = \mathbf{g}^i = \mathbf{e}_i \qquad \Longleftrightarrow \qquad \text{Cartesian coordinate system.}$$

Any vector $\mathbf{v}$ in $\mathcal{E}^3$ can be expressed in terms of either a covariant basis or a contravariant basis, i.e.,

$$\mathbf{v} = v^i\,\mathbf{g}_i = v^1\mathbf{g}_1 + v^2\,\mathbf{g}_2 + v^3\,\mathbf{g}_3 \tag{2.112a}$$

or

$$\mathbf{v} = v_i\,\mathbf{g}^i = v_1\,\mathbf{g}^1 + v_2\,\mathbf{g}^2 + v_3\,\mathbf{g}^3. \tag{2.112b}$$

Similarly, any tensor $\mathbf{T}$ in $\mathcal{L}$ can be be expressed in terms of either a covariant basis or a contravariant basis, i.e.,

$$\mathbf{T} = T^{ij}\,\mathbf{g}_i \otimes \mathbf{g}_j \tag{2.113a}$$

or

$$\mathbf{T} = T_{ij}\,\mathbf{g}^i \otimes \mathbf{g}^j. \tag{2.113b}$$

In curvilinear coordinates it is important to keep track of whether the indices are superscripts or subscripts. The components v^i and v_i of vector $\mathbf{v}$ are in general different, since $\mathbf{g}_i$ and $\mathbf{g}^i$ are generally different.

When using curvilinear components, one can contract indices (i.e., sum on repeated indices) only on a superscript and a subscript, never on two superscripts or two subscripts. It is only for Cartesian coordinate systems, in which $\mathbf{g}_i = \mathbf{g}^i$, that the distinction between v^i and v_i disappears, and all indices can be written as subscripts, and summation can be performed on two subscripts.

2.6.2 PHYSICAL COMPONENTS

Since $\mathbf{g}_i$ and $\mathbf{g}^i$ in general have dimensions, the components v^i and v_i of the vector $\mathbf{v}$ do not in general have the dimensions of $\mathbf{v}$ itself. For instance, if we regard $\mathbf{v}$ as the velocity at position $\mathbf{x}$, then, in our cylindrical polar example, v^1 and v^3 have the dimensions of velocity (length/time) since $\mathbf{g}_1$ and $\mathbf{g}_3$ are dimensionless, but v^2 has dimensions of 1/time since $\mathbf{g}_2$ has dimensions of length. Similarly, v_1 and v_3 are dimensionless, but v_2 has dimensions of $(\text{length})^2/\text{time}$.

From v^i or v_i we can produce the **physical components** of velocity $\mathbf{v}$. Physical components of a vector or tensor have the same dimensional units as the vector or tensor itself. They are obtained by first defining unit vectors along the directions of $\mathbf{g}_i$ or $\mathbf{g}^i$. Physical components are usually employed only in orthogonal coordinate systems.

PROBLEM 2.60

For the cylindrical polar coordinate system (r, θ, z), determine the relations between the physical components v_r, v_θ, v_z, the covariant components v_1, v_2, v_3, and the contravariant components v^1, v^2, v^3 of the vector $\mathbf{v}$.

Solution

We define the unit vectors $\mathbf{e}_r, \mathbf{e}_\theta, \mathbf{e}_z$ along the directions of $\mathbf{g}_1, \mathbf{g}_2, \mathbf{g}_3$:

$$\begin{aligned} \mathbf{e}_r &\equiv \cos\theta\,\mathbf{e}_1 + \sin\theta\,\mathbf{e}_2 = \mathbf{g}_1, \\ \mathbf{e}_\theta &\equiv -\sin\theta\,\mathbf{e}_1 + \cos\theta\,\mathbf{e}_2 = \frac{\mathbf{g}_2}{|\mathbf{g}_2|} = \frac{1}{r}\mathbf{g}_2, \\ \mathbf{e}_z &\equiv \mathbf{e}_3 = \mathbf{g}_3. \end{aligned} \tag{a}$$

Then

$$\begin{aligned} \mathbf{g}^1 &= \mathbf{g}_1 = \mathbf{e}_r, \\ \mathbf{g}^2 &= \frac{1}{r^2}\mathbf{g}_2 = \frac{1}{r}\mathbf{e}_\theta, \\ \mathbf{g}^3 &= \mathbf{g}_3 = \mathbf{e}_z. \end{aligned} \tag{b}$$

The vector $\mathbf{v}$ can then be expressed in terms of the physical components v_r, v_θ, v_z along the unit vectors $\mathbf{e}_r, \mathbf{e}_\theta, \mathbf{e}_z$:

$$\mathbf{v} = v_r\,\mathbf{e}_r + v_\theta\,\mathbf{e}_\theta + v_z\,\mathbf{e}_z. \tag{c}$$

Note that in physical components, as in Cartesian components (which are a special case of physical components), we can dispense with the distinction between superscripts and subscripts. Using the relations (b) to compare (c) with (2.112a) and (2.112b), we see that

$$v_r = v^1 = v_1, \qquad v_\theta = rv^2 = \frac{1}{r}v_2, \qquad v_z = v^3 = v_3. \tag{d}$$

PROBLEM 2.61

For the cylindrical polar coordinate system, determine the relations between the physical components and the contravariant components of the tensor **T**.

Solution

We write the tensor **T** in terms of its contravariant components by expanding (2.113a):

$$\mathbf{T} = T^{11}\,\mathbf{g}_1 \otimes \mathbf{g}_1 + T^{12}\,\mathbf{g}_1 \otimes \mathbf{g}_2 + T^{13}\,\mathbf{g}_1 \otimes \mathbf{g}_3 + T^{21}\,\mathbf{g}_2 \otimes \mathbf{g}_1 + T^{22}\,\mathbf{g}_2 \otimes \mathbf{g}_2$$

$$+ T^{23}\,\mathbf{g}_2 \otimes \mathbf{g}_3 + T^{31}\,\mathbf{g}_3 \otimes \mathbf{g}_1 + T^{32}\,\mathbf{g}_3 \otimes \mathbf{g}_2 + T^{33}\,\mathbf{g}_3 \otimes \mathbf{g}_3. \quad \text{(a)}$$

The tensor **T** can also be written in terms of its physical cylindrical polar components as

$$\mathbf{T} = T_{rr}\,\mathbf{e}_r \otimes \mathbf{e}_r + T_{r\theta}\,\mathbf{e}_r \otimes \mathbf{e}_\theta + T_{rz}\,\mathbf{e}_r \otimes \mathbf{e}_z + T_{\theta r}\,\mathbf{e}_\theta \otimes \mathbf{e}_r + T_{\theta\theta}\,\mathbf{e}_\theta \otimes \mathbf{e}_\theta$$

$$+ T_{\theta z}\,\mathbf{e}_\theta \otimes \mathbf{e}_z + T_{zr}\,\mathbf{e}_z \otimes \mathbf{e}_r + T_{z\theta}\,\mathbf{e}_z \otimes \mathbf{e}_\theta + T_{zz}\,\mathbf{e}_z \otimes \mathbf{e}_z. \quad \text{(b)}$$

Using the relations (b) in Problem 2.60 between $\mathbf{e}_r$, $\mathbf{e}_\theta$, $\mathbf{e}_z$ and $\mathbf{g}_1$, $\mathbf{g}_2$, $\mathbf{g}_3$, we compare (a) and (b) and conclude that

$$T^{11} = T_{rr}, \qquad T^{12} = \frac{1}{r}\,T_{r\theta}, \qquad T^{13} = T_{rz},$$

$$T^{21} = \frac{1}{r}\,T_{\theta r}, \qquad T^{22} = \frac{1}{r^2}\,T_{\theta\theta}, \qquad T^{23} = \frac{1}{r}\,T_{\theta z}, \quad \text{(c)}$$

$$T^{31} = T_{zr}, \qquad T^{32} = \frac{1}{r}\,T_{z\theta}, \qquad T^{33} = T_{zz}.$$

2.6.3 SPATIAL DERIVATIVES: COVARIANT DIFFERENTIATION

Since the basis vectors $\mathbf{g}_i$ are not fixed in space, their spatial derivatives are not zero. Consider the change of the basis vector $\mathbf{g}_i$ as one proceeds along the θ^j coordinate curve. This change in magnitude and direction, denoted by

$$\frac{\partial \mathbf{g}_i}{\partial \theta^j} = \text{change of } \mathbf{g}_i \text{ along } \theta^j \text{ curve},$$

is itself a vector, and therefore can be expressed in terms of the basis $\{\mathbf{g}_i\}$:

$$\frac{\partial \mathbf{g}_i}{\partial \theta^j} = \Gamma^k_{ij}\,\mathbf{g}_k = \Gamma^1_{ij}\,\mathbf{g}_1 + \Gamma^2_{ij}\,\mathbf{g}_2 + \Gamma^3_{ij}\,\mathbf{g}_3. \quad (2.114)$$

The coefficients Γ^k_{ij} in (2.114) are called **Christoffel symbols of the second kind**. From the definition (2.114), it follows that

$$\frac{\partial \mathbf{g}^i}{\partial \theta^j} = -\Gamma^i_{kj}\,\mathbf{g}^k. \tag{2.115}$$

2.6.3.1 Gradient and divergence of a vector

The spatial derivative of a vector-valued function $\mathbf{v}$ of position $\mathbf{x}$, when expressed with respect to the basis vectors $\mathbf{g}_i$, must take into account the possible spatial dependence of both the coefficients v^i and the basis $\{\mathbf{g}_i\}$:

$$\begin{aligned}
\text{grad}\,\mathbf{v} = \frac{\partial \mathbf{v}}{\partial \mathbf{x}} &= \frac{\partial}{\partial \mathbf{x}}\left(v^i\,\mathbf{g}_i\right)\\
&= \frac{\partial}{\partial \theta^j}\left(v^i\,\mathbf{g}_i\right)\otimes \mathbf{g}^j\\
&= \frac{\partial v^i}{\partial \theta^j}\,\mathbf{g}_i \otimes \mathbf{g}^j + v^i\,\frac{\partial \mathbf{g}_i}{\partial \theta^j}\otimes \mathbf{g}^j\\
&= \frac{\partial v^i}{\partial \theta^j}\,\mathbf{g}_i \otimes \mathbf{g}^j + v^i\,\Gamma^k_{ij}\,\mathbf{g}_k \otimes \mathbf{g}^j \qquad (2.116)\\
&= \frac{\partial v^i}{\partial \theta^j}\,\mathbf{g}_i \otimes \mathbf{g}^j + v^m\,\Gamma^i_{mj}\,\mathbf{g}_i \otimes \mathbf{g}^j\\
&= \left(\frac{\partial v^i}{\partial \theta^j} + \Gamma^i_{mj}\,v^m\right)\mathbf{g}_i \otimes \mathbf{g}^j.
\end{aligned}$$

The coefficient $\dfrac{\partial v^i}{\partial \theta^j} + \Gamma^i{}_{mj}\,v^m$ is called the covariant derivative of v^i, and is often denoted by $v^i|_j$ or $v^i||_j$. Similarly, it can be shown that

$$\text{grad}\,\mathbf{v} = \frac{\partial \mathbf{v}}{\partial \mathbf{x}} = \left(\frac{\partial v_i}{\partial \theta^j} - \Gamma^m_{ij}\,v_m\right)\mathbf{g}^i \otimes \mathbf{g}^j. \tag{2.117}$$

With use of definition (2.97), the divergence of a vector $\mathbf{v}$ in curvilinear coordinates is given by

$$\begin{aligned}
\text{div}\,\mathbf{v} &\equiv \text{tr}\,(\text{grad}\,\mathbf{v})\\
&= \left(\frac{\partial v^i}{\partial \theta^j} + \Gamma^i_{mj}\,v^m\right)\text{tr}\left(\mathbf{g}_i \otimes \mathbf{g}^j\right)\\
&= \left(\frac{\partial v^i}{\partial \theta^j} + \Gamma^i_{mj}\,v^m\right)\mathbf{g}_i \cdot \mathbf{g}^j \qquad (2.118)\\
&= \frac{\partial v^i}{\partial \theta^i} + \Gamma^i_{mi}\,v^m.
\end{aligned}$$

In can be shown (refer to Problem 2.63) that for the special case of cylindrical polar coordinates (r, θ, z), we have

$$\text{grad}\,\mathbf{v} = \begin{bmatrix} \dfrac{\partial v_r}{\partial r} & \dfrac{1}{r}\dfrac{\partial v_r}{\partial \theta} - \dfrac{v_\theta}{r} & \dfrac{\partial v_r}{\partial z} \\ \dfrac{\partial v_\theta}{\partial r} & \dfrac{1}{r}\dfrac{\partial v_\theta}{\partial \theta} + \dfrac{v_r}{r} & \dfrac{\partial v_\theta}{\partial z} \\ \dfrac{\partial v_z}{\partial r} & \dfrac{1}{r}\dfrac{\partial v_z}{\partial \theta} & \dfrac{\partial v_z}{\partial z} \end{bmatrix} \tag{2.119}$$

and (refer to Problem 2.64)

$$\text{div}\,\mathbf{v} = \frac{\partial v_r}{\partial r} + \frac{v_r}{r} + \frac{1}{r}\frac{\partial v_\theta}{\partial \theta} + \frac{\partial v_z}{\partial z}. \tag{2.120}$$

In spherical coordinates (r, θ, ϕ), where θ is the polar angle and ϕ is the azimuthal angle, we have

$$\text{grad}\,\mathbf{v} = \begin{bmatrix} \dfrac{\partial v_r}{\partial r} & \dfrac{1}{r}\dfrac{\partial v_r}{\partial \theta} - \dfrac{v_\theta}{r} & \dfrac{1}{r\sin\theta}\dfrac{\partial v_r}{\partial \phi} - \dfrac{v_\phi}{r} \\ \dfrac{\partial v_\theta}{\partial r} & \dfrac{1}{r}\dfrac{\partial v_\theta}{\partial \theta} + \dfrac{v_r}{r} & \dfrac{1}{r\sin\theta}\dfrac{\partial v_\theta}{\partial \phi} - \dfrac{\cot\theta}{r}v_\phi \\ \dfrac{\partial v_\phi}{\partial r} & \dfrac{1}{r}\dfrac{\partial v_\phi}{\partial \theta} & \dfrac{1}{r\sin\theta}\dfrac{\partial v_\phi}{\partial \phi} + \dfrac{v_r}{r} + \dfrac{\cot\theta}{r}v_\theta \end{bmatrix} \tag{2.121}$$

and

$$\text{div}\,\mathbf{v} = \frac{\partial v_r}{\partial r} + \frac{2v_r}{r} + \frac{1}{r}\frac{\partial v_\theta}{\partial \theta} + \frac{\cot\theta}{r}v_\theta + \frac{1}{r\sin\theta}\frac{\partial v_\phi}{\partial \phi}. \tag{2.122}$$

PROBLEM 2.62

Determine the Christoffel symbols Γ_{ij}^k in the cylindrical polar coordinate system (r, θ, z).

Solution

Recall from (c) in Problem 2.64 that the covariant basis of a cylindrical polar coordinate system $(\theta^1 = r, \theta^2 = \theta, \theta^3 = z)$ is

$$\mathbf{g}_1 = \cos\theta\,\mathbf{e}_1 + \sin\theta\,\mathbf{e}_2, \qquad \mathbf{g}_2 = -r\sin\theta\,\mathbf{e}_1 + r\cos\theta\,\mathbf{e}_2, \qquad \mathbf{g}_3 = \mathbf{e}_3.$$

It follows that

$$\frac{\partial \mathbf{g}_1}{\partial \theta^1} = \frac{\partial}{\partial r}(\cos\theta\,\mathbf{e}_1 + \sin\theta\,\mathbf{e}_2) = \mathbf{0}$$

so, by (2.114),

$$\Gamma_{11}^1 = \Gamma_{11}^2 = \Gamma_{11}^3 = 0;$$

$$\frac{\partial \mathbf{g}_1}{\partial \theta^2} = \frac{\partial}{\partial \theta}(\cos\theta\,\mathbf{e}_1 + \sin\theta\,\mathbf{e}_2) = -\sin\theta\,\mathbf{e}_1 + \cos\theta\,\mathbf{e}_2 = \mathbf{e}_\theta = \frac{1}{r}\mathbf{g}_2,$$

so

$$\Gamma^1_{12} = \Gamma^3_{12} = 0, \qquad \Gamma^2_{12} = \frac{1}{r};$$

$$\frac{\partial \mathbf{g}_2}{\partial \theta^1} = \frac{\partial}{\partial r}(-r \sin\theta\, \mathbf{e}_1 + r\cos\theta\, \mathbf{e}_2) = -\sin\theta\, \mathbf{e}_1 + \cos\theta\, \mathbf{e}_2 = \frac{1}{r}\mathbf{g}_2,$$

so

$$\Gamma^1_{21} = \Gamma^3_{21} = 0, \qquad \Gamma^2_{21} = \frac{1}{r};$$

and

$$\frac{\partial \mathbf{g}_2}{\partial \theta^2} = \frac{\partial}{\partial \theta}(-r \sin\theta\, \mathbf{e}_1 + r\cos\theta\, \mathbf{e}_2) = -r(\cos\theta\, \mathbf{e}_1 + \sin\theta\, \mathbf{e}_2) = -r\,\mathbf{g}_1,$$

so

$$\Gamma^1_{22} = -r, \qquad \Gamma^2_{22} = \Gamma^3_{22} = 0.$$

Also,

$$\frac{\partial \mathbf{g}_1}{\partial \theta^3} = \frac{\partial \mathbf{g}_2}{\partial \theta^3} = \frac{\partial \mathbf{g}_3}{\partial \theta^1} = \frac{\partial \mathbf{g}_3}{\partial \theta^2} = \frac{\partial \mathbf{g}_3}{\partial \theta^3} = \mathbf{0},$$

so

$$\Gamma^i_{3k} = \Gamma^i_{j3} = 0.$$

Therefore, the only nonzero Christoffel symbols in the cylindrical polar coordinate system are

$$\Gamma^2_{12} = \Gamma^2_{21} = \frac{1}{r}, \qquad \Gamma^1_{22} = -r. \tag{a}$$

PROBLEM 2.63

Determine grad $\mathbf{v}$ in cylindrical polar coordinates.

Solution

Recall from (2.116) that

$$\text{grad}\,\mathbf{v} = \left(\frac{\partial v^i}{\partial \theta^j} + \Gamma^i_{mj} v^m\right)\mathbf{g}_i \otimes \mathbf{g}^j.$$

Then, using (b) and (d) in Problem 2.66 and (a) in Problem 2.68 we can determine the expression for grad $\mathbf{v}$ in physical cylindrical polar coordinates:

$$(i,j = 1,1) \quad \Rightarrow \quad \left(\frac{\partial v^1}{\partial \theta^1} + 0\right)\mathbf{g}_1 \otimes \mathbf{g}^1 = \frac{\partial v_r}{\partial r}\mathbf{e}_r \otimes \mathbf{e}_r,$$

$$(i,j = 1,2) \quad \Rightarrow \quad \left(\frac{\partial v^1}{\partial \theta^2} + \Gamma^1_{22} v^2\right)\mathbf{g}_1 \otimes \mathbf{g}^2 = \left[\frac{\partial v_r}{\partial \theta} - r\left(\frac{v_\theta}{r}\right)\right]\mathbf{e}_r \otimes \frac{1}{r}\mathbf{e}_\theta$$

$$= \left(\frac{1}{r}\frac{\partial v_r}{\partial \theta} - \frac{v_\theta}{r}\right)\mathbf{e}_r \otimes \mathbf{e}_\theta,$$

$$(i,j=1,3) \quad \Rightarrow \quad \left(\frac{\partial v^1}{\partial \theta^3} + 0\right)\mathbf{g}_1 \otimes \mathbf{g}^3 = \frac{\partial v_r}{\partial z}\mathbf{e}_r \otimes \mathbf{e}_z,$$

$$(i,j=2,1) \quad \Rightarrow \quad \left(\frac{\partial v^2}{\partial \theta^1} + \Gamma^2_{21}v^2\right)\mathbf{g}_2 \otimes \mathbf{g}^1 = \left[\frac{\partial}{\partial r}\left(\frac{v_\theta}{r}\right) + \frac{1}{r}\left(\frac{v_\theta}{r}\right)\right] r\,\mathbf{e}_\theta \otimes \mathbf{e}_r$$

$$= \left(\frac{1}{r}\frac{\partial v_\theta}{\partial r} - \frac{v_\theta}{r^2} + \frac{v_\theta}{r^2}\right) r\,\mathbf{e}_\theta \otimes \mathbf{e}_r$$

$$= \frac{\partial v_\theta}{\partial r}\mathbf{e}_\theta \otimes \mathbf{e}_r,$$

$$(i,j=2,2) \quad \Rightarrow \quad \left(\frac{\partial v^2}{\partial \theta^2} + \Gamma^2_{12}v^1\right)\mathbf{g}_2 \otimes \mathbf{g}^2 = \left[\frac{\partial}{\partial \theta}\left(\frac{v_\theta}{r}\right) + \frac{1}{r}v_r\right] r\,\mathbf{e}_\theta \otimes \frac{1}{r}\mathbf{e}_\theta$$

$$= \left(\frac{1}{r}\frac{\partial v_\theta}{\partial \theta} + \frac{v_r}{r}\right)\mathbf{e}_\theta \otimes \mathbf{e}_\theta,$$

$$(i,j=2,3) \quad \Rightarrow \quad \left(\frac{\partial v^2}{\partial \theta^3} + 0\right)\mathbf{g}_2 \otimes \mathbf{g}^3 = \frac{\partial}{\partial z}\left(\frac{v_\theta}{r}\right) r\,\mathbf{e}_\theta \otimes \mathbf{e}_z = \frac{\partial v_\theta}{\partial z}\mathbf{e}_\theta \otimes \mathbf{e}_z,$$

$$(i,j=3,1) \quad \Rightarrow \quad \left(\frac{\partial v^3}{\partial \theta^1} + 0\right)\mathbf{g}_3 \otimes \mathbf{g}^1 = \frac{\partial v_z}{\partial r}\mathbf{e}_z \otimes \mathbf{e}_r,$$

$$(i,j=3,2) \quad \Rightarrow \quad \left(\frac{\partial v^3}{\partial \theta^2} + 0\right)\mathbf{g}_3 \otimes \mathbf{g}^2 = \frac{1}{r}\frac{\partial v_z}{\partial \theta}\mathbf{e}_z \otimes \mathbf{e}_\theta,$$

$$(i,j=3,3) \quad \Rightarrow \quad \left(\frac{\partial v^3}{\partial \theta^3} + 0\right)\mathbf{g}_3 \otimes \mathbf{g}^3 = \frac{\partial v_z}{\partial z}\mathbf{e}_z \otimes \mathbf{e}_z.$$

Thus, we have

$$\operatorname{grad}\mathbf{v} = \frac{\partial v_r}{\partial r}\mathbf{e}_r \otimes \mathbf{e}_r + \left(\frac{1}{r}\frac{\partial v_r}{\partial \theta} - \frac{v_\theta}{r}\right)\mathbf{e}_r \otimes \mathbf{e}_\theta + \frac{\partial v_r}{\partial z}\mathbf{e}_r \otimes \mathbf{e}_z$$

$$+ \frac{\partial v_\theta}{\partial r}\mathbf{e}_\theta \otimes \mathbf{e}_r + \left(\frac{1}{r}\frac{\partial v_\theta}{\partial \theta} + \frac{v_r}{r}\right)\mathbf{e}_\theta \otimes \mathbf{e}_\theta + \frac{\partial v_\theta}{\partial z}\mathbf{e}_\theta \otimes \mathbf{e}_z$$

$$+ \frac{\partial v_z}{\partial r}\mathbf{e}_z \otimes \mathbf{e}_r + \frac{1}{r}\frac{\partial v_z}{\partial \theta}\mathbf{e}_z \otimes \mathbf{e}_\theta + \frac{\partial v_z}{\partial z}\mathbf{e}_z \otimes \mathbf{e}_z$$

or, in matrix form,

$$[\text{grad}\,\mathbf{v}] = \begin{bmatrix} \dfrac{\partial v_r}{\partial r} & \dfrac{1}{r}\dfrac{\partial v_r}{\partial \theta} - \dfrac{v_\theta}{r} & \dfrac{\partial v_r}{\partial z} \\ \dfrac{\partial v_\theta}{\partial r} & \dfrac{1}{r}\dfrac{\partial v_\theta}{\partial \theta} + \dfrac{v_r}{r} & \dfrac{\partial v_\theta}{\partial z} \\ \dfrac{\partial v_z}{\partial r} & \dfrac{1}{r}\dfrac{\partial v_z}{\partial \theta} & \dfrac{\partial v_z}{\partial z} \end{bmatrix}.$$

PROBLEM 2.64

Determine div **v** in cylindrical polar coordinates.

Solution

Recall from (2.118) that

$$\text{div}\,\mathbf{v} = \frac{\partial v^i}{\partial \theta^i} + \Gamma^i_{mi}\, v^m.$$

Expanding this, we have

$$\text{div}\,\mathbf{v} = \frac{\partial v^1}{\partial \theta^1} + \frac{\partial v^2}{\partial \theta^2} + \frac{\partial v^3}{\partial \theta^3} + \Gamma^1_{11}\, v^1 + \Gamma^2_{12}\, v^1 + \cdots.$$

Then, with use of (d) in Problem 2.66 and (a) in Problem 2.68 this becomes

$$\text{div}\,\mathbf{v} = \frac{\partial v_r}{\partial r} + \frac{\partial}{\partial \theta}\left(\frac{v_\theta}{r}\right) + \frac{\partial v_z}{\partial z} + \frac{1}{r}v_r = \frac{\partial v_r}{\partial r} + \frac{v_r}{r} + \frac{1}{r}\frac{\partial v_\theta}{\partial \theta} + \frac{\partial v_z}{\partial z}.$$

2.6.3.2 Divergence of a tensor

Recall the definition (2.98) of the divergence of a tensor **T**:

$$(\text{div}\,\mathbf{T}) \cdot \mathbf{a} = \text{div}\left(\mathbf{T}^{\mathrm{T}}\mathbf{a}\right)$$

for any vector **a** in $\mathcal{E}^3$. For **a** = constant, we have

$$\frac{\partial \mathbf{a}}{\partial \mathbf{x}} = \mathbf{0},$$

or, in curvilinear coordinates,

$$\frac{\partial}{\partial \theta^j}\left(a_i\, \mathbf{g}^i\right) \otimes \mathbf{g}^j = \left(\frac{\partial a_i}{\partial \theta^j} - \Gamma^m_{ij}\, a_m\right) \mathbf{g}^i \otimes \mathbf{g}^j = \mathbf{0},$$

which implies

$$\frac{\partial a_i}{\partial \theta^j} = \Gamma^m_{ij}\, a_m \tag{2.123}$$

for $\mathbf{a}$ = constant. The right-hand side of (2.98) becomes in curvilinear coordinates

$$\begin{aligned}
\operatorname{div}\left(\mathbf{T}^{\mathrm{T}}\mathbf{a}\right) &= \operatorname{div}\left[\left(T^{ji}\,\mathbf{g}_i \otimes \mathbf{g}_j\right) a_k\,\mathbf{g}^k\right] \\
&= \operatorname{div}\left[T^{ji}\,a_k\left(\mathbf{g}_j \cdot \mathbf{g}^k\right)\mathbf{g}_i\right] \\
&= \operatorname{div}\left(T^{ji}\,a_j\,\mathbf{g}_i\right) \\
&= \frac{\partial}{\partial\theta^i}\left(T^{ji}a_j\right) + \Gamma^i_{mi}\,T^{jm}\,a_j \\
&= \frac{\partial T^{ji}}{\partial\theta^i}\,a_j + T^{ji}\,\frac{\partial a_j}{\partial\theta^i} + \Gamma^i_{mi}\,T^{jm}\,a_j.
\end{aligned}$$

With use of (2.123), this becomes

$$\operatorname{div}\left(\mathbf{T}^{\mathrm{T}}\mathbf{a}\right) = \frac{\partial T^{ji}}{\partial\theta^i}\,a_j + T^{ji}\,\Gamma^m_{ji}\,a_m + \Gamma^i_{mi}\,T^{jm}\,a_j = \left(\frac{\partial T^{ji}}{\partial\theta^i} + \Gamma^j_{mi}\,T^{mi} + \Gamma^i_{mi}\,T^{jm}\right)a_j,$$

so (2.98) in curvilinear coordinates is

$$(\operatorname{div}\mathbf{T})\cdot\mathbf{g}^j\,a_j = \left(\frac{\partial T^{ji}}{\partial\theta^i} + \Gamma^j_{mi}\,T^{mi} + \Gamma^i_{mi}\,T^{jm}\right)a_j.$$

Therefore,

$$\operatorname{div}\mathbf{T} = \left(\frac{\partial T^{ij}}{\partial\theta^j} + \Gamma^i_{mj}\,T^{mj} + \Gamma^j_{mj}\,T^{im}\right)\mathbf{g}_i. \tag{2.124}$$

It can be shown (refer to Problem 2.65) that for the special case of cylindrical polar coordinates (r,θ,z), (2.124) becomes

$$\begin{aligned}
\operatorname{div}\mathbf{T} &= \left[\frac{\partial T_{rr}}{\partial r} + \frac{1}{r}\frac{\partial T_{r\theta}}{\partial\theta} + \frac{\partial T_{rz}}{\partial z} + \frac{1}{r}\left(T_{rr} - T_{\theta\theta}\right)\right]\mathbf{e}_r \\
&+ \left[\frac{\partial T_{\theta r}}{\partial r} + \frac{1}{r}\frac{\partial T_{\theta\theta}}{\partial\theta} + \frac{\partial T_{\theta z}}{\partial z} + \frac{1}{r}\left(T_{r\theta} + T_{\theta r}\right)\right]\mathbf{e}_\theta \qquad (2.125) \\
&+ \left[\frac{\partial T_{zr}}{\partial r} + \frac{1}{r}\frac{\partial T_{z\theta}}{\partial\theta} + \frac{\partial T_{zz}}{\partial z} + \frac{1}{r}T_{zr}\right]\mathbf{e}_z.
\end{aligned}$$

In spherical coordinates (r,θ,ϕ), we have

$$\begin{aligned}
\operatorname{div}\mathbf{T} &= \left[\frac{\partial T_{rr}}{\partial r} + \frac{1}{r}\frac{\partial T_{r\theta}}{\partial\theta} + \frac{1}{r\sin\theta}\frac{\partial T_{r\phi}}{\partial\phi} + \frac{1}{r}\left(2T_{rr} - T_{\theta\theta} - T_{\phi\phi}\right) + \frac{\cot\theta}{r}T_{r\theta}\right]\mathbf{e}_r \\
&+ \left[\frac{\partial T_{\theta r}}{\partial r} + \frac{1}{r}\frac{\partial T_{\theta\theta}}{\partial\theta} + \frac{1}{r\sin\theta}\frac{\partial T_{\theta\phi}}{\partial\phi} + \frac{1}{r}\left(T_{r\theta} + 2T_{\theta r}\right) + \frac{\cot\theta}{r}\left(T_{\theta\theta} - T_{\phi\phi}\right)\right]\mathbf{e}_\theta \\
&\qquad (2.126) \\
&+ \left[\frac{\partial T_{\phi r}}{\partial r} + \frac{1}{r}\frac{\partial T_{\phi\theta}}{\partial\theta} + \frac{1}{r\sin\theta}\frac{\partial T_{\phi\phi}}{\partial\phi} + \frac{1}{r}\left(T_{r\phi} + 2T_{\phi r}\right) + \frac{\cot\theta}{r}\left(T_{\theta\phi} + T_{\phi\theta}\right)\right]\mathbf{e}_\phi.
\end{aligned}$$

PROBLEM 2.65

Determine div **T** in cylindrical polar coordinates.

Solution

It follows from (2.124) that

$$\operatorname{div}\mathbf{T} = \left(\frac{\partial T^{11}}{\partial\theta^1} + \frac{\partial T^{12}}{\partial\theta^2} + \frac{\partial T^{13}}{\partial\theta^3} + \Gamma^1_{22}T^{22} + \Gamma^2_{12}T^{11} + \dots\right)\mathbf{g}_1$$

$$+\left(\frac{\partial T^{21}}{\partial\theta^1} + \frac{\partial T^{22}}{\partial\theta^2} + \frac{\partial T^{23}}{\partial\theta^3} + \Gamma^2_{12}T^{12} + \Gamma^2_{21}T^{21} + \Gamma^2_{12}T^{21} + \dots\right)\mathbf{g}_2$$

$$+\left(\frac{\partial T^{31}}{\partial\theta^1} + \frac{\partial T^{32}}{\partial\theta^2} + \frac{\partial T^{33}}{\partial\theta^3} + \Gamma^2_{12}T^{31} + \dots\right)\mathbf{g}_3.$$

Using (b) in Problem 2.66, (c) in Problem 2.67, and (a) in Problem 2.68 we have

$$\operatorname{div}\mathbf{T} = \left[\frac{\partial T_{rr}}{\partial r} + \frac{\partial}{\partial\theta}\left(\frac{1}{r}T_{r\theta}\right) + \frac{\partial T_{rz}}{\partial z} - r\frac{1}{r^2}T_{\theta\theta} + \frac{1}{r}T_{rr}\right]\mathbf{e}_r$$

$$+\left[\frac{\partial}{\partial r}\left(\frac{1}{r}T_{\theta r}\right) + \frac{\partial}{\partial\theta}\left(\frac{1}{r^2}T_{\theta\theta}\right) + \frac{\partial}{\partial z}\left(\frac{1}{r}T_{\theta z}\right) + \frac{1}{r}\left(\frac{1}{r}T_{r\theta} + \frac{2}{r}T_{\theta r}\right)\right] r\,\mathbf{e}_\theta$$

$$+\left[\frac{\partial T_{zr}}{\partial r} + \frac{\partial}{\partial\theta}\left(\frac{1}{r}T_{z\theta}\right) + \frac{\partial T_{zz}}{\partial z} + \frac{1}{r}T_{zr}\right]\mathbf{e}_z$$

$$= \left[\frac{\partial T_{rr}}{\partial r} + \frac{1}{r}\frac{\partial T_{r\theta}}{\partial\theta} + \frac{\partial T_{rz}}{\partial z} + \frac{1}{r}(T_{rr} - T_{\theta\theta})\right]\mathbf{e}_r$$

$$+\left[\frac{\partial T_{\theta r}}{\partial r} + \frac{1}{r}\frac{\partial T_{\theta\theta}}{\partial\theta} + \frac{\partial T_{\theta z}}{\partial z} + \frac{1}{r}(T_{r\theta} + T_{\theta r})\right]\mathbf{e}_\theta$$

$$+\left[\frac{\partial T_{zr}}{\partial r} + \frac{1}{r}\frac{\partial T_{z\theta}}{\partial\theta} + \frac{\partial T_{zz}}{\partial z} + \frac{1}{r}T_{zr}\right]\mathbf{e}_z.$$

PART

II

Kinematics, Kinetics, and the Fundamental Laws of Mechanics and Thermodynamics

CHAPTER

Kinematics: Motion and Deformation

3

3.1 BODY, CONFIGURATION, MOTION, DISPLACEMENT

A **body** $\mathcal{B}$ is a set of **particles.** A representative particle is designated as Y, where Y is the **name** of the particle (e.g., Steve, Bob, or the color red); see Figure 3.1(a). Body $\mathcal{B}$ is seen only in its configurations. A **configuration** of $\mathcal{B}$ is the region of $\mathcal{E}^3$ (Euclidean 3-space) occupied at a particular instant by the body.

Consider the configuration of body $\mathcal{B}$ at time t, called the **present configuration,** in which $\mathcal{B}$ occupies an *open volume* $\mathcal{R}$ of $\mathcal{E}^3$ bounded by a *closed surface* $\partial\mathcal{R}$; see Figure 3.1(b). (If, for instance, the body $\mathcal{B}$ is an orange, the region occupied by the flesh constitutes an open volume, and the region occupied the peel constitutes a closed surface.) Let $\mathbf{x}$ be the position vector of the place occupied by the representative particle Y at time t; see Figure 3.1(b). Then

$$\mathbf{x} = \bar{\chi}\,(Y,t). \tag{3.1}$$

The mapping (3.1) is called the **motion** of body $\mathcal{B}$. Note that we distinguish the *function* $\bar{\chi}\,(Y,t)$ from its *value* (or output) $\mathbf{x}$. The mapping $\bar{\chi}$ is assumed to be differentiable as many times as necessary with respect to both Y and t. We also make the important assumption that the motion (3.1) is invertible for each t, so the function exists.

$$Y = \bar{\chi}^{-1}(\mathbf{x},t) \tag{3.2}$$

One particular configuration κ of body $\mathcal{B}$ may be selected as the **reference configuration,** and then the body and its motion may be referred to this fixed configuration. A common practice is to choose κ to be the **initial configuration,** i.e., the configuration occupied by $\mathcal{B}$ at time $t = 0$. However, the reference configuration need not be the initial configuration; in fact, it can even be a configuration that the body $\mathcal{B}$ never actually attains. Since the selection of the reference configuration κ is taken from an infinite number of possible configurations of $\mathcal{B}$, a motion has infinitely many different referential descriptions, all equally valid.

In the reference configuration κ, the body $\mathcal{B}$ occupies an *open volume* $\mathcal{R}_\mathrm{R}$ of $\mathcal{E}^3$ bounded by a *closed surface* $\partial\mathcal{R}_\mathrm{R}$; see Figure 3.1(c). Let $\mathbf{X}$ be the position vector of the place occupied by the representative particle Y in this reference configuration.

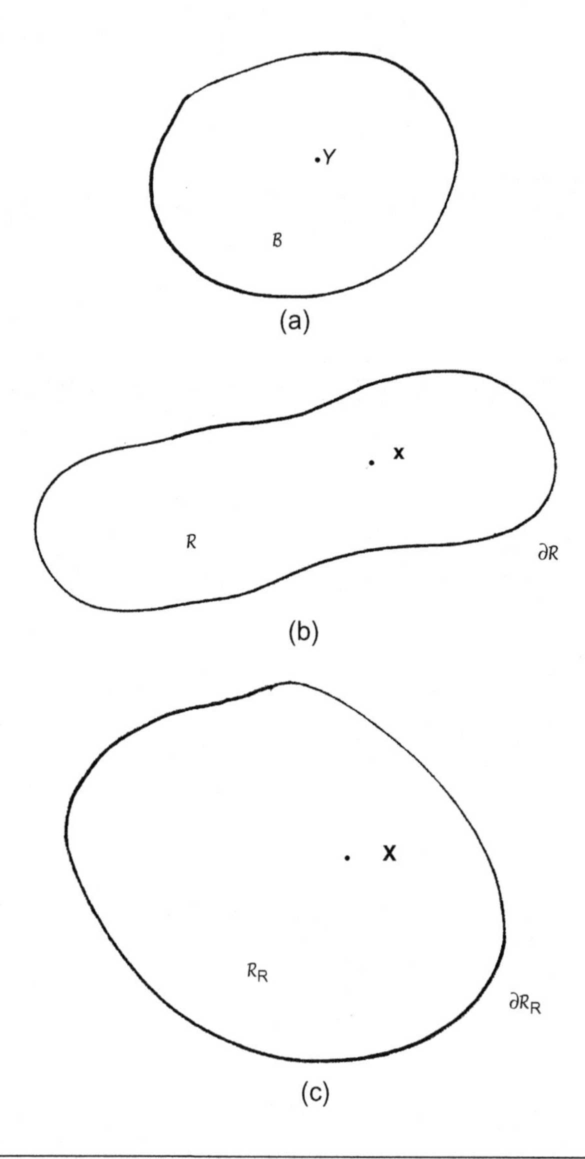

FIGURE 3.1

The body $\mathcal{B}$, its present configuration $\mathcal{R}$, and its reference configuration $\mathcal{R}_R$.

The mapping from Y to $\mathbf{X}$ is written

$$\mathbf{X} = \kappa\,(Y). \tag{3.3}$$

The mapping (3.3), called the **reference map,** is assumed to be differentiable as many times as desired, and invertible. The inverse mapping is

$$Y = \kappa^{-1}(\mathbf{X}). \tag{3.4}$$

Using (3.4), we can rewrite the motion (3.1) as

$$\mathbf{x} = \bar{\chi}\left[\kappa^{-1}(\mathbf{X}), t\right] = {}_{\kappa}\chi(\mathbf{X}, t). \tag{3.5}$$

Equation (3.5) serves as a definition of the mapping ${}_{\kappa}\chi$. It should be noted that ${}_{\kappa}\chi$ depends on the choice of reference configuration κ. If only one reference configuration is adopted, then the subscript κ is generally omitted, i.e.,

$$\mathbf{x} = \chi(\mathbf{X}, t), \tag{3.6}$$

and the dependence of χ on a particular reference configuration κ is understood.

The function (3.1) is called the **material description** of motion. In the material description, we deal directly with the particle Y of $\mathcal{B}$. The description of motion (3.6) is called the **referential description** of motion. In the referential description, the particle Y of $\mathcal{B}$ is labeled by the position $\mathbf{X}$ it occupied when $\mathcal{B}$ was in its reference configuration κ.

The **displacement u** of the particle Y is defined as the difference between its reference and present positions, i.e.,

$$\mathbf{u} = \mathbf{x} - \mathbf{X}; \tag{3.7}$$

see Figure 3.2.

Consider a vector quantity **f** related to the motion of body $\mathcal{B}$. (Our use of a vector quantity is arbitrary; the discussion that follows is equally valid for a scalar quantity ϕ or a tensor quantity **A**.) In the **material description**, the quantity **f** is expressed as a function of particle Y and time t, i.e.,

$$\mathbf{f} = \bar{\mathbf{f}}(Y, t). \tag{3.8}$$

In the **referential description, f** is expressed as a function of the position **X** occupied by Y in the reference configuration κ, and time t. To see this we use (3.4) in (3.8):

$$\mathbf{f} = \bar{\mathbf{f}}(Y, t) = \bar{\mathbf{f}}(\kappa^{-1}(\mathbf{X}), t) = \hat{\mathbf{f}}(\mathbf{X}, t). \tag{3.9}$$

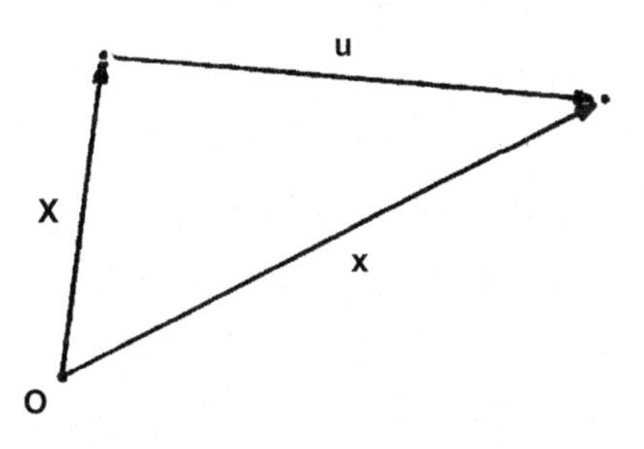

FIGURE 3.2

The displacement of a particle.

The quantity $\mathbf{f}$ may also be expressed as a function of the position $\mathbf{x}$ occupied at time t by the particle Y, and time t. To see this we use (3.2) in (3.8):

$$\mathbf{f} = \bar{\mathbf{f}}\,(Y,t) = \bar{\mathbf{f}}\left(\bar{\chi}^{-1}(\mathbf{x},t),t\right) = \tilde{\mathbf{f}}\,(\mathbf{x},t). \tag{3.10}$$

In (3.10) we have expressed $\mathbf{f}$ in terms of the present position $\mathbf{x}$ of Y, and time t. Such a description is called the **spatial description**. Referential and spatial descriptions are also referred to as **Lagrangian** and **Eulerian** descriptions, respectively.

To review: A vector quantity $\mathbf{f}$ (or a scalar quantity ϕ or a tensor quantity $\mathbf{A}$) related to the motion of body $\mathcal{B}$ may be expressed in any of three different descriptions:

$$\begin{aligned} \mathbf{f} &= \bar{\mathbf{f}}\,(Y,t) && \text{(material)} \\ &= \hat{\mathbf{f}}\,(\mathbf{X},t) && \text{(referential)} \\ &= \tilde{\mathbf{f}}\,(\mathbf{x},t) && \text{(spatial).} \end{aligned}$$

This freedom is due to our assumptions that the mappings (3.1) and (3.3) are invertible. In the referential description, each particle is labeled by its reference position; in the spatial description, each particle is labeled by its present position. We note that the material description is seldom used in continuum mechanics since it becomes intractable to find a different name for each particle in the body; the other two representations, however, are useful and will be employed.

PROBLEM 3.1

In a particular motion of body $\mathcal{B}$, the reference configuration is chosen to be the configuration at time $t = 0$, i.e., the initial configuration. At $t = 0$, the red particle is at (1, 1, 1), the blue particle is at (0, 0, 0), and the white particle is at (−1, −1, −1). At $t = 5$ (the present configuration), the red particle is at (2, 2, 2), the blue particle is at (1, 1, 1), and the white particle is at (3, 3, 3), and their temperatures are 5°, 0°, and 10°, respectively; see Figure 3.3.

Solution

(a) The equation $\mathbf{x} = \bar{\chi}\,(Y,t)$ answers the question: "Where is particle Y at time t?" That is, the input to the function $\bar{\chi}\,(Y,t)$ is the particle name Y and the time t, and the output is the present position $\mathbf{x}$ of the particle at time t. For instance, $\bar{\chi}\,(\text{red},5) = (2,2,2)$ and $\bar{\chi}\,(\text{blue},5) = (1,1,1)$.

(b) The equation $Y = \bar{\chi}^{-1}(\mathbf{x},t)$ answers the question: "Which particle is at $\mathbf{x}$ at time t?" That is, the input to the function $\bar{\chi}^{-1}(\mathbf{x},t)$ is the time t and the present position $\mathbf{x}$ of the particle at time t, and the output is the name Y of the particle. For instance, $\bar{\chi}^{-1}((1,1,1),5) = \text{blue}$.

(c) The equation $\mathbf{X} = \kappa\,(Y)$ answers the question: "Where is particle Y in the reference configuration?" That is, the input to the function $\kappa\,(Y)$ is the particle name Y, and the output is the reference position $\mathbf{X}$ of the particle. For instance, $\kappa\,(\text{red}) = (1,1,1)$.

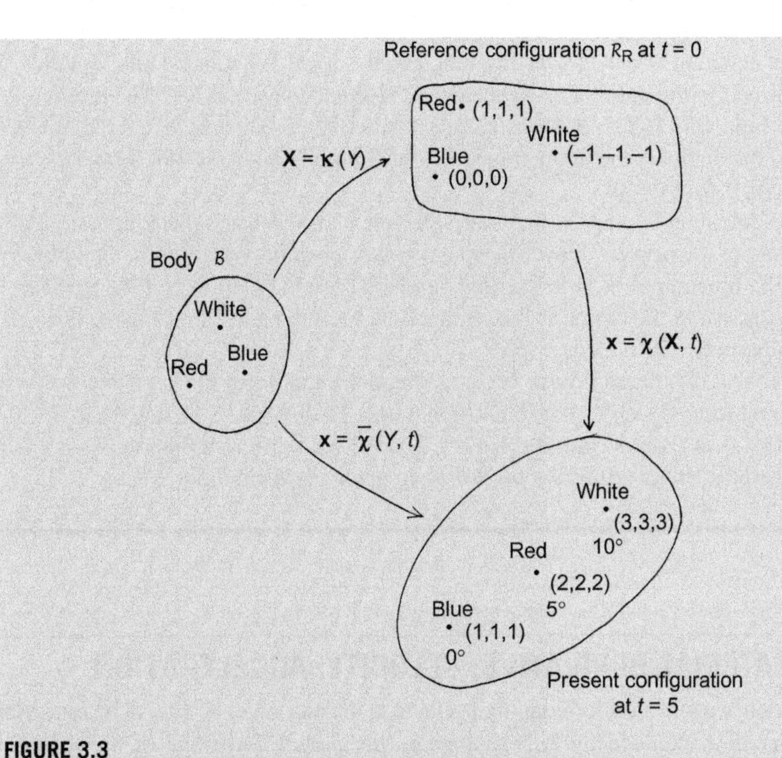

FIGURE 3.3

Schematic illustrating the positions and temperatures of the red, white, and blue particles in the reference and present configurations of the body at $t = 0$ and $t = 5$, respectively.

(d) The equation $Y = \kappa^{-1}(\mathbf{X})$ answers the question: "Which particle is at $\mathbf{X}$ in the reference configuration?" That is, the input to the function $\kappa^{-1}(\mathbf{X})$ is the reference position $\mathbf{X}$ of the particle, and the output is the particle name Y. For instance, $\kappa^{-1}(-1, -1, -1) = \text{white}$.

(e) The equation $\mathbf{x} = \chi\,(\mathbf{X}, t)$ answers the question: "Where is the particle that *was* at $\mathbf{X}$ in the reference configuration *now* located at time t?" That is, the input to the function $\chi\,(\mathbf{X}, t)$ is the reference position $\mathbf{X}$ of the particle and the time t, and the output is the present position $\mathbf{x}$ of the particle at time t. For instance, $\chi\,((1, 1, 1), 5) = (2, 2, 2)$. To see this, note that

$$\chi\,((1, 1, 1), 5) = \chi\,(\kappa\,(\text{red}), 5) = \bar{\chi}\,(\text{red}, 5) = (2, 2, 2).$$

(f) Let Θ be the temperature. The equation $\Theta = \bar{\Theta}\,(Y, t)$ answers the question: "What is the temperature of particle Y at time t?" The input to the function $\bar{\Theta}\,(Y, t)$ is the particle name Y and the time t, and the output is the temperature Θ of the particle at time t. For instance, $\bar{\Theta}\,(\text{white}, 5) = 10°$.

(g) The equation $\Theta = \hat{\Theta}(\mathbf{X}, t)$ answers the question: "What is the temperature at time t of the particle Y that occupied reference position $\mathbf{X}$?" The input to the function $\hat{\Theta}(\mathbf{X}, t)$ is the reference position $\mathbf{X}$ of the particle and the time t, and the output is the temperature Θ of the particle at time t. For instance, $\hat{\Theta}((0,0,0),5) = 0°$.

(h) The equation $\Theta = \tilde{\Theta}(\mathbf{x}, t)$ answers the question: "What is the temperature at time t of the particle Y that occupies present position $\mathbf{x}$ at time t?" The input to the function $\tilde{\Theta}(\mathbf{x}, t)$ is the time t and the present position $\mathbf{x}$ of the particle at time t, and the output is the temperature Θ of the particle at time t. For instance, $\tilde{\Theta}((2,2,2),5) = 5°$.

(i) Note that the material form $\bar{\Theta}(Y, t)$, the referential form $\hat{\Theta}(\mathbf{X}, t)$, and the spatial form $\tilde{\Theta}(\mathbf{x}, t)$ are different functions. To illustrate this, note that $\hat{\Theta}((1,1,1),5) = 5°$ but $\tilde{\Theta}((1,1,1),5) = 0°$, i.e., different functions generally provide different output when given identical input.

3.2 MATERIAL DERIVATIVE, VELOCITY, ACCELERATION

Consider once again a vector quantity $\mathbf{f}$ related to the motion of body $\mathcal{B}$. The **material derivative** of $\mathbf{f}$, denoted by $\dot{\mathbf{f}}$, is defined as the partial derivative of the material description (3.8) of $\mathbf{f}$ with respect to time t, holding the label Y fixed:

$$\dot{\mathbf{f}} = \frac{\partial}{\partial t}\bar{\mathbf{f}}(Y, t). \tag{3.11}$$

The quantity $\dot{\mathbf{f}}$ can be described as the time rate change of $\mathbf{f}$ following a particle. It follows that the **velocity v** and **acceleration a** of the particle Y are defined by

$$\mathbf{v} = \dot{\mathbf{x}} = \frac{\partial}{\partial t}\bar{\chi}(Y, t), \qquad \mathbf{a} = \dot{\mathbf{v}} = \frac{\partial}{\partial t}\bar{\mathbf{v}}(Y, t) = \frac{\partial^2}{\partial t^2}\bar{\chi}(Y, t). \tag{3.12}$$

We now determine the referential and spatial descriptions of the material derivative $\dot{\mathbf{f}}$, which was defined in its impractical material description in (3.11). Using (3.3), (3.9), (3.11), and the chain rule (2.95), we have

$$\begin{aligned}
\dot{\mathbf{f}} &= \frac{\partial}{\partial t}\bar{\mathbf{f}}(Y, t) \\
&= \frac{\partial}{\partial t}\hat{\mathbf{f}}(\mathbf{X}, t) + \frac{\partial}{\partial \mathbf{X}}\hat{\mathbf{f}}(\mathbf{X}, t)\,\frac{\partial \mathbf{X}}{\partial t} \\
&= \frac{\partial}{\partial t}\hat{\mathbf{f}}(\mathbf{X}, t) + \frac{\partial}{\partial \mathbf{X}}\hat{\mathbf{f}}(\mathbf{X}, t)\,\frac{\partial}{\partial t}\kappa(Y) \\
&= \frac{\partial}{\partial t}\hat{\mathbf{f}}(\mathbf{X}, t).
\end{aligned}$$

Therefore, in the referential description, $\dot{\mathbf{f}}$ is

$$\dot{\mathbf{f}} = \frac{\partial}{\partial t}\hat{\mathbf{f}}(\mathbf{X}, t), \tag{3.13}$$

so the referential forms of the velocity $\mathbf{v}$ and acceleration $\mathbf{a}$ are

$$\mathbf{v} = \dot{\mathbf{x}} = \frac{\partial}{\partial t}\chi(\mathbf{X}, t), \qquad \mathbf{a} = \dot{\mathbf{v}} = \frac{\partial}{\partial t}\hat{\mathbf{v}}(\mathbf{X}, t) = \frac{\partial^2}{\partial t^2}\chi(\mathbf{X}, t). \tag{3.14}$$

These forms give the velocity and acceleration at time t of a particle Y that occupied position $\mathbf{X}$ in the reference configuration.

To obtain the spatial description of $\dot{\mathbf{f}}$, we use (3.1), (3.10), (3.11), (3.12), and the chain rule (2.95):

$$\begin{aligned}
\dot{\mathbf{f}} &= \frac{\partial}{\partial t}\bar{\mathbf{f}}(Y, t) \\
&= \frac{\partial}{\partial t}\tilde{\mathbf{f}}(\mathbf{x}, t) + \frac{\partial}{\partial \mathbf{x}}\tilde{\mathbf{f}}(\mathbf{x}, t)\frac{\partial \mathbf{x}}{\partial t} \\
&= \frac{\partial}{\partial t}\tilde{\mathbf{f}}(\mathbf{x}, t) + \frac{\partial}{\partial \mathbf{x}}\tilde{\mathbf{f}}(\mathbf{x}, t)\frac{\partial}{\partial t}\bar{\chi}(Y, t) \\
&= \frac{\partial}{\partial t}\tilde{\mathbf{f}}(\mathbf{x}, t) + \frac{\partial}{\partial \mathbf{x}}\tilde{\mathbf{f}}(\mathbf{x}, t)\,\mathbf{v}.
\end{aligned}$$

Therefore, the spatial description of $\dot{\mathbf{f}}$ is

$$\dot{\mathbf{f}} = \frac{\partial}{\partial t}\tilde{\mathbf{f}}(\mathbf{x}, t) + \frac{\partial}{\partial \mathbf{x}}\tilde{\mathbf{f}}(\mathbf{x}, t)\,\mathbf{v}, \tag{3.15}$$

so the spatial form of the acceleration $\mathbf{a}$ is

$$\mathbf{a} = \dot{\mathbf{v}} = \frac{\partial}{\partial t}\tilde{\mathbf{v}}(\mathbf{x}, t) + \frac{\partial}{\partial \mathbf{x}}\tilde{\mathbf{v}}(\mathbf{x}, t)\,\mathbf{v}. \tag{3.16}$$

This form gives the acceleration at time t of a particle Y that occupies position $\mathbf{x}$ at time t.

Similarly, for a scalar quantity ϕ, we have material, referential, and spatial descriptions of the material derivative:

$$\dot{\phi} = \frac{\partial}{\partial t}\bar{\phi}(Y, t), \qquad \dot{\phi} = \frac{\partial}{\partial t}\hat{\phi}(\mathbf{X}, t), \qquad \dot{\phi} = \frac{\partial}{\partial t}\tilde{\phi}(\mathbf{x}, t) + \frac{\partial}{\partial \mathbf{x}}\tilde{\phi}(\mathbf{x}, t)\cdot\mathbf{v}, \tag{3.17}$$

and, for a tensor quantity $\mathbf{A}$,

$$\dot{\mathbf{A}} = \frac{\partial}{\partial t}\bar{\mathbf{A}}(Y, t), \qquad \dot{\mathbf{A}} = \frac{\partial}{\partial t}\hat{\mathbf{A}}(\mathbf{X}, t), \qquad \dot{\mathbf{A}} = \frac{\partial}{\partial t}\tilde{\mathbf{A}}(\mathbf{x}, t) + \frac{\partial}{\partial \mathbf{x}}\tilde{\mathbf{A}}(\mathbf{x}, t)\,\mathbf{v}. \tag{3.18}$$

For a scalar-, vector-, or tensor-valued function of position and time associated with the motion of body $\mathcal{B}$, there are four possible partial derivatives, depending on if the function is considered in its referential form or its spatial form:

$$\dot{\phi} = \frac{\partial}{\partial t}\hat{\phi}(\mathbf{X}, t), \quad \phi' = \frac{\partial}{\partial t}\tilde{\phi}(\mathbf{x}, t), \quad \operatorname{Grad}\phi = \frac{\partial}{\partial \mathbf{X}}\hat{\phi}(\mathbf{X}, t), \quad \operatorname{grad}\phi = \frac{\partial}{\partial \mathbf{x}}\tilde{\phi}(\mathbf{x}, t),$$

$$\dot{\mathbf{f}} = \frac{\partial}{\partial t}\hat{\mathbf{f}}(\mathbf{X},t), \quad \mathbf{f}' = \frac{\partial}{\partial t}\tilde{\mathbf{f}}(\mathbf{x},t), \quad \text{Grad}\,\mathbf{f} = \frac{\partial}{\partial \mathbf{X}}\hat{\mathbf{f}}(\mathbf{X},t), \quad \text{grad}\,\mathbf{f} = \frac{\partial}{\partial \mathbf{x}}\tilde{\mathbf{f}}(\mathbf{x},t), \tag{3.19}$$

$$\dot{\mathbf{A}} = \frac{\partial}{\partial t}\hat{\mathbf{A}}(\mathbf{X},t), \quad \mathbf{A}' = \frac{\partial}{\partial t}\tilde{\mathbf{A}}(\mathbf{x},t), \quad \text{Grad}\,\mathbf{A} = \frac{\partial}{\partial \mathbf{X}}\hat{\mathbf{A}}(\mathbf{X},t), \quad \text{grad}\,\mathbf{A} = \frac{\partial}{\partial \mathbf{x}}\tilde{\mathbf{A}}(\mathbf{x},t).$$

We employ this notation throughout the remainder of the book. In words, $\dot{\mathbf{f}}$ denotes the partial derivative of the referential description of $\mathbf{f}$ with respect to time (this is a restatement of (3.13)); $\mathbf{f}'$ denotes the partial derivative of the spatial description of $\mathbf{f}$ with respect to time; Grad $\mathbf{f}$ denotes the gradient of the referential description of $\mathbf{f}$; and grad $\mathbf{f}$ denotes the gradient of the spatial description of $\mathbf{f}$. Also, Div $\mathbf{f}$ and div $\mathbf{f}$ are the divergence of $\mathbf{f}$ when considered in its referential and spatial descriptions, respectively.

With use of the notation introduced in (3.19), the spatial forms of $\dot{\phi}$, $\dot{\mathbf{f}}$, and $\dot{\mathbf{A}}$ (refer to (3.15), $(3.17)_3$, and $(3.18)_3$) become

$$\dot{\phi} = \phi' + \text{grad}\,\phi \cdot \mathbf{v}, \qquad \dot{\mathbf{f}} = \mathbf{f}' + (\text{grad}\,\mathbf{f})\,\mathbf{v}, \qquad \dot{\mathbf{A}} = \mathbf{A}' + (\text{grad}\,\mathbf{A})\,\mathbf{v}. \tag{3.20}$$

Note that to relate the operations Grad and grad, we have, for instance,

$$\text{Grad}\,\phi = \mathbf{F}^{\mathrm{T}}\,\text{grad}\,\phi, \qquad \text{Grad}\,\mathbf{f} = (\text{grad}\,\mathbf{f})\,\mathbf{F}, \tag{3.21}$$

where $\mathbf{F} = \text{Grad}\,\mathbf{x} = \partial\mathbf{x}/\partial\mathbf{X}$ is called the deformation gradient (there is more on this in Section 3.3).

The following forms of the product rule can be verified:

$$\dot{\overline{\phi\mathbf{f}}} = \dot{\phi}\mathbf{f} + \phi\dot{\mathbf{f}}, \qquad \dot{\overline{\mathbf{f}\cdot\mathbf{g}}} = \dot{\mathbf{f}}\cdot\mathbf{g} + \mathbf{f}\cdot\dot{\mathbf{g}}, \qquad \dot{\overline{\mathbf{f}\times\mathbf{g}}} = \dot{\mathbf{f}}\times\mathbf{g} + \mathbf{f}\times\dot{\mathbf{g}},$$

$$\dot{\overline{\mathbf{A}\mathbf{f}}} = \dot{\mathbf{A}}\mathbf{f} + \mathbf{A}\dot{\mathbf{f}}, \qquad \dot{\overline{\mathbf{A}\mathbf{B}}} = \dot{\mathbf{A}}\mathbf{B} + \mathbf{A}\dot{\mathbf{B}}, \qquad \dot{\overline{\mathbf{A}\cdot\mathbf{B}}} = \dot{\mathbf{A}}\cdot\mathbf{B} + \mathbf{A}\cdot\dot{\mathbf{B}}, \tag{3.22}$$

where ϕ is a scalar, $\mathbf{f}$ and $\mathbf{g}$ are vectors, and $\mathbf{A}$ and $\mathbf{B}$ are tensors. We also have

$$\dot{\overline{\mathbf{A}+\mathbf{B}}} = \dot{\mathbf{A}} + \dot{\mathbf{B}}, \qquad \dot{\overline{\mathbf{A}^{\mathrm{T}}}} = \left(\dot{\mathbf{A}}\right)^{\mathrm{T}}. \tag{3.23}$$

PROBLEM 3.2

Consider a motion $\mathbf{x} = \chi(\mathbf{X},t)$ whose Cartesian components are

$$x_1 = \mathrm{e}^{t} X_1, \qquad x_2 = \mathrm{e}^{-2t} X_2, \qquad x_3 = (1+t)\,X_3.$$

(a) Invert the motion to obtain $\mathbf{X} = \chi^{-1}(\mathbf{x},t)$.
(b) Calculate the referential and spatial forms of the displacement $\mathbf{u}$.
(c) Determine the referential and spatial forms of the velocity $\mathbf{v}$.
(d) Calculate the referential and spatial forms of the acceleration $\mathbf{a}$.

Solution

(a) Upon inverting the motion $\mathbf{x} = \boldsymbol{\chi}(\mathbf{X}, t)$, we obtain

$$X_1 = \mathrm{e}^{-t}x_1, \qquad X_2 = \mathrm{e}^{2t}x_2, \qquad X_3 = \frac{x_3}{1+t}.$$

(b) Recall that the displacement is defined in (3.7); in Cartesian component form, we have

$$u_1 = x_1 - X_1, \qquad u_2 = x_2 - X_2, \qquad u_3 = x_3 - X_3.$$

It follows that the referential form of the displacement is

$$\hat{u}_1 = x_1 - X_1 = \mathrm{e}^{t}X_1 - X_1 = (\mathrm{e}^{t} - 1)X_1,$$

$$\hat{u}_2 = x_2 - X_2 = \mathrm{e}^{-2t}X_2 - X_2 = (\mathrm{e}^{-2t} - 1)X_2,$$

$$\hat{u}_3 = x_3 - X_3 = (1+t)X_3 - X_3 = tX_3,$$

while the spatial form is

$$\tilde{u}_1 = x_1 - X_1 = x_1 - \mathrm{e}^{-t}x_1 = (1 - \mathrm{e}^{-t})x_1,$$

$$\tilde{u}_2 = x_2 - X_2 = x_2 - \mathrm{e}^{2t}x_2 = (1 - \mathrm{e}^{2t})x_2,$$

$$\tilde{u}_3 = x_3 - X_3 = x_3 - \frac{x_3}{1+t} = \frac{t}{1+t}x_3.$$

(c) The referential form of the velocity $\hat{\mathbf{v}}(\mathbf{X}, t)$ is obtained using $(3.14)_1$:

$$\hat{v}_1 = \frac{\partial x_1(X_1, X_2, X_3, t)}{\partial t} = \frac{\partial}{\partial t}(\mathrm{e}^{t}X_1) = \mathrm{e}^{t}X_1,$$

$$\hat{v}_2 = \frac{\partial x_2(X_1, X_2, X_3, t)}{\partial t} = \frac{\partial}{\partial t}(\mathrm{e}^{-2t}X_2) = -2\mathrm{e}^{-2t}X_2,$$

$$\hat{v}_3 = \frac{\partial x_3(X_1, X_2, X_3, t)}{\partial t} = \frac{\partial}{\partial t}\left[(1+t)X_3\right] = X_3.$$

The spatial form of the velocity $\tilde{\mathbf{v}}(\mathbf{x}, t)$ is obtained by substituting the inverted motion

$$X_1 = \mathrm{e}^{-t}x_1, \qquad X_2 = \mathrm{e}^{2t}x_2, \qquad X_3 = \frac{x_3}{1+t}$$

into $\hat{v}_1$, $\hat{v}_2$, and $\hat{v}_3$, yielding

$$\tilde{v}_1 = \mathrm{e}^{t}X_1 = \mathrm{e}^{t}(\mathrm{e}^{-t}x_1) = x_1,$$

$$\tilde{v}_2 = -2\mathrm{e}^{-2t}X_2 = -2\mathrm{e}^{-2t}(\mathrm{e}^{2t}x_2) = -2x_2,$$

$$\tilde{v}_3 = X_3 = \frac{x_3}{1+t}.$$

(d) The referential form of the acceleration $\hat{\mathbf{a}}(\mathbf{X},t)$ is deduced using $(3.14)_2$:

$$\hat{a}_1 = \frac{\partial \hat{v}_1(X_1,X_2,X_3,t)}{\partial t} = \frac{\partial}{\partial t}\left(e^t X_1\right) = e^t X_1,$$

$$\hat{a}_2 = \frac{\partial \hat{v}_2(X_1,X_2,X_3,t)}{\partial t} = \frac{\partial}{\partial t}\left(-2e^{-2t}X_2\right) = 4e^{-2t}X_2,$$

$$\hat{a}_3 = \frac{\partial \hat{v}_3(X_1,X_2,X_3,t)}{\partial t} = \frac{\partial}{\partial t}\left(X_3\right) = 0.$$

The spatial form of the acceleration $\tilde{\mathbf{a}}(\mathbf{x},t)$ can be obtained by substituting the inverted motion into $\hat{a}_1$, $\hat{a}_2$, and $\hat{a}_3$:

$$\tilde{a}_1 = e^t X_1 = e^t\left(e^{-t}x_1\right) = x_1,$$

$$\tilde{a}_2 = 4e^{-2t}X_2 = 4e^{-2t}\left(e^{2t}x_2\right) = 4x_2,$$

$$\tilde{a}_3 = 0.$$

Alternatively, the spatial form of the acceleration $\tilde{\mathbf{a}}(\mathbf{x},t)$ can be found using (3.16), whose Cartesian component form is

$$\tilde{a}_i = \frac{\partial \tilde{v}_i(\mathbf{x},t)}{\partial t} + \frac{\partial \tilde{v}_i(\mathbf{x},t)}{\partial x_j}\tilde{v}_j(\mathbf{x},t).$$

Then,

$$\tilde{a}_1 = \frac{\partial \tilde{v}_1}{\partial t} + \frac{\partial \tilde{v}_1}{\partial x_1}\tilde{v}_1 + \frac{\partial \tilde{v}_1}{\partial x_2}\tilde{v}_2 + \frac{\partial \tilde{v}_1}{\partial x_3}\tilde{v}_3 = 0 + (1)(x_1) + 0 + 0 = x_1,$$

$$\tilde{a}_2 = \frac{\partial \tilde{v}_2}{\partial t} + \frac{\partial \tilde{v}_2}{\partial x_1}\tilde{v}_1 + \frac{\partial \tilde{v}_2}{\partial x_2}\tilde{v}_2 + \frac{\partial \tilde{v}_2}{\partial x_3}\tilde{v}_3 = 0 + 0 + (-2)(-2x_2) + 0 = 4x_2,$$

$$\tilde{a}_3 = \frac{\partial \tilde{v}_3}{\partial t} + \frac{\partial \tilde{v}_3}{\partial x_1}\tilde{v}_1 + \frac{\partial \tilde{v}_3}{\partial x_2}\tilde{v}_2 + \frac{\partial \tilde{v}_3}{\partial x_3}\tilde{v}_3 = -\frac{x_3}{(1+t)^2} + 0 + 0 + \frac{x_3}{(1+t)^2} = 0.$$

EXERCISES

1. In Cartesian component notation, verify the following forms of the product rule:

   (a) $\dot{\overline{\phi\mathbf{f}}} = \dot{\phi}\mathbf{f} + \phi\dot{\mathbf{f}}$.

   (b) $\dot{\overline{\mathbf{f}\cdot\mathbf{g}}} = \dot{\mathbf{f}}\cdot\mathbf{g} + \mathbf{f}\cdot\dot{\mathbf{g}}$.

   (c) $\dot{\overline{\mathbf{f}\times\mathbf{g}}} = \dot{\mathbf{f}}\times\mathbf{g} + \mathbf{f}\times\dot{\mathbf{g}}$.

   (d) $\dot{\overline{\mathbf{A}\mathbf{f}}} = \dot{\mathbf{A}}\mathbf{f} + \mathbf{A}\dot{\mathbf{f}}$.

(e) $\overline{\mathbf{AB}}^{\cdot} = \dot{\mathbf{A}}\mathbf{B} + \mathbf{A}\dot{\mathbf{B}}$.
(f) $\overline{\mathbf{A} \cdot \mathbf{B}}^{\cdot} = \dot{\mathbf{A}} \cdot \mathbf{B} + \mathbf{A} \cdot \dot{\mathbf{B}}$.

2. In Cartesian component notation, verify that $\overline{\mathbf{A} + \mathbf{B}}^{\cdot} = \dot{\mathbf{A}} + \dot{\mathbf{B}}$.

3. Prove in direct notation that $\overline{\mathbf{A}^{\mathrm{T}}}^{\cdot} = (\dot{\mathbf{A}})^{\mathrm{T}}$.

3.3 DEFORMATION AND STRAIN

A fundamental postulate of classical continuum mechanics is that the response at a particle of the continuum to a deformation at a particular time depends only on the deformation of a *small neighborhood* of the particle *up to* that time. We will therefore devote a sizeable amount of space to (1) defining what is meant by the small neighborhood of a particle, (2) defining what is meat by deformation, and (3) describing the kinematical quantities that characterize deformation.

3.3.1 DEFORMATION GRADIENT

The **small neighborhood** of a particle Y consists of particle Y and all of its nearest neighbors. Suppose Y occupies position $\mathbf{X}$ in the reference configuration $\mathcal{R}_{\mathrm{R}}$ and position $\mathbf{x}$ in the present configuration $\mathcal{R}$; refer to Figure 3.1. A vector $\mathrm{d}\mathbf{X}$ is a filament of infinitesimal length $\mathrm{d}S$ in the reference configuration, radiating from $\mathbf{X}$ in the direction along unit vector $\mathbf{N}$:

$$\mathrm{d}\mathbf{X} = \mathbf{N}\,\mathrm{d}S. \tag{3.24}$$

The line element $\mathrm{d}\mathbf{X}$ connects particle Y to one of its nearest neighbors in the reference configuration; see Figure 3.4. The small neighborhood of particle Y in the reference configuration $\mathcal{R}_{\mathrm{R}}$ is the set of all $\mathrm{d}\mathbf{X}$s radiating from position $\mathbf{X}$.

The motion $\mathbf{x} = \chi(\mathbf{X}, t)$ is defined for all particles in the body, i.e., for all $\mathbf{X}$ in $\mathcal{R}_{\mathrm{R}}$. During this motion, the line element $\mathrm{d}\mathbf{X}$ will be deformed—in general stretched (or compressed) and rotated, but not bent ($\mathrm{d}\mathbf{X}$ is too short to be bent)—into a line element $\mathrm{d}\mathbf{x}$ of length $\mathrm{d}s$, radiating from the new position $\mathbf{x}$ of particle Y in the direction along unit vector $\mathbf{n}$:

$$\mathrm{d}\mathbf{x} = \mathbf{n}\,\mathrm{d}s; \tag{3.25}$$

see Figure 3.4.

Recall from Section 3.1 that the **motion** $\mathbf{x} = \chi(\mathbf{X}, t)$ of the body maps each particle Y from its reference position $\mathbf{X}$ to its present position $\mathbf{x}$. Since the motion is defined for all particles in the body, the **deformation**, or change in local geometry, of the small neighborhood of particle Y can be found by evaluating the function $\chi(\mathbf{X}, t)$ on the particle and all of its nearest neighbors. However, a more tractable approach to quantifying the deformation of this small neighborhood involves defining a deformation measure that can be evaluated at Y *alone*, so the need to track the

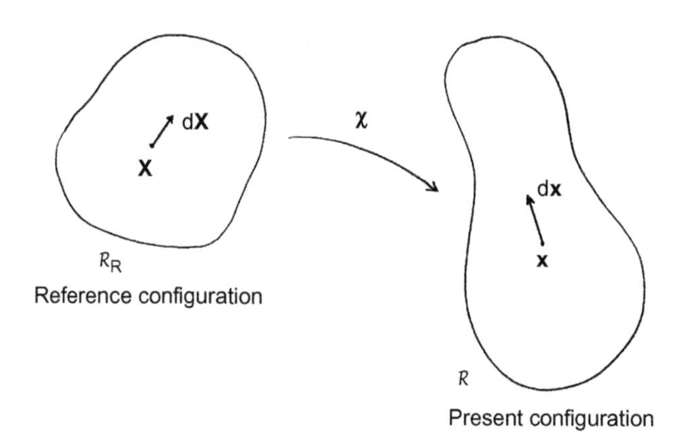

FIGURE 3.4

During the motion of the body, line element d**X** in the reference configuration stretches and rotates into line element d**x** in the present configuration.

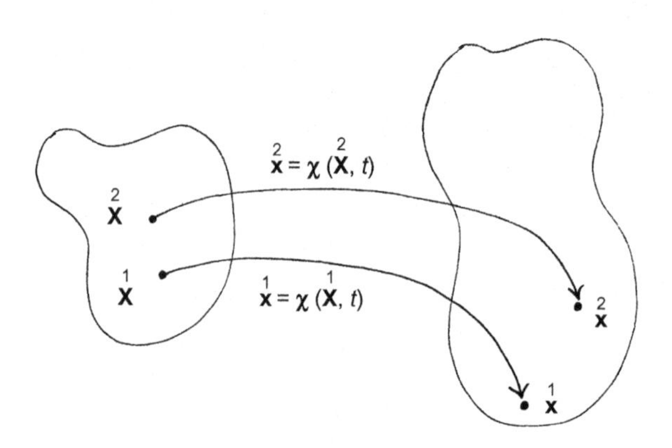

FIGURE 3.5

The motion of two nearest neighbors from their reference positions $\overset{1}{\mathbf{X}}$ and $\overset{2}{\mathbf{X}}$ to their present positions $\overset{1}{\mathbf{x}}$ and $\overset{2}{\mathbf{x}}$.

motion of all of the nearest neighbors of *Y* is obviated. This is accomplished through the use of a Taylor series.

A Taylor series is a mathematical tool that expresses the value of a function at one point in terms of the value of the function and its derivatives at another point. Suppose two particles that occupy points $\overset{1}{\mathbf{X}}$ and $\overset{2}{\mathbf{X}}$ in the reference configuration are mapped to points $\overset{1}{\mathbf{x}}$ and $\overset{2}{\mathbf{x}}$ in the present configuration (see Figure 3.5). A Taylor series expansion at $\overset{1}{\mathbf{X}}$ gives

$$\overset{2}{\mathbf{x}} = \chi\left(\overset{2}{\mathbf{X}}, t\right) = \chi\left(\overset{1}{\mathbf{X}}, t\right) + \left[\frac{\partial \chi}{\partial \mathbf{X}}\left(\overset{1}{\mathbf{X}}, t\right)\right]\left(\overset{2}{\mathbf{X}} - \overset{1}{\mathbf{X}}\right)$$

$$+ \frac{1}{2}\left[\frac{\partial^2 \chi}{\partial \mathbf{X}^2}\left(\overset{1}{\mathbf{X}}, t\right)\right]\left(\overset{2}{\mathbf{X}} - \overset{1}{\mathbf{X}}\right) \otimes \left(\overset{2}{\mathbf{X}} - \overset{1}{\mathbf{X}}\right) + \ldots. \tag{3.26}$$

If $\overset{1}{\mathbf{X}}$ and $\overset{2}{\mathbf{X}}$ are nearest neighbors, then the higher-order terms are negligible, leaving

$$\overset{2}{\mathbf{x}} = \overset{1}{\mathbf{x}} + \left[\frac{\partial \chi}{\partial \mathbf{X}}\left(\overset{1}{\mathbf{X}}, t\right)\right]\left(\overset{2}{\mathbf{X}} - \overset{1}{\mathbf{X}}\right).$$

Upon replacing $\overset{1}{\mathbf{X}}$ with $\mathbf{X}$, $\overset{2}{\mathbf{X}} - \overset{1}{\mathbf{X}}$ with $d\mathbf{X}$, and $\overset{2}{\mathbf{x}} - \overset{1}{\mathbf{x}}$ with $d\mathbf{x}$, we have

$$d\mathbf{x} = \left[\frac{\partial \chi}{\partial \mathbf{X}}(\mathbf{X}, t)\right] d\mathbf{X}.$$

We label

$$\mathbf{F} \equiv \frac{\partial \chi(\mathbf{X}, t)}{\partial \mathbf{X}} = \text{Grad}\, \chi = \text{Grad}\, \mathbf{x}, \tag{3.27}$$

so

$$d\mathbf{x} = \mathbf{F}\, d\mathbf{X}, \tag{3.28}$$

where $\mathbf{F}$ is called the **deformation gradient**.

Using definition (3.28), one can verify that $\mathbf{F}$ satisfies requirements (2.7) and is thus a tensor. Hence, $\mathbf{F}$ *linearly maps* each line element $d\mathbf{X}$ radiating from $\mathbf{X}$ in the reference configuration into a line element $d\mathbf{x}$ radiating from $\mathbf{x}$ in the present configuration. ($d\mathbf{X}$ is not bent when it is mapped by $\mathbf{F}$ to $d\mathbf{x}$.) Therefore $\mathbf{F}$ describes the mapping of the small neighborhood of $\mathbf{X}$ to the small neighborhood of $\mathbf{x}$, and thus quantifies the changes in the relative positions of nearest neighbors in the body; see Figure 3.6. In other words, $\mathbf{F}$, a function evaluated only at the point $\mathbf{X}$, is a measure of the deformation, or change of local geometry, in the neighborhood of $\mathbf{X}$ as it moves to $\mathbf{x}$.

Since $\mathbf{F}$ is a linear map, if the small neighborhood of $\mathbf{X}$ in the reference configuration is a sphere, then the deformation maps it into an ellipsoid in the present

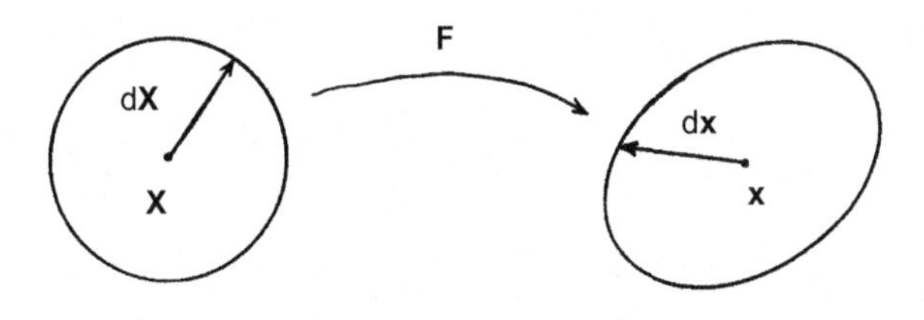

FIGURE 3.6

The deformation of a small neighborhood of **X** from a sphere into an ellipsoid.

configuration; see Figure 3.6. (If a spherical volume in the reference configuration deforms into a shape other than an ellipsoid, then it is too large to qualify as a *small* neighborhood, thereby invalidating our truncation of the series (3.26).) Note that although $\mathbf{F}$ is a tensor, which is by definition a linear map, the deformation is not restricted to be "linear," in the sense of infinitesimal. The stretch and rotation of the map from $\mathrm{d}\mathbf{X}$ to $\mathrm{d}\mathbf{x}$ can be large.

The tensor $\mathbf{F}$, in general, is a function of position and time, i.e.,

$$\mathbf{F} = \hat{\mathbf{F}}(\mathbf{X}, t) = \tilde{\mathbf{F}}(\mathbf{x}, t). \tag{3.29}$$

If $\mathbf{F}$ is independent of position, i.e., $\mathbf{F} = \mathbf{F}(t)$, the deformation is said to be *homogeneous.* If $\mathbf{F}$ is independent of time, i.e., $\mathbf{F} = \hat{\mathbf{F}}(\mathbf{X}) = \tilde{\mathbf{F}}(\mathbf{x})$, the deformation is said to be *static.*

EXERCISES

1. Prove that the deformation gradient $\mathbf{F}$ is a tensor.

3.3.2 STRETCH, ROTATION, GREEN'S DEFORMATION TENSOR, CAUCHY DEFORMATION TENSOR

The deformation gradient $\mathbf{F}$ contains the knowledge of what happens, *both* **stretch** *and* **rotation**, to *all* elements $\mathrm{d}\mathbf{X}$ radiating from $\mathbf{X}$ in the reference configuration, as they deform to $\mathrm{d}\mathbf{x}$ in the present configuration at time t. It is worthwhile separating the two concepts of stretch and rotation, which are combined in $\mathbf{F}$, since, as we shall see in later sections, sometimes only one or the other is physically important. We now produce functions of $\mathbf{F}$ that isolate stretch or rotation. We will also identify *particular* elements $\mathrm{d}\mathbf{X}$ and $\mathrm{d}\mathbf{x}$, of the infinitely many radiating from $\mathbf{X}$ and $\mathbf{x}$, which have special significance in understanding the deformation.

In general, the line element $\mathrm{d}\mathbf{X}$ undergoes both stretch and rotation in the deformation (3.28). The ratio

$$\frac{\mathrm{d}s}{\mathrm{d}S} = \lambda \tag{3.30}$$

of the length of $\mathrm{d}\mathbf{x}$ to the length of $\mathrm{d}\mathbf{X}$ is called the **stretch** λ of the line element. Note that λ is always positive and

$$\begin{aligned} &\lambda > 1 \Rightarrow \text{the line element lengthens,} \\ &\lambda < 1 \Rightarrow \text{the line element shortens,} \\ &\lambda = 1 \Rightarrow \text{the length of the line element remains the same.} \end{aligned} \tag{3.31}$$

Suppose we are interested only in the stretch, and not the rotation, that *will happen* to a particular line element $\mathrm{d}\mathbf{X}$ along direction $\mathbf{N}$. Using (3.24) and (3.25), we can rewrite (3.28) as

$$\mathbf{n}\,\mathrm{d}s = \mathbf{F}\mathbf{N}\mathrm{d}S,$$

or, using the definition of stretch (3.30),

$$\lambda \mathbf{n} = \mathbf{F}\mathbf{N}. \tag{3.32}$$

By taking the inner product of (3.32) with itself, we can show that (refer to Problem 3.3)

$$\lambda^2 = \mathbf{N} \cdot \mathbf{C}\mathbf{N} = \mathbf{C} \cdot (\mathbf{N} \otimes \mathbf{N}), \tag{3.33}$$

where

$$\mathbf{C} = \mathbf{F}^{\mathrm{T}}\mathbf{F} \tag{3.34}$$

is called **Green's deformation tensor** (or the **right Cauchy-Green deformation tensor**). It can be shown (refer to Problem 3.4) that **C** is symmetric and positive definite. From relation (3.33), we see that **C** contains information about the stretch, and not the rotation, that *will happen* to a line element $\mathrm{d}\mathbf{X}$ along direction **N** in the reference configuration.

Suppose instead we are interested only in the stretch, and not the rotation, that *has happened* to a particular $\mathrm{d}\mathbf{x}$ along direction **n** in the present configuration. Recall that we assumed in Section 3.1 that the motion $\mathbf{x} = \bar{\chi}(\mathrm{Y}, t)$ is invertible for all particles Y and times t. Then, using (3.2) and (3.3), we obtain

$$\mathbf{X} = \kappa\,(Y) = \kappa\left(\bar{\chi}^{-1}(\mathbf{x}, t)\right) = \chi^{-1}(\mathbf{x}, t), \tag{3.35}$$

so the motion $\mathbf{x} = \chi\,(\mathbf{X}, t)$ is invertible. From linear algebra, it follows that

$$J \equiv \det \mathbf{F} = \det\left(\frac{\partial \mathbf{x}}{\partial \mathbf{X}}\right) \neq 0. \tag{3.36}$$

We refer to $J \equiv \det \mathbf{F}$ as the **Jacobian** of the deformation gradient, which, as we will see later, is related to the local **dilatation** or volume change. Inequality (3.36) implies that the tensor $\mathbf{F}^{-1}$ exists, i.e., $\mathbf{F} = \operatorname{Grad} \mathbf{x}$ is invertible:

$$\mathrm{d}\mathbf{X} = \mathbf{F}^{-1}\,\mathrm{d}\mathbf{x}. \tag{3.37}$$

Using (3.24) and (3.25), we can rewrite (3.37) as

$$\mathbf{N}\,\mathrm{d}S = \mathbf{F}^{-1}\,\mathbf{n}\,ds,$$

or

$$\frac{1}{\lambda}\mathbf{N} = \mathbf{F}^{-1}\,\mathbf{n}. \tag{3.38}$$

By taking the inner product of (3.38) with itself, we can show that

$$\frac{1}{\lambda^2} = \mathbf{n} \cdot \mathbf{c}\mathbf{n} = \mathbf{c} \cdot (\mathbf{n} \otimes \mathbf{n}), \tag{3.39}$$

where

$$\mathbf{c} = \mathbf{F}^{-\mathrm{T}}\,\mathbf{F}^{-1} \tag{3.40}$$

is called the **Cauchy deformation tensor**. It can be verified that **c** is symmetric and positive definite. From relation (3.39), we see that **c** contains information about the stretch, and not the rotation, that *has happened* to a line element $\mathbf{dx}$ along direction $\mathbf{n}$ in the present configuration.

Therefore, in summary, to find the stretch of a line element without concern for its rotation, if one has knowledge of the shape of the reference configuration (e.g., a specimen in a tensile test), use (3.33), and if one has knowledge of the shape of the deformed configuration (e.g., an inflated inner tube or beach ball), use (3.39).

PROBLEM 3.3

Prove in direct notation that $\lambda^2 = \mathbf{N} \cdot \mathbf{CN} = \mathbf{C} \cdot (\mathbf{N} \otimes \mathbf{N})$.

Solution

Upon taking the inner product of (3.32) with itself, we have

$$\lambda^2 \mathbf{n} \cdot \mathbf{n} = \mathbf{FN} \cdot \mathbf{FN}.$$

Since $\mathbf{n}$ is a unit vector, $\mathbf{n} \cdot \mathbf{n} = 1$. It follows that

$$\lambda^2 = \mathbf{N} \cdot \mathbf{F}^{\mathrm{T}}(\mathbf{FN}) = \mathbf{N} \cdot (\mathbf{F}^{\mathrm{T}}\mathbf{F})\mathbf{N},$$

where we have used definitions (2.11) and (2.13). Thus, since $\mathbf{C} = \mathbf{F}^{\mathrm{T}}\mathbf{F}$,

$$\lambda^2 = \mathbf{N} \cdot \mathbf{CN}.$$

Result (2.43) then implies that

$$\lambda^2 = \mathbf{C} \cdot (\mathbf{N} \otimes \mathbf{N}).$$

PROBLEM 3.4

In direct notation, prove that the right Cauchy-Green deformation tensor $\mathbf{C}$ is symmetric and positive definite.

Solution

We have

$$\mathbf{C}^{\mathrm{T}} = (\mathbf{F}^{\mathrm{T}}\mathbf{F})^{\mathrm{T}} = \mathbf{F}^{\mathrm{T}}\mathbf{F}^{\mathrm{TT}} = \mathbf{F}^{\mathrm{T}}\mathbf{F} = \mathbf{C},$$

so, according to definition (2.15), $\mathbf{C}$ is symmetric. It follows from (3.33) and positivity of λ that

$$\mathbf{N} \cdot \mathbf{CN} > 0,$$

with the unit vector $\mathbf{N}$ arbitrary and nonzero, so by (2.53) $\mathbf{C}$ is positive definite.

EXERCISES

1. Prove in direct notation that $\frac{1}{\lambda^2} = \mathbf{n} \cdot \mathbf{c}\,\mathbf{n} = \mathbf{c} \cdot (\mathbf{n} \otimes \mathbf{n})$.
2. Prove in direct notation that the Cauchy deformation tensor **c** is symmetric and positive definite.

3.3.3 POLAR DECOMPOSITION, STRETCH TENSORS, ROTATION TENSOR

We return to our discussion of the deformation gradient **F**. Since **F** is invertible (see (3.37)), it can be decomposed using the polar decomposition theorem (refer to Section 2.3) into the form

$$\mathbf{F} = \mathbf{R}\mathbf{U} \tag{3.41a}$$

or

$$\mathbf{F} = \mathbf{V}\mathbf{R}, \tag{3.41b}$$

where **R** is an orthogonal tensor (it will be shown in Sections 3.6 and 4.12 that $\det \mathbf{F} > 0$, so **R** is in fact proper orthogonal), and **U** and **V** are symmetric, positive-definite tensors. For physical reasons to be revealed shortly, the tensors **R**, **U**, and **V** are called the **rotation tensor**, **right stretch tensor**, and **left stretch tensor**, respectively.

The effect of the multiplicative decompositions (3.41a) and (3.41b) is to replace the linear transformation (3.28) by either of two pairs of linear transformations:

$$\mathrm{d}\mathbf{x} = \mathbf{F}\,\mathrm{d}\mathbf{X} = (\mathbf{R}\mathbf{U})\,\mathrm{d}\mathbf{X} = \mathbf{R}\,(\mathbf{U}\,\mathrm{d}\mathbf{X}),$$

i.e.,

$$\mathrm{d}\mathbf{X}^* = \mathbf{U}\,\mathrm{d}\mathbf{X} \quad \text{followed by} \quad \mathrm{d}\mathbf{x} = \mathbf{R}\,\mathrm{d}\mathbf{X}^*, \tag{3.42a}$$

or

$$\mathrm{d}\mathbf{x} = \mathbf{F}\,\mathrm{d}\mathbf{X} = (\mathbf{V}\mathbf{R})\,\mathrm{d}\mathbf{X} = \mathbf{V}(\mathbf{R}\,\mathrm{d}\mathbf{X}),$$

i.e.,

$$\mathrm{d}\mathbf{x}^* = \mathbf{R}\,\mathrm{d}\mathbf{X} \quad \text{followed by} \quad \mathrm{d}\mathbf{x} = \mathbf{V}\,\mathrm{d}\mathbf{x}^*; \tag{3.42b}$$

see Figure 3.7. We emphasize that (3.42a) and (3.42b) involve sequential (or serial) mappings, a distinguishing feature of multiplicative decompositions. Physically, it can be shown that (3.42a) represents stretch followed by *pure* rotation (refer to Problem 3.6), while (3.42b) represents *pure* rotation followed by stretch (refer to Problem 3.7).

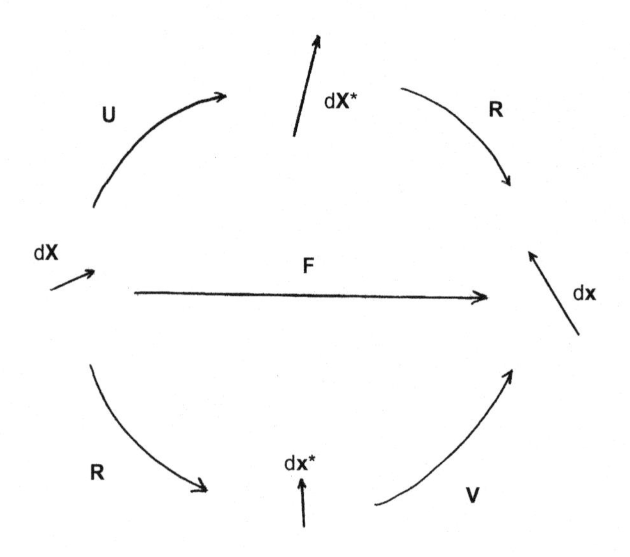

FIGURE 3.7

The polar decomposition of the deformation $d\mathbf{x} = \mathbf{F}d\mathbf{X}$ (center) into $d\mathbf{x} = \mathbf{R}(\mathbf{U}d\mathbf{X})$ (top) and $d\mathbf{x} = \mathbf{V}(\mathbf{R}d\mathbf{X})$ (bottom).

PROBLEM 3.5

Prove in direct notation that $\mathbf{C} = \mathbf{U}^2$.

Solution

From definition (3.34) and the polar decomposition (3.41a), we have

$$\mathbf{C} = \mathbf{F}^T\mathbf{F} = (\mathbf{RU})^T(\mathbf{RU}).$$

It follows from result $(2.14)_2$ that

$$\mathbf{C} = \mathbf{U}^T\mathbf{R}^T\mathbf{RU}.$$

Recall that the rotation tensor $\mathbf{R}$ is orthogonal, so $\mathbf{R}^T\mathbf{R} = \mathbf{I}$ (refer to (2.51)). It follows that

$$\mathbf{C} = \mathbf{U}^T\mathbf{IU} = \mathbf{U}^T\mathbf{U} = \mathbf{UU} \equiv \mathbf{U}^2$$

since $\mathbf{U}$ is symmetric.

PROBLEM 3.6

Prove in direct notation that decomposition (3.42a) represents stretch followed by *pure* rotation. That is, show that the deformation $(3.42a)_1$ of $d\mathbf{X}$ into $d\mathbf{X}^*$ involves stretching (and perhaps rotation), and the deformation $(3.42a)_2$ of $d\mathbf{X}^*$ into $d\mathbf{x}$ is rotation without stretch (i.e., pure rotation).

Solution

To prove this, we demonstrate that the intermediate element $\mathbf{dX}^*$ has the same length $\mathrm{d}s$ as the deformed element $\mathbf{dx}$; this, in turn, implies that the part of the deformation (3.42a) described by the map $\mathbf{U}$ contains all of the stretch of the total deformation from $\mathbf{dX}$ to $\mathbf{dx}$, so the part of the deformation (3.42a) described by the map $\mathbf{R}$ is pure rotation. To begin, we have

$$|\mathbf{dX}^*|^2 = \mathbf{dX}^* \cdot \mathbf{dX}^* = \mathbf{U}\,\mathbf{dX} \cdot \mathbf{U}\,\mathbf{dX}.$$

Using definitions (2.11) and (2.13), and noting that $\mathbf{U}$ is symmetric, we find that

$$|\mathbf{dX}^*|^2 = \mathbf{dX} \cdot \mathbf{U}^{\mathrm{T}}(\mathbf{U}\,\mathbf{dX}) = \mathbf{dX} \cdot \left(\mathbf{U}^{\mathrm{T}}\mathbf{U}\right)\mathbf{dX} = \mathbf{dX} \cdot \mathbf{U}^2\,\mathbf{dX}.$$

It follows that

$$|\mathbf{dX}^*|^2 = \mathbf{N} \cdot \mathbf{U}^2\mathbf{N}\mathrm{d}S^2 = \mathbf{N} \cdot \mathbf{C}\mathbf{N}\mathrm{d}S^2 = \lambda^2\mathrm{d}S^2 = (\lambda\mathrm{d}S)^2 = \mathrm{d}s^2,$$

where we have used (3.24), (3.30), (3.33), and $\mathbf{C} = \mathbf{U}^2$ (refer to Problem 3.5). We then conclude that

$$|\mathbf{dX}^*| = |\mathbf{dx}| = \mathrm{d}s.$$

PROBLEM 3.7

Prove in direct notation that decomposition (3.42b) represents *pure* rotation followed by stretch. That is, show that the deformation $(3.42\mathrm{b})_1$ of $\mathbf{dX}$ into $\mathbf{dx}^*$ is pure rotation (rotation without stretch), and the deformation $(3.42\mathrm{b})_2$ of $\mathbf{dx}^*$ into $\mathbf{dx}$ involves stretching (and perhaps rotation).

Solution

To prove this, we demonstrate that the intermediate element $\mathbf{dx}^*$ has the same length $\mathrm{d}S$ as the undeformed element $\mathbf{dX}$; this, in turn, implies that the part of the deformation (3.42b) described by the map $\mathbf{R}$ is only rotation, and all of the stretch of the total deformation from $\mathbf{dX}$ to $\mathbf{dx}$ is contained in the map $\mathbf{V}$. To begin, we have

$$|\mathbf{dx}^*|^2 = \mathbf{dx}^* \cdot \mathbf{dx}^* = \mathbf{R}\,\mathbf{dX} \cdot \mathbf{R}\,\mathbf{dX}.$$

Since $\mathbf{R}$ is orthogonal, it follows from definition (2.50) that

$$|\mathbf{dx}^*|^2 = \mathbf{R}\,\mathbf{dX} \cdot \mathbf{R}\,\mathbf{dX} = \mathbf{dX} \cdot \mathbf{dX} = |\mathbf{dX}|^2,$$

so

$$|\mathbf{dx}^*| = |\mathbf{dX}| = \mathrm{d}S.$$

3.3.4 PRINCIPAL STRETCHES AND PRINCIPAL DIRECTIONS

Recall the sequential mappings (3.42a) and (3.42b) that arise as a result of the polar decomposition of $\mathbf{F}$, and refer to Figure 3.7. We showed in Problems 3.6 and 3.7 that the maps $\mathrm{d}\mathbf{x}^* = \mathbf{R}\,\mathrm{d}\mathbf{X}$ and $\mathrm{d}\mathbf{x} = \mathbf{R}\,\mathrm{d}\mathbf{X}^*$ are pure rotations since they involve no length change. The maps $\mathrm{d}\mathbf{X}^* = \mathbf{U}\,\mathrm{d}\mathbf{X}$ and $\mathrm{d}\mathbf{x} = \mathbf{V}\,\mathrm{d}\mathbf{x}^*$, on the other hand, are called stretches since they incorporate all of the length change of the line element, from length $\mathrm{d}S$ of $\mathrm{d}\mathbf{X}$ to length $\mathrm{d}s = \lambda \mathrm{d}S$ of $\mathrm{d}\mathbf{x}$. *In general, these latter two maps also involve a change of direction (or rotation), in addition to this stretch.*

When $\mathrm{d}\mathbf{X}$ is in certain directions, however, the deformation $(3.42\text{a})_1$ from $\mathrm{d}\mathbf{X}$ to $\mathrm{d}\mathbf{X}^* = \mathbf{U}\,\mathrm{d}\mathbf{X}$ is *pure* stretch, i.e., $\mathrm{d}\mathbf{X}^*$ is parallel to $\mathrm{d}\mathbf{X}$. Similarly, when $\mathrm{d}\mathbf{x}^* = \mathbf{R}\,\mathrm{d}\mathbf{X}$ is in certain directions, the deformation $(3.42\text{b})_2$ from $\mathrm{d}\mathbf{x}^*$ to $\mathrm{d}\mathbf{x} = \mathbf{V}\,\mathrm{d}\mathbf{x}^*$ is *pure* stretch, i.e., $\mathrm{d}\mathbf{x}$ is parallel to $\mathrm{d}\mathbf{x}^*$.

3.3.4.1 Directions of pure stretch in the map U

Recall the map $\mathrm{d}\mathbf{X}^* = \mathbf{U}\,\mathrm{d}\mathbf{X}$, where $\mathbf{U}$ is the right stretch tensor. The condition that a particular element $\mathrm{d}\overset{1}{\mathbf{X}}$ along direction $\overset{1}{\mathbf{N}}$ undergoes pure stretch in the map $\mathbf{U}$ is

$$\mathrm{d}\overset{1}{\mathbf{X}}{}^* = \mathbf{U}\,\mathrm{d}\overset{1}{\mathbf{X}} = \overset{1}{\lambda}\mathrm{d}\overset{1}{\mathbf{X}},$$

or, equivalently,

$$\mathbf{U}\overset{1}{\mathbf{N}} = \overset{1}{\lambda}\,\overset{1}{\mathbf{N}}.$$

Therefore, if $\mathrm{d}\overset{1}{\mathbf{X}}$ is along a direction $\overset{1}{\mathbf{N}}$ that undergoes pure stretch in the map $\mathbf{U}$, then it is along an eigenvector of $\mathbf{U}$, in which case the direction $\overset{1}{\mathbf{N}}$ is an eigenvector of $\mathbf{U}$, and the stretch $\overset{1}{\lambda}$ that $\mathrm{d}\overset{1}{\mathbf{X}}$ undergoes is an eigenvalue of $\mathbf{U}$ (refer to Section 2.3).

Since $\mathbf{U}$ is symmetric and positive definite, there always exist at least three mutually perpendicular eigenvectors $\overset{1}{\mathbf{N}}, \overset{2}{\mathbf{N}}, \overset{3}{\mathbf{N}}$ of $\mathbf{U}$, with corresponding positive eigenvalues $\overset{1}{\lambda}, \overset{2}{\lambda}, \overset{3}{\lambda}$. Therefore, there exist *at least three* directions of *pure stretch* in the map $\mathbf{U}$, which we will label $\overset{1}{\mathbf{N}}, \overset{2}{\mathbf{N}}, \overset{3}{\mathbf{N}}$, with corresponding stretches $\overset{1}{\lambda}, \overset{2}{\lambda}, \overset{3}{\lambda}$. We call $\overset{1}{\mathbf{N}}, \overset{2}{\mathbf{N}}, \overset{3}{\mathbf{N}}$ the **principal directions**, and $\overset{1}{\lambda}, \overset{2}{\lambda}, \overset{3}{\lambda}$ the **principal stretches**. These three mutually perpendicular unit vectors $\overset{1}{\mathbf{N}}, \overset{2}{\mathbf{N}}, \overset{3}{\mathbf{N}}$ in the directions of pure stretch form an orthonormal basis. With respect to this basis, the matrix of tensor $\mathbf{U}$ is diagonal, with components equal to the principal stretches:

$$\mathbf{U} = \overset{1}{\lambda}\overset{1}{\mathbf{N}} \otimes \overset{1}{\mathbf{N}} + \overset{2}{\lambda}\overset{2}{\mathbf{N}} \otimes \overset{2}{\mathbf{N}} + \overset{3}{\lambda}\overset{3}{\mathbf{N}} \otimes \overset{3}{\mathbf{N}} = \sum_{i=1}^{3} \overset{i}{\lambda}\overset{i}{\mathbf{N}} \otimes \overset{i}{\mathbf{N}}. \tag{3.43}$$

Again, refer to Section 2.3.

With this understanding of the principal directions (directions of pure stretch) and principal stretches of $\mathbf{U}$, we can now give the following description of the deformation $\mathrm{d}\mathbf{x} = \mathbf{F}\mathrm{d}\mathbf{X}$, as decomposed into $\mathrm{d}\mathbf{x} = \mathbf{R}(\mathbf{U}\mathrm{d}\mathbf{X})$:

Suppose the small neighborhood of $\mathbf{X}$ *is a sphere of radius* $\mathrm{d}S$. *The deformation* $\mathbf{F}$ *takes this sphere into an ellipsoid (see Figure 3.6). In the decomposition* $\mathbf{F} = \mathbf{RU}$, *the sphere is first stretched into an ellipsoid; in this deformation there are (at least) three line elements* $\mathrm{d}\overset{1}{\mathbf{X}}, \mathrm{d}\overset{2}{\mathbf{X}}, \mathrm{d}\overset{3}{\mathbf{X}}$ *that do not rotate. These elements become the principal axes* $\overset{1}{\lambda}\mathrm{d}\overset{1}{\mathbf{X}}, \overset{2}{\lambda}\mathrm{d}\overset{2}{\mathbf{X}}, \overset{3}{\lambda}\mathrm{d}\overset{3}{\mathbf{X}}$ *of the ellipsoid (see Figure 3.8). The rotation* $\mathbf{R}$ *then rigidly rotates this ellipsoid, so* $\overset{1}{\mathbf{N}}$ *(the direction of* $\mathrm{d}\overset{1}{\mathbf{X}}$*) rotates to* $\overset{1}{\mathbf{n}}$, $\overset{2}{\mathbf{N}}$ *rotates to* $\overset{2}{\mathbf{n}}$, *and* $\overset{3}{\mathbf{N}}$ *rotates to* $\overset{3}{\mathbf{n}}$. *Since* $\mathbf{R}$ *is a rigid rotation,* $\overset{1}{\mathbf{n}}, \overset{2}{\mathbf{n}}, \overset{3}{\mathbf{n}}$ *is an orthonormal basis for* $\mathcal{E}^3$. *Therefore,*

$$\mathbf{R} = \overset{1}{\mathbf{n}} \otimes \overset{1}{\mathbf{N}} + \overset{2}{\mathbf{n}} \otimes \overset{2}{\mathbf{N}} + \overset{3}{\mathbf{n}} \otimes \overset{3}{\mathbf{N}} = \sum_{i=1}^{3} \overset{i}{\mathbf{n}} \otimes \overset{i}{\mathbf{N}}. \tag{3.44}$$

The final product of the composition of these two maps is an ellipsoid with principal axes

$$\mathrm{d}\overset{1}{\mathbf{x}} = \overset{1}{\lambda}\overset{1}{\mathbf{n}}\mathrm{d}S, \qquad \mathrm{d}\overset{2}{\mathbf{x}} = \overset{2}{\lambda}\overset{2}{\mathbf{n}}\mathrm{d}S, \qquad \mathrm{d}\overset{3}{\mathbf{x}} = \overset{3}{\lambda}\overset{3}{\mathbf{n}}\mathrm{d}S.$$

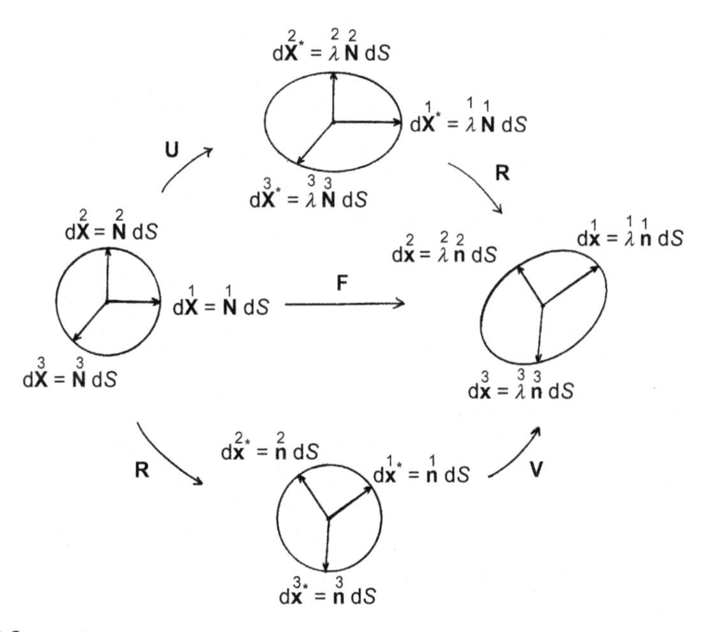

FIGURE 3.8

Polar decomposition: principal directions.

3.3.4.2 Directions of pure stretch in the map V

Suppose $d\overset{1}{\mathbf{X}}$ is along a direction of pure stretch in $\mathbf{U}$, i.e., $d\overset{1}{\mathbf{X}}$ is along an eigenvector of $\mathbf{U}$. Then,

$$d\overset{1}{\mathbf{x}} = \mathbf{F}\, d\overset{1}{\mathbf{X}} = \mathbf{R}\,(\mathbf{U}\, d\overset{1}{\mathbf{X}}) = \mathbf{R}\,(\overset{1}{\lambda} d\overset{1}{\mathbf{X}}) = \overset{1}{\lambda}(\mathbf{R}\, d\overset{1}{\mathbf{X}}).$$

We also have

$$d\overset{1}{\mathbf{x}} = \mathbf{F}\, d\overset{1}{\mathbf{X}} = \mathbf{V}(\mathbf{R}\, d\overset{1}{\mathbf{X}}).$$

Thus,

$$\mathbf{V}(\mathbf{R}\, d\overset{1}{\mathbf{X}}) = \overset{1}{\lambda}(\mathbf{R}\, d\overset{1}{\mathbf{X}}),$$

so $d\overset{1}{\mathbf{x}}{}^* = \mathbf{R}\, d\overset{1}{\mathbf{X}}$ is along an eigenvector of the left stretch tensor $\mathbf{V}$. As such, the deformation $\mathbf{V}$ takes $\mathbf{R}\, d\overset{1}{\mathbf{X}}$ to $\overset{1}{\lambda}\mathbf{R}\, d\overset{1}{\mathbf{X}}$, so it stretches without rotating. Therefore, the rotation $\mathbf{R}$ rotates the principal stretch directions $\overset{1}{\mathbf{N}}, \overset{2}{\mathbf{N}}, \overset{3}{\mathbf{N}}$ of $\mathbf{U}$ into the principal stretch directions $\overset{1}{\mathbf{n}}, \overset{2}{\mathbf{n}}, \overset{3}{\mathbf{n}}$ of $\mathbf{V}$; see Figure 3.8. Furthermore, the principal stretches $\overset{1}{\lambda}$, $\overset{2}{\lambda}, \overset{3}{\lambda}$ of $\mathbf{V}$ are the principal stretches $\overset{1}{\lambda}, \overset{2}{\lambda}, \overset{3}{\lambda}$ of $\mathbf{U}$. Hence, $\mathbf{V}$ can be expressed as

$$\mathbf{V} = \overset{1}{\lambda}\overset{1}{\mathbf{n}} \otimes \overset{1}{\mathbf{n}} + \overset{2}{\lambda}\overset{2}{\mathbf{n}} \otimes \overset{2}{\mathbf{n}} + \overset{3}{\lambda}\overset{3}{\mathbf{n}} \otimes \overset{3}{\mathbf{n}} = \sum_{i=1}^{3} \overset{i}{\lambda}\overset{i}{\mathbf{n}} \otimes \overset{i}{\mathbf{n}}. \tag{3.45}$$

We can now give the following description of the deformation $d\mathbf{x} = \mathbf{F}\, d\mathbf{X}$, as decomposed into $d\mathbf{x} = \mathbf{V}(\mathbf{R}\, d\mathbf{X})$:

Suppose the small neighborhood of $\mathbf{X}$ *is a sphere of radius* d*S. The deformation* $\mathbf{F}$ *takes this sphere into an ellipsoid (see Figure 3.6). In the decomposition* $\mathbf{F} = \mathbf{VR}$, *the sphere is first rotated rigidly, and is then stretched into the final ellipsoid. The initial rotation is such that the final stretch is accomplished without any change of direction in the principal axes of the ellipsoid; see Figure 3.8.*

Note that $\mathbf{R}$ is a rigid rotation of the small neighborhood of $\mathbf{X}$, not of the body as a whole; $\mathbf{R}$ can be a function of position, so different parts of the body rotate different amounts.

3.3.5 OTHER MEASURES OF DEFORMATION AND STRAIN

Recall Green's deformation tensor

$$\mathbf{C} = \mathbf{F}^{\mathrm{T}}\mathbf{F}$$

and the Cauchy deformation tensor

$$\mathbf{c} = \mathbf{F}^{-\mathrm{T}}\mathbf{F}^{-1},$$

which were defined in Section 3.3.2. Also recall that

$$\mathbf{C} = \mathbf{F}^{\mathrm{T}}\mathbf{F} = (\mathbf{RU})^{\mathrm{T}}(\mathbf{RU}) = \mathbf{U}^{\mathrm{T}}\mathbf{R}^{\mathrm{T}}\mathbf{RU} = \mathbf{U}^{\mathrm{T}}\mathbf{I}\,\mathbf{U} = \mathbf{U}^{\mathrm{T}}\mathbf{U} = \mathbf{UU} \equiv \mathbf{U}^2. \tag{3.46}$$

We now define

$$\mathbf{B} = \mathbf{F}\mathbf{F}^{\mathrm{T}}, \tag{3.47}$$

called the **Finger deformation tensor** (or the **left Cauchy-Green deformation tensor**). It can be shown that

$$\mathbf{B} = \mathbf{V}\mathbf{V} \equiv \mathbf{V}^2. \tag{3.48}$$

Also, we can verify that

$$\mathbf{c}\mathbf{B} = \mathbf{B}\mathbf{c} = \mathbf{I},$$

so

$$\mathbf{B} = \mathbf{c}^{-1}. \tag{3.49}$$

Suppose $\overset{1}{\mathbf{N}}$ is an eigenvector of $\mathbf{U}$, with corresponding eigenvalue $\overset{1}{\lambda}$. Then

$$\mathbf{C}\overset{1}{\mathbf{N}} = \mathbf{U}\,\mathbf{U}\overset{1}{\mathbf{N}} = \mathbf{U}(\overset{1}{\lambda}\overset{1}{\mathbf{N}}) = \overset{1}{\lambda}\mathbf{U}\overset{1}{\mathbf{N}} = \left(\overset{1}{\lambda}\right)^2 \overset{1}{\mathbf{N}}.$$

Therefore, the eigenvectors of $\mathbf{C}$ are the eigenvectors of $\mathbf{U}$, and the eigenvalues of $\mathbf{C}$ are the squares of the principal stretches. It follows that

$$\mathbf{C} = \left(\overset{1}{\lambda}\right)^2 \overset{1}{\mathbf{N}} \otimes \overset{1}{\mathbf{N}} + \left(\overset{2}{\lambda}\right)^2 \overset{2}{\mathbf{N}} \otimes \overset{2}{\mathbf{N}} + \left(\overset{3}{\lambda}\right)^2 \overset{3}{\mathbf{N}} \otimes \overset{3}{\mathbf{N}} = \sum_{i=1}^{3} \left(\overset{i}{\lambda}\right)^2 \overset{i}{\mathbf{N}} \otimes \overset{i}{\mathbf{N}}. \tag{3.50}$$

Similarly, it can be shown that $\mathbf{V}$ and $\mathbf{B}$ have the same eigenvectors $\overset{1}{\mathbf{n}}$, $\overset{2}{\mathbf{n}}$, $\overset{3}{\mathbf{n}}$, and that the eigenvalues of $\mathbf{B}$ are the squares of the principal stretches, so

$$\mathbf{B} = \left(\overset{1}{\lambda}\right)^2 \overset{1}{\mathbf{n}} \otimes \overset{1}{\mathbf{n}} + \left(\overset{2}{\lambda}\right)^2 \overset{2}{\mathbf{n}} \otimes \overset{2}{\mathbf{n}} + \left(\overset{3}{\lambda}\right)^2 \overset{3}{\mathbf{n}} \otimes \overset{3}{\mathbf{n}} = \sum_{i=1}^{3} \left(\overset{i}{\lambda}\right)^2 \overset{i}{\mathbf{n}} \otimes \overset{i}{\mathbf{n}}. \tag{3.51}$$

All of the measures of deformation we have so far defined, namely, $\mathbf{F}$, $\mathbf{F}^{-1}$, $\mathbf{C}$, $\mathbf{c}$, $\mathbf{B}$, $\mathbf{R}$, $\mathbf{U}$, and $\mathbf{V}$, reduce to the identity tensor in the absence of deformation:

$$\mathbf{x} = \mathbf{X} \quad \Rightarrow \quad \mathbf{F} = \mathbf{F}^{-1} = \mathbf{C} = \mathbf{c} = \mathbf{B} = \mathbf{R} = \mathbf{U} = \mathbf{V} = \mathbf{I}.$$

We now define two additional measures,

$$\mathbf{E} = \frac{1}{2}(\mathbf{C} - \mathbf{I}), \tag{3.52}$$

called the **Lagrangian strain tensor** (or **Green-Lagrange strain tensor**), and

$$\mathbf{e} = \frac{1}{2}(\mathbf{I} - \mathbf{c}), \tag{3.53}$$

called the **Eulerian strain tensor** (or **Euler-Almansi strain tensor**). These strain tensors vanish in the absence of deformation:

$$\mathbf{x} = \mathbf{X} \quad \Rightarrow \quad \mathbf{E} = \mathbf{e} = \mathbf{0}.$$

Recall that $\mathbf{C}$, $\mathbf{c}$, $\mathbf{U}$, and $\mathbf{V}$ are symmetric; it can be shown that $\mathbf{B}$, $\mathbf{E}$, and $\mathbf{e}$ are also symmetric.

The deformation and strain tensors presented heretofore can be expressed in terms of the displacement $\mathbf{u} = \mathbf{x} - \mathbf{X}$. For instance, we have

$$\begin{aligned}\mathbf{C} &= (\text{Grad}\,\mathbf{x})^{\mathrm{T}}\,\text{Grad}\,\mathbf{x}\\ &= \left[\text{Grad}\,(\mathbf{X}+\mathbf{u})\right]^{\mathrm{T}}\text{Grad}\,(\mathbf{X}+\mathbf{u})\\ &= (\mathbf{I}+\text{Grad}\,\mathbf{u})^{\mathrm{T}}\,(\mathbf{I}+\text{Grad}\,\mathbf{u})\\ &= \mathbf{I}+\text{Grad}\,\mathbf{u}+(\text{Grad}\,\mathbf{u})^{\mathrm{T}}+(\text{Grad}\,\mathbf{u})^{\mathrm{T}}\,\text{Grad}\,\mathbf{u}\end{aligned}$$

and

$$\mathbf{E} = \frac{1}{2}(\mathbf{C}-\mathbf{I}) = \frac{1}{2}\left[\text{Grad}\,\mathbf{u}+(\text{Grad}\,\mathbf{u})^{\mathrm{T}}+(\text{Grad}\,\mathbf{u})^{\mathrm{T}}\,\text{Grad}\,\mathbf{u}\right]. \tag{3.54}$$

Similarly,

$$\mathbf{c} = \mathbf{I} - \text{grad}\,\mathbf{u} - (\text{grad}\,\mathbf{u})^{\mathrm{T}} + (\text{grad}\,\mathbf{u})^{\mathrm{T}}\,\text{grad}\,\mathbf{u}$$

and

$$\mathbf{e} = \frac{1}{2}(\mathbf{I}-\mathbf{c}) = \frac{1}{2}\left[\text{grad}\,\mathbf{u}+(\text{grad}\,\mathbf{u})^{\mathrm{T}}-(\text{grad}\,\mathbf{u})^{\mathrm{T}}\,\text{grad}\,\mathbf{u}\right]. \tag{3.55}$$

Equations (3.54) and (3.55) are referred to as *strain-displacement relations*.

PROBLEM 3.8

Given the motion

$$x_1 = -2X_1 + X_2, \qquad x_2 = -X_1 - 2X_2, \qquad x_3 = X_3,$$

determine matrix representations of

(a) the deformation gradient **F**, Green's deformation tensor **C**, the Finger deformation tensor **B**, and the Cauchy deformation tensor **c**;
(b) the Lagrangian strain tensor **E** and the Eulerian strain tensor **e**;
(c) the right stretch tensor **U**, the left stretch tensor **V**, and the rotation tensor **R**.

Solution

(a) We have $\mathbf{x} = x_i\mathbf{e}_i$ and $\mathbf{X} = X_A\mathbf{e}_A$, where $\mathbf{e}_i$ and $\mathbf{e}_A$ are the Cartesian basis vectors associated with the present and reference configurations, respectively. (Note that lowercase subscripts pertain to the present configuration, and uppercase subscripts pertain to the reference configuration.) It then follows from (3.27) that the Cartesian components of the deformation gradient are $F_{iA} = \partial x_i/\partial X_A$, which in turn implies that

$$\mathbf{F} = F_{iA}\,\mathbf{e}_i \otimes \mathbf{e}_A = \frac{\partial x_i}{\partial X_A}\,\mathbf{e}_i \otimes \mathbf{e}_A.$$

Thus, using (2.34), we have

$$[\mathbf{F}] = \begin{bmatrix} \dfrac{\partial x_1}{\partial X_1} & \dfrac{\partial x_1}{\partial X_2} & \dfrac{\partial x_1}{\partial X_3} \\ \dfrac{\partial x_2}{\partial X_1} & \dfrac{\partial x_2}{\partial X_2} & \dfrac{\partial x_2}{\partial X_3} \\ \dfrac{\partial x_3}{\partial X_1} & \dfrac{\partial x_3}{\partial X_2} & \dfrac{\partial x_3}{\partial X_3} \end{bmatrix} = \begin{bmatrix} -2 & 1 & 0 \\ -1 & -2 & 0 \\ 0 & 0 & 1 \end{bmatrix}.$$

Note that the deformation is *homogeneous* since the components of **F** are independent of position **X**, and *static* since the components of **F** are independent of time t. Then, since $\mathbf{C} = \mathbf{F}^T\mathbf{F}$, it follows from (2.35) and (2.37) that

$$[\mathbf{C}] = [\mathbf{F}]^T[\mathbf{F}] = \begin{bmatrix} -2 & -1 & 0 \\ 1 & -2 & 0 \\ 0 & 0 & 1 \end{bmatrix} \begin{bmatrix} -2 & 1 & 0 \\ -1 & -2 & 0 \\ 0 & 0 & 1 \end{bmatrix} = \begin{bmatrix} 5 & 0 & 0 \\ 0 & 5 & 0 \\ 0 & 0 & 1 \end{bmatrix}.$$

Similarly,

$$[\mathbf{B}] = [\mathbf{F}][\mathbf{F}]^T = \begin{bmatrix} -2 & 1 & 0 \\ -1 & -2 & 0 \\ 0 & 0 & 1 \end{bmatrix} \begin{bmatrix} -2 & -1 & 0 \\ 1 & -2 & 0 \\ 0 & 0 & 1 \end{bmatrix} = \begin{bmatrix} 5 & 0 & 0 \\ 0 & 5 & 0 \\ 0 & 0 & 1 \end{bmatrix},$$

and

$$[\mathbf{c}] = [\mathbf{F}]^{-T}[\mathbf{F}]^{-1} = \begin{bmatrix} -\frac{2}{5} & \frac{1}{5} & 0 \\ -\frac{1}{5} & -\frac{2}{5} & 0 \\ 0 & 0 & 1 \end{bmatrix} \begin{bmatrix} -\frac{2}{5} & -\frac{1}{5} & 0 \\ \frac{1}{5} & -\frac{2}{5} & 0 \\ 0 & 0 & 1 \end{bmatrix} = \begin{bmatrix} \frac{1}{5} & 0 & 0 \\ 0 & \frac{1}{5} & 0 \\ 0 & 0 & 1 \end{bmatrix}.$$

Note that, as expected, $[\mathbf{c}]^{-1} = [\mathbf{B}]$.

(b) Recall that $\mathbf{E} = \frac{1}{2}(\mathbf{C} - \mathbf{I})$; it follows from (2.36) that

$$[\mathbf{E}] = \frac{1}{2}([\mathbf{C}] - [\mathbf{I}]) = \frac{1}{2}\left(\begin{bmatrix} 5 & 0 & 0 \\ 0 & 5 & 0 \\ 0 & 0 & 1 \end{bmatrix} - \begin{bmatrix} 1 & 0 & 0 \\ 0 & 1 & 0 \\ 0 & 0 & 1 \end{bmatrix} \right) = \begin{bmatrix} 2 & 0 & 0 \\ 0 & 2 & 0 \\ 0 & 0 & 0 \end{bmatrix}.$$

Similarly,

$$[\mathbf{e}] = \frac{1}{2}([\mathbf{I}] - [\mathbf{c}]) = \frac{1}{2}\left(\begin{bmatrix} 1 & 0 & 0 \\ 0 & 1 & 0 \\ 0 & 0 & 1 \end{bmatrix} - \begin{bmatrix} \frac{1}{5} & 0 & 0 \\ 0 & \frac{1}{5} & 0 \\ 0 & 0 & 1 \end{bmatrix} \right) = \begin{bmatrix} \frac{2}{5} & 0 & 0 \\ 0 & \frac{2}{5} & 0 \\ 0 & 0 & 0 \end{bmatrix}.$$

(c) To calculate the right stretch tensor **U**, we use the relationship $\mathbf{U}^2 \equiv \mathbf{U}\mathbf{U} = \mathbf{C}$. Since $[\mathbf{C}]$ is a diagonal matrix in this example, $[\mathbf{U}]$ is obtained by taking the square root of each of the elements of $[\mathbf{C}]$, i.e.,

$$[\mathbf{U}] = [\sqrt{\mathbf{C}}] = \begin{bmatrix} \sqrt{5} & 0 & 0 \\ 0 & \sqrt{5} & 0 \\ 0 & 0 & 1 \end{bmatrix}.$$

Similarly,

$$[\mathbf{V}] = [\sqrt{\mathbf{B}}] = \begin{bmatrix} \sqrt{5} & 0 & 0 \\ 0 & \sqrt{5} & 0 \\ 0 & 0 & 1 \end{bmatrix}.$$

(To calculate the square root of a matrix with *nonzero off-diagonal terms*, the matrix is first re-expressed in terms of its basis of eigenvectors, which results in a diagonal representation of the matrix. The square root of this diagonal matrix is calculated as shown above, i.e., by taking the square root of the diagonal elements. The resulting matrix is then rotated back to the original Cartesian basis.)

Finally, we use the relationship $\mathbf{R} = \mathbf{F}\mathbf{U}^{-1}$ to calculate the rotation tensor:

$$[\mathbf{R}] = [\mathbf{F}][\mathbf{U}]^{-1} = \begin{bmatrix} -2 & 1 & 0 \\ -1 & -2 & 0 \\ 0 & 0 & 1 \end{bmatrix} \begin{bmatrix} \frac{1}{\sqrt{5}} & 0 & 0 \\ 0 & \frac{1}{\sqrt{5}} & 0 \\ 0 & 0 & 1 \end{bmatrix} = \begin{bmatrix} -\frac{2}{\sqrt{5}} & \frac{1}{\sqrt{5}} & 0 \\ -\frac{1}{\sqrt{5}} & -\frac{2}{\sqrt{5}} & 0 \\ 0 & 0 & 1 \end{bmatrix}.$$

Note that since $\mathbf{R}$ is orthogonal,

$$\det \mathbf{R} = \det [\mathbf{R}] = \left(-\frac{2}{\sqrt{5}}\right)^2 - \left(\frac{1}{\sqrt{5}}\right)\left(-\frac{1}{\sqrt{5}}\right) = 1,$$

as expected.

PROBLEM 3.9

Given the motion

$$x_1 = 2X_2, \qquad x_2 = 5X_3, \qquad x_3 = X_1,$$

(a) calculate the principal stretches $\overset{1}{\lambda}$, $\overset{2}{\lambda}$, and $\overset{3}{\lambda}$ (common to both $\mathbf{U}$ and $\mathbf{V}$);

(b) calculate the principal directions $\overset{1}{\mathbf{N}}, \overset{2}{\mathbf{N}}, \overset{3}{\mathbf{N}}$ of $\mathbf{U}$;

(c) calculate the principal directions of $\overset{1}{\mathbf{n}}, \overset{2}{\mathbf{n}}, \overset{3}{\mathbf{n}}$ of $\mathbf{V}$;

(d) verify that the principal directions of $\mathbf{U}$ and $\mathbf{V}$ differ by a rotation, i.e.,

$$\overset{1}{\mathbf{n}} = \mathbf{R}\overset{1}{\mathbf{N}}, \qquad \overset{2}{\mathbf{n}} = \mathbf{R}\overset{2}{\mathbf{N}}, \qquad \overset{3}{\mathbf{n}} = \mathbf{R}\overset{3}{\mathbf{N}}.$$

Solution

First, we compute the deformation gradient

$$[\mathbf{F}] = \begin{bmatrix} \dfrac{\partial x_1}{\partial X_1} & \dfrac{\partial x_1}{\partial X_2} & \dfrac{\partial x_1}{\partial X_3} \\ \dfrac{\partial x_2}{\partial X_1} & \dfrac{\partial x_2}{\partial X_2} & \dfrac{\partial x_2}{\partial X_3} \\ \dfrac{\partial x_3}{\partial X_1} & \dfrac{\partial x_3}{\partial X_2} & \dfrac{\partial x_3}{\partial X_3} \end{bmatrix} = \begin{bmatrix} 0 & 2 & 0 \\ 0 & 0 & 5 \\ 1 & 0 & 0 \end{bmatrix},$$

Green's deformation tensor

$$[\mathbf{C}] = [\mathbf{F}]^{\mathrm{T}}[\mathbf{F}] = \begin{bmatrix} 0 & 0 & 1 \\ 2 & 0 & 0 \\ 0 & 5 & 0 \end{bmatrix} \begin{bmatrix} 0 & 2 & 0 \\ 0 & 0 & 5 \\ 1 & 0 & 0 \end{bmatrix} = \begin{bmatrix} 1 & 0 & 0 \\ 0 & 4 & 0 \\ 0 & 0 & 25 \end{bmatrix},$$

the Finger deformation tensor

$$[\mathbf{B}] = [\mathbf{F}][\mathbf{F}]^{\mathrm{T}} = \begin{bmatrix} 0 & 2 & 0 \\ 0 & 0 & 5 \\ 1 & 0 & 0 \end{bmatrix} \begin{bmatrix} 0 & 0 & 1 \\ 2 & 0 & 0 \\ 0 & 5 & 0 \end{bmatrix} = \begin{bmatrix} 4 & 0 & 0 \\ 0 & 25 & 0 \\ 0 & 0 & 1 \end{bmatrix},$$

the right stretch tensor

$$[\mathbf{U}] = [\sqrt{\mathbf{C}}] = \begin{bmatrix} 1 & 0 & 0 \\ 0 & 2 & 0 \\ 0 & 0 & 5 \end{bmatrix},$$

the left stretch tensor

$$[\mathbf{V}] = [\sqrt{\mathbf{B}}] = \begin{bmatrix} 2 & 0 & 0 \\ 0 & 5 & 0 \\ 0 & 0 & 1 \end{bmatrix},$$

and the rotation tensor

$$[\mathbf{R}] = [\mathbf{F}][\mathbf{U}]^{-1} = \begin{bmatrix} 0 & 2 & 0 \\ 0 & 0 & 5 \\ 1 & 0 & 0 \end{bmatrix} \begin{bmatrix} 1 & 0 & 0 \\ 0 & \frac{1}{2} & 0 \\ 0 & 0 & \frac{1}{5} \end{bmatrix} = \begin{bmatrix} 0 & 1 & 0 \\ 0 & 0 & 1 \\ 1 & 0 & 0 \end{bmatrix}.$$

Note that $\det \mathbf{R} = 1$, as expected.

(a) The principal strains $\overset{1}{\lambda}$, $\overset{2}{\lambda}$, and $\overset{3}{\lambda}$ are the eigenvalues of $\mathbf{U}$ (and $\mathbf{V}$), and are obtained by solving the characteristic equation (refer to (2.71))

$$\det(\mathbf{U} - \lambda\mathbf{I}) = \begin{vmatrix} 1-\lambda & 0 & 0 \\ 0 & 2-\lambda & 0 \\ 0 & 0 & 5-\lambda \end{vmatrix} = 0,$$

which gives

$$(1-\lambda)(2-\lambda)(5-\lambda)=0.$$

Thus,

$$\overset{1}{\lambda}=1, \qquad \overset{2}{\lambda}=2, \qquad \overset{3}{\lambda}=5.$$

(b) The principal directions $\overset{1}{\mathbf{N}}, \overset{2}{\mathbf{N}}, \overset{3}{\mathbf{N}}$ of $\mathbf{U}$ are found by solving the eigenvalue problem

$$(\mathbf{U}-\lambda\mathbf{I})\mathbf{N}=\mathbf{0}$$

for $\overset{1}{\lambda}=1$, $\overset{2}{\lambda}=2$, and $\overset{3}{\lambda}=5$, respectively. For instance, in the case of $\overset{1}{\lambda}=1$, we have

$$(\mathbf{U}-\overset{1}{\lambda}\mathbf{I})\overset{1}{\mathbf{N}}=\mathbf{0},$$

i.e.,

$$\begin{bmatrix} 0 & 0 & 0 \\ 0 & 1 & 0 \\ 0 & 0 & 4 \end{bmatrix} \begin{bmatrix} \overset{1}{N_1} \\ \overset{1}{N_2} \\ \overset{1}{N_3} \end{bmatrix} = \begin{bmatrix} 0 \\ 0 \\ 0 \end{bmatrix},$$

which implies that

$$\left[\overset{1}{\mathbf{N}}\right] = \begin{bmatrix} \overset{1}{N_1} \\ \overset{1}{N_2} \\ \overset{1}{N_3} \end{bmatrix} = \begin{bmatrix} 1 \\ 0 \\ 0 \end{bmatrix}.$$

In a similar fashion, we find that

$$\left[\overset{2}{\mathbf{N}}\right] = \begin{bmatrix} \overset{2}{N_1} \\ \overset{2}{N_2} \\ \overset{2}{N_3} \end{bmatrix} = \begin{bmatrix} 0 \\ 1 \\ 0 \end{bmatrix}, \qquad \left[\overset{3}{\mathbf{N}}\right] = \begin{bmatrix} \overset{3}{N_1} \\ \overset{3}{N_2} \\ \overset{3}{N_3} \end{bmatrix} = \begin{bmatrix} 0 \\ 0 \\ 1 \end{bmatrix}.$$

(c) The principal directions $\overset{1}{\mathbf{n}}, \overset{2}{\mathbf{n}}, \overset{3}{\mathbf{n}}$ of $\mathbf{V}$ are found by solving the eigenvalue problem

$$(\mathbf{V}-\lambda\mathbf{I})\mathbf{n}=\mathbf{0}$$

for $\overset{1}{\lambda}=1$, $\overset{2}{\lambda}=2$, and $\overset{3}{\lambda}=5$, respectively. (Recall that $\mathbf{U}$ and $\mathbf{V}$ have the same eigenvalues.) For instance, in the case of $\overset{1}{\lambda}=1$, we have

$$(\mathbf{V}-\overset{1}{\lambda}\mathbf{I})\overset{1}{\mathbf{n}}=\mathbf{0},$$

i.e.,

$$\begin{bmatrix} 1 & 0 & 0 \\ 0 & 4 & 0 \\ 0 & 0 & 0 \end{bmatrix} \begin{bmatrix} \overset{1}{n}_1 \\ \overset{1}{n}_2 \\ \overset{1}{n}_3 \end{bmatrix} = \begin{bmatrix} 0 \\ 0 \\ 0 \end{bmatrix},$$

which implies that

$$\left[\overset{1}{\mathbf{n}}\right] = \begin{bmatrix} \overset{1}{n}_1 \\ \overset{1}{n}_2 \\ \overset{1}{n}_3 \end{bmatrix} = \begin{bmatrix} 0 \\ 0 \\ 1 \end{bmatrix}.$$

In a similar fashion, we find that

$$\left[\overset{2}{\mathbf{n}}\right] = \begin{bmatrix} \overset{2}{n}_1 \\ \overset{2}{n}_2 \\ \overset{2}{n}_3 \end{bmatrix} = \begin{bmatrix} 1 \\ 0 \\ 0 \end{bmatrix}, \qquad \left[\overset{3}{\mathbf{n}}\right] = \begin{bmatrix} \overset{3}{n}_1 \\ \overset{3}{n}_2 \\ \overset{3}{n}_3 \end{bmatrix} = \begin{bmatrix} 0 \\ 1 \\ 0 \end{bmatrix}.$$

(d) We now verify that the principal directions of **U** and **V** differ by a rotation, i.e.,

$$\overset{1}{\mathbf{n}} = \mathbf{R}\overset{1}{\mathbf{N}}, \qquad \overset{2}{\mathbf{n}} = \mathbf{R}\overset{2}{\mathbf{N}}, \qquad \overset{3}{\mathbf{n}} = \mathbf{R}\overset{3}{\mathbf{N}}.$$

We have

$$\left[\overset{1}{\mathbf{n}}\right] = \left[\mathbf{R}\right]\left[\overset{1}{\mathbf{N}}\right] = \begin{bmatrix} 0 & 1 & 0 \\ 0 & 0 & 1 \\ 1 & 0 & 0 \end{bmatrix} \begin{bmatrix} 1 \\ 0 \\ 0 \end{bmatrix} = \begin{bmatrix} 0 \\ 0 \\ 1 \end{bmatrix},$$

$$\left[\overset{2}{\mathbf{n}}\right] = \left[\mathbf{R}\right]\left[\overset{2}{\mathbf{N}}\right] = \begin{bmatrix} 0 & 1 & 0 \\ 0 & 0 & 1 \\ 1 & 0 & 0 \end{bmatrix} \begin{bmatrix} 0 \\ 1 \\ 0 \end{bmatrix} = \begin{bmatrix} 1 \\ 0 \\ 0 \end{bmatrix},$$

and

$$\left[\overset{3}{\mathbf{n}}\right] = \left[\mathbf{R}\right]\left[\overset{3}{\mathbf{N}}\right] = \begin{bmatrix} 0 & 1 & 0 \\ 0 & 0 & 1 \\ 1 & 0 & 0 \end{bmatrix} \begin{bmatrix} 0 \\ 0 \\ 1 \end{bmatrix} = \begin{bmatrix} 0 \\ 1 \\ 0 \end{bmatrix},$$

which are self-consistent with our results from part (c), as expected.

EXERCISES

1. Prove in direct notation that $\mathbf{B} = \mathbf{V}^2$.

2. In direct notation, verify that $\mathbf{cB} = \mathbf{Bc} = \mathbf{I}$, so $\mathbf{B} = \mathbf{c}^{-1}$.

3. Show that $\mathbf{V}$ and $\mathbf{B}$ have the same eigenvectors $\overset{1}{\mathbf{n}}, \overset{2}{\mathbf{n}}, \overset{3}{\mathbf{n}}$, and that the eigenvalues of $\mathbf{B}$ are the squares of the principal stretches.

4. Prove in direct notation that $\mathbf{B}$, $\mathbf{E}$, and $\mathbf{e}$ are symmetric.

3.4 VELOCITY GRADIENT, RATE OF DEFORMATION TENSOR, VORTICITY TENSOR

The **velocity gradient** $\mathbf{L}$ is defined as the gradient of the spatial description of the velocity $\mathbf{v}$, i.e.,

$$\mathbf{L} = \operatorname{grad} \mathbf{v} = \frac{\partial}{\partial \mathbf{x}} \tilde{\mathbf{v}}(\mathbf{x}, t). \tag{3.56}$$

Following (2.17), the velocity gradient may be expressed as the sum of a symmetric tensor $\mathbf{D}$ and a skew tensor $\mathbf{W}$, i.e.,

$$\mathbf{L} = \mathbf{D} + \mathbf{W}, \tag{3.57}$$

where

$$\mathbf{D} = \frac{1}{2}(\mathbf{L} + \mathbf{L}^{\mathrm{T}}), \qquad \mathbf{W} = \frac{1}{2}(\mathbf{L} - \mathbf{L}^{\mathrm{T}}). \tag{3.58}$$

$\mathbf{D}$ and $\mathbf{W}$ are called the **rate of deformation tensor** and the **vorticity tensor**, respectively. Note that (3.57) is an *additive* (or parallel) decomposition of the velocity gradient $\mathbf{L}$; contrast this with the *multiplicative* (or serial) decomposition of the deformation gradient $\mathbf{F}$ presented in Section 3.3.3.

Recall that there is a one-to-one correspondence between vectors and skew tensors (refer to (2.65)). Thus, given the skew vorticity tensor $\mathbf{W}$ in (3.58), there exists a unique vector $\mathbf{w}$ such that

$$\mathbf{W}\mathbf{s} = \mathbf{w} \times \mathbf{s}$$

for any vector $\mathbf{s}$. The vector $\mathbf{w}$ is the *axial vector* of the vorticity tensor $\mathbf{W}$, and may be expressed as

$$\mathbf{w} = \frac{1}{2} \operatorname{curl} \mathbf{v} = \frac{1}{2} \boldsymbol{\omega}. \tag{3.59}$$

In (3.59), $\boldsymbol{\omega}$ is called the **vorticity vector**, and the velocity $\mathbf{v}$ is regarded as a function of $\mathbf{x}$ and t (spatial description).

It can be shown (refer to Problems 3.10–3.12) that the material derivatives of $\mathbf{F}$, $\mathbf{C}$, $\mathbf{E}$, and J are

$$\dot{\mathbf{F}} = \mathbf{L}\mathbf{F}, \qquad \dot{\mathbf{C}} = 2\dot{\mathbf{E}} = 2\mathbf{F}^{\mathrm{T}}\mathbf{D}\mathbf{F}, \qquad \dot{J} = \overline{\det \mathbf{F}}^{\,\cdot} = J \operatorname{div} \mathbf{v}. \tag{3.60}$$

We can also verify that

$$\dot{\mathbf{F}} = \operatorname{Grad} \mathbf{v}, \qquad \overline{\mathbf{F}^{-1}}^{\,\cdot} = -\mathbf{F}^{-1}\mathbf{L}, \qquad \operatorname{div} \mathbf{v} = \operatorname{tr} \mathbf{D}. \tag{3.61}$$

Recall from Section 3.3.2 that λ is the stretch of the line element $d\mathbf{x}$ in the current configuration (whose direction is along unit normal $\mathbf{n}$), which was formerly line element $d\mathbf{X}$ in the reference configuration (whose direction was along unit normal $\mathbf{N}$). It can be shown (refer to Problem 3.13) that

$$\frac{\dot{\lambda}}{\lambda} = \dot{\overline{\ln \lambda}} = \mathbf{n} \cdot \mathbf{D}\mathbf{n} = \mathbf{D} \cdot (\mathbf{n} \otimes \mathbf{n}). \quad (3.62)$$

Result (3.62) indicates the physical significance of $\mathbf{D}$ as a measure of the rate of stretching and rate of shearing of line elements: diagonal components of $\mathbf{D}$ give the logarithmic rates of stretching of line elements instantaneously aligned with the basis directions; off-diagonal components of $\mathbf{D}$ give the rate of shearing of line elements instantaneously aligned with perpendicular basis directions.

If $\mathbf{n}^*$ is an eigenvector of $\mathbf{D}$, then it can be shown (refer to Problem 3.14) that

$$\mathbf{D}\mathbf{n}^* = \frac{\dot{\lambda}^*}{\lambda^*}\mathbf{n}^*, \quad (3.63)$$

so the associated eigenvalue of $\mathbf{D}$ is $\dot{\lambda}^*/\lambda^*$. Also, we can verify (refer to Problem 3.14) that

$$\dot{\mathbf{n}}^* = \mathbf{W}\mathbf{n}^* = \mathbf{w} \times \mathbf{n}^*. \quad (3.64)$$

It is seen from (3.64) that the axial vector $\mathbf{w}$ is the angular velocity of the line element that is along an eigenvector of $\mathbf{D}$, so $\mathbf{W}$ has the physical significance of being the rate of rotation of the small neighborhood. Motions for which $\mathbf{W}$ (or $\mathbf{w}$) are zero are called *irrotational motions.*

Note that the velocity gradient $\mathbf{L}$, the rate of deformation tensor $\mathbf{D}$, and the vorticity tensor $\mathbf{W}$ depend only on the present configuration, and have no connection to any reference or previous configuration. This is in contrast to the deformation measures $\mathbf{F}$, $\mathbf{F}^{-1}$, $\mathbf{C}$, $\mathbf{B}$, $\mathbf{c}$, $\mathbf{E}$, and $\mathbf{e}$ of Section 3.3, which all describe some feature of the present configuration relative to a reference configuration.

Loosely, the response of a "solid" depends on the deformation of the continuum away from some reference (usually stress-free) configuration, whereas the response of a "fluid" depends only on the flow configuration at that instant. Therefore, as we shall see, the response functions (e.g., the relations between stress and strain) for a "solid" involve deformation measures like $\mathbf{F}$, and the response functions for a "fluid" involve deformation measures like $\mathbf{L}$. The response functions of rate-dependent materials (e.g., viscoelastic materials) involve both $\mathbf{F}$ and $\mathbf{L}$ (through $\dot{\mathbf{F}} = \mathbf{L}\mathbf{F}$; such materials have both "solid-like" and "liquid-like" characteristics).

PROBLEM 3.10

In direct notation, prove that $\dot{\mathbf{F}} = \mathbf{L}\mathbf{F}$.

Solution

$$\dot{\mathbf{F}} = \frac{\partial}{\partial t}\hat{\mathbf{F}}(\mathbf{X}, t) \qquad \text{(description (3.13))}$$

$$= \frac{\partial}{\partial t}\left(\frac{\partial \chi(\mathbf{X},t)}{\partial \mathbf{X}}\right) \qquad \text{(definition (3.27))}$$

$$= \frac{\partial}{\partial \mathbf{X}}\left(\frac{\partial \chi(\mathbf{X},t)}{\partial t}\right) \qquad \text{(continuity of } \chi(\mathbf{X},t)\text{)}$$

$$= \frac{\partial \hat{\mathbf{v}}(\mathbf{X},t)}{\partial \mathbf{X}} \qquad \text{(definition } (3.14)_1\text{)}$$

$$= \frac{\partial \tilde{\mathbf{v}}(\mathbf{x},t)}{\partial \mathbf{x}} \frac{\partial \chi(\mathbf{X},t)}{\partial \mathbf{X}} \qquad \text{(switch to Eulerian description of } \mathbf{v}\text{)}$$

$$= \mathbf{LF} \qquad \text{(definitions (3.27) and (3.56)).}$$

PROBLEM 3.11

Prove in direct notation that $\dot{\mathbf{C}} = 2\dot{\mathbf{E}} = 2\mathbf{F}^{\mathrm{T}}\mathbf{DF}$.

Solution

To begin, we have

$$\dot{\mathbf{C}} = \overline{\mathbf{F}^{\mathrm{T}}\mathbf{F}}^{\boldsymbol{\cdot}} = \overline{\mathbf{F}^{\mathrm{T}}}^{\boldsymbol{\cdot}}\mathbf{F} + \mathbf{F}^{\mathrm{T}}\dot{\mathbf{F}} = (\dot{\mathbf{F}})^{\mathrm{T}}\mathbf{F} + \mathbf{F}^{\mathrm{T}}\dot{\mathbf{F}},$$

where results $(3.22)_5$ and $(3.23)_2$ have been used. Then,

$$(\dot{\mathbf{F}})^{\mathrm{T}}\mathbf{F} + \mathbf{F}^{\mathrm{T}}\dot{\mathbf{F}} = (\mathbf{LF})^{\mathrm{T}}\mathbf{F} + \mathbf{F}^{\mathrm{T}}\mathbf{LF} = \mathbf{F}^{\mathrm{T}}\mathbf{L}^{\mathrm{T}}\mathbf{F} + \mathbf{F}^{\mathrm{T}}\mathbf{LF} = 2\mathbf{F}^{\mathrm{T}}\left[\frac{1}{2}\left(\mathbf{L}^{\mathrm{T}} + \mathbf{L}\right)\right]\mathbf{F} = 2\mathbf{F}^{\mathrm{T}}\mathbf{DF},$$

so

$$\dot{\mathbf{C}} = 2\mathbf{F}^{\mathrm{T}}\mathbf{DF}.$$

Also,

$$\dot{\mathbf{E}} = \frac{1}{2}\overline{(\mathbf{C} - \mathbf{I})}^{\boldsymbol{\cdot}} = \frac{1}{2}\dot{\mathbf{C}} = \mathbf{F}^{\mathrm{T}}\mathbf{DF}.$$

PROBLEM 3.12

In direct notation, show that $\dot{J} = J \operatorname{div} \mathbf{v}$.

Solution

We have

$$\dot{J} = \overline{\det \mathbf{F}}^{\boldsymbol{\cdot}} = \frac{\mathrm{d}}{\mathrm{d}\mathbf{F}}(\det \mathbf{F}) \cdot \dot{\mathbf{F}},$$

which follows from use of the chain rule $(2.95)_1$. Recall from Problem 2.48 that

$$\frac{d}{d\mathbf{A}}(\det\mathbf{A}) \cdot \mathbf{S} = (\det\mathbf{A})\text{tr}(\mathbf{S}\mathbf{A}^{-1})$$

for any tensor $\mathbf{S}$; it follows from this result (by setting $\mathbf{A} = \mathbf{F}$ and $\mathbf{S} = \dot{\mathbf{F}}$) that

$$\frac{d}{d\mathbf{F}}(\det\mathbf{F}) \cdot \dot{\mathbf{F}} = (\det\mathbf{F})\,\text{tr}(\dot{\mathbf{F}}\mathbf{F}^{-1}) = J\,\text{tr}(\mathbf{L}\mathbf{F}\mathbf{F}^{-1}) = J\,\text{tr}(\mathbf{L}\mathbf{I}) = J\,\text{tr}\,\mathbf{L}.$$

Thus,

$$\dot{J} = J\,\text{tr}\,\mathbf{L} = J\,\text{tr}(\text{grad}\,\mathbf{v}) = J\,\text{div}\,\mathbf{v}.$$

PROBLEM 3.13

Prove in direct notation that $\dot{\lambda}/\lambda = \mathbf{D} \cdot (\mathbf{n} \otimes \mathbf{n})$.

Solution

To begin, we take the material derivative of (3.33), which leads to

$$2\lambda\dot{\lambda} = \dot{\mathbf{C}} \cdot (\mathbf{N} \otimes \mathbf{N}),$$

where we have used result $(3.22)_6$; note that $\overline{\mathbf{N} \otimes \mathbf{N}}^{\,\boldsymbol{\cdot}} = \mathbf{0}$. Now, working with the right-hand side, we have

$$\dot{\mathbf{C}} \cdot (\mathbf{N} \otimes \mathbf{N}) = 2\mathbf{F}^{\mathrm{T}}\mathbf{D}\mathbf{F} \cdot (\mathbf{N} \otimes \mathbf{N}) = 2\mathbf{N} \cdot (\mathbf{F}^{\mathrm{T}}\mathbf{D}\mathbf{F})\mathbf{N},$$

where we have used results (2.43) and $(3.60)_2$. Definition (2.13) and results (2.43) and (3.32) then imply that

$$2\mathbf{N} \cdot \mathbf{F}^{\mathrm{T}}(\mathbf{D}\mathbf{F}\mathbf{N}) = 2\,(\mathbf{F}\mathbf{N}) \cdot \mathbf{D}\,(\mathbf{F}\mathbf{N}) = 2\,(\lambda\mathbf{n}) \cdot \mathbf{D}\,(\lambda\mathbf{n}) = 2\lambda^2(\mathbf{n} \cdot \mathbf{D}\mathbf{n}) = 2\lambda^2\,\mathbf{D} \cdot (\mathbf{n} \otimes \mathbf{n}).$$

Thus,

$$2\lambda\dot{\lambda} = 2\lambda^2\,\mathbf{D} \cdot (\mathbf{n} \otimes \mathbf{n}),$$

or, equivalently,

$$\frac{\dot{\lambda}}{\lambda} = \mathbf{D} \cdot (\mathbf{n} \otimes \mathbf{n}).$$

PROBLEM 3.14

If $\mathbf{n}^*$ is an eigenvector of $\mathbf{D}$, prove in direct notation that

(a) $\mathbf{D}\mathbf{n}^* = (\dot{\lambda}^*/\lambda^*)\,\mathbf{n}^*$, so the associated eigenvalue is $\dot{\lambda}^*/\lambda^*$, and
(b) $\dot{\mathbf{n}}^* = \mathbf{W}\mathbf{n}^* = \mathbf{w} \times \mathbf{n}^*$.

Solution

(a) Recall from (3.62) that

$$\mathbf{n}\cdot\mathbf{D}\mathbf{n}=\frac{\dot{\lambda}}{\lambda},$$

where λ is the stretch associated with the line element oriented along direction $\mathbf{n}$ in the present configuration. Since $\mathbf{n}$ is a unit vector, $\mathbf{n}\cdot\mathbf{n}=1$, and we have

$$\mathbf{n}\cdot\mathbf{D}\mathbf{n}=\frac{\dot{\lambda}}{\lambda}(\mathbf{n}\cdot\mathbf{n}).$$

Properties (2.6) of the inner product then allow us to write

$$\left(\mathbf{D}\mathbf{n}-\frac{\dot{\lambda}}{\lambda}\mathbf{n}\right)\cdot\mathbf{n}=0,$$

from which the eigenvalue problem for $\mathbf{D}$ follows:

$$\mathbf{D}\mathbf{n}=\frac{\dot{\lambda}}{\lambda}\mathbf{n}.$$

If $\mathbf{n}=\mathbf{n}^*$ is an eigenvector of $\mathbf{D}$, then it satisfies the above equation, i.e.,

$$\mathbf{D}\mathbf{n}^*=\frac{\dot{\lambda}^*}{\lambda^*}\mathbf{n}^*,$$

and the associated eigenvalue is $\dot{\lambda}^*/\lambda^*$.

(b) Recall from (3.32) that

$$\lambda\mathbf{n}=\mathbf{F}\mathbf{N}.$$

Taking the material derivative of both sides, i.e.,

$$\dot{\overline{\lambda\mathbf{n}}}=\dot{\overline{\mathbf{F}\mathbf{N}}},$$

leads to

$$\dot{\lambda}\mathbf{n}+\lambda\dot{\mathbf{n}}=\dot{\mathbf{F}}\mathbf{N}+\mathbf{F}\dot{\mathbf{N}},$$

where we have used (3.22). Noting that $\dot{\mathbf{N}}=\mathbf{0}$, we find from (3.32) and $(3.60)_1$ that

$$\dot{\mathbf{n}}=\mathbf{L}\mathbf{n}-\frac{\dot{\lambda}}{\lambda}\mathbf{n}.$$

If $\mathbf{n}=\mathbf{n}^*$ is an eigenvector of $\mathbf{D}$, with associated eigenvalue $\dot{\lambda}^*/\lambda^*$, then

$$\dot{\mathbf{n}}^*=\mathbf{L}\mathbf{n}^*-\frac{\dot{\lambda}^*}{\lambda^*}\mathbf{n}^*=\mathbf{L}\mathbf{n}^*-\mathbf{D}\mathbf{n}^*=\left[\mathbf{L}-\frac{1}{2}(\mathbf{L}+\mathbf{L}^{\mathrm{T}})\right]\mathbf{n}^*=\frac{1}{2}(\mathbf{L}-\mathbf{L}^{\mathrm{T}})\mathbf{n}^*=\mathbf{W}\mathbf{n}^*.$$

The vorticity tensor $\mathbf{W}$ is a skew tensor, and for every skew tensor there exists a unique axial vector $\mathbf{w}$ such that $\mathbf{Ws} = \mathbf{w} \times \mathbf{s}$ for any vector $\mathbf{s}$. Thus,

$$\dot{\mathbf{n}}^* = \mathbf{W}\mathbf{n}^* = \mathbf{w} \times \mathbf{n}^*.$$

EXERCISES

1. Prove in direct notation that $\dot{\mathbf{F}} = \text{Grad}\,\mathbf{v}$.
2. In direct notation, show that $\dot{\overline{\mathbf{F}^{-1}}} = -\mathbf{F}^{-1}\mathbf{L}$.
3. Prove in direct notation that $\text{div}\,\mathbf{v} = \text{tr}\,\mathbf{D}$.

3.5 MATERIAL POINT, MATERIAL LINE, MATERIAL SURFACE, MATERIAL VOLUME

Consider a body $\mathcal{B}$ with arbitrary subset $\mathcal{S}$. The body $\mathcal{B}$ occupies open region $\mathcal{R}_\text{R}$ of $\mathcal{E}^3$ in the reference configuration (with closed boundary $\partial\mathcal{R}_\text{R}$), and open region $\mathcal{R}$ of $\mathcal{E}^3$ in the present configuration (with closed boundary $\partial\mathcal{R}$). The subset $\mathcal{S}$ occupies open region $\mathcal{P}_\text{R} \subset \mathcal{R}_\text{R}$ in the reference configuration (with closed boundary $\partial\mathcal{P}_\text{R}$), and open region $\mathcal{P} \subset \mathcal{R}$ in the present configuration (with closed boundary $\partial\mathcal{P}$); see Figure 3.9.

A particle Y of body $\mathcal{B}$ is a **material point**; Y occupies location $\mathbf{X}$ in the reference configuration and location $\mathbf{x}$ in the present configuration (refer to Figure 3.1).

A **material line** is a curve that consists of the same set of particles for all time. Similarly, a **material surface** is a surface that consists of the same collection of particles for all time. It is defined in the form

$$\hat{f}(\mathbf{X}) = 0, \tag{3.65}$$

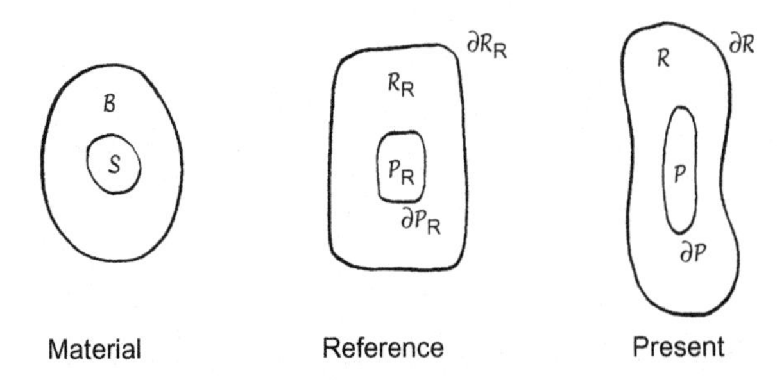

FIGURE 3.9

Body $\mathcal{B}$ with arbitrary subset $\mathcal{S}$ in the reference configuration and present configuration.

i.e., a scalar-valued function of reference location $\mathbf{X}$ equals zero. Note that the function $\hat{f}$ has no time dependence. The closed boundary $\partial\mathcal{P}$ of subset $\mathcal{S}$ is a material surface.

Recall that we have assumed the motion $\mathbf{x} = \chi(\mathbf{X}, t)$ is invertible. Therefore,

$$\hat{f}(\mathbf{X}) = \hat{f}\left(\chi^{-1}(\mathbf{x}, t)\right) = \tilde{f}(\mathbf{x}, t),$$

so the material surface with Lagrangian description $\hat{f}(\mathbf{X}) = 0$ also has an Eulerian description:

$$\tilde{f}(\mathbf{x}, t) = 0. \tag{3.66}$$

However, not all Eulerian surfaces of the form $\tilde{g}(\mathbf{x}, t) = 0$ are material surfaces. To be a material surface, any particle on the surface $\tilde{g}(\mathbf{x}, t) = 0$ at one time must stay on the surface for all time. Hence, to represent a material surface, the function $\tilde{g}(\mathbf{x}, t)$ must satisfy

$$\dot{g} = \frac{\partial}{\partial t}\hat{g}(\mathbf{X}, t) = \frac{\partial}{\partial t}\tilde{g}(\mathbf{x}, t) + \mathbf{v} \cdot \frac{\partial}{\partial \mathbf{x}}\tilde{g}(\mathbf{x}, t) = g' + \mathbf{v} \cdot \operatorname{grad} g = 0.$$

Therefore, we can state *Lagrange's criterion for a material surface:*

$$\tilde{f}(\mathbf{x}, t) \text{ is a material surface} \quad \Longleftrightarrow \quad \dot{f} = f' + \mathbf{v} \cdot \operatorname{grad} f = 0. \tag{3.67}$$

A **material volume** is a set of particles within a closed material surface, and therefore contains the same set of particles for all time. The region $\mathcal{P}$ occupied by subset $\mathcal{S}$ is a material volume.

3.6 VOLUME ELEMENTS AND SURFACE ELEMENTS IN VOLUME AND SURFACE INTEGRATIONS

Consider three line elements $\overset{1}{d\mathbf{X}}, \overset{2}{d\mathbf{X}}, \overset{3}{d\mathbf{X}}$ at point $\mathbf{X}$ that form a right-handed system in the reference configuration; see Figure 3.10. The **volume element** dV in the reference configuration is the volume of the parallelepiped formed by these three line elements, i.e.,

$$dV = \overset{1}{d\mathbf{X}} \cdot \overset{2}{d\mathbf{X}} \times \overset{3}{d\mathbf{X}} = \left[\overset{1}{d\mathbf{X}}\,\overset{2}{d\mathbf{X}}\,\overset{3}{d\mathbf{X}}\right], \tag{3.68}$$

where brackets denote a scalar triple product (refer to (2.62)). During the motion of the body, the line elements $\overset{1}{d\mathbf{X}}, \overset{2}{d\mathbf{X}}, \overset{3}{d\mathbf{X}}$ at point $\mathbf{X}$ in the reference configuration deform into line elements $\overset{1}{d\mathbf{x}}, \overset{2}{d\mathbf{x}}, \overset{3}{d\mathbf{x}}$ at point $\mathbf{x}$ in the present configuration, i.e.,

$$\overset{1}{d\mathbf{x}} = \mathbf{F}\,\overset{1}{d\mathbf{X}}, \qquad \overset{2}{d\mathbf{x}} = \mathbf{F}\,\overset{2}{d\mathbf{X}}, \qquad \overset{3}{d\mathbf{x}} = \mathbf{F}\,\overset{3}{d\mathbf{X}}; \tag{3.69}$$

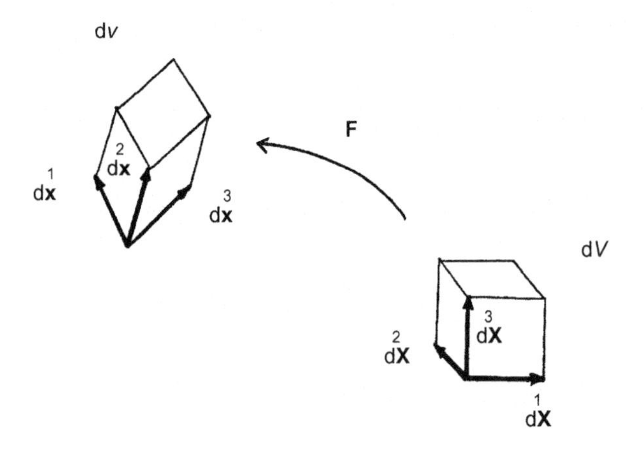

FIGURE 3.10

The volume elements dV and dv.

see Figure 3.10. The volume element dv in the present configuration is the volume of the parallelepiped formed by the deformed line elements $\overset{1}{d\mathbf{x}}$, $\overset{2}{d\mathbf{x}}$, $\overset{3}{d\mathbf{x}}$, i.e.,

$$dv = \overset{1}{d\mathbf{x}} \cdot \overset{2}{d\mathbf{x}} \times \overset{3}{d\mathbf{x}} = \left[\overset{1}{d\mathbf{x}}\, \overset{2}{d\mathbf{x}}\, \overset{3}{d\mathbf{x}} \right]. \tag{3.70}$$

Note that the parallelepipeds formed by $\overset{1}{d\mathbf{X}}$, $\overset{2}{d\mathbf{X}}$, $\overset{3}{d\mathbf{X}}$ and $\overset{1}{d\mathbf{x}}$, $\overset{2}{d\mathbf{x}}$, $\overset{3}{d\mathbf{x}}$ are the same material volume. It can be verified (refer to Problem 3.15) that

$$dv = J\, dV, \tag{3.71}$$

where $J = \det \mathbf{F}$ is the **Jacobian** of the deformation gradient. Hence, J quantifies the local *volume change* experienced by an infinitesimal parallelepiped during the deformation of the body. It follows from (3.36) and (3.71) that $J > 0$. The **dilatation** (or *normalized* volume change) is defined as

$$\Delta = \frac{dv - dV}{dV} = J - 1. \tag{3.72}$$

It can be shown (refer to Problem 3.16) that

$$\dot{\overline{dv}} = (\operatorname{div} \mathbf{v}) dv. \tag{3.73}$$

A motion that is volume preserving (i.e., $dv = dV$ for all $\mathbf{x}$ and t) is said to be **isochoric**. From (3.71)–(3.73), we deduce that a motion is isochoric if and only if

$$J = 1,$$

or, equivalently,

$$\operatorname{div} \mathbf{v} = \operatorname{tr} \mathbf{D} = 0.$$

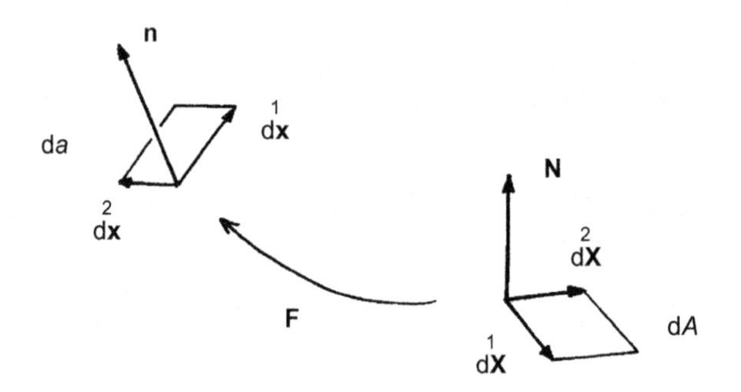

FIGURE 3.11

The area elements dA and da.

Consider now two nonparallel line elements $d\overset{1}{\mathbf{X}}$, $d\overset{2}{\mathbf{X}}$ at $\mathbf{X}$ in the reference configuration, which deform during the motion into $d\overset{1}{\mathbf{x}}$, $d\overset{2}{\mathbf{x}}$ at $\mathbf{x}$ in the present configuration. The **area element** dA in the reference configuration is the area of the parallelogram defined by $d\overset{1}{\mathbf{X}}$ and $d\overset{2}{\mathbf{X}}$. The outward unit normal $\mathbf{N}$ to area element dA is coincident with the direction of $d\overset{1}{\mathbf{X}} \times d\overset{2}{\mathbf{X}}$; see Figure 3.11. Hence,

$$\mathbf{N}\,dA = d\overset{1}{\mathbf{X}} \times d\overset{2}{\mathbf{X}}. \tag{3.74}$$

The area element da in the present configuration is the area of the parallelogram defined by $d\overset{1}{\mathbf{x}}$ and $d\overset{2}{\mathbf{x}}$. Its outward unit normal $\mathbf{n}$ is coincident with the direction of $d\overset{1}{\mathbf{x}} \times d\overset{2}{\mathbf{x}}$; see Figure 3.11. Hence,

$$\mathbf{n}\,da = d\overset{1}{\mathbf{x}} \times d\overset{2}{\mathbf{x}}. \tag{3.75}$$

Note that the two parallelograms are the same material surface. Since

$$d\overset{1}{\mathbf{x}} = \mathbf{F}\,d\overset{1}{\mathbf{X}}, \qquad d\overset{2}{\mathbf{x}} = \mathbf{F}\,d\overset{2}{\mathbf{X}},$$

it can be shown (refer to Problem 3.17) that

$$\mathbf{F}^{\mathrm{T}}\mathbf{n}\,da = J\,\mathbf{N}\,dA. \tag{3.76}$$

Also, we have (refer to Problem 3.18)

$$\dot{\overline{\mathbf{n}\,da}} = \left[(\operatorname{div}\mathbf{v})\mathbf{n} - \mathbf{L}^{\mathrm{T}}\mathbf{n}\right]da. \tag{3.77}$$

PROBLEM 3.15

In direct notation, prove that $\mathrm{d}v = J\,\mathrm{d}V$.

Solution

It follows from (3.68) and (3.70) that

$$\frac{\mathrm{d}v}{\mathrm{d}V} = \frac{\left[\overset{1}{\mathrm{d}\mathbf{x}}\,\overset{2}{\mathrm{d}\mathbf{x}}\,\overset{3}{\mathrm{d}\mathbf{x}}\right]}{\left[\overset{1}{\mathrm{d}\mathbf{X}}\,\overset{2}{\mathrm{d}\mathbf{X}}\,\overset{3}{\mathrm{d}\mathbf{X}}\right]}.$$

Then, using (2.64) and (3.69), we have

$$\frac{\mathrm{d}v}{\mathrm{d}V} = \frac{\left[\mathbf{F}\overset{1}{\mathrm{d}\mathbf{X}}\,\mathbf{F}\overset{2}{\mathrm{d}\mathbf{X}}\,\mathbf{F}\overset{3}{\mathrm{d}\mathbf{X}}\right]}{\left[\overset{1}{\mathrm{d}\mathbf{X}}\,\overset{2}{\mathrm{d}\mathbf{X}}\,\overset{3}{\mathrm{d}\mathbf{X}}\right]} = \det\mathbf{F} \equiv J,$$

so $\mathrm{d}v = J\,\mathrm{d}V$.

PROBLEM 3.16

In direct notation, show that $\dot{\overline{\mathrm{d}v}} = (\operatorname{div}\mathbf{v})\,\mathrm{d}v$.

Solution

Using result (3.71) and the product rule, we have

$$\dot{\overline{\mathrm{d}v}} = \dot{\overline{J\,\mathrm{d}V}} = \dot{J}\,\mathrm{d}V + J\,\dot{\overline{\mathrm{d}V}} = \dot{J}\,\mathrm{d}V,$$

noting that $\dot{\overline{dV}} = 0$. Then,

$$\dot{J}\,\mathrm{d}V = J\,(\operatorname{div}\mathbf{v})\,\mathrm{d}V = (\operatorname{div}\mathbf{v})\,\mathrm{d}v,$$

where we have used $(3.60)_3$ and (3.71), so

$$\dot{\overline{\mathrm{d}v}} = (\operatorname{div}\mathbf{v})\,\mathrm{d}v.$$

PROBLEM 3.17

Verify in direct notation that $\mathbf{F}^{\mathrm{T}}\mathbf{n}\,\mathrm{d}a = J\,\mathbf{N}\,\mathrm{d}A$.

Solution

It follows from (2.62), (2.63), (3.68), and (3.74) that

$$\mathrm{d}V = \left[\overset{1}{\mathrm{d}\mathbf{X}}\,\overset{2}{\mathrm{d}\mathbf{X}}\,\overset{3}{\mathrm{d}\mathbf{X}}\right] = \left[\overset{3}{\mathrm{d}\mathbf{X}}\,\overset{1}{\mathrm{d}\mathbf{X}}\,\overset{2}{\mathrm{d}\mathbf{X}}\right] = \overset{3}{\mathrm{d}\mathbf{X}}\cdot\overset{1}{\mathrm{d}\mathbf{X}}\times\overset{2}{\mathrm{d}\mathbf{X}} = \overset{3}{\mathrm{d}\mathbf{X}}\cdot\mathbf{N}\mathrm{d}A.$$

Similarly, it follows from (2.13), (2.62), (2.63), (3.69), (3.70), and (3.75) that

$$\mathrm{d}v = \left[\overset{1}{\mathrm{d}\mathbf{x}}\, \overset{2}{\mathrm{d}\mathbf{x}}\, \overset{3}{\mathrm{d}\mathbf{x}}\right] = \left[\overset{3}{\mathrm{d}\mathbf{x}}\, \overset{1}{\mathrm{d}\mathbf{x}}\, \overset{2}{\mathrm{d}\mathbf{x}}\right] = \overset{3}{\mathrm{d}\mathbf{x}} \cdot \overset{1}{\mathrm{d}\mathbf{x}} \times \overset{2}{\mathrm{d}\mathbf{x}} = \mathbf{F}\overset{3}{\mathrm{d}\mathbf{X}} \cdot \mathbf{n}\,\mathrm{d}a = \overset{3}{\mathrm{d}\mathbf{X}} \cdot \mathbf{F}^{\mathrm{T}}\mathbf{n}\,\mathrm{d}a.$$

Then, $\mathrm{d}v = J\,\mathrm{d}V$ implies that

$$\overset{3}{\mathrm{d}\mathbf{X}} \cdot \left(\mathbf{F}^{\mathrm{T}}\mathbf{n}\,\mathrm{d}a - J\,\mathbf{N}\mathrm{d}A\right) = 0.$$

We note that $\overset{3}{\mathrm{d}\mathbf{X}}$ is independent of $\mathbf{F}$, $\mathbf{n}\,\mathrm{d}a$, and $\mathbf{N}\,\mathrm{d}A$, so

$$\mathbf{F}^{\mathrm{T}}\mathbf{n}\,\mathrm{d}a = J\,\mathbf{N}\,\mathrm{d}A.$$

PROBLEM 3.18

Show in direct notation that $\dot{\overline{\mathbf{n}\,\mathrm{d}a}} = \left[(\mathrm{div}\,\mathbf{v})\mathbf{n} - \mathbf{L}^{\mathrm{T}}\mathbf{n}\right]\mathrm{d}a$.

Solution

We begin by taking the material derivative of (3.76), i.e.,

$$\dot{\overline{\mathbf{F}^{\mathrm{T}}\mathbf{n}\,\mathrm{d}a}} = \dot{\overline{J\,\mathbf{N}\,\mathrm{d}A}}.$$

Results (3.22) and (3.23) then imply that

$$\left(\dot{\mathbf{F}}\right)^{\mathrm{T}}\mathbf{n}\,\mathrm{d}a + \mathbf{F}^{\mathrm{T}}\,\dot{\overline{\mathbf{n}\,\mathrm{d}a}} = \dot{J}\,\mathbf{N}\,\mathrm{d}A,$$

noting that $\dot{\overline{\mathbf{N}\mathrm{d}A}} = \mathbf{0}$. It follows from $(2.14)_2$ and (3.60) that

$$\mathbf{F}^{\mathrm{T}}\mathbf{L}^{\mathrm{T}}\mathbf{n}\,\mathrm{d}a + \mathbf{F}^{\mathrm{T}}\,\dot{\overline{\mathbf{n}\,\mathrm{d}a}} = J\,(\mathrm{div}\,\mathbf{v})\,\mathbf{N}\,\mathrm{d}A,$$

or, after some tensor algebra,

$$\dot{\overline{\mathbf{n}\,\mathrm{d}a}} = (\mathrm{div}\,\mathbf{v})\left(J\mathbf{F}^{-\mathrm{T}}\mathbf{N}\,\mathrm{d}A\right) - \mathbf{L}^{\mathrm{T}}\mathbf{n}\,\mathrm{d}a.$$

Finally, use of (3.76) leads to

$$\dot{\overline{\mathbf{n}\,\mathrm{d}a}} = (\mathrm{div}\,\mathbf{v})\mathbf{n}\,\mathrm{d}a - \mathbf{L}^{\mathrm{T}}\mathbf{n}\,\mathrm{d}a.$$

CHAPTER

The Fundamental Laws of Thermomechanics

4

4.1 MASS

Consider again the configuration of body $\mathcal{B}$ at time t (i.e., the present configuration) in which $\mathcal{B}$ occupies the region $\mathcal{R}$ of $\mathcal{E}^3$ bounded by the closed surface $\partial\mathcal{R}$; a part $\mathcal{S}$ of body $\mathcal{B}$ occupies a region $\mathcal{P} \subseteq \mathcal{R}$ bounded by a closed surface $\partial\mathcal{P}$; refer to Figure 3.9. Each part $\mathcal{S}$ of body $\mathcal{B}$ at each instant of time is assumed to be endowed with a nonnegative measure $\mathcal{M}(\mathcal{S}, t)$, called the **mass** of part $\mathcal{S}$.

We define the **mass density** ρ at particle Y **in the present configuration** by

$$\rho = \bar{\rho}\,(Y, t) = \lim_{V \to 0} \frac{\mathcal{M}(\mathcal{S}, t)}{V}, \tag{4.1}$$

where $V = V(\mathcal{P})$ is the volume of the region $\mathcal{P}$ occupied by subset $\mathcal{S}$ as it collapses to particle Y. Since V is positive and $\mathcal{M}(\mathcal{S}, t)$ is nonnegative, density ρ is nonnegative.

In continuum mechanics, it is assumed that the measure $\mathcal{M}(\mathcal{S}, t)$ is absolutely continuous, i.e., in every configuration a part $\mathcal{S}$ occupying a sufficiently small volume has arbitrarily small mass $\mathcal{M}(\mathcal{S}, t)$. Thus, concentrated point, line, and surface masses are excluded, and the limit (4.1) always exists.

The mass of part $\mathcal{S}$ of body $\mathcal{B}$ and the mass of body $\mathcal{B}$ itself at time t can be expressed in terms of density ρ by

$$\mathcal{M}(\mathcal{S}, t) = \int_{\mathcal{P}} \rho \, \mathrm{d}v, \qquad \mathcal{M}(\mathcal{B}, t) = \int_{\mathcal{R}} \rho \, \mathrm{d}v, \tag{4.2}$$

where $\mathrm{d}v$ is an element of volume in the present configuration (refer to Section 3.6).

We define the **mass density** ρ_{R} at particle Y **in the reference configuration** by

$$\rho_{\mathrm{R}} = \bar{\rho}_{\mathrm{R}}\,(Y) = \lim_{V_{\mathrm{R}} \to 0} \frac{\mathcal{M}(\mathcal{S}_{\mathrm{R}})}{V_{\mathrm{R}}}, \tag{4.3}$$

where $V_{\mathrm{R}} = V_{\mathrm{R}}(\mathcal{P}_{\mathrm{R}})$ is the volume of region $\mathcal{P}_{\mathrm{R}}$ occupied by subset $\mathcal{S}$ in the reference configuration, and $\mathcal{M}(\mathcal{S}_{\mathrm{R}})$ is the mass of $\mathcal{S}$ in the reference configuration, as $\mathcal{S}$ collapses to particle Y.

The mass of part $\mathcal{S}$ of body $\mathcal{B}$ and the mass of body $\mathcal{B}$ itself in the reference configuration can be expressed in terms of the reference density ρ_{R} by

$$\mathcal{M}(\mathcal{S}_{\mathrm{R}}) = \int_{\mathcal{P}_{\mathrm{R}}} \rho_{\mathrm{R}} \, \mathrm{d}V, \qquad \mathcal{M}(\mathcal{B}_{\mathrm{R}}) = \int_{\mathcal{R}_{\mathrm{R}}} \rho_{\mathrm{R}} \, \mathrm{d}V, \tag{4.4}$$

where dV is an element of volume in the reference configuration (refer to Section 3.6).

The densities ρ and ρ_{R}, defined in (4.1) and (4.3), respectively, both have material, Lagrangian, and Eulerian descriptions:

$$\rho = \bar{\rho}(Y,t) = \bar{\rho}(\kappa^{-1}(\mathbf{X}),t) = \hat{\rho}(\mathbf{X},t) = \hat{\rho}(\chi^{-1}(\mathbf{x},t),t) = \tilde{\rho}(\mathbf{x},t),$$

$$\rho_{\mathrm{R}} = \bar{\rho}_{\mathrm{R}}(Y) = \bar{\rho}_{\mathrm{R}}(\kappa^{-1}(\mathbf{X})) = \hat{\rho}_{\mathrm{R}}(\mathbf{X}) = \hat{\rho}_{\mathrm{R}}(\chi^{-1}(\mathbf{x},t)) = \tilde{\rho}_{\mathrm{R}}(\mathbf{x},t).$$

Note that since the material and Lagrangian descriptions of ρ_{R} have no time dependence,

$$\dot{\rho}_{\mathrm{R}} = \frac{\partial}{\partial t}\bar{\rho}_{\mathrm{R}}(Y) = \frac{\partial}{\partial t}\hat{\rho}_{\mathrm{R}}(\mathbf{X}) = 0. \tag{4.5}$$

4.2 FORCES AND MOMENTS, LINEAR AND ANGULAR MOMENTUM

Again, consider the part $\mathcal{S}$ of body $\mathcal{B}$ that occupies region $\mathcal{P}$ bounded by surface $\partial\mathcal{P}$ in the present configuration at time t. We denote the **resultant external force** acting on part $\mathcal{S}$ of the body by $\mathbf{f}$:

$\mathbf{f} = \mathbf{f}(\mathcal{S},t)$, the resultant external force acting on part $\mathcal{S}$ at time t,

and the **resultant external moment** about the origin (point $\mathbf{0}$) by $\mathbf{M_o}$:

$\mathbf{M_o} = \mathbf{M_o}(\mathcal{S},t,\mathbf{0})$, the resultant external moment acting on part $\mathcal{S}$ at time t about the origin $\mathbf{0}$.

We *assume* that the resultant external force $\mathbf{f}$ acting on part $\mathcal{S}$ at time t is composed of a **body force** $\mathbf{f}_{\mathrm{b}}$ and a **contact force** $\mathbf{f}_{\mathrm{c}}$:

$$\mathbf{f} = \mathbf{f}_{\mathrm{b}} + \mathbf{f}_{\mathrm{c}}. \tag{4.6}$$

Additionally, we make the following *smoothness assumptions:*

$$\mathbf{f}_{\mathrm{b}} = \mathbf{f}_{\mathrm{b}}(\mathcal{S},t) = \int_{\mathcal{P}} \mathbf{b}\,\rho\,dv, \qquad \mathbf{f}_{\mathrm{c}} = \mathbf{f}_{\mathrm{c}}(\mathcal{S},t) = \int_{\partial\mathcal{P}} \mathbf{t}\,da, \tag{4.7}$$

where

$\mathbf{b} = \tilde{\mathbf{b}}(\mathbf{x},t)$, the **specific body force**, i.e., the body force per unit mass

and

$\mathbf{t} = \tilde{\mathbf{t}}(\mathbf{x}, t$, geometry of surface), the **traction**, i.e., the contact force per unit area of the present configuration.

The geometry of the surface includes orientation, curvature, etc. The smoothness assumptions demand that $\mathbf{b}$ and $\mathbf{t}$ are bounded, continuous functions of space and time. We have already invoked a smoothness assumption on mass in the previous section, namely,

$$\mathcal{M}(\mathcal{S},t) = \int_{\mathcal{P}} \rho \, \mathrm{d}v,$$

where density ρ is a bounded, continuous function of space and time. With this smoothness assumption on mass, we have excluded point, line, and surface masses; our smoothness assumptions on forces exclude concentrated forces.

We *assume* there are no distributed moment fields and no concentrated moments (we later relax this assumption in Chapter 9 to accommodate the effects of electromagnetism), so the resultant external moment $\mathbf{M_o}$ on part $\mathcal{S}$ at time t about the origin $\mathbf{0}$ comes from only $\mathbf{t}$ and $\mathbf{b}$:

$$\mathbf{M_o}\,(\mathcal{S},t,\mathbf{0}) = \int_{\mathcal{P}} (\mathbf{x}-\mathbf{0}) \times \mathbf{b}\rho \, \mathrm{d}v + \int_{\partial\mathcal{P}} (\mathbf{x}-\mathbf{0}) \times \mathbf{t} \, \mathrm{d}a. \tag{4.8}$$

Because of our smoothness assumption on mass, the **linear momentum** $\mathcal{L}$ of part $\mathcal{S}$ at time t and the **angular momentum** $\mathbf{H_o}$ of part $\mathcal{S}$ about the origin at time t are also smooth:

$$\mathcal{L} = \mathcal{L}\,(\mathcal{S},t) = \int_{\mathcal{P}} \mathbf{v}\rho \, \mathrm{d}v, \qquad \mathbf{H_o} = \mathbf{H_o}\,(\mathcal{S},t,\mathbf{0}) = \int_{\mathcal{P}} (\mathbf{x}-\mathbf{0}) \times \mathbf{v}\rho \, \mathrm{d}v. \tag{4.9}$$

4.3 EQUATIONS OF MOTION (MECHANICAL CONSERVATION LAWS)

We *postulate* the following equations of motion:

- The mass of every subset $\mathcal{S}$ of the body remains constant throughout the motion, or, equivalently, the rate of change of the mass of $\mathcal{S}$ is zero (**conservation of mass**).
- The rate of change of linear momentum of $\mathcal{S}$ is equal to the resultant force acting on $\mathcal{S}$ (**balance of linear momentum**).
- The rate of change of angular momentum of $\mathcal{S}$ about the origin is equal to the resultant moment acting on $\mathcal{S}$ about the origin (**balance of angular momentum**).

Mathematically, these equations of motion can be expressed in **material form** as

$$\frac{\mathrm{d}}{\mathrm{d}t}\mathcal{M}(\mathcal{S},t) = 0 \qquad \text{or} \qquad \mathcal{M}(\mathcal{S}) = \text{independent of } t = \mathcal{M}(\mathcal{S}_{\mathrm{R}}), \tag{4.10a}$$

$$\frac{\mathrm{d}}{\mathrm{d}t}\mathcal{L}(\mathcal{S},t) = \mathbf{f}(\mathcal{S},t), \tag{4.10b}$$

$$\frac{\mathrm{d}}{\mathrm{d}t}\mathbf{H_o}(\mathcal{S},t,\mathbf{0}) = \mathbf{M_o}(\mathcal{S},t,\mathbf{0}) \tag{4.10c}$$

for *arbitrary* part $\mathcal{S}$ of body $\mathcal{B}$, and *all* time t. We emphasize that the equations of motion (4.10a)–(4.10c) are global; i.e., they are valid not only for the body as a whole, but also for every arbitrary subset of the body. (The reader is already familiar with this requirement: as an example, in a static truss, not only is the entire structure in equilibrium, but so is each joint and each member.)

The smoothness assumptions discussed in Sections 4.1 and 4.2 can then be used to express the equations of motion $(4.10a)_1$, (4.10b), and (4.10c) in **Eulerian integral form**:

$$\frac{\mathrm{d}}{\mathrm{d}t}\int_{\mathcal{P}} \rho \,\mathrm{d}v = 0, \tag{4.11a}$$

$$\frac{\mathrm{d}}{\mathrm{d}t}\int_{\mathcal{P}} \mathbf{v}\rho \,\mathrm{d}v = \int_{\mathcal{P}} \mathbf{b}\rho \,\mathrm{d}v + \int_{\partial\mathcal{P}} \mathbf{t}\,\mathrm{d}a, \tag{4.11b}$$

$$\frac{\mathrm{d}}{\mathrm{d}t}\int_{\mathcal{P}} \mathbf{x}\times\mathbf{v}\rho \,\mathrm{d}v = \int_{\mathcal{P}} \mathbf{x}\times\mathbf{b}\rho \,\mathrm{d}v + \int_{\partial\mathcal{P}} \mathbf{x}\times\mathbf{t}\,\mathrm{d}a \tag{4.11c}$$

for *arbitrary* material volume $\mathcal{P}$ in the present configuration $\mathcal{R}$ of body $\mathcal{B}$, and *all* time t. To perform the integrations in (4.11a)–(4.11c) over areas and volumes in the present configuration, the functions ρ, $\mathbf{v}$, $\mathbf{b}$, and $\mathbf{t}$ must be in their Eulerian forms, i.e., functions of the independent variables $\mathbf{x}$ and t.

4.4 THE FIRST LAW OF THERMODYNAMICS (CONSERVATION OF ENERGY)

To complete the set of fundamental laws of thermomechanics, we now *postulate* the law of conservation of energy, also known as the first law of thermodynamics. In general:

> *The rate of change of the total energy of any part $\mathcal{S}$ of the body is equal to the rate of mechanical work generated by the resultant external force acting on $\mathcal{S}$ plus the rate of all other energies that enter or leave $\mathcal{S}$ (such as heat energy, chemical energy, or electromagnetic energy).*

In this chapter, we specialize to thermomechanical systems, so the only other energy entering or leaving part $\mathcal{S}$ is heat. (This restriction is later relaxed in Chapter 9 to accommodate electromagnetic sources of energy.) It follows, then, that the law of conservation of energy in *material form* is

$$\frac{\mathrm{d}}{\mathrm{d}t} T(\mathcal{S},t) = R(\mathcal{S},t) + H(\mathcal{S},t) \tag{4.12}$$

for *arbitrary* subset $\mathcal{S}$ of body $\mathcal{B}$ and *all* time t, where

$T = T(\mathcal{S}, t)$, the **total energy** of part $\mathcal{S}$ at time t,

$R = R(\mathcal{S}, t)$, the **rate of work** done on part $\mathcal{S}$ at time t by the resultant external force $\mathbf{f}$,

$H = H(\mathcal{S}, t)$, the **rate of heat energy** *entering* part $\mathcal{S}$ at time t.

We emphasize that (4.12) is valid not only for the body as a whole, but also for every subset of the body.

The assumption (4.6) that the resultant external force $\mathbf{f}$ on $\mathcal{S}$ is the sum of a body force $\mathbf{f}_\mathrm{b}$ and a contact force $\mathbf{f}_\mathrm{c}$ implies that

$$R = R_\mathrm{b} + R_\mathrm{c}, \tag{4.13}$$

and the smoothness assumptions on $\mathbf{f}_\mathrm{b}$ and $\mathbf{f}_\mathrm{c}$ (refer to (4.7)) further imply that

$$R_\mathrm{b}(\mathcal{S}, t) = \int_{\mathcal{P}} \mathbf{b} \cdot \mathbf{v}\, \rho \,\mathrm{d}v \qquad \text{(the rate of work of body forces)}, \tag{4.14a}$$

$$R_\mathrm{c}(\mathcal{S}, t) = \int_{\partial\mathcal{P}} \mathbf{t} \cdot \mathbf{v} \,\mathrm{d}a \qquad \text{(the rate of work of contact forces)}, \tag{4.14b}$$

where $\mathcal{P}$ is the region of $\mathcal{E}^3$ occupied by part $\mathcal{S}$ in the present configuration at time t, with boundary $\partial\mathcal{P}$.

Our smoothness assumption $(4.2)_1$ on mass implies that the energy of motion, or **kinetic energy,** of part $\mathcal{S}$ at time t is given by

$$K(\mathcal{S}, t) = \int_{\mathcal{P}} \frac{1}{2} \mathbf{v} \cdot \mathbf{v}\, \rho \,\mathrm{d}v. \tag{4.15}$$

We *assume* the existence of an **internal energy** E of part $\mathcal{S}$ of the body, such that

$$T \text{ (total energy)} = K + E \text{ (kinetic energy } + \text{ internal energy)}, \tag{4.16}$$

and further make the *smoothness assumption*

$$E(\mathcal{S}, t) = \int_{\mathcal{P}} \varepsilon\, \rho \,\mathrm{d}v, \tag{4.17}$$

where

$\varepsilon = \tilde{\varepsilon}(\mathbf{x}, t)$, the **specific internal energy**, i.e., the internal energy per unit mass,

is a bounded, continuous function (so we have excluded point, line, and surface concentrations of internal energy).

We *assume* that the rate of heat energy H entering part $\mathcal{S}$ at time t is composed of two parts, that entering throughout the volume $\mathcal{P}$ and that flowing through the surface $\partial\mathcal{P}$:

$$H(\mathcal{S}, t) = \int_{\mathcal{P}} r \rho \,\mathrm{d}v - \int_{\partial\mathcal{P}} h \,\mathrm{d}a, \tag{4.18}$$

where

$r = \tilde{r}(\mathbf{x}, t)$, the **specific heat supply rate**, i.e., the heat energy absorbed per unit mass per unit time,

and

$h = \tilde{h}(\mathbf{x}, t, \text{geometry of surface})$, the **heat flux rate**, i.e., the heat flow *out of* $\partial\mathcal{P}$ per unit area of the present configuration per unit time.

Physically, (4.18) represents heat transfer due to radiation (first term) and conduction (second term). Again, we make the *smoothness assumption* that r and h are bounded, continuous functions. The minus sign appears in (4.18) because of the sign convention that h is positive when heat flows *out of* $\partial\mathcal{P}$, i.e., through the surface $\partial\mathcal{P}$ in the direction of its outward unit normal $\mathbf{n}$.

Using (4.13)–(4.18), we can express the first law of thermodynamics (4.12) in *Eulerian integral form:*

$$\frac{\mathrm{d}}{\mathrm{d}t}\int_{\mathcal{P}} \frac{1}{2}\mathbf{v}\cdot\mathbf{v}\,\rho\,\mathrm{d}v + \frac{\mathrm{d}}{\mathrm{d}t}\int_{\mathcal{P}} \varepsilon\,\rho\,\mathrm{d}v = \int_{\mathcal{P}} \mathbf{b}\cdot\mathbf{v}\,\rho\,\mathrm{d}v + \int_{\partial\mathcal{P}} \mathbf{t}\cdot\mathbf{v}\,\mathrm{d}a + \int_{\mathcal{P}} r\,\rho\,\mathrm{d}v - \int_{\partial\mathcal{P}} h\,\mathrm{d}a \quad (4.19)$$

for *arbitrary* material volume $\mathcal{P}$ in the present configuration $\mathcal{R}$ of the body for all time t.

4.5 THE TRANSPORT AND LOCALIZATION THEOREMS

In this section, we present the transport theorem and the localization theorem. As will soon be evident, these are essential tools for rigorously deriving local (or pointwise) statements of the integral conservation laws developed in the previous two sections.

4.5.1 THE TRANSPORT THEOREM

The transport theorem allows us to take the time derivative of a volume integral whose region of integration $\mathcal{P}$ changes with time. It is the three-dimensional analog of Leibniz's rule:

$$\frac{\mathrm{d}}{\mathrm{d}t}\int_{a(t)}^{b(t)} f(x,t)\,\mathrm{d}x = \int_{a(t)}^{b(t)} \frac{\partial f(x,t)}{\partial t}\,\mathrm{d}x + f(b(t),t)\,\frac{\mathrm{d}b(t)}{\mathrm{d}t} - f(a(t),t)\,\frac{\mathrm{d}a(t)}{\mathrm{d}t}.$$

Let $\mathcal{S}$ be an arbitrary part (or subset) of the body $\mathcal{B}$ that occupies a region $\mathcal{P}_{\mathrm{R}}$, with closed boundary $\partial\mathcal{P}_{\mathrm{R}}$, in a fixed reference configuration, and occupies region $\mathcal{P}$, with closed boundary $\partial\mathcal{P}$, in the present configuration at time t. Let ϕ be any continuous (in space and time) scalar-, vector-, or tensor-valued function, with the representations

$$\phi = \tilde{\phi}(\mathbf{x},t) = \tilde{\phi}(\chi(\mathbf{X},t),t) = \hat{\phi}(\mathbf{X},t).$$

Then

$$\frac{\mathrm{d}}{\mathrm{d}t}\int_{\mathcal{P}} \tilde{\phi}(\mathbf{x},t)\,\mathrm{d}v = \int_{\mathcal{P}} (\dot{\phi} + \phi\,\mathrm{div}\,\mathbf{v})\,\mathrm{d}v. \tag{4.20}$$

It can be shown that an alternative form of (4.20) is

$$\frac{\mathrm{d}}{\mathrm{d}t}\int_{\mathcal{P}} \tilde{\phi}(\mathbf{x},t)\,\mathrm{d}v = \int_{\mathcal{P}} \phi'\,\mathrm{d}v + \int_{\partial\mathcal{P}} \phi\mathbf{v}\cdot\mathbf{n}\,\mathrm{d}a, \tag{4.21}$$

where $\mathbf{n}$ is the outward unit normal to the surface $\partial\mathcal{P}$. Refer to Section 3.2 for the definitions of $\dot{\phi}$ and ϕ'. (Note that the function $\tilde{\phi}(\mathbf{x},t)$ must be continuous for $\dot{\phi}$ and ϕ' to make sense.)

PROBLEM 4.1

Prove the transport theorem. That is, show that

$$\frac{\mathrm{d}}{\mathrm{d}t}\int_{\mathcal{P}} \tilde{\phi}(\mathbf{x},t)\,\mathrm{d}v = \int_{\mathcal{P}} (\dot{\phi} + \phi\,\mathrm{div}\,\mathbf{v})\,\mathrm{d}v.$$

Solution

We begin by using $\mathrm{d}v = J\,\mathrm{d}V$ (refer to (3.71)) and a change of independent variable from $\mathbf{x}$ to $\mathbf{X}$ to convert the Eulerian integration to a Lagrangian integration:

$$\frac{\mathrm{d}}{\mathrm{d}t}\int_{\mathcal{P}} \tilde{\phi}(\mathbf{x},t)\,\mathrm{d}v = \frac{\mathrm{d}}{\mathrm{d}t}\int_{\mathcal{P}_R} \hat{\phi}(\mathbf{X},t)\,\hat{J}(\mathbf{X},t)\,\mathrm{d}V.$$

Since the region of integration $\mathcal{P}_R$ is fixed and independent of time t, it follows that

$$\frac{\mathrm{d}}{\mathrm{d}t}\int_{\mathcal{P}_R} \hat{\phi}(\mathbf{X},t)\,\hat{J}(\mathbf{X},t)\,\mathrm{d}V = \int_{\mathcal{P}_R} \frac{\partial}{\partial t}\left[\hat{\phi}(\mathbf{X},t)\,\hat{J}(\mathbf{X},t)\right]\mathrm{d}V.$$

The product rule, the definition (3.13) of the material derivative, and result $(3.60)_3$ then imply that

$$\begin{aligned}
\int_{\mathcal{P}_R} \frac{\partial}{\partial t}\left[\hat{\phi}(\mathbf{X},t)\,\hat{J}(\mathbf{X},t)\right]\mathrm{d}V &= \int_{\mathcal{P}_R}\left[\frac{\partial}{\partial t}\hat{\phi}(\mathbf{X},t)\,\hat{J}(\mathbf{X},t) + \hat{\phi}(\mathbf{X},t)\frac{\partial}{\partial t}\hat{J}(\mathbf{X},t)\right]\mathrm{d}V \\
&= \int_{\mathcal{P}_R}\left[\frac{\partial}{\partial t}\hat{\phi}(\mathbf{X},t)\,\hat{J}(\mathbf{X},t) + \hat{\phi}(\mathbf{X},t)\,\dot{J}\right]\mathrm{d}V \\
&= \int_{\mathcal{P}_R}\left[\frac{\partial}{\partial t}\hat{\phi}(\mathbf{X},t)\,\hat{J}(\mathbf{X},t) + \hat{\phi}(\mathbf{X},t)\,\hat{J}(\mathbf{X},t)\,\mathrm{div}\,\mathbf{v}\right]\mathrm{d}V
\end{aligned}$$

$$= \int_{\mathcal{P}_R} \left[\frac{\partial}{\partial t} \hat{\phi}(\mathbf{X}, t) + \hat{\phi}(\mathbf{X}, t) \operatorname{div} \mathbf{v} \right] \hat{J}(\mathbf{X}, t) \, \mathrm{d}V$$

$$= \int_{\mathcal{P}_R} \left[\dot{\phi} + \hat{\phi}(\mathbf{X}, t) \operatorname{div} \mathbf{v} \right] \hat{J}(\mathbf{X}, t) \, \mathrm{d}V.$$

We once again use (3.71) and a change of independent variable from $\mathbf{X}$ to $\mathbf{x}$, this time to convert the Lagrangian integration back to an Eulerian integration:

$$\int_{\mathcal{P}_R} \left[\dot{\phi} + \hat{\phi}(\mathbf{X}, t) \operatorname{div} \mathbf{v} \right] \hat{J}(\mathbf{X}, t) \, \mathrm{d}V = \int_{\mathcal{P}} \left[\dot{\phi} + \tilde{\phi}(\mathbf{x}, t) \operatorname{div} \mathbf{v} \right] \mathrm{d}v.$$

Thus, we conclude that

$$\frac{\mathrm{d}}{\mathrm{d}t} \int_{\mathcal{P}} \tilde{\phi}(\mathbf{x}, t) \, \mathrm{d}v = \int_{\mathcal{P}} (\dot{\phi} + \phi \operatorname{div} \mathbf{v}) \, \mathrm{d}v.$$

EXERCISES

1. Prove Leibniz's rule

$$\frac{\mathrm{d}}{\mathrm{d}t} \int_{a(t)}^{b(t)} \tilde{f}(x, t) \, \mathrm{d}x = \int_{a(t)}^{b(t)} \frac{\partial \tilde{f}(x, t)}{\partial t} \mathrm{d}x + \tilde{f}(b(t), t) \frac{\mathrm{d}b(t)}{\mathrm{d}t} - \tilde{f}(a(t), t) \frac{\mathrm{d}a(t)}{\mathrm{d}t}$$

as the one-dimensional analog of the three-dimensional transport theorem. (Hint: First, define a one-dimensional motion $x = \chi(X, t)$, and define the limits A and B such that $\chi(A, t) = a(t)$ and $\chi(B, t) = b(t)$. Proceed as in the proof of the three-dimensional transport theorem shown in Problem 4.1.)

2. Using the definition of the material derivative and the divergence theorem, show that

$$\int_{\mathcal{P}} (\dot{\phi} + \phi \operatorname{div} \mathbf{v}) \, \mathrm{d}v = \int_{\mathcal{P}} \phi' \, \mathrm{d}v + \int_{\partial \mathcal{P}} \phi \mathbf{v} \cdot \mathbf{n} \, \mathrm{d}a.$$

Then show that this result implies that

$$\frac{\mathrm{d}}{\mathrm{d}t} \int_{\mathcal{P}} \tilde{\phi}(\mathbf{x}, t) \, \mathrm{d}v = \int_{\mathcal{P}} \phi' \, \mathrm{d}v + \int_{\partial \mathcal{P}} \phi \mathbf{v} \cdot \mathbf{n} \, \mathrm{d}a,$$

which is an alternative form of the transport theorem.

4.5.2 THE LOCALIZATION THEOREM

If ϕ is a continuous scalar- or tensor-valued field in $\mathcal{R}$ and

$$\int_{\mathcal{P}} \phi \, \mathrm{d}v = 0 \tag{4.22}$$

for *any* part $\mathcal{P} \subseteq \mathcal{R}$, then it is necessary and sufficient that

$$\phi = 0 \tag{4.23}$$

in $\mathcal{R}$. Said differently, if (4.22) holds for any arbitrary subset of the body, then the integrand ϕ vanishes everywhere throughout the body, and vice versa.

PROBLEM 4.2

Prove the localization theorem.

Solution

That (4.23) implies (4.22) is trivial. To show that (4.22) implies (4.23), we first recall from real analysis the definition of continuity:

A function $\phi(\mathbf{x}, t)$ is continuous in a region $\mathcal{R}$ if for every $\mathbf{x} \in \mathcal{R}$ and every $\epsilon > 0$ there exists a $\delta > 0$ such that

$$|\mathbf{x} - {}_0\mathbf{x}| < \delta \quad \Rightarrow \quad |\phi(\mathbf{x}, t) - \phi({}_0\mathbf{x}, t)| < \epsilon.$$

In what follows, we verify that (4.22) implies (4.23) using this definition of continuity and proof by contrapositive.

Suppose not (4.23), i.e., there is a point ${}_0\mathbf{x} \in \mathcal{R}$ at which $\phi_0 = \phi({}_0\mathbf{x}, t) \neq 0$. Assume first that $\phi_0 > 0$. Since ϕ is continuous on $\mathcal{R}$, we are free to select $\epsilon = \phi_0/2$, and there must exist a $\delta > 0$ such that

$$|\mathbf{x} - {}_0\mathbf{x}| < \delta \quad \Rightarrow \quad |\phi(\mathbf{x}, t) - \phi_0| < \frac{\phi_0}{2}. \tag{a}$$

Let $\mathcal{P}_\delta$ be the region of all $\mathbf{x} \in \mathcal{R}$ such that $|\mathbf{x} - {}_0\mathbf{x}| < \delta$, and let V_δ be the volume of this region, so

$$V_\delta = \int_{\mathcal{P}_\delta} \mathrm{d}v > 0.$$

From (a) it follows that

$$\phi(\mathbf{x}, t) > \frac{\phi_0}{2} \text{ in } \mathcal{P}_\delta \quad \Rightarrow \quad \int_{\mathcal{P}_\delta} \phi \,\mathrm{d}v > \frac{\phi_0}{2} \int_{\mathcal{P}_\delta} \mathrm{d}v = \frac{1}{2}\phi_0 V_\delta > 0. \tag{b}$$

Now assume $\phi_0 < 0$. By continuity of ϕ, there exists a $\delta > 0$ such that

$$|\mathbf{x} - {}_0\mathbf{x}| < \delta \quad \Rightarrow \quad |\phi(\mathbf{x}, t) - \phi_0| < -\frac{\phi_0}{2} \quad \Rightarrow \quad \phi(\mathbf{x}, t) < \frac{\phi_0}{2} \text{ in } \mathcal{P}_\delta,$$

and therefore

$$\int_{\mathcal{P}_\delta} \phi \,\mathrm{d}v < \frac{\phi_0}{2} \int_{\mathcal{P}_\delta} \mathrm{d}v = \frac{1}{2}\phi_0 V_\delta < 0. \tag{c}$$

From (b) and (c) we see that, for either $\phi_0 > 0$ or $\phi_0 < 0$, we can find a $\mathcal{P}_\delta \subseteq \mathcal{R}$ for which (4.22) is not satisfied. Hence, we have shown

$$\text{not } (4.23) \Rightarrow \text{not } (4.22), \quad \text{or equivalently,} \quad (4.22) \Rightarrow (4.23).$$

4.6 CAUCHY STRESS TENSOR, HEAT FLUX VECTOR

Recall from Section 4.2 that the traction $\mathbf{t}$ acting on the surface $\partial\mathcal{P}$ is, in general, a function of position $\mathbf{x}$, time t, and the geometry of the surface, i.e.,

$$\mathbf{t} = \tilde{\mathbf{t}}(\mathbf{x}, t, \text{geometry of surface}). \tag{4.24}$$

The geometry of the surface includes orientation, curvature, and so on. We now restrict our attention to a particular type of contact force, namely, one such that the traction $\mathbf{t}$ is the same for all like-oriented surfaces with a common tangent plane at $\mathbf{x}$ and t. Therefore, the dependence of $\mathbf{t}$ on the geometry of the surface $\partial\mathcal{P}$ is only through the outward unit normal $\mathbf{n}$, so (4.24) becomes

$$\mathbf{t} = \tilde{\mathbf{t}}(\mathbf{x}, t, \mathbf{n}). \tag{4.25}$$

(An example of a traction that is not of the form (4.25) is surface tension, which depends on a measure of the curvature of the surface.)

We further assume that $\mathbf{t}$ is a *continuous* function of $\mathbf{x}$, t, and $\mathbf{n}$. Then, as a consequence of the conservation laws of mass (4.11a) and linear momentum (4.11b), and our assumptions that $\ddot{\mathbf{x}}$, $\mathbf{b}$, and ρ are bounded, it can be shown (refer to Problem 4.3) that the dependence of $\mathbf{t}$ on $\mathbf{n}$ in (4.25) is in fact linear, i.e.,

$$\tilde{\mathbf{t}}(\mathbf{x}, t, \mathbf{n}) = \tilde{\mathbf{T}}(\mathbf{x}, t)\mathbf{n} \qquad \text{or} \qquad \mathbf{t} = \mathbf{T}\mathbf{n}. \tag{4.26}$$

The tensor $\mathbf{T}$ in (4.26) is called the **Cauchy stress tensor.** Note that the components of the Cauchy stress $\mathbf{T}$ are defined by

$$T_{ij} = \mathbf{t}(\mathbf{e}_j) \cdot \mathbf{e}_i, \tag{4.27}$$

where $\mathbf{t}(\mathbf{e}_j)$ is the traction on a surface whose unit normal is $\mathbf{n} = \mathbf{e}_j$. This definition, along with result (4.26), implies that

$$T_{ij} = \mathbf{e}_i \cdot \mathbf{T}\mathbf{e}_j, \tag{4.28}$$

which is consistent with definition (2.29) of the Cartesian components of a tensor.

Recall from Section 4.4 that the heat flux h out of the surface $\partial\mathcal{P}$ is, in general, a function of position $\mathbf{x}$, time t, and the geometry of the surface, i.e.,

$$h = \tilde{h}(\mathbf{x}, t, \text{geometry of surface}). \tag{4.29}$$

We restrict our attention to a particular type of heat flow through the boundary $\partial\mathcal{P}$, namely, one such that the heat flux h is the same for all like-oriented surfaces with a common tangent plane at $\mathbf{x}$ and t, so (4.29) becomes

$$h = \tilde{h}(\mathbf{x}, t, \mathbf{n}). \tag{4.30}$$

We further assume that h is a *continuous* function of $\mathbf{x}$, t, and $\mathbf{n}$. Then, as a consequence of the conservation laws of mass (4.11a), linear momentum (4.11b), and energy (4.19), and our boundedness assumptions, it can be shown (refer to Problem 4.4) that the dependence of h on $\mathbf{n}$ in (4.30) must be linear, i.e.,

$$\tilde{h}(\mathbf{x}, t, \mathbf{n}) = \tilde{\mathbf{q}}(\mathbf{x}, t) \cdot \mathbf{n} \qquad \text{or} \qquad h = \mathbf{q} \cdot \mathbf{n}. \tag{4.31}$$

The vector $\mathbf{q}$ in (4.31) is called the **heat flux vector.**

PROBLEM 4.3

Prove that $\mathbf{t} = \mathbf{Tn}$.

Solution

Consider an arbitrary part $\mathcal{S}$ of body $\mathcal{B}$ that occupies a region $\mathcal{P}$ in the present configuration at time t. Let $\mathcal{P}$ be divided into two regions, $\mathcal{P}_1$ and $\mathcal{P}_2$, separated by a surface σ (see Figure 4.1). Define $\partial\mathcal{P}'$ and $\partial\mathcal{P}''$ so that $\partial\mathcal{P} = \partial\mathcal{P}' \cup \partial\mathcal{P}''$. Then, $\mathcal{P} = \mathcal{P}_1 \cup \mathcal{P}_2$, $\partial\mathcal{P} = \partial\mathcal{P}' \cup \partial\mathcal{P}''$, $\partial\mathcal{P}_1 = \partial\mathcal{P}' \cup \sigma$, $\partial\mathcal{P}_2 = \partial\mathcal{P}'' \cup \sigma$.

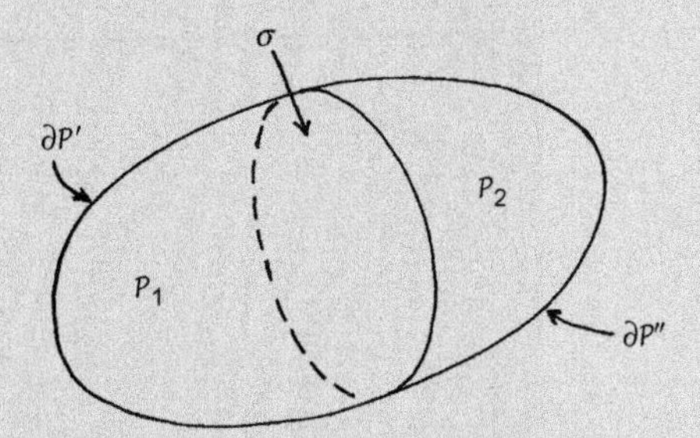

FIGURE 4.1

Schematic illustrating the division of $\mathcal{P}$ into two regions, $\mathcal{P}_1$ and $\mathcal{P}_2$, separated by a surface σ.

Applying balance of linear momentum to the three regions $\mathcal{P}$, $\mathcal{P}_1$, and $\mathcal{P}_2$, we obtain

$$\frac{\mathrm{d}}{\mathrm{d}t}\int_{\mathcal{P}} \dot{\mathbf{x}}\rho \,\mathrm{d}v = \int_{\mathcal{P}} \mathbf{b}\rho \,\mathrm{d}v + \int_{\partial\mathcal{P}} \mathbf{t}(\mathbf{n}) \,\mathrm{d}a, \tag{a}$$

$$\frac{\mathrm{d}}{\mathrm{d}t}\int_{\mathcal{P}_1} \dot{\mathbf{x}}\rho \,\mathrm{d}v = \int_{\mathcal{P}_1} \mathbf{b}\rho \,\mathrm{d}v + \int_{\partial\mathcal{P}'\cup\sigma} \mathbf{t}(\mathbf{n}) \,\mathrm{d}a, \tag{b}$$

$$\frac{\mathrm{d}}{\mathrm{d}t}\int_{\mathcal{P}_2} \dot{\mathbf{x}}\rho \,\mathrm{d}v = \int_{\mathcal{P}_2} \mathbf{b}\rho \mathrm{d}v + \int_{\partial\mathcal{P}''\cup\sigma} \mathbf{t}(\mathbf{n}) \,\mathrm{d}a. \tag{c}$$

Adding (b) and (c) then subtracting (a) gives

$$\int_{\sigma} [\mathbf{t}(\mathbf{n}) + \mathbf{t}(-\mathbf{n})] \,\mathrm{d}a = \mathbf{0}. \tag{d}$$

In (d) we have noted that the outward normal of σ, when considered as a part of the boundary of $\mathcal{P}_1$, is directly opposed to the outward normal of σ when considered as a part of the boundary of $\mathcal{P}_2$. Assuming the traction is a continuous function of $\mathbf{x}$ and $\mathbf{n}$, so that we can use the two-dimensional localization theorem, (d) implies

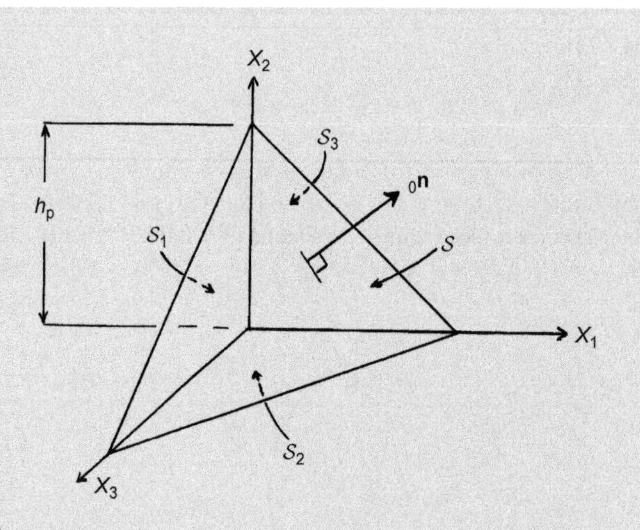

FIGURE 4.2

A tetrahedron of height h_p. Sides $\mathcal{S}_1$, $\mathcal{S}_2$, and $\mathcal{S}_3$ have outward unit normals $-\mathbf{e}_1$, $-\mathbf{e}_2$, and $-\mathbf{e}_3$, respectively, while side $\mathcal{S}$ has outward unit normal $_o\mathbf{n}$.

$$\mathbf{t}(\mathbf{n}) = -\mathbf{t}(-\mathbf{n}). \tag{e}$$

(Traction vectors on opposite sides of the same surface are equal in magnitude, and opposite in direction.)

Consider now a family of similar tetrahedra $\mathcal{T}$ with heights h_p and a common vertex at some point $_o\mathbf{x}$ (see Figure 4.2). Sides i are perpendicular to the x_i directions and have outward normals $-\mathbf{e}_i$. The remaining side has outward normal $_o\mathbf{n}$. From geometry, areas $\mathcal{S}$, $\mathcal{S}_1$, $\mathcal{S}_2$, and $\mathcal{S}_3$ are related by

$$\mathcal{S}_i = \mathcal{S}\,({}_o\mathbf{n} \cdot \mathbf{e}_i) = \mathcal{S}\,n_i. \tag{f}$$

The volumes of the tetrahedra are

$$V = \frac{1}{3}\, h_p\, \mathcal{S}. \tag{g}$$

It is assumed that each of the tetrahedra lies completely within the region $\mathcal{R}$ occupied by $\mathcal{B}$ in the present configuration. By virtue of the transport theorem (4.20) and the *local form* of conservation of mass (presented later in Section 4.8), (a) becomes

$$\int_{\mathcal{P}} \ddot{\mathbf{x}}\rho\, \mathrm{d}v = \int_{\mathcal{P}} \mathbf{b}\rho\, \mathrm{d}v + \int_{\partial\mathcal{P}} \mathbf{t}(\mathbf{n})\, \mathrm{d}a. \tag{h}$$

Applying (h) to a tetrahedron $\mathcal{T}$ gives

$$\int_{\mathcal{T}} (\ddot{\mathbf{x}} - \mathbf{b})\,\rho\,\mathrm{d}v = \int_{\Sigma \mathcal{S}_i} \mathbf{t}(-\mathbf{e}_i)\,\mathrm{d}a + \int_{\mathcal{S}} \mathbf{t}({}_{\mathrm{o}}\mathbf{n})\,\mathrm{d}a. \tag{i}$$

With use of (e), (i) becomes

$$\int_{\mathcal{T}} (\ddot{\mathbf{x}} - \mathbf{b})\,\rho\,\mathrm{d}v = \int_{\mathcal{S}} \mathbf{t}({}_{\mathrm{o}}\mathbf{n})\,\mathrm{d}a - \int_{\Sigma \mathcal{S}_i} \mathbf{t}(\mathbf{e}_i)\,\mathrm{d}a. \tag{j}$$

Recall that the specific body force $\mathbf{b}$ and density ρ are assumed bounded throughout $\mathcal{R}$. The acceleration $\ddot{\mathbf{x}}$ is also assumed bounded. Then,

$$\begin{aligned}
\left| \int_{\mathcal{T}} (\ddot{\mathbf{x}} - \mathbf{b})\,\rho\,\mathrm{d}v \right| &\le \int_{\mathcal{T}} \left|(\ddot{\mathbf{x}} - \mathbf{b})\right| \rho\,\mathrm{d}v && \text{(by a theorem of analysis)} \\
&= \int_{\mathcal{T}} K\,\mathrm{d}v && \text{(by boundedness)} \\
&= K \int_{\mathcal{T}} \mathrm{d}v \\
&= K\,\frac{\mathcal{S} h_{\mathrm{p}}}{3} && \text{(using (g)),} \qquad \text{(k)}
\end{aligned}$$

where K is some number. Assuming that $\mathbf{t}$ is a continuous function of $\mathbf{x}$ and t, the mean value theorem and (f) imply that

$$\int_{\mathcal{S}} \mathbf{t}({}_{\mathrm{o}}\mathbf{n})\,\mathrm{d}a - \int_{\Sigma \mathcal{S}_i} \mathbf{t}(\mathbf{e}_i)\,\mathrm{d}a \;=\; \mathbf{t}^*({}_{\mathrm{o}}\mathbf{n})\,\mathcal{S} - \mathbf{t}^*(\mathbf{e}_i)\,\mathcal{S}_i \;=\; \left[\mathbf{t}^*({}_{\mathrm{o}}\mathbf{n}) - \mathbf{t}^*(\mathbf{e}_i)\,n_i\right]\mathcal{S}, \tag{l}$$

where $\mathbf{t}^*({}_{\mathrm{o}}\mathbf{n})$ and $\mathbf{t}^*(\mathbf{e}_i)$ stand for some specific interior values of the traction vectors on the faces $\mathcal{S}$ and $\mathcal{S}_i$, respectively. From (j) to (l),

$$\begin{aligned}
\frac{1}{3} K \mathcal{S} h_{\mathrm{p}} &\ge \left| \int_{\mathcal{T}} (\ddot{\mathbf{x}} - \mathbf{b})\,\rho\,\mathrm{d}v \right| \\
&= \left| \int_{\mathcal{S}} \mathbf{t}({}_{\mathrm{o}}\mathbf{n})\,\mathrm{d}a - \int_{\Sigma \mathcal{S}_i} \mathbf{t}(\mathbf{e}_i)\,\mathrm{d}a \right| \\
&= \left| \mathbf{t}^*({}_{\mathrm{o}}\mathbf{n}) - \mathbf{t}^*(\mathbf{e}_i)\,n_i \right| \mathcal{S},
\end{aligned}$$

so

$$\left| \mathbf{t}^*({}_{\mathrm{o}}\mathbf{n}) - \mathbf{t}^*(\mathbf{e}_i)\,n_i \right| \le \frac{1}{3} K h_{\mathrm{p}}.$$

As we consider smaller and smaller tetrahedra, $h_{\text{p}} \to 0$, and

$$\lim_{h_{\text{p}} \to 0} \left| \mathbf{t}^*({}_{\mathbf{o}}\mathbf{n}) - \mathbf{t}^*(\mathbf{e}_i)\, n_i \right| = 0, \tag{m}$$

where $\mathbf{t}^*({}_{\mathbf{o}}\mathbf{n})$ and $\mathbf{t}^*(\mathbf{e}_i)$ are evaluated at the point ${}_{\mathbf{o}}\mathbf{x}$, which is the common vertex of the family of tetrahedra. From (m), we see that in the limit $h_{\text{p}} \to 0$, $\mathbf{t}^*({}_{\mathbf{o}}\mathbf{n}) - \mathbf{t}^*(\mathbf{e}_i)\, n_i$ must be the zero vector, i.e.,

$$\mathbf{t}^*({}_{\mathbf{o}}\mathbf{n}) = \mathbf{t}^*(\mathbf{e}_i)_{\mathbf{o}}\, n_i, \tag{n}$$

where $\mathbf{t}^*(\mathbf{e}_i)_{\mathbf{o}}$ denotes the value of $\mathbf{t}^*(\mathbf{e}_i)$ at the point ${}_{\mathbf{o}}\mathbf{x}$. Since (n) must hold at any point ${}_{\mathbf{o}}\mathbf{x}$ and corresponding to any direction ${}_{\mathbf{o}}\mathbf{n}$, without ambiguity, we may delete the star, replace ${}_{\mathbf{o}}\mathbf{x}$ and ${}_{\mathbf{o}}\mathbf{n}$ by $\mathbf{x}$ and $\mathbf{n}$, and write

$$\mathbf{t}(\mathbf{n}) = \mathbf{t}(\mathbf{e}_i)\, n_i. \tag{o}$$

We define T_{ki} by

$$T_{ki} = \mathbf{t}(\mathbf{e}_i) \cdot \mathbf{e}_k,$$

and denote the components of $\mathbf{t}(\mathbf{n})$ by

$$t_k = \mathbf{t} \cdot \mathbf{e}_k.$$

Taking the inner product of (o) with the unit vector $\mathbf{e}_k$ thus gives

$$t_k = \mathbf{t}(\mathbf{e}_i)\, n_i \cdot \mathbf{e}_k = T_{ki}\, n_i. \tag{p}$$

Equation (p) is the Cartesian component form of

$$\mathbf{t} = \mathbf{T}\mathbf{n} \qquad \text{or} \qquad \mathbf{t}(\mathbf{x}, t, \mathbf{n}) = \mathbf{T}(\mathbf{x}, t)\, \mathbf{n}.$$

PROBLEM 4.4

Show that $h = \mathbf{q} \cdot \mathbf{n}$.

Solution

Consider once again the regions $\mathcal{P}$, $\mathcal{P}_1$, and $\mathcal{P}_2$ shown in Figure 4.1. Applying the first law of thermodynamics to these three regions, we have

$$\frac{\mathrm{d}}{\mathrm{d}t} \int_{\mathcal{P}} \left(\frac{\dot{\mathbf{x}} \cdot \dot{\mathbf{x}}}{2} + \varepsilon \right) \rho\, \mathrm{d}v = \int_{\mathcal{P}} (\mathbf{b} \cdot \mathbf{v} + r)\, \rho\, \mathrm{d}v + \int_{\partial \mathcal{P}} [\mathbf{t}(\mathbf{n}) \cdot \mathbf{v} - h(\mathbf{n})]\, \mathrm{d}a, \tag{a}$$

$$\frac{\mathrm{d}}{\mathrm{d}t} \int_{\mathcal{P}_1} \left(\frac{\dot{\mathbf{x}} \cdot \dot{\mathbf{x}}}{2} + \varepsilon \right) \rho\, \mathrm{d}v = \int_{\mathcal{P}_1} (\mathbf{b} \cdot \mathbf{v} + r)\, \rho\, \mathrm{d}v + \int_{\partial \mathcal{P}' \cup \sigma} [\mathbf{t}(\mathbf{n}) \cdot \mathbf{v} - h(\mathbf{n})]\, \mathrm{d}a, \tag{b}$$

$$\frac{\mathrm{d}}{\mathrm{d}t} \int_{\mathcal{P}_2} \left(\frac{\dot{\mathbf{x}} \cdot \dot{\mathbf{x}}}{2} + \varepsilon \right) \rho\, \mathrm{d}v = \int_{\mathcal{P}_2} (\mathbf{b} \cdot \mathbf{v} + r)\, \rho\, \mathrm{d}v + \int_{\partial \mathcal{P}'' \cup \sigma} [\mathbf{t}(\mathbf{n}) \cdot \mathbf{v} - h(\mathbf{n})]\, \mathrm{d}a. \tag{c}$$

Adding (b) and (c) then subtracting (a) gives

$$\int_{\sigma} [\mathbf{t}(\mathbf{n}) + \mathbf{t}(-\mathbf{n})] \cdot \mathbf{v} \, da - \int_{\sigma} [h(\mathbf{n}) + h(-\mathbf{n})] \, da = 0. \tag{d}$$

In (d) we have noted that the outward normal of σ when considered as a part of $\partial\mathcal{P}_1$ is directly opposed to the outward normal of σ when considered as a part of $\partial\mathcal{P}_2$. The result

$$\mathbf{t}(\mathbf{n}) = -\mathbf{t}(-\mathbf{n})$$

from Problem 4.3 implies that (d) becomes

$$\int_{\sigma} [h(\mathbf{n}) + h(-\mathbf{n})] \, da = 0.$$

Assuming continuity of h, it then follows that

$$h(\mathbf{n}) = -h(-\mathbf{n}). \tag{e}$$

(The heat leaving one part equals the heat entering the other part.)

By virtue of the transport theorem (4.20) and the *local form* of conservation of mass (presented later in Section 4.8), (a) reduces to

$$\int_{\mathcal{P}} (\mathbf{v} \cdot \dot{\mathbf{v}} + \dot{\varepsilon}) \rho \, dv = \int_{\mathcal{P}} (\mathbf{b} \cdot \mathbf{v} + r) \rho \, dv + \int_{\partial\mathcal{P}} [\mathbf{t}(\mathbf{n}) \cdot \mathbf{v} - h(\mathbf{n})] \, da.$$

Applying this to the tetrahedron $\mathcal{T}$ in Figure 4.2 gives

$$\int_{\mathcal{T}} (\mathbf{v} \cdot \dot{\mathbf{v}} + \dot{\varepsilon} - \mathbf{b} \cdot \mathbf{v} - r) \rho \, dv = \int_{\Sigma \mathcal{S}_i} [\mathbf{t}(-\mathbf{e}_i) \cdot \mathbf{v} - h(-\mathbf{e}_i)] \, da$$

$$+ \int_{\mathcal{S}} [\mathbf{t}({}_o\mathbf{n}) \cdot \mathbf{v} - h({}_o\mathbf{n})] \, da,$$

or, using (e),

$$\int_{\mathcal{T}} (\mathbf{v} \cdot \dot{\mathbf{v}} + \dot{\varepsilon} - \mathbf{b} \cdot \mathbf{v} - r) \rho \, dv = \int_{\Sigma \mathcal{S}_i} [-\mathbf{t}(\mathbf{e}_i) \cdot \mathbf{v} + h(\mathbf{e}_i)] \, da$$

$$+ \int_{\mathcal{S}} [\mathbf{t}({}_o\mathbf{n}) \cdot \mathbf{v} - h({}_o\mathbf{n})] \, da. \tag{f}$$

We assume that ε, $\dot{\varepsilon}$, $\mathbf{v}$, $\dot{\mathbf{v}}$, and r are bounded, so

$$\left| \int_{\mathcal{T}} (\mathbf{v} \cdot \dot{\mathbf{v}} + \dot{\varepsilon} - \mathbf{b} \cdot \mathbf{v} - r) \rho \, dv \right| \leq M \frac{\mathcal{S} h_p}{3} \tag{g}$$

for some number M. Therefore, from (f), (g), and the mean value theorem, we have

$$\frac{1}{3}MSh_p \geq \left| \int_{\Sigma S_i} [-\mathbf{t}(\mathbf{e}_i)\cdot\mathbf{v} + h(\mathbf{e}_i)]\,da + \int_{S} [\mathbf{t}({}_o\mathbf{n})\cdot\mathbf{v} - h({}_o\mathbf{n})]\,da \right|$$

$$= \left| [\mathbf{t}^*({}_o\mathbf{n})\cdot\mathbf{v}^* - h^*({}_o\mathbf{n})]S + [-\mathbf{t}^*(\mathbf{e}_i)\cdot\mathbf{v}^{*(i)} + h^*(\mathbf{e}_i)]S_i \right|$$

$$= \left| \mathbf{t}^*({}_o\mathbf{n})\cdot\mathbf{v}^* - \mathbf{t}^*(\mathbf{e}_i)\,n_i\cdot\mathbf{v}^{*(i)} - h^*({}_o\mathbf{n}) + h^*(\mathbf{e}_i)\,n_i \right| S,$$

where $\mathbf{t}^*({}_o\mathbf{n})$, $\mathbf{v}^*$, and $h^*({}_o\mathbf{n})$ stand for some specific interior values on face S, and $\mathbf{t}^*(\mathbf{e}_i)$, $\mathbf{v}^{*(i)}$, and $h^*(\mathbf{e}_i)$ stand for some specific values on faces S_i. Then

$$\lim_{h_p\to 0} \left| \mathbf{t}^*({}_o\mathbf{n})\cdot\mathbf{v}^* - \mathbf{t}^*(\mathbf{e}_i)\,n_i\cdot\mathbf{v}^{*(i)} - h^*({}_o\mathbf{n}) + h^*(\mathbf{e}_i)\,n_i \right| = 0,$$

so in the limit,

$$\mathbf{t}^*({}_o\mathbf{n})\cdot\mathbf{v}^* - \mathbf{t}^*(\mathbf{e}_i)\,n_i\cdot\mathbf{v}^{*(i)} - h^*({}_o\mathbf{n}) + h^*(\mathbf{e}_i)\,n_i = 0. \tag{h}$$

In the limit, $\mathbf{v}^* = \mathbf{v}^{*(i)} = \mathbf{v}({}_o\mathbf{x}, t)$, and so (h) becomes

$$[\mathbf{t}^*({}_o\mathbf{n}) - \mathbf{t}^*(\mathbf{e}_i)\,n_i]\cdot\mathbf{v}^* - h^*({}_o\mathbf{n}) + h^*(\mathbf{e}_i)\,n_i = 0. \tag{i}$$

Recall from Problem 4.3 that

$$\mathbf{t}^*({}_o\mathbf{n}) = \mathbf{t}^*(\mathbf{e}_i)_o\,n_i.$$

As a consequence, (i) becomes

$$h^*({}_o\mathbf{n}) = h^*(\mathbf{e}_i)\,n_i \quad \text{or} \quad h(\mathbf{n}) = h(\mathbf{e}_i)\,n_i. \tag{j}$$

We define q_i by

$$q_i = h(\mathbf{e}_i)$$

so that (j) is written as

$$h(\mathbf{n}) = q_i\,n_i. \tag{k}$$

Equation (k) is the Cartesian component form of

$$h = \mathbf{q}\cdot\mathbf{n} \quad \text{or} \quad h(\mathbf{x}, t, \mathbf{n}) = \mathbf{q}(\mathbf{x}, t)\cdot\mathbf{n}.$$

4.7 THE ENERGY THEOREM AND STRESS POWER

Recall from Section 4.3 the Eulerian representations of the mechanical integral conservation laws:

$$\frac{d}{dt}\int_{\mathcal{P}} \rho\,dv = 0, \tag{4.32a}$$

$$\frac{d}{dt}\int_{\mathcal{P}} \mathbf{v}\rho \, dv = \int_{\mathcal{P}} \mathbf{b}\rho \, dv + \int_{\partial\mathcal{P}} \mathbf{Tn} \, da, \tag{4.32b}$$

$$\frac{d}{dt}\int_{\mathcal{P}} \mathbf{x}\times\mathbf{v}\rho \, dv = \int_{\mathcal{P}} \mathbf{x}\times\mathbf{b}\rho \, dv + \int_{\partial\mathcal{P}} \mathbf{x}\times\mathbf{Tn} \, da, \tag{4.32c}$$

where we have used $\mathbf{t} = \mathbf{Tn}$ (result (4.26)). As a consequence of these mechanical conservation laws, it can be shown (refer to Problem 4.7) that the rate of work of all external forces acting on $\mathcal{P}$ and its boundary $\partial\mathcal{P}$ minus the rate of increase of the kinetic energy is equal to the **total stress power** of $\mathcal{P}$:

$$\int_{\mathcal{P}} \mathbf{b}\cdot\mathbf{v}\rho \, dv + \int_{\partial\mathcal{P}} \mathbf{t}\cdot\mathbf{v} \, da - \frac{d}{dt}\int_{\mathcal{P}} \frac{1}{2}\mathbf{v}\cdot\mathbf{v}\rho \, dv = \int_{\mathcal{P}} \mathbf{T}\cdot\mathbf{D} \, dv. \tag{4.33}$$

This result, called the **energy theorem**, is a mechanical result: it follows only from the mechanical laws (4.32a)–(4.32c) and *does not depend on conservation of energy*. The scalar quantity $\mathbf{T}\cdot\mathbf{D}$ in (4.33) is called the **stress power** P, i.e.,

$$\mathrm{P} = \mathbf{T}\cdot\mathbf{D}. \tag{4.34}$$

4.8 LOCAL FORMS OF THE CONSERVATION LAWS

We now obtain the **local**, or **pointwise**, forms of the Eulerian representations of the thermomechanical integral conservation laws developed in Sections 4.3 and 4.4:

$$\frac{d}{dt}\int_{\mathcal{P}} \rho \, dv = 0, \tag{4.35a}$$

$$\frac{d}{dt}\int_{\mathcal{P}} \mathbf{v}\rho \, dv = \int_{\mathcal{P}} \mathbf{b}\rho \, dv + \int_{\partial\mathcal{P}} \mathbf{Tn} \, da, \tag{4.35b}$$

$$\frac{d}{dt}\int_{\mathcal{P}} \mathbf{x}\times\mathbf{v}\rho \, dv = \int_{\mathcal{P}} \mathbf{x}\times\mathbf{b}\rho \, dv + \int_{\partial\mathcal{P}} \mathbf{x}\times\mathbf{Tn} \, da, \tag{4.35c}$$

$$\frac{d}{dt}\int_{\mathcal{P}} \left(\frac{1}{2}\mathbf{v}\cdot\mathbf{v} + \varepsilon\right)\rho \, dv = \int_{\mathcal{P}} (\mathbf{b}\cdot\mathbf{v} + r)\rho \, dv + \int_{\partial\mathcal{P}} (\mathbf{Tn}\cdot\mathbf{v} - \mathbf{q}\cdot\mathbf{n}) \, da, \tag{4.35d}$$

where we have used $\mathbf{t} = \mathbf{Tn}$ and $h = \mathbf{q}\cdot\mathbf{n}$ (results (4.26) and (4.31)).

We have made suitable smoothness assumptions on all of the terms in the integrands of (4.35a)–(4.35d) to allow us to convert the area integrals to volume integrals using the divergence theorem (2.104), and to take the time derivatives inside the volume integrals using the transport theorem (4.20). By combining the volume integrals, each of (4.35a)–(4.35d) is put in the form

$$\int_{\mathcal{P}} \phi \, dv = 0.$$

Since each of these integrals holds for arbitrary part $\mathcal{P}$ of the present configuration and we have made suitable continuity assumptions, use of the localization theorem from Section 4.5.2 yields

$$\dot{\rho} + \rho \operatorname{div} \mathbf{v} = 0, \tag{4.36a}$$

$$\operatorname{div} \mathbf{T} + \rho \mathbf{b} = \rho \dot{\mathbf{v}}, \tag{4.36b}$$

$$\mathbf{T} = \mathbf{T}^{\mathrm{T}}, \tag{4.36c}$$

$$\rho \dot{\varepsilon} = \mathbf{T} \cdot \mathbf{D} + \rho r - \operatorname{div} \mathbf{q}; \tag{4.36d}$$

refer to Problems 4.5 and 4.6. Equations (4.36a)–(4.36d) are the local Eulerian forms of the conservation laws for mass, linear momentum, angular momentum, and energy, respectively. Note that div denotes the divergence calculated with respect to the present configuration (refer to Section 3.2).

PROBLEM 4.5

Starting with the Eulerian (or spatial) statement of balance of linear momentum in integral form,

$$\frac{\mathrm{d}}{\mathrm{d}t} \int_{\mathcal{P}} \rho \mathbf{v} \, \mathrm{d}v = \int_{\mathcal{P}} \rho \mathbf{b} \, \mathrm{d}v + \int_{\partial \mathcal{P}} \mathbf{t} \, \mathrm{d}a,$$

derive the corresponding pointwise form

$$\rho \dot{\mathbf{v}} = \rho \mathbf{b} + \operatorname{div} \mathbf{T}.$$

Solution

We begin with the Eulerian integral form of balance of linear momentum, i.e.,

$$\frac{\mathrm{d}}{\mathrm{d}t} \int_{\mathcal{P}} \rho \mathbf{v} \, \mathrm{d}v = \int_{\mathcal{P}} \rho \mathbf{b} \, \mathrm{d}v + \int_{\partial \mathcal{P}} \mathbf{t} \, \mathrm{d}a.$$

We first consider the left-hand side of this integral equation:

$$\frac{\mathrm{d}}{\mathrm{d}t} \int_{\mathcal{P}} \rho \mathbf{v} \, \mathrm{d}v = \int_{\mathcal{P}} \left(\dot{\overline{\rho \mathbf{v}}} + \rho \mathbf{v} \operatorname{div} \mathbf{v} \right) \mathrm{d}v \qquad \text{(transport theorem (4.20))}$$

$$= \int_{\mathcal{P}} \left[\rho \dot{\mathbf{v}} + \left(\dot{\rho} + \rho \operatorname{div} \mathbf{v} \right) \mathbf{v} \right] \mathrm{d}v \qquad \text{(product rule (3.22)}_1\text{)}.$$

Note that div denotes the divergence calculated with respect to the present configuration. It then follows from the local form of conservation of mass, $\dot{\rho} + \rho \operatorname{div} \mathbf{v} = 0$, that

$$\frac{\mathrm{d}}{\mathrm{d}t}\int_{\mathcal{P}} \rho\mathbf{v}\,\mathrm{d}v = \int_{\mathcal{P}} \rho\dot{\mathbf{v}}\,\mathrm{d}v.$$

We now consider the second term on the right-hand side of the integral form of balance of linear momentum:

$$\begin{aligned}\int_{\partial\mathcal{P}} \mathbf{t}\,\mathrm{d}a &= \int_{\partial\mathcal{P}} \mathbf{Tn}\,\mathrm{d}a && \text{(result (4.26))}\\ &= \int_{\mathcal{P}} \operatorname{div}\mathbf{T}\,\mathrm{d}v && \text{(divergence theorem } (2.104)_3).\end{aligned}$$

Using these results in the original equation, we arrive at

$$\int_{\mathcal{P}} \left(\rho\dot{\mathbf{v}} - \rho\mathbf{b} - \operatorname{div}\mathbf{T}\right)\mathrm{d}v = \mathbf{0}.$$

Note that the integrand is continuous, and $\mathcal{P}$ is arbitrary. Subsequent application of the localization theorem from Section 4.5.2 gives

$$\rho\dot{\mathbf{v}} - \rho\mathbf{b} - \operatorname{div}\mathbf{T} = \mathbf{0},$$

or

$$\rho\dot{\mathbf{v}} = \rho\mathbf{b} + \operatorname{div}\mathbf{T}.$$

PROBLEM 4.6

Starting with the Eulerian (or spatial) statement of balance of angular momentum in integral form,

$$\frac{\mathrm{d}}{\mathrm{d}t}\int_{\mathcal{P}} \mathbf{x}\times\rho\mathbf{v}\,\mathrm{d}v = \int_{\mathcal{P}} \mathbf{x}\times\rho\mathbf{b}\,\mathrm{d}v + \int_{\partial\mathcal{P}} \mathbf{x}\times\mathbf{t}\,\mathrm{d}a,$$

derive the corresponding pointwise form

$$\mathbf{T} = \mathbf{T}^{\mathrm{T}}.$$

Solution

We begin with the Eulerian integral form of balance of angular momentum, i.e.,

$$\frac{\mathrm{d}}{\mathrm{d}t}\int_{\mathcal{P}} \mathbf{x}\times\rho\mathbf{v}\,\mathrm{d}v = \int_{\mathcal{P}} \mathbf{x}\times\rho\mathbf{b}\,\mathrm{d}v + \int_{\partial\mathcal{P}} \mathbf{x}\times\mathbf{t}\,\mathrm{d}a.$$

We first consider the left-hand side of this integral equation:

$$\frac{\mathrm{d}}{\mathrm{d}t}\int_{\mathcal{P}} \mathbf{x}\times\rho\mathbf{v}\,\mathrm{d}v = \int_{\mathcal{P}} \left[\overline{\mathbf{x}\times\rho\mathbf{v}}^{\cdot} + (\mathbf{x}\times\rho\mathbf{v})\,\mathrm{div}\,\mathbf{v}\right]\mathrm{d}v \qquad \text{(transport theorem (4.20))}$$

$$= \int_{\mathcal{P}} \left[\mathbf{x}\times\rho\dot{\mathbf{v}} + \rho(\mathbf{v}\times\mathbf{v}) + \mathbf{x}\times\left(\dot{\rho} + \rho\,\mathrm{div}\,\mathbf{v}\right)\mathbf{v}\right]\mathrm{d}v \text{ (product rule (3.22))}$$

$$= \int_{\mathcal{P}} \mathbf{x}\times\rho\dot{\mathbf{v}}\,\mathrm{d}v \qquad \text{(conservation of mass (4.36a)).}$$

Next, we consider the second term on the right-hand side of the integral form of balance of angular momentum:

$$\int_{\partial\mathcal{P}} \mathbf{x}\times\mathbf{t}\,\mathrm{d}a = \int_{\partial\mathcal{P}} \mathbf{x}\times(\mathbf{Tn})\,\mathrm{d}a \qquad \text{(result (4.26))}$$

$$= \int_{\mathcal{P}} (\mathbf{x}\times\mathrm{div}\,\mathbf{T} + \boldsymbol{\tau})\,\mathrm{d}v \qquad \text{(tensor calculus result).}$$

Note that

$$\left(\mathbf{T}-\mathbf{T}^{\mathrm{T}}\right)\mathbf{a} = \boldsymbol{\tau}\times\mathbf{a}$$

for any vector $\mathbf{a}$ in $\mathcal{E}^3$. In other words, $\boldsymbol{\tau}$ is the axial vector corresponding to the skew tensor $\mathbf{T}-\mathbf{T}^{\mathrm{T}}$ (refer to (2.65)).

Using these results in the original equation, we arrive at

$$\int_{\mathcal{P}} \left[\mathbf{x}\times\left(\rho\mathbf{b} + \mathrm{div}\,\mathbf{T} - \rho\dot{\mathbf{v}}\right) + \boldsymbol{\tau}\right]\mathrm{d}v = \mathbf{0}.$$

By virtue of the local form of balance of linear momentum (4.36b), this reduces to

$$\int_{\mathcal{P}} \boldsymbol{\tau}\,\mathrm{d}v = \mathbf{0}.$$

Since the integrand is continuous and $\mathcal{P}$ is arbitrary, the localization theorem implies that

$$\boldsymbol{\tau} = \mathbf{0}.$$

It then follows that

$$\boldsymbol{\tau}\times\mathbf{a} = \mathbf{0}$$

for any vector $\mathbf{a}$ in $\mathcal{E}^3$. Since $\boldsymbol{\tau}$ is the axial vector corresponding to the skew tensor $\mathbf{T}-\mathbf{T}^{\mathrm{T}}$, this becomes

$$\left(\mathbf{T}-\mathbf{T}^{\mathrm{T}}\right)\mathbf{a} = \mathbf{0}.$$

Since $\mathbf{a}$ is arbitrary, and $\mathbf{T}-\mathbf{T}^{\mathrm{T}}$ is independent of $\mathbf{a}$, we conclude that

$$\mathbf{T}-\mathbf{T}^{\mathrm{T}} = \mathbf{0} \quad \text{or} \quad \mathbf{T} = \mathbf{T}^{\mathrm{T}}.$$

PROBLEM 4.7

Verify that the Eulerian (or spatial) form of the energy theorem

$$\int_{\mathcal{P}} \mathbf{b} \cdot \mathbf{v}\rho \, \mathrm{d}v + \int_{\partial\mathcal{P}} \mathbf{t} \cdot \mathbf{v} \, \mathrm{d}a - \frac{\mathrm{d}}{\mathrm{d}t} \int_{\mathcal{P}} \frac{1}{2} \mathbf{v} \cdot \mathbf{v}\rho \, \mathrm{d}v = \int_{\mathcal{P}} \mathbf{T} \cdot \mathbf{D} \, \mathrm{d}v$$

is a consequence of the mechanical conservation laws of mass, linear momentum, and angular momentum.

Solution

To begin, we consider the second term on the left-hand side of the energy theorem:

$$\begin{aligned}
\int_{\partial\mathcal{P}} \mathbf{t} \cdot \mathbf{v} \, \mathrm{d}a &= \int_{\partial\mathcal{P}} \mathbf{T}\mathbf{n} \cdot \mathbf{v} \, \mathrm{d}a && \text{(result (4.26))} \\
&= \int_{\partial\mathcal{P}} \mathbf{T}^{\mathrm{T}}\mathbf{v} \cdot \mathbf{n} \, \mathrm{d}a && \text{(definition (2.13))} \\
&= \int_{\mathcal{P}} \operatorname{div}\left(\mathbf{T}^{\mathrm{T}}\mathbf{v}\right) \mathrm{d}v && \text{(divergence theorem (2.104)}_2\text{)} \\
&= \int_{\mathcal{P}} (\mathbf{T} \cdot \operatorname{grad} \mathbf{v} + \mathbf{v} \cdot \operatorname{div} \mathbf{T}) \, \mathrm{d}v && \text{(result (2.99)}_4\text{).}
\end{aligned}$$

Note that this result is consistent with (2.105). Also note that grad and div are the gradient and divergence calculated with respect to the present configuration. Next, considering the third term on the left-hand side of the energy theorem, we have

$$\begin{aligned}
\frac{\mathrm{d}}{\mathrm{d}t} \int_{\mathcal{P}} \frac{1}{2} \mathbf{v} \cdot \mathbf{v}\rho \, \mathrm{d}v &= \int_{\mathcal{P}} \left(\frac{1}{2} \dot{\overline{\mathbf{v} \cdot \mathbf{v}\rho}} + \frac{1}{2} \mathbf{v} \cdot \mathbf{v}\rho \operatorname{div} \mathbf{v} \right) \mathrm{d}v && \text{(transport theorem (4.20))} \\
&= \int_{\mathcal{P}} \left[\rho\dot{\mathbf{v}} \cdot \mathbf{v} + \frac{1}{2} \mathbf{v} \cdot \mathbf{v} \left(\dot{\rho} + \rho \operatorname{div} \mathbf{v} \right) \right] \mathrm{d}v && \text{(product rule (3.22))} \\
&= \int_{\mathcal{P}} \rho\dot{\mathbf{v}} \cdot \mathbf{v} \, \mathrm{d}v && \text{(conservation of mass (4.36a)).}
\end{aligned}$$

After assembling these results, the left-hand side of the energy theorem becomes

$$\int_{\mathcal{P}} \left[\left(\rho\mathbf{b} + \operatorname{div} \mathbf{T} - \rho\dot{\mathbf{v}} \right) \cdot \mathbf{v} + \mathbf{T} \cdot \operatorname{grad} \mathbf{v} \right] \mathrm{d}v.$$

By virtue of the local form of conservation of linear momentum (4.36b), this further reduces to

$$\int_{\mathcal{P}} \mathbf{T} \cdot \operatorname{grad} \mathbf{v} \, \mathrm{d}v.$$

Then,

$$\begin{aligned}
\int_{\mathcal{P}} \mathbf{T} \cdot \operatorname{grad} \mathbf{v} \, \mathrm{d}v &= \int_{\mathcal{P}} \mathbf{T} \cdot \mathbf{L} \, \mathrm{d}v && \text{(definition (3.56))} \\
&= \int_{\mathcal{P}} \mathbf{T} \cdot (\mathbf{D} + \mathbf{W}) \, \mathrm{d}v && \text{(decomposition (3.57))} \\
&= \int_{\mathcal{P}} (\mathbf{T} \cdot \mathbf{D} + \mathbf{T} \cdot \mathbf{W}) \, \mathrm{d}v && \text{(property (2.6)}_3\text{)} \\
&= \int_{\mathcal{P}} \mathbf{T} \cdot \mathbf{D} \, \mathrm{d}v && \text{(result (2.44)),}
\end{aligned}$$

where, in the last step, we have exploited the local form of conservation of angular momentum (4.36c), which demands that $\mathbf{T}$ is symmetric; recall that $\mathbf{W}$ is skew by construction. (Note that we were able to employ the distributive property $(2.6)_3$ of the inner product in the next-to-last step since we demonstrated in Problem 2.35 that the set of all second-order tensors is an inner product space.) Thus, we conclude that

$$\int_{\mathcal{P}} \mathbf{b} \cdot \mathbf{v} \rho \, \mathrm{d}v + \int_{\partial \mathcal{P}} \mathbf{t} \cdot \mathbf{v} \, \mathrm{d}a - \frac{\mathrm{d}}{\mathrm{d}t} \int_{\mathcal{P}} \frac{1}{2} \mathbf{v} \cdot \mathbf{v} \rho \, \mathrm{d}v = \int_{\mathcal{P}} \mathbf{T} \cdot \mathbf{D} \, \mathrm{d}v.$$

EXERCISES

1. Starting with the Eulerian (or spatial) statement of conservation of mass in integral form,

$$\frac{\mathrm{d}}{\mathrm{d}t} \int_{\mathcal{P}} \rho \, \mathrm{d}v = 0,$$

derive the corresponding pointwise form

$$\dot{\rho} + \rho \operatorname{div} \mathbf{v} = 0.$$

2. Starting with the Eulerian (or spatial) statement of the first law of thermodynamics (or conservation of energy) in integral form,

$$\frac{d}{dt}\int_{\mathcal{P}} \frac{1}{2}\mathbf{v}\cdot\mathbf{v}\rho\, dv + \frac{d}{dt}\int_{\mathcal{P}} \varepsilon\rho\, dv = \int_{\mathcal{P}} \mathbf{b}\cdot\mathbf{v}\rho\, dv + \int_{\partial\mathcal{P}} \mathbf{t}\cdot\mathbf{v}\, da + \int_{\mathcal{P}} r\rho\, dv - \int_{\partial\mathcal{P}} h\, da,$$

derive the corresponding pointwise form

$$\rho\dot{\varepsilon} = \mathbf{T}\cdot\mathbf{D} + \rho r - \operatorname{div}\mathbf{q}.$$

4.9 LAGRANGIAN FORMS OF THE INTEGRAL CONSERVATION LAWS

In Sections 4.2–4.8, the traction vector $\mathbf{t}$ and heat flux h acting on $\partial\mathcal{P}$ at time t are measured per unit area of surface $\partial\mathcal{P}$ *in the present configuration.* In view of the invertibility of the mapping $\mathbf{x} = \chi(\mathbf{X}, t)$, all surfaces in the region $\mathcal{R}$ occupied by the body in the present configuration can be mapped into corresponding surfaces in the region $\mathcal{R}_R$ occupied by the body in the reference configuration. For some purposes, it is more convenient to measure the traction and heat flux acting on $\partial\mathcal{P}$ per unit area of surfaces in $\mathcal{R}_R$. To illustrate this, consider the following example:

> *In a uniaxial tension test in the laboratory, the reference configuration is taken to be the initial, unstressed configuration of the specimen. The area A_o of a particular cross section in the reference configuration is easily and safely measured. During the test, the total tension $\mathbf{f}(t)$ on the specimen is measured. Assuming the traction at the cross section is uniform over its current area $A(t)$, the traction vector $\mathbf{t}$ is $\mathbf{f}(t)/A(t)$. However, it is difficult, and perhaps more dangerous, to measure the changing area $A(t)$ as the specimen is necking. Hence, it is advantageous to define the traction at time t measured per unit area of the reference configuration, which is $\mathbf{f}(t)/A_o$.*

To recap, the traction $\mathbf{t}$ and heat flux h are defined with respect to surfaces *in the present configuration*: $\mathbf{t}$ is the contact force per unit area in the present configuration, and h is the heat energy flowing through the surface per unit time per unit area in the present configuration. Owing to the invertibility of the motion, however, we can label material surfaces by either their present location or their reference location. For any subset $\mathcal{S}$ of the body, corresponding to the volume $\mathcal{P}$ with boundary $\partial\mathcal{P}$ it occupies in the present configuration is a volume $\mathcal{P}_R$ with boundary $\partial\mathcal{P}_R$ it occupies in the reference configuration (see Figure 4.3).

Therefore, each element of the material surface bounding the subset $\mathcal{S}$ has two possible and interchangeable labels: da in the present configuration and dA in the reference configuration. Both the size and the orientation, given by outward normal $\mathbf{n}$ in the present configuration and outward normal $\mathbf{N}$ in the reference configuration, are subject to alteration by the deformation induced by the motion (refer to Figure 3.11). As discussed in Section 3.6 (refer to (3.76)), $\mathbf{n}$ and $\mathbf{N}$ are related by

$$\mathbf{F}^{\mathrm{T}}\mathbf{n}\, da = J\,\mathbf{N}\, dA.$$

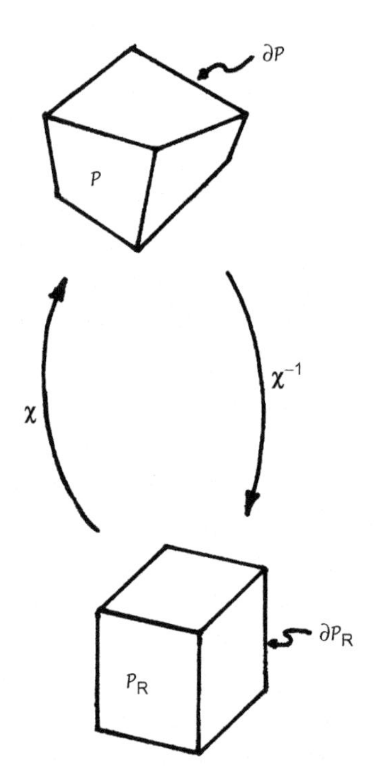

FIGURE 4.3

The reference and present configurations of part $\mathcal{S}$ of the body.

(In our uniaxial tension example, the deformation gradient $\mathbf{F}$ is diagonal, so the orientations of dA and da are the same (i.e., $\mathbf{n} = \mathbf{N}$), and only the size of the area changes.)

The contact force $\mathbf{f}_c(\mathcal{S}, t)$ acting on part $\mathcal{S}$ of the body at time t always acts on the present boundary $\partial\mathcal{P}$ of part $\mathcal{S}$, and the heat flux $H(\mathcal{S}, t)$ out of $\mathcal{S}$ at time t always flows through the present boundary $\partial\mathcal{P}$, but we are free to label this material boundary by its reference location $\partial\mathcal{P}_R$ for measurement purposes. We denote the traction vector acting on $\partial\mathcal{P}$ in the present configuration, but measured per unit area of the corresponding surface $\partial\mathcal{P}_R$ in the reference configuration, by $\mathbf{p}$. In the uniaxial tension example discussed earlier in this section,

$$\mathbf{t} = \frac{\mathbf{f}(t)}{A(t)}, \qquad \mathbf{p} = \frac{\mathbf{f}(t)}{A_o}.$$

Note that in the expressions for both $\mathbf{t}$ and $\mathbf{p}$, the force in the numerator is evaluated at the present time t. The denominator in the expression for $\mathbf{t}$ is also evaluated at the present time t, while the denominator in the expression for $\mathbf{p}$ is fixed. Therefore, for the contact force $\mathbf{f}_c$ acting on part $\mathcal{S}$ at time t, we have

$$\mathbf{f}_c(\mathcal{S}, t) = \int_{\text{boundary of } \mathcal{S}} d\mathbf{f}_c = \int_{\partial\mathcal{P}} \mathbf{t}\, da = \int_{\partial\mathcal{P}_R} \mathbf{p}\, dA. \tag{4.37}$$

The tractions $\mathbf{t}$ and $\mathbf{p}$ in (4.37) are not merely different descriptions of the same vector, but are different vectors altogether, since in general $da \neq dA$ and $\mathbf{n} \neq \mathbf{N}$. We can always represent $\mathbf{t}$ and $\mathbf{p}$, as we can any quantity related to the motion, in either their Eulerian or Lagrangian forms, i.e.,

$$\mathbf{t} = \tilde{\mathbf{t}}(\mathbf{x}, t, \mathbf{n}) = \hat{\mathbf{t}}(\mathbf{X}, t, \mathbf{N}), \qquad \mathbf{p} = \tilde{\mathbf{p}}(\mathbf{x}, t, \mathbf{n}) = \hat{\mathbf{p}}(\mathbf{X}, t, \mathbf{N}).$$

Although the definitions in (4.37) indicate that it is more natural to consider $\mathbf{t}$ in its Eulerian form $\tilde{\mathbf{t}}$ and $\mathbf{p}$ in its Lagrangian form $\hat{\mathbf{p}}$, it is important to realize that $\mathbf{p}$ is not just another title for $\mathbf{t}$; $\hat{\mathbf{p}}$ and $\hat{\mathbf{t}}$ are different functions of $\mathbf{X}$, t, and $\mathbf{N}$.

With this notation, we now present Lagrangian statements of the equations of motion and the first law of thermodynamics.

4.9.1 MASS, FORCES, MOMENTS, LINEAR AND ANGULAR MOMENTUM

Recall from Section 4.1 that the mass of part $\mathcal{S}$ can be expressed in material, spatial, and referential forms,

$$\mathcal{M}(\mathcal{S}) = \underbrace{\int_{\mathcal{S}} dm}_{\text{material}} = \underbrace{\int_{\mathcal{P}} \rho\, dv}_{\text{spatial}} = \underbrace{\int_{\mathcal{P}_R} \rho_R\, dV}_{\text{referential}}, \tag{4.38}$$

as can the linear momentum of part $\mathcal{S}$ at time t,

$$\mathcal{L}(\mathcal{S}, t) = \underbrace{\int_{\mathcal{S}} \mathbf{v}\, dm}_{\text{material}} = \underbrace{\int_{\mathcal{P}} \mathbf{v}\rho\, dv}_{\text{spatial}} = \underbrace{\int_{\mathcal{P}_R} \mathbf{v}\rho_R\, dV}_{\text{referential}}, \tag{4.39}$$

the angular momentum about the origin,

$$\mathbf{H_o}(\mathcal{S}, t, \mathbf{0}) = \int_{\mathcal{S}} \mathbf{x} \times \mathbf{v}\, dm = \int_{\mathcal{P}} \mathbf{x} \times \mathbf{v}\rho\, dv = \int_{\mathcal{P}_R} \mathbf{x} \times \mathbf{v}\rho_R\, dV, \tag{4.40}$$

the body force,

$$\mathbf{f}_b(\mathcal{S}, t) = \int_{\mathcal{S}} \mathbf{b}\, dm = \int_{\mathcal{P}} \mathbf{b}\rho\, dv = \int_{\mathcal{P}_R} \mathbf{b}\rho_R\, dV, \tag{4.41}$$

the contact force,

$$\mathbf{f}_c(\mathcal{S}, t) = \int_{\text{boundary of } \mathcal{S}} d\mathbf{f}_c = \int_{\partial\mathcal{P}} \mathbf{t}\, da = \int_{\partial\mathcal{P}_R} \mathbf{p}\, dA,$$

which is a restatement of (4.37), and the resultant moment about the origin,

$$\mathbf{M_o}(\mathcal{S}, t, \mathbf{0}) = \int_{\mathcal{S}} \mathbf{x} \times \mathbf{b}\, dm + \int_{\text{boundary of } \mathcal{S}} \mathbf{x} \times d\mathbf{f}_c$$

$$= \int_{\mathcal{P}} \mathbf{x} \times \mathbf{b}\rho \, \mathrm{d}v + \int_{\partial\mathcal{P}} \mathbf{x} \times \mathbf{t} \, \mathrm{d}a$$

$$= \int_{\mathcal{P}_{\mathrm{R}}} \mathbf{x} \times \mathbf{b}\rho_{\mathrm{R}} \, \mathrm{d}V + \int_{\partial\mathcal{P}_{\mathrm{R}}} \mathbf{x} \times \mathbf{p} \, \mathrm{d}A. \tag{4.42}$$

Note that to go from the spatial forms to the referential forms, only one new quantity (namely, $\mathbf{p}$) had to be defined for the area integrals in (4.37) and (4.42).

To integrate the spatial forms, functions of position and time in the integrands are considered in their Eulerian representations, e.g.,

$$\mathbf{f}_{\mathrm{b}}\,(\mathcal{S}, t) = \int_{\mathcal{P}} \mathbf{b}\rho \, \mathrm{d}v = \int_{\mathcal{P}(t)} \tilde{\mathbf{b}}(\mathbf{x}, t)\, \tilde{\rho}(\mathbf{x}, t) \, \mathrm{d}v,$$

$$\mathbf{H}_{\mathbf{o}}(\mathcal{S}, t, \mathbf{0}) = \int_{\mathcal{P}} \mathbf{x} \times \mathbf{v}\rho \, \mathrm{d}v = \int_{\mathcal{P}(t)} \mathbf{x} \times \tilde{\mathbf{v}}(\mathbf{x}, t)\, \tilde{\rho}(\mathbf{x}, t) \, \mathrm{d}v.$$

On the other hand, to integrate the referential forms, functions of position and time in the integrands are considered in their Lagrangian representations, e.g.,

$$\mathbf{f}_{\mathrm{b}}\,(\mathcal{S}, t) = \int_{\mathcal{P}_{\mathrm{R}}} \mathbf{b}\rho_{\mathrm{R}} \, \mathrm{d}V = \int_{\mathcal{P}_{\mathrm{R}}} \hat{\mathbf{b}}(\mathbf{X}, t)\, \hat{\rho}_{\mathrm{R}}(\mathbf{X}) \, \mathrm{d}V,$$

$$\mathbf{H}_{\mathbf{o}}(\mathcal{S}, t, \mathbf{0}) = \int_{\mathcal{P}_{\mathrm{R}}} \mathbf{x} \times \mathbf{v}\rho_{\mathrm{R}} \, \mathrm{d}V = \int_{\mathcal{P}_{\mathrm{R}}} \chi(\mathbf{X}, t) \times \hat{\mathbf{v}}(\mathbf{X}, t)\, \hat{\rho}_{\mathrm{R}}(\mathbf{X}) \, \mathrm{d}V.$$

4.9.2 CONSERVATION OF MASS, LINEAR MOMENTUM, AND ANGULAR MOMENTUM

With use of (4.37)–(4.42), the material forms $(4.10\mathrm{a})_2$, (4.10b), (4.10c) of the equations of motion from Section 4.3 become, in Lagrangian form,

$$\int_{\mathcal{P}} \rho \, \mathrm{d}v = \int_{\mathcal{P}_{\mathrm{R}}} \rho_{\mathrm{R}} \, \mathrm{d}V,$$

$$\frac{\mathrm{d}}{\mathrm{d}t} \int_{\mathcal{P}_{\mathrm{R}}} \mathbf{v}\rho_{\mathrm{R}} \, \mathrm{d}V = \int_{\mathcal{P}_{\mathrm{R}}} \mathbf{b}\rho_{\mathrm{R}} \, \mathrm{d}V + \int_{\partial\mathcal{P}_{\mathrm{R}}} \mathbf{p} \, \mathrm{d}A,$$

$$\frac{\mathrm{d}}{\mathrm{d}t} \int_{\mathcal{P}_{\mathrm{R}}} \mathbf{x} \times \mathbf{v}\rho_{\mathrm{R}} \, \mathrm{d}V = \int_{\mathcal{P}_{\mathrm{R}}} \mathbf{x} \times \mathbf{b}\rho_{\mathrm{R}} \, \mathrm{d}V + \int_{\partial\mathcal{P}_{\mathrm{R}}} \mathbf{x} \times \mathbf{p} \, \mathrm{d}A.$$

Compare these with the corresponding Eulerian forms given in (4.11a)–(4.11c).

4.9.3 FIRST LAW OF THERMODYNAMICS

Returning our attention to thermodynamics, we can express the kinetic energy of part $\mathcal{S}$ at time t in material, spatial, and referential forms,

$$K(\mathcal{S},t) = \underbrace{\int_{\mathcal{S}} \frac{1}{2}\mathbf{v}\cdot\mathbf{v}\,\mathrm{d}m}_{\text{material}} = \underbrace{\int_{\mathcal{P}} \frac{1}{2}\mathbf{v}\cdot\mathbf{v}\,\rho\,\mathrm{d}v}_{\text{spatial}} = \underbrace{\int_{\mathcal{P}_\mathrm{R}} \frac{1}{2}\mathbf{v}\cdot\mathbf{v}\,\rho_\mathrm{R}\,\mathrm{d}V}_{\text{referential}}, \tag{4.43}$$

and similarly the internal energy,

$$E(\mathcal{S},t) = \int_{\mathcal{S}} \varepsilon\,\mathrm{d}m = \int_{\mathcal{P}} \varepsilon\,\rho\,\mathrm{d}v = \int_{\mathcal{P}_\mathrm{R}} \varepsilon\,\rho_\mathrm{R}\,\mathrm{d}V. \tag{4.44}$$

We denote the heat flux flowing through $\partial\mathcal{P}$ in the present configuration, but measured per unit area of $\partial\mathcal{P}_\mathrm{R}$ in the reference configuration, by h_R. Therefore, the rate of heat energy *entering* part $\mathcal{S}$ at time t, in spatial and referential forms, is

$$H(\mathcal{S},t) = \int_{\mathcal{P}} r\rho\,\mathrm{d}v - \int_{\partial\mathcal{P}} h\,\mathrm{d}a = \int_{\mathcal{P}_\mathrm{R}} r\rho_\mathrm{R}\,\mathrm{d}V - \int_{\partial\mathcal{P}_\mathrm{R}} h_\mathrm{R}\,\mathrm{d}A. \tag{4.45}$$

Similarly, the rate of work done by the resultant external force, in spatial and referential forms, is

$$R(\mathcal{S},t) = \int_{\mathcal{P}} \mathbf{b}\cdot\mathbf{v}\,\rho\,\mathrm{d}v + \int_{\partial\mathcal{P}} \mathbf{t}\cdot\mathbf{v}\,\mathrm{d}a = \int_{\mathcal{P}_\mathrm{R}} \mathbf{b}\cdot\mathbf{v}\,\rho_\mathrm{R}\,\mathrm{d}V + \int_{\partial\mathcal{P}_\mathrm{R}} \mathbf{p}\cdot\mathbf{v}\,\mathrm{d}A. \tag{4.46}$$

With use of (4.16) and (4.43)–(4.46), the material form (4.12) of conservation of energy (i.e., the first law of thermodynamics) becomes, in Lagrangian form,

$$\frac{\mathrm{d}}{\mathrm{d}t}\int_{\mathcal{P}_\mathrm{R}} \left(\frac{1}{2}\mathbf{v}\cdot\mathbf{v}+\varepsilon\right)\rho_\mathrm{R}\,\mathrm{d}V = \int_{\mathcal{P}_\mathrm{R}} (\mathbf{b}\cdot\mathbf{v}+r)\,\rho_\mathrm{R}\,\mathrm{d}V + \int_{\partial\mathcal{P}_\mathrm{R}} (\mathbf{p}\cdot\mathbf{v}-h_\mathrm{R})\,\mathrm{d}A.$$

4.9.4 SUMMARY

As a summary of our results, the Lagrangian integral forms of the equations of motion and the first law of thermodynamics are

$$\int_{\mathcal{P}} \rho\,\mathrm{d}v = \int_{\mathcal{P}_\mathrm{R}} \rho_\mathrm{R}\,\mathrm{d}V, \tag{4.47a}$$

$$\frac{\mathrm{d}}{\mathrm{d}t}\int_{\mathcal{P}_\mathrm{R}} \mathbf{v}\rho_\mathrm{R}\,\mathrm{d}V = \int_{\mathcal{P}_\mathrm{R}} \mathbf{b}\,\rho_\mathrm{R}\,\mathrm{d}V + \int_{\partial\mathcal{P}_\mathrm{R}} \mathbf{p}\,\mathrm{d}A, \tag{4.47b}$$

$$\frac{\mathrm{d}}{\mathrm{d}t}\int_{\mathcal{P}_\mathrm{R}} \mathbf{x}\times\mathbf{v}\rho_\mathrm{R}\,\mathrm{d}V = \int_{\mathcal{P}_\mathrm{R}} \mathbf{x}\times\mathbf{b}\,\rho_\mathrm{R}\,\mathrm{d}V + \int_{\partial\mathcal{P}_\mathrm{R}} \mathbf{x}\times\mathbf{p}\,\mathrm{d}A, \tag{4.47c}$$

$$\frac{\mathrm{d}}{\mathrm{d}t}\int\limits_{\mathcal{P}_{\mathrm{R}}}\left(\frac{1}{2}\mathbf{v}\cdot\mathbf{v}+\varepsilon\right)\rho_{\mathrm{R}}\,\mathrm{d}V=\int\limits_{\mathcal{P}_{\mathrm{R}}}(\mathbf{b}\cdot\mathbf{v}+r)\,\rho_{\mathrm{R}}\,\mathrm{d}V+\int\limits_{\partial\mathcal{P}_{\mathrm{R}}}(\mathbf{p}\cdot\mathbf{v}-h_{\mathrm{R}})\,\mathrm{d}A. \tag{4.47d}$$

We emphasize that to perform integrations over areas and volumes in the reference configuration, the functions in the integrands must be in their Lagrangian forms, i.e., functions of the independent variables $\mathbf{X}$ and t.

4.10 PIOLA-KIRCHHOFF STRESS TENSORS, REFERENTIAL HEAT FLUX VECTOR

The traction vector $\mathbf{p}$ and heat flux h_{R} are assumed to depend on the geometry of the surface $\partial\mathcal{P}_{\mathrm{R}}$ only through the outward unit normal $\mathbf{N}$, i.e.,

$$\mathbf{p}=\hat{\mathbf{p}}(\mathbf{X},t,\mathbf{N}),\qquad h_{\mathrm{R}}=\hat{h}_{\mathrm{R}}(\mathbf{X},t,\mathbf{N}). \tag{4.48}$$

Analogously to what was done in Section 4.6, we can show that a consequence of the conservation laws (4.47a)–(4.47d) is that the dependence of $\mathbf{p}$ and h_{R} on $\mathbf{N}$ must be linear, so

$$\mathbf{p}=\mathbf{P}\mathbf{N},\qquad h_{\mathrm{R}}=\mathbf{q}_{\mathrm{R}}\cdot\mathbf{N}. \tag{4.49}$$

In (4.49), $\mathbf{P}$ is called the **first Piola-Kirchhoff stress tensor** (or **nonsymmetric Piola-Kirchhoff stress tensor**) and $\mathbf{q}_{\mathrm{R}}$ is the **referential heat flux vector**. We define a tensor $\mathbf{S}$, called the **second Piola-Kirchhoff stress tensor** (or **symmetric Piola-Kirchhoff stress tensor**), by

$$\mathbf{P}=\mathbf{F}\mathbf{S}. \tag{4.50}$$

It remains for us to demonstrate that $\mathbf{S}$ is indeed symmetric and $\mathbf{P}$ is indeed nonsymmetric.

4.10.1 RELATIONS BETWEEN SPATIAL AND REFERENTIAL QUANTITIES

We now relate $\mathbf{p}$ to $\mathbf{t}$, $\mathbf{P}$ and $\mathbf{S}$ to $\mathbf{T}$, h_{R} to h, and $\mathbf{q}_{\mathrm{R}}$ to $\mathbf{q}$. Recall from Section 4.9 that for arbitrary part $\mathcal{S}$ of the body,

$$\mathbf{f}_c(\mathcal{S},t)=\int\limits_{\partial\mathcal{P}}\mathbf{t}\,\mathrm{d}a=\int\limits_{\partial\mathcal{P}_{\mathrm{R}}}\mathbf{p}\,\mathrm{d}A,$$

so, using $\mathbf{t}=\mathbf{T}\mathbf{n}$ and $\mathbf{p}=\mathbf{P}\mathbf{N}$, we obtain

$$\int\limits_{\partial\mathcal{P}}\mathbf{T}\mathbf{n}\,\mathrm{d}a=\int\limits_{\partial\mathcal{P}_{\mathrm{R}}}\mathbf{P}\mathbf{N}\,\mathrm{d}A.$$

Using (3.76) and a change of independent variable from $\mathbf{X}$ to $\mathbf{x}$, we convert the Lagrangian integration to an Eulerian integration:

$$\int\limits_{\partial\mathcal{P}}\left(\mathbf{T}-\frac{1}{J}\mathbf{P}\mathbf{F}^{\mathrm{T}}\right)\mathbf{n}\,\mathrm{d}a=\mathbf{0}.$$

By the two-dimensional localization theorem, since $\partial\mathcal{P}$ is arbitrary and the integrand is continuous,

$$\left(\mathbf{T} - \frac{1}{J}\mathbf{P}\mathbf{F}^{\mathrm{T}}\right)\mathbf{n} = \mathbf{0}.$$

Further, because $\mathbf{T} - \frac{1}{J}\mathbf{P}\mathbf{F}^{\mathrm{T}}$ does not depend on $\mathbf{n}$,

$$J\mathbf{T} = \mathbf{P}\mathbf{F}^{\mathrm{T}} \qquad \text{or} \qquad \mathbf{P} = J\mathbf{T}\mathbf{F}^{-\mathrm{T}}. \tag{4.51}$$

Subsequent use of (4.50) in (4.51) implies that

$$J\mathbf{T} = \mathbf{F}\mathbf{S}\mathbf{F}^{\mathrm{T}}. \tag{4.52}$$

In a similar fashion, we can also show that

$$J\mathbf{q} = \mathbf{F}\,\mathbf{q}_{\mathrm{R}} \qquad \text{or} \qquad \mathbf{q}_{\mathrm{R}} = J\mathbf{F}^{-1}\mathbf{q}. \tag{4.53}$$

EXERCISES

1. Prove that $\mathbf{p} = \mathbf{P}\mathbf{N}$.
2. Show that $h_{\mathrm{R}} = \mathbf{q}_{\mathrm{R}} \cdot \mathbf{N}$.
3. Starting with

$$\int_{\partial\mathcal{P}} h\,\mathrm{d}a = \int_{\partial\mathcal{P}_{\mathrm{R}}} h_{\mathrm{R}}\,\mathrm{d}A,$$

prove that

$$\mathbf{q}_{\mathrm{R}} = J\mathbf{F}^{-1}\mathbf{q}.$$

4.11 THE LAGRANGIAN FORM OF THE ENERGY THEOREM

Recall from Section 4.9.2 the Lagrangian representations of the conservation laws of mass, linear momentum, and angular momentum:

$$\int_{\mathcal{P}} \rho\,\mathrm{d}v = \int_{\mathcal{P}_{\mathrm{R}}} \rho_{\mathrm{R}}\,\mathrm{d}V,$$

$$\frac{\mathrm{d}}{\mathrm{d}t}\int_{\mathcal{P}_{\mathrm{R}}} \mathbf{v}\rho_{\mathrm{R}}\,\mathrm{d}V = \int_{\mathcal{P}_{\mathrm{R}}} \mathbf{b}\,\rho_{\mathrm{R}}\,\mathrm{d}V + \int_{\partial\mathcal{P}_{\mathrm{R}}} \mathbf{P}\mathbf{N}\,\mathrm{d}A,$$

$$\frac{\mathrm{d}}{\mathrm{d}t}\int_{\mathcal{P}_{\mathrm{R}}} \mathbf{x}\times\mathbf{v}\rho_{\mathrm{R}}\,\mathrm{d}V = \int_{\mathcal{P}_{\mathrm{R}}} \mathbf{x}\times\mathbf{b}\,\rho_{\mathrm{R}}\,\mathrm{d}V + \int_{\partial\mathcal{P}_{\mathrm{R}}} \mathbf{x}\times\mathbf{P}\mathbf{N}\,\mathrm{d}A,$$

where we have used $\mathbf{p} = \mathbf{PN}$ (result $(4.49)_1$). Similar to what we did in Section 4.7, as a consequence of these mechanical conservation laws, we can show that

$$\int_{\mathcal{P}_R} \rho_R \mathbf{b} \cdot \mathbf{v}\, dV + \int_{\partial\mathcal{P}_R} \mathbf{p} \cdot \mathbf{v}\, dA - \frac{d}{dt}\int_{\mathcal{P}_R} \frac{1}{2}\rho_R \mathbf{v} \cdot \mathbf{v}\, dV = \int_{\mathcal{P}_R} \mathbf{P} \cdot \mathrm{Grad}\, \mathbf{v}\, dV. \tag{4.54}$$

This result is the Lagrangian form of the **energy theorem**; compare it with (4.33), the Eulerian form of the energy theorem.

4.12 LOCAL CONSERVATION LAWS IN LAGRANGIAN FORM

In Section 4.8 we obtained the Eulerian representations of the local conservation laws by using the localization theorem on the Eulerian representations of the integral conservation laws. The same procedure is performed here, except that we now involve the Lagrangian representations.

From Section 4.9 we recall the Lagrangian representations of the integral conservation laws:

$$\int_{\mathcal{P}} \rho\, dv = \int_{\mathcal{P}_R} \rho_R\, dV, \tag{4.55a}$$

$$\frac{d}{dt}\int_{\mathcal{P}_R} \mathbf{v}\rho_R\, dV = \int_{\mathcal{P}_R} \mathbf{b}\rho_R\, dV + \int_{\partial\mathcal{P}_R} \mathbf{PN}\, dA, \tag{4.55b}$$

$$\frac{d}{dt}\int_{\mathcal{P}_R} \mathbf{x} \times \mathbf{v}\rho_R\, dV = \int_{\mathcal{P}_R} \mathbf{x} \times \mathbf{b}\rho_R\, dV + \int_{\partial\mathcal{P}_R} \mathbf{x} \times \mathbf{PN}\, dA, \tag{4.55c}$$

$$\frac{d}{dt}\int_{\mathcal{P}_R} \left(\frac{1}{2}\mathbf{v} \cdot \mathbf{v} + \varepsilon\right)\rho_R\, dV = \int_{\mathcal{P}_R} (\mathbf{b} \cdot \mathbf{v} + r)\,\rho_R\, dV + \int_{\partial\mathcal{P}_R} (\mathbf{PN} \cdot \mathbf{v} - \mathbf{q}_R \cdot \mathbf{N})\, dA, \tag{4.55d}$$

where we have used $\mathbf{p} = \mathbf{PN}$ and $h_R = \mathbf{q}_R \cdot \mathbf{N}$ (results $(4.49)_1$ and $(4.49)_2$). After use of the divergence theorem, taking the time derivatives inside the integrals, and use of the localization theorem, one can show that the local forms of (4.55a)–(4.55d) are

$$\rho J = \rho_R, \tag{4.56a}$$

$$\mathrm{Div}\, \mathbf{P} + \rho_R\, \mathbf{b} = \rho_R\, \dot{\mathbf{v}}, \tag{4.56b}$$

$$\mathbf{PF}^T = \mathbf{FP}^T \quad \text{or} \quad \mathbf{S} = \mathbf{S}^T, \tag{4.56c}$$

$$\rho_R\, \dot{\varepsilon} = \mathbf{P} \cdot \mathrm{Grad}\, \mathbf{v} + \rho_R\, r - \mathrm{Div}\, \mathbf{q}_R; \tag{4.56d}$$

refer to Problems 4.8 and 4.9. Note that Div and Grad are the divergence and gradient calculated with respect to the reference configuration (refer to Section 3.2).

Equations (4.56a)–(4.56d) are the local Lagrangian forms of the conservation laws for mass, linear momentum, angular momentum, and energy, respectively, and are an alternative set to the local spatial forms given by (4.36a)–(4.36d). The Eulerian forms (4.36a)–(4.36d) are most useful for problems in which the present (loaded) configuration is known, or for a fluid, which has no concept of a reference configuration. The Lagrangian forms (4.56a)–(4.56d) are most useful when the reference (unloaded) configuration is known, as is often the case with solids.

From (3.36) and conservation of mass (4.56a), we see that J must be positive, i.e., $J > 0$. This is consistent with our findings in Section 3.6. Also, as a consequence of balance of angular momentum (4.56c), we see that the first Piola-Kirchhoff stress tensor $\mathbf{P}$ is indeed not symmetric, and the second Piola-Kirchhoff stress tensor $\mathbf{S}$ is in fact symmetric.

PROBLEM 4.8

Starting with the Lagrangian (or referential) statement of balance of linear momentum in integral form,

$$\frac{\mathrm{d}}{\mathrm{d}t}\int_{\mathcal{P}_R} \rho_R \mathbf{v}\, \mathrm{d}V = \int_{\mathcal{P}_R} \rho_R \mathbf{b}\, \mathrm{d}V + \int_{\partial\mathcal{P}_R} \mathbf{p}\, \mathrm{d}A,$$

derive the corresponding pointwise form

$$\rho_R \dot{\mathbf{v}} = \rho_R \mathbf{b} + \mathrm{Div}\, \mathbf{P}.$$

Solution

An important feature of the Lagrangian integral conservation laws (4.55a)–(4.55d) is that the region $\mathcal{P}_R$ occupied by subset $\mathcal{S}$ in the reference configuration is fixed. Hence, the region of integration $\mathcal{P}_R$ does not change with time, so time derivatives of Lagrangian volume integrals can be directly passed inside the integrals. (Conversely, with the Eulerian integral conservation laws, the region of integration $\mathcal{P}$ changes with time; the transport theorem (4.20) is needed to take time derivatives of these time-varying volume integrals.)

Here, we begin with the Lagrangian integral form of balance of linear momentum, i.e.,

$$\frac{\mathrm{d}}{\mathrm{d}t}\int_{\mathcal{P}_R} \rho_R \mathbf{v}\, \mathrm{d}V = \int_{\mathcal{P}_R} \rho_R \mathbf{b}\, \mathrm{d}V + \int_{\partial\mathcal{P}_R} \mathbf{p}\, \mathrm{d}A.$$

We first consider the left-hand side of this integral equation:

$$\frac{\mathrm{d}}{\mathrm{d}t}\int_{\mathcal{P}_R} \rho_R \mathbf{v}\, \mathrm{d}V = \int_{\mathcal{P}_R} \overline{\rho_R \dot{\mathbf{v}}}\, \mathrm{d}V \qquad (\mathcal{P}_R \text{ is fixed})$$

$$= \int_{\mathcal{P}_R} (\dot{\rho}_R \mathbf{v} + \rho_R \dot{\mathbf{v}}) \, dV \qquad \text{(product rule (3.22)}_1\text{)}$$

$$= \int_{\mathcal{P}_R} \rho_R \dot{\mathbf{v}} \, dV \qquad (\rho_R \text{ is independent of time}).$$

We next consider the second term on the right-hand side of the integral form of balance of linear momentum:

$$\int_{\partial \mathcal{P}_R} \mathbf{p} \, dA = \int_{\partial \mathcal{P}_R} \mathbf{PN} \, dA \qquad \text{(result (4.49)}_1\text{)}$$

$$= \int_{\mathcal{P}_R} \text{Div} \, \mathbf{P} \, dV \qquad \text{(divergence theorem (2.104)}_3\text{)}.$$

Note that Div is the divergence calculated with respect to the reference configuration. Use of these results in the original equation leads to

$$\int_{\mathcal{P}_R} (\rho_R \dot{\mathbf{v}} - \rho_R \mathbf{b} - \text{Div} \, \mathbf{P}) \, dV = \mathbf{0}.$$

Since the integrand is continuous and the region of integration $\mathcal{P}_R$ is arbitrary, it follows from the localization theorem that the integrand vanishes everywhere in the reference configuration of the body; i.e.,

$$\rho_R \dot{\mathbf{v}} - \rho_R \mathbf{b} - \text{Div} \, \mathbf{P} = \mathbf{0},$$

or

$$\rho_R \dot{\mathbf{v}} = \rho_R \mathbf{b} + \text{Div} \, \mathbf{P}$$

everywhere in $\mathcal{R}_R$.

PROBLEM 4.9

Starting with the Lagrangian (or referential) statement of the first law of thermodynamics (or conservation of energy) in integral form,

$$\frac{d}{dt} \int_{\mathcal{P}_R} \left(\frac{1}{2} \mathbf{v} \cdot \mathbf{v} + \varepsilon \right) \rho_R \, dV = \int_{\mathcal{P}_R} (\mathbf{b} \cdot \mathbf{v} + r) \rho_R \, dV + \int_{\partial \mathcal{P}_R} (\mathbf{p} \cdot \mathbf{v} - h_R) \, dA,$$

derive the corresponding pointwise form

$$\rho_R \dot{\varepsilon} = \mathbf{P} \cdot \text{Grad} \, \mathbf{v} + \rho_R r - \text{Div} \, \mathbf{q}_R.$$

Solution

We begin with the Lagrangian integral form of the first law of thermodynamics, i.e.,

$$\frac{\mathrm{d}}{\mathrm{d}t}\int_{\mathcal{P}_\mathrm{R}}\left(\frac{1}{2}\mathbf{v}\cdot\mathbf{v}+\varepsilon\right)\rho_\mathrm{R}\,\mathrm{d}V=\int_{\mathcal{P}_\mathrm{R}}(\mathbf{b}\cdot\mathbf{v}+r)\,\rho_\mathrm{R}\,\mathrm{d}V+\int_{\partial\mathcal{P}_\mathrm{R}}(\mathbf{p}\cdot\mathbf{v}-h_\mathrm{R})\,\mathrm{d}A.$$

Subsequent use of (4.54), the Lagrangian form of the energy theorem, simplifies the above statement of the first law by consolidating the mechanical energy contributions into a single term. Consequently, we have

$$\frac{\mathrm{d}}{\mathrm{d}t}\int_{\mathcal{P}_\mathrm{R}}\varepsilon\,\rho_\mathrm{R}\,\mathrm{d}V=\int_{\mathcal{P}_\mathrm{R}}\mathbf{P}\cdot\operatorname{Grad}\mathbf{v}\,\mathrm{d}V+\int_{\mathcal{P}_\mathrm{R}}r\,\rho_\mathrm{R}\,\mathrm{d}V-\int_{\partial\mathcal{P}_\mathrm{R}}h_\mathrm{R}\,\mathrm{d}A.$$

Working with the left-hand side of this integral equation, we have

$$\begin{aligned}\frac{\mathrm{d}}{\mathrm{d}t}\int_{\mathcal{P}_\mathrm{R}}\varepsilon\,\rho_\mathrm{R}\,\mathrm{d}V&=\int_{\mathcal{P}_\mathrm{R}}\overline{\varepsilon\,\rho_\mathrm{R}}^{\,\boldsymbol{\cdot}}\,\mathrm{d}V &&(\mathcal{P}_\mathrm{R}\text{ is fixed})\\&=\int_{\mathcal{P}_\mathrm{R}}\left(\varepsilon\,\dot{\rho}_\mathrm{R}+\rho_\mathrm{R}\,\dot{\varepsilon}\right)\mathrm{d}V &&(\text{product rule})\\&=\int_{\mathcal{P}_\mathrm{R}}\rho_\mathrm{R}\,\dot{\varepsilon}\,\mathrm{d}V &&(\rho_\mathrm{R}\text{ is independent of time}).\end{aligned}$$

Now, working with the third term on the right-hand side of this integral equation, we have

$$\begin{aligned}\int_{\partial\mathcal{P}_\mathrm{R}}h_\mathrm{R}\,\mathrm{d}A&=\int_{\partial\mathcal{P}_\mathrm{R}}\mathbf{q}_\mathrm{R}\cdot\mathbf{N}\,\mathrm{d}A &&(\text{result }(4.49)_2)\\&=\int_{\mathcal{P}_\mathrm{R}}\operatorname{Div}\mathbf{q}_\mathrm{R}\,\mathrm{d}V &&(\text{divergence theorem }(2.104)_2).\end{aligned}$$

Assembling the above results, we have

$$\int_{\mathcal{P}_\mathrm{R}}\left(\rho_\mathrm{R}\,\dot{\varepsilon}-\mathbf{P}\cdot\operatorname{Grad}\mathbf{v}-\rho_\mathrm{R}\,r+\operatorname{Div}\mathbf{q}_\mathrm{R}\right)\mathrm{d}V=0.$$

Since the integrand is continuous and the region of integration $\mathcal{P}_\mathrm{R}$ is arbitrary, it follows from the localization theorem that

$$\rho_\mathrm{R}\,\dot{\varepsilon}-\mathbf{P}\cdot\operatorname{Grad}\mathbf{v}-\rho_\mathrm{R}\,r+\operatorname{Div}\mathbf{q}_\mathrm{R}=0,$$

or

$$\rho_\mathrm{R}\,\dot{\varepsilon}=\mathbf{P}\cdot\operatorname{Grad}\mathbf{v}+\rho_\mathrm{R}\,r-\operatorname{Div}\mathbf{q}_\mathrm{R}.$$

EXERCISES

1. Starting with the Lagrangian (or referential) statement of conservation of mass in integral form,

$$\int_{\mathcal{P}} \rho \, \mathrm{d}v = \int_{\mathcal{P}_\mathrm{R}} \rho_\mathrm{R} \, \mathrm{d}V,$$

derive the corresponding pointwise form

$$\rho J = \rho_\mathrm{R}.$$

2. Starting with the Lagrangian (or referential) statement of balance of angular momentum in integral form,

$$\frac{\mathrm{d}}{\mathrm{d}t} \int_{\mathcal{P}_\mathrm{R}} \mathbf{x} \times \rho_\mathrm{R} \mathbf{v} \, \mathrm{d}V = \int_{\mathcal{P}_\mathrm{R}} \mathbf{x} \times \rho_\mathrm{R} \mathbf{b} \, \mathrm{d}V + \int_{\partial \mathcal{P}_\mathrm{R}} \mathbf{x} \times \mathbf{p} \, \mathrm{d}A,$$

derive the corresponding pointwise form

$$\mathbf{F}\mathbf{P}^\mathrm{T} = \mathbf{P}\mathbf{F}^\mathrm{T},$$

and show that this implies that

$$\mathbf{S} = \mathbf{S}^\mathrm{T}.$$

(You may use the result

$$\int_{\partial \mathcal{P}_\mathrm{R}} \mathbf{x} \times (\mathbf{P}\mathbf{N}) \, \mathrm{d}A = \int_{\mathcal{P}_\mathrm{R}} (\mathbf{x} \times \mathrm{Div}\, \mathbf{P} + \boldsymbol{\tau}) \, \mathrm{d}V$$

from tensor calculus without proving it, where

$$\left(\mathbf{F}\mathbf{P}^\mathrm{T} - \mathbf{P}\mathbf{F}^\mathrm{T}\right) \mathbf{a} = \boldsymbol{\tau} \times \mathbf{a}$$

for any vector $\mathbf{a}$ in $\mathcal{E}^3$.)

3. Verify that the Lagrangian (or referential) form of the energy theorem

$$\int_{\mathcal{P}_\mathrm{R}} \rho_\mathrm{R} \mathbf{b} \cdot \mathbf{v} \, \mathrm{d}V + \int_{\partial \mathcal{P}_\mathrm{R}} \mathbf{p} \cdot \mathbf{v} \, \mathrm{d}A - \frac{\mathrm{d}}{\mathrm{d}t} \int_{\mathcal{P}_\mathrm{R}} \frac{1}{2} \rho_\mathrm{R} \mathbf{v} \cdot \mathbf{v} \, \mathrm{d}V = \int_{\mathcal{P}_\mathrm{R}} \mathbf{P} \cdot \mathrm{Grad}\, \mathbf{v} \, \mathrm{d}V$$

is a consequence of the Lagrangian forms of the mechanical conservation laws of mass, linear momentum, and angular momentum.

4.13 THE SECOND LAW OF THERMODYNAMICS

Recall that the *first law of thermodynamics* states that the rate of change of the total (kinetic plus internal) energy of a part $\mathcal{S}$ of a body $\mathcal{B}$ is equal to the rate of mechanical work generated by the resultant external force acting on $\mathcal{S}$ plus the rate at which heat enters $\mathcal{S}$, i.e.,

$$\frac{\mathrm{d}}{\mathrm{d}t}\int_{\mathcal{P}}\left(\frac{1}{2}\mathbf{v}\cdot\mathbf{v}+\varepsilon\right)\rho\,\mathrm{d}v=\int_{\mathcal{P}}\mathbf{b}\cdot\mathbf{v}\rho\,\mathrm{d}v+\int_{\partial\mathcal{P}}\mathbf{t}\cdot\mathbf{v}\,\mathrm{d}a+\int_{\mathcal{P}}r\rho\,\mathrm{d}v-\int_{\partial\mathcal{P}}h\,\mathrm{d}a.$$

It has been observed that the transformation of mechanical energy to heat energy (and the converse) in the first law is not arbitrary, but rather subject to definite restrictions. These restrictions are collectively called the **second law of thermodynamics.** Three of the restrictions of the second law are as follows:

(1) A heat engine cannot convert heat energy solely to mechanical energy in the absence of other effects (i.e., a heat engine cannot be 100% thermally efficient).
(2) A process in which friction changes mechanical energy into heat cannot be reversed.
(3) Heat cannot spontaneously pass from a body of lower temperature to one of higher temperature.

Each of these restrictions gives different aspects of the second law. Of course, to be usable, the second law must be put into mathematical language. *The ideal mathematical statement would apply all consequences of the second law to all materials.* One form that demands *most of the consequences for most materials* is the **Clausius-Duhem inequality**.

A quantity $\mathcal{N}(\mathcal{S},t)$ called the **entropy** of part $\mathcal{S}$ at time t is introduced. The Clausius-Duhem inequality states that mechanical forces and deformation can only tend to increase the entropy of $\mathcal{S}$. Said differently, the rate of change of the entropy of part $\mathcal{S}$ at time t is greater than or equal to the rate of entropy generation at time t *due to the radiative heat supply* minus the rate of entropy loss at time t *due to the outward heat flux.* Mathematically, this can be expressed in *material form* as

$$\frac{\mathrm{d}}{\mathrm{d}t}\mathcal{N}(\mathcal{S},t)\geq\mathcal{R}(\mathcal{S},t)-\mathcal{H}(\mathcal{S},t) \tag{4.57}$$

for *arbitrary* subset $\mathcal{S}$ of body $\mathcal{B}$ and *all* time t, where $\mathcal{N}(\mathcal{S},t)$ is the entropy of part $\mathcal{S}$ at time t, $\mathcal{R}(\mathcal{S},t)$ is the rate of entropy production at time t due to radiative heat absorbed by $\mathcal{S}$, and $\mathcal{H}(\mathcal{S},t)$ is the rate of entropy loss at time t due to heat flow out through the boundary of $\mathcal{S}$.

We emphasize that inequality (4.57) is valid not only for the body as a whole, but also for every subset of the body.

The entropy $\mathcal{N}(\mathcal{S},t)$ of subset $\mathcal{S}$ at time t can be expressed in spatial and referential forms,

$$\mathcal{N}(\mathcal{S},t)=\underbrace{\int_{\mathcal{P}}\eta\,\rho\,\mathrm{d}v}_{\text{spatial}}=\underbrace{\int_{\mathcal{P}_\mathrm{R}}\eta\,\rho_\mathrm{R}\,\mathrm{d}V}_{\text{referential}}, \tag{4.58}$$

as can the rate of entropy production $\mathcal{R}(\mathcal{S},t)$ at time t due to heat absorbed by $\mathcal{S}$,

$$\mathcal{R}(\mathcal{S},t)=\underbrace{\int_{\mathcal{P}}\frac{r}{\Theta}\,\rho\,\mathrm{d}v}_{\text{spatial}}=\underbrace{\int_{\mathcal{P}_\mathrm{R}}\frac{r}{\Theta}\,\rho_\mathrm{R}\,\mathrm{d}V}_{\text{referential}} \tag{4.59}$$

and the rate of entropy loss $\mathcal{H}(\mathcal{S}, t)$ at time t due to heat flow out through the boundary of $\mathcal{S}$,

$$\mathcal{H}(\mathcal{S}, t) = \underbrace{\int_{\partial\mathcal{P}} \frac{h}{\Theta}\, da}_{\text{spatial}} = \underbrace{\int_{\partial\mathcal{P}_R} \frac{h_R}{\Theta}\, dA}_{\text{referential}}. \tag{4.60}$$

Appearing in (4.58)–(4.60) are the **specific entropy** (or entropy per unit mass)

$$\eta = \hat{\eta}(\mathbf{X}, t) = \tilde{\eta}(\mathbf{x}, t)$$

and the **absolute temperature**

$$\Theta = \hat{\Theta}(\mathbf{X}, t) = \tilde{\Theta}(\mathbf{x}, t) \; > \; 0.$$

Both η and Θ are assumed to be continuous functions of space and time.

Using (4.58)–(4.60), we can express the Clausius-Duhem inequality (4.57) in either Eulerian integral form,

$$\frac{d}{dt}\int_{\mathcal{P}} \eta\rho\, dv \geq \int_{\mathcal{P}} \frac{r}{\Theta}\, \rho\, dv - \int_{\partial\mathcal{P}} \frac{h}{\Theta}\, da, \tag{4.61a}$$

or Lagrangian integral form,

$$\frac{d}{dt}\int_{\mathcal{P}_R} \eta\rho_R\, dV \geq \int_{\mathcal{P}_R} \frac{r}{\Theta}\, \rho_R\, dV - \int_{\partial\mathcal{P}_R} \frac{h_R}{\Theta}\, dA. \tag{4.61b}$$

It can be verified that the local versions of (4.61a) and (4.61b) are

$$\rho\dot{\eta} \geq \rho\frac{r}{\Theta} - \operatorname{div}\left(\frac{\mathbf{q}}{\Theta}\right) \tag{4.62a}$$

and

$$\rho_R\,\dot{\eta} \geq \rho_R\frac{r}{\Theta} - \operatorname{Div}\left(\frac{\mathbf{q}_R}{\Theta}\right), \tag{4.62b}$$

respectively; refer to Problem 4.10. It can also be shown that the local Eulerian and Lagrangian forms of conservation of energy

$$\rho\dot{\varepsilon} = \mathbf{T}\cdot\mathbf{D} + \rho r - \operatorname{div}\mathbf{q}, \qquad \rho_R\,\dot{\varepsilon} = \mathbf{P}\cdot\operatorname{Grad}\mathbf{v} + \rho_R\, r - \operatorname{Div}\mathbf{q}_R$$

can be used to eliminate r in (4.62a) and (4.62b), respectively, leading to the local Eulerian and Lagrangian forms of the **reduced Clausius-Duhem inequality**:

$$\rho\Theta\dot{\eta} - \rho\dot{\varepsilon} + \mathbf{T}\cdot\mathbf{D} - \frac{1}{\Theta}\mathbf{q}\cdot\mathbf{g} \geq 0, \tag{4.63a}$$

$$\rho_R\Theta\dot{\eta} - \rho_R\dot{\varepsilon} + \mathbf{P}\cdot\operatorname{Grad}\mathbf{v} - \frac{1}{\Theta}\mathbf{q}_R\cdot\mathbf{g}_R \geq 0; \tag{4.63b}$$

refer to Problem 4.11. Note that

$$\mathbf{g} = \operatorname{grad}\Theta, \qquad \mathbf{g}_R = \operatorname{Grad}\Theta \tag{4.64}$$

are the spatial and referential temperature gradients, related by

$$\mathbf{g}_{R} = \mathbf{F}^{T}\mathbf{g}; \tag{4.65}$$

refer to (3.21).

We remark that the Clausius-Duhem inequality is open to the following criticisms:

(1) It is not clear that physical ideas such as the three restrictions listed at the beginning of this section are included in the Clausius-Duhem inequality. (We will see in the upcoming chapters that some of these physical ideas fall out of the Clausius-Duhem inequality in special thermomechanical processes.)

(2) The Clausius-Duhem inequality is unable to rule out unacceptable behavior in some materials. It is therefore incomplete as a statement of all the physical ideas that are inclusively called "the second law of thermodynamics."

Hence, statements of the second law that either supplement or replace the Clausius-Duhem inequality have been presented by numerous authors, e.g., [3–7]. For instance, an approach championed by Green and Naghdi [7] involves a separation of the Clausius-Duhem inequality into an entropy balance law and isolated statements of second law inequalities. We also mention an approach to the thermodynamic treatment of continua recently set forth by Rajagopal and Srinivasa [8, 9] that appeals to the maximization of the rate of entropy production.

Nevertheless, despite it shortcomings, the Clausius-Duhem inquality remains popular, in large part, because (a) it demands most of the consequences of the second law for most materials, (b) it is compact and straightforward to employ, and (c) it has proven to be useful in many practical applications, as witnessed in Chapters 5–9.

PROBLEM 4.10

Starting with the Eulerian (or spatial) statement of the Clausius-Duhem inequality in integral form,

$$\frac{\mathrm{d}}{\mathrm{d}t}\int_{\mathcal{P}} \eta\rho \,\mathrm{d}v \geq \int_{\mathcal{P}} \frac{r}{\Theta}\rho \,\mathrm{d}v - \int_{\partial\mathcal{P}} \frac{h}{\Theta}\, da,$$

derive the corresponding pointwise form

$$\rho\dot{\eta} \geq \rho\frac{r}{\Theta} - \operatorname{div}\left(\frac{\mathbf{q}}{\Theta}\right).$$

Solution

We begin with the Eulerian integral form of the Clausius-Duhem inequality, i.e.,

$$\frac{\mathrm{d}}{\mathrm{d}t}\int_{\mathcal{P}} \eta\rho \,\mathrm{d}v \geq \int_{\mathcal{P}} \frac{r}{\Theta}\rho \,\mathrm{d}v - \int_{\partial\mathcal{P}} \frac{h}{\Theta}\, da.$$

We first consider the left-hand side of the inequality:

$$\frac{\mathrm{d}}{\mathrm{d}t}\int_{\mathcal{P}} \eta\rho\, \mathrm{d}v = \int_{\mathcal{P}} \left(\dot{\overline{\eta\rho}} + \eta\rho \operatorname{div}\mathbf{v}\right) \mathrm{d}v \qquad \text{(transport theorem (4.20))}$$

$$= \int_{\mathcal{P}} \Big[\rho\dot{\eta} + \eta\left(\dot{\rho} + \rho \operatorname{div}\mathbf{v}\right)\Big] \mathrm{d}v \qquad \text{(product rule)}$$

$$= \int_{\mathcal{P}} \rho\dot{\eta}\, \mathrm{d}v \qquad \text{(conservation of mass (4.36a)).}$$

Next, we consider the second term on the right-hand side of the inequality:

$$\int_{\partial\mathcal{P}} \frac{h}{\Theta}\, \mathrm{d}a = \int_{\partial\mathcal{P}} \frac{\mathbf{q}}{\Theta} \cdot \mathbf{n}\, \mathrm{d}a \qquad \text{(result (4.31))}$$

$$= \int_{\mathcal{P}} \operatorname{div}\left(\frac{\mathbf{q}}{\Theta}\right) \mathrm{d}v \qquad \text{(divergence theorem (2.104)}_2\text{).}$$

Use of these results in the original inequality leads to

$$\int_{\mathcal{P}} \left[\rho\dot{\eta} - \rho\frac{r}{\Theta} + \operatorname{div}\left(\frac{\mathbf{q}}{\Theta}\right)\right] \mathrm{d}v \geq 0.$$

Since the integrand is continuous and the region of integration $\mathcal{P}$ is arbitrary, it follows from the localization theorem (extended to inequalities) that

$$\rho\dot{\eta} - \rho\frac{r}{\Theta} + \operatorname{div}\left(\frac{\mathbf{q}}{\Theta}\right) \geq 0,$$

or

$$\rho\dot{\eta} \geq \rho\frac{r}{\Theta} - \operatorname{div}\left(\frac{\mathbf{q}}{\Theta}\right).$$

PROBLEM 4.11

Show that the local Eulerian form of conservation of energy

$$\rho\dot{\varepsilon} = \mathbf{T}\cdot\mathbf{D} + \rho r - \operatorname{div}\mathbf{q}$$

can be used to eliminate r in the local Eulerian form of the Clausius-Duhem inequality

$$\rho\dot{\eta} \geq \rho\frac{r}{\Theta} - \operatorname{div}\left(\frac{\mathbf{q}}{\Theta}\right)$$

to give the reduced Clausius-Duhem inequality

$$\rho\Theta\dot{\eta} - \rho\dot{\varepsilon} + \mathbf{T}\cdot\mathbf{D} - \frac{1}{\Theta}\mathbf{q}\cdot\mathbf{g} \geq 0.$$

Solution

Multiplying the Clausius-Duhem inequality by the absolute temperature $\Theta > 0$ gives

$$\rho\Theta\dot{\eta} - \rho r + \Theta\,\mathrm{div}\left(\frac{\mathbf{q}}{\Theta}\right) \geq 0.$$

Using conservation of energy to eliminate ρr, we obtain the reduced Clausius-Duhem inequality:

$$\rho\Theta\dot{\eta} - \rho\dot{\varepsilon} + \mathbf{T}\cdot\mathbf{D} - \mathrm{div}\,\mathbf{q} + \Theta\,\mathrm{div}\left(\frac{\mathbf{q}}{\Theta}\right) \geq 0.$$

The fifth term can be rewritten as

$$\mathrm{div}\left(\frac{\mathbf{q}}{\Theta}\right) = \frac{1}{\Theta}\,\mathrm{div}\,\mathbf{q} + \mathbf{q}\cdot\mathrm{grad}\left(\frac{1}{\Theta}\right) \qquad \text{(result } (2.99)_2\text{)}$$

$$= \frac{1}{\Theta}\,\mathrm{div}\,\mathbf{q} - \frac{1}{\Theta^2}\,\mathbf{q}\cdot\mathrm{grad}\,\Theta \qquad \text{(result } (2.99)_5\text{)}$$

$$= \frac{1}{\Theta}\,\mathrm{div}\,\mathbf{q} - \frac{1}{\Theta^2}\,\mathbf{q}\cdot\mathbf{g} \qquad \text{(definition (4.64)),}$$

leading to a simplified form of the reduced Clausius-Duhem inequality:

$$\rho\Theta\dot{\eta} - \rho\dot{\varepsilon} + \mathbf{T}\cdot\mathbf{D} - \frac{1}{\Theta}\mathbf{q}\cdot\mathbf{g} \geq 0.$$

EXERCISES

1. Starting with the Lagrangian (or referential) statement of the Clausius-Duhem inequality in integral form,

$$\frac{\mathrm{d}}{\mathrm{d}t}\int_{\mathcal{P}_\mathrm{R}} \eta\rho_\mathrm{R}\,\mathrm{d}V \geq \int_{\mathcal{P}_\mathrm{R}} \frac{r}{\Theta}\,\rho_\mathrm{R}\,\mathrm{d}V - \int_{\partial\mathcal{P}_\mathrm{R}} \frac{h_\mathrm{R}}{\Theta}\,\mathrm{d}A,$$

derive the corresponding pointwise form

$$\rho_\mathrm{R}\,\dot{\eta} \geq \rho_\mathrm{R}\,\frac{r}{\Theta} - \mathrm{Div}\left(\frac{\mathbf{q}_\mathrm{R}}{\Theta}\right).$$

2. Show that the local Lagrangian form of conservation of energy

$$\rho_\mathrm{R}\,\dot{\varepsilon} = \mathbf{P}\cdot\mathrm{Grad}\,\mathbf{v} + \rho_\mathrm{R}\,r - \mathrm{Div}\,\mathbf{q}_\mathrm{R}$$

can be used to eliminate r in the local Lagrangian form of the Clausius-Duhem inequality

$$\rho_R\, \dot{\eta} \geq \rho_R\, \frac{r}{\Theta} - \mathrm{Div}\left(\frac{\mathbf{q}_R}{\Theta}\right)$$

to give the reduced Clausius-Duhem inequality

$$\rho_R\, \Theta \dot{\eta} - \rho_R \dot{\varepsilon} + \mathbf{P} \cdot \mathrm{Grad}\,\mathbf{v} - \frac{1}{\Theta}\, \mathbf{q}_R \cdot \mathbf{g}_R \geq 0.$$

PART

Constitutive Modeling

CHAPTER

Constitutive Modeling in Mechanics and Thermomechanics

5

PART I: MECHANICS

5.1 FUNDAMENTAL LAWS, CONSTITUTIVE EQUATIONS, A WELL-POSED INITIAL-VALUE BOUNDARY-VALUE PROBLEM

Recall from Chapter 4 that the field equations, i.e., those equations that must be satisfied at each location and time in the domain of interest, are, in the mechanical theory,

$$\rho' + \mathbf{v} \cdot \operatorname{grad} \rho + \rho \operatorname{div} \mathbf{v} = 0, \tag{5.1}$$

$$\operatorname{div} \mathbf{T} + \rho \mathbf{b} = \rho\big[\mathbf{v}' + (\operatorname{grad} \mathbf{v})\,\mathbf{v}\big], \tag{5.2}$$

$$\mathbf{T} = \mathbf{T}^{\mathrm{T}}. \tag{5.3}$$

Equations (5.1)–(5.3) are the local Eulerian forms of conservation of mass, conservation of linear momentum, and conservation of angular momentum, respectively. In this form, the density ρ, velocity $\mathbf{v}$, and Cauchy stress $\mathbf{T}$ are scalar, vector, and tensor functions, respectively, of present position $\mathbf{x}$ and time t. Note that in the mechanical theory (sometimes referred to as the **isothermal**, or constant temperature, theory), we do not concern ourselves with the thermal quantities of temperature Θ, entropy η, heat flux h, and heat supply r, nor the first and second laws of thermodynamics. Recall from Chapter 3 the definitions of the partial derivatives ρ', $\operatorname{grad} \rho$, $\operatorname{grad} \mathbf{v}$, $\operatorname{div} \mathbf{v}$, and $\operatorname{div} \mathbf{T}$ appearing in (5.1)–(5.3), e.g.,

$$\rho' = \frac{\partial}{\partial t}\tilde{\rho}(\mathbf{x}, t), \qquad \operatorname{grad} \rho = \frac{\partial}{\partial \mathbf{x}}\tilde{\rho}(\mathbf{x}, t), \qquad \operatorname{grad} \mathbf{v} = \frac{\partial}{\partial \mathbf{x}}\tilde{\mathbf{v}}(\mathbf{x}, t).$$

The field equations (5.1)–(5.3) must be satisfied at all locations $\mathbf{x}$ in the open region $\mathcal{R}$ of $\mathcal{E}^3$ occupied by the body in the present configuration at each time t, for all time t in the interval of interest. In Cartesian component notation, (5.1)–(5.3) consist of seven coupled nonlinear partial differential equations (PDEs) for the 13 scalar functions $\rho, v_1, v_2, v_3, T_{11}, T_{12}, T_{13}, T_{21}, T_{22}, T_{23}, T_{31}, T_{32}$, and T_{33} of the independent variables x_1, x_2, x_3, and t. To close this system of field equations (i.e., to have the

Fundamentals of Continuum Mechanics

same number of unknowns as equations, which is necessary for a *well-posed* initial-value boundary-value problem formulation), there must be six additional scalar equations.

These six scalar equations, called **constitutive equations**, depend on the physical behavior of the particular material being modeled. (In contrast, the field equations (5.1)–(5.3) are valid for all continua.) In general, different materials will have different constitutive equations. The constitutive equations proposed for a specific material or class of materials must be grounded on physical experiments.

In the Eulerian formulation, the six scalar constitutive equations are a single tensor equation relating the stress $\mathbf{T}$ to the motion χ (and possibly rates, gradients, or the history of the motion). In general, the stress $\mathbf{T}$ at time t can depend on the *entire* history of motion up to time t; for practical models, however, this dependence is reduced to manageable limits, while still including enough dependence to be physically useful. The tensorial constitutive equation defines the response of the continuum to the motion, and is therefore also called the **response function** of the continuum. In general, different materials have different responses to the same motion, and are thus characterized by different constitutive equations.

The tensor constitutive equation relating $\mathbf{T}$ to the motion must be consistent with conservation of angular momentum (5.3), so it must be symmetric in $\mathbf{T}$. Therefore, although the tensor constitutive equation corresponds to nine scalar equations, only six of these are independent, since we must have $T_{21} = T_{12}$, $T_{31} = T_{13}$, and $T_{32} = T_{23}$. By counting equations, we see that these six independent constitutive equations are the field equations necessary to close the set (5.1)–(5.3): the six constitutive equations plus (5.1)–(5.3) are 13 coupled nonlinear PDEs for the 13 functions of x_1, x_2, x_3, and t listed previously.

With the constitutive equations given and the set of equations closed, to complete a well-posed initial-value boundary-value problem formulation, it is necessary to specify the appropriate initial conditions and boundary conditions. The initial conditions and boundary conditions necessary in a particular problem depend on the differential operators present in the constitutive equation, as well as the differential operators appearing in the field equations (5.1) and (5.2). For instance, since first-order time derivatives of ρ and $\mathbf{v}$ appear in (5.1) and (5.2), it is necessary to give initial values of ρ and $\mathbf{v}$ throughout the domain. Initial conditions on ρ' and $\mathbf{v}'$, etc., will be necessary if there are higher-order time derivatives in the constitutive equations; see Table 5.1.

The Lagrangian formulation of the mechanical theory differs from the Eulerian formulation in that (1) it has different unknowns, (2) it has different differential operators in the field equations and hence different initial and boundary conditions, and (3) the concept of a reference configuration requires that information be supplied about the reference state. The independent variables in the Lagrangian formulation are reference position $\mathbf{X}$ and time t; see Table 5.2.

Table 5.1 Eulerian Formulation of the Mechanical Initial-Value Boundary-Value Problem

Domain	All $\mathbf{x} \in \mathcal{R}(t)$, the region occupied by the body in the present configuration at time t
Time interval	$t_o \leq t \leq t_f$
Governing equations	$\rho' + \mathbf{v} \cdot \operatorname{grad} \rho + \rho \operatorname{div} \mathbf{v} = 0$ $\operatorname{div} \mathbf{T} + \rho \mathbf{b} = \rho[\mathbf{v}' + (\operatorname{grad} \mathbf{v})\, \mathbf{v}]$ $\mathbf{T} = \breve{\mathbf{T}}(\text{motion})$ satisfying $\mathbf{T} = \mathbf{T}^{\mathrm{T}}$
Specified functions	$\mathbf{b} = \tilde{\mathbf{b}}(\mathbf{x}, t)$ specified for all $\mathbf{x} \in \mathcal{R}(t)$, $t_o \leq t \leq t_f$
Unknowns	$\tilde{\rho}(\mathbf{x}, t)$, $\tilde{\mathbf{v}}(\mathbf{x}, t)$, $\tilde{\mathbf{T}}(\mathbf{x}, t)$
Initial conditions	Specified at $t = t_o$ for all $\mathbf{x} \in \mathcal{R}(t_o)$
Boundary conditions	Specified on $\mathbf{x} \in \partial\mathcal{R}(t)$ for $t_o \leq t \leq t_f$

Table 5.2 Lagrangian Formulation of the Mechanical Initial-Value Boundary-Value Problem

Domain	All $\mathbf{X} \in \mathcal{R}_R$, the region occupied by the body in the reference configuration
Time interval	$t_o \leq t \leq t_f$
Governing equations	$\rho J = \rho_R$ $\operatorname{Div} \mathbf{P} + \rho_R \mathbf{b} = \rho_R \ddot{\chi}$ $\mathbf{P} = \breve{\mathbf{P}}(\text{motion})$ satisfying $\mathbf{P}\mathbf{F}^{\mathrm{T}} = \mathbf{F}\mathbf{P}^{\mathrm{T}}$
Specified functions	$\mathbf{b} = \hat{\mathbf{b}}(\mathbf{X}, t)$ specified for all $\mathbf{X} \in \mathcal{R}_R$, $t_o \leq t \leq t_f$ $\rho_R = \hat{\rho}_R(\mathbf{X})$ specified for all $\mathbf{X} \in \mathcal{R}_R$
Unknowns	$\hat{\rho}(\mathbf{X}, t)$, $\chi(\mathbf{X}, t)$, $\hat{\mathbf{P}}(\mathbf{X}, t)$
Initial conditions	Specified at $t = t_o$ for all $\mathbf{X} \in \mathcal{R}_R$
Boundary conditions	Specified on $\mathbf{X} \in \partial\mathcal{R}_R$ for $t_o \leq t \leq t_f$

EXERCISES

1. Write the field equations (5.1)–(5.3) in indicial notation. Then fully expand them into their Cartesian component forms. Verify that the resulting system of equations consists of seven scalar PDEs for 13 scalar unknowns.

5.2 RESTRICTIONS ON THE CONSTITUTIVE EQUATIONS

As discussed in Section 5.1, the tensor constitutive equation relating the Cauchy stress $\mathbf{T}$ to the motion χ must satisfy conservation of angular momentum. Additionally, it must satisfy invariance requirements, and perhaps some material symmetry conditions. These restrictions will prove useful. For instance, in Chapter 7, we will show how these restrictions allow us to reduce a proposed form of the constitutive equation for viscous fluids. Such reductions are welcomed since they allow the fluid to be characterized with fewer laboratory experiments.

5.2.1 INVARIANCE UNDER SUPERPOSED RIGID BODY MOTIONS

In this book, we require that the constitutive equation for the Cauchy stress $\mathbf{T}$ be invariant under all possible **superposed rigid body motions (SRBMs)** of the body. (Refer to Appendix A for other notions of invariance.) Invariance under SRBMs demands that if two motions of a body composed of the same material differ only by a SRBM, then physically the internal response generated in the two motions must be the same, apart from orientation. *The response due to the SRBM should be zero.* If it is not, the proposed constitutive equation is not physical.

5.2.1.1 Superposed rigid body motions

Consider two motions, given by the functions $\mathbf{x} = \chi(\mathbf{X}, t)$ and $\mathbf{x}^+ = \chi^+(\mathbf{X}, t^+)$, both referred to the same reference configuration. A particle at reference position $\mathbf{X}$ is mapped by motion χ to position $\mathbf{x}$ at time t; this same particle at reference position $\mathbf{X}$ is mapped by motion χ^+ to position $\mathbf{x}^+$ at time t^+. Another particle, at position $\mathbf{Y}$ in the reference configuration, is mapped by motion χ to position $\mathbf{y}$ at time t, and by motion χ^+ to position $\mathbf{y}^+$ at time t^+; see Figure 5.1. The motions χ and χ^+ differ by a SRBM if

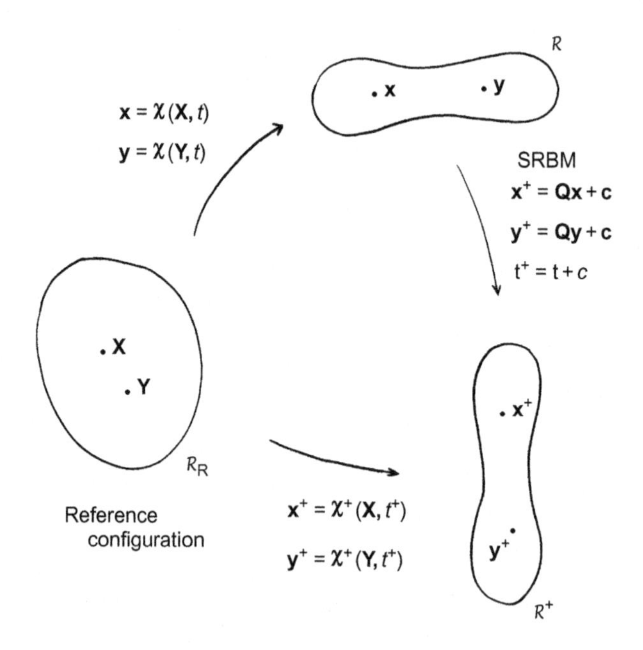

FIGURE 5.1

Two motions χ and χ^+ that differ by a SRBM.

(1) the elapsed time between an arbitrary pair of instants (or events) t_1 and t_2, and t_1^+ and t_2^+, is the same,
(2) the order in which two distinct instants t_1 and t_2, and t_1^+ and t_2^+, occur is the same, and
(3) the distance between an arbitrary pair of particles $\mathbf{x}$ and $\mathbf{y}$, and $\mathbf{x}^+$ and $\mathbf{y}^+$, at any particular instant is the same.

Necessary and sufficient conditions that the two motions $\mathbf{x} = \chi(\mathbf{X}, t)$ and $\mathbf{x}^+ = \chi^+(\mathbf{X}, t^+)$ differ by a SRBM are that $\mathbf{x}^+$ and t^+ are related to $\mathbf{x}$ and t by

$$\mathbf{x}^+ = \mathbf{Q}(t)\,\mathbf{x} + \mathbf{c}(t), \tag{5.4a}$$

$$t^+ = t + c, \tag{5.4b}$$

where $\mathbf{Q}(t)$ is a proper orthogonal tensor-valued function of time t, $\mathbf{c}(t)$ is a vector-valued function of time t, and c is a scalar constant. (Refer to Section 2.2.6 for a discussion of proper orthogonal tensors.) Said differently,

$$\text{motions } \mathbf{x} = \chi(\mathbf{X}, t) \text{ and } \mathbf{x}^+ = \chi^+(\mathbf{X}, t^+) \text{ differ by a SRBM}$$

$$\text{if and only if} \tag{5.5}$$

$$\mathbf{x}^+ = \mathbf{Q}(t)\,\mathbf{x} + \mathbf{c}(t) \quad \text{and} \quad t^+ = t + c;$$

refer to Problem 5.1. Physically, the proper orthogonal tensor $\mathbf{Q}(t)$ can be interpreted as a rigid body rotation, and the vector $\mathbf{c}(t)$ can be interpreted as a rigid body translation. Note that the linear equation (5.4a) can be inverted, i.e.,

$$\mathbf{x} = \mathbf{Q}^{\mathrm{T}}(t)\mathbf{x}^+ - \mathbf{Q}^{\mathrm{T}}(t)\mathbf{c}(t),$$

so

$$\frac{\partial \mathbf{x}}{\partial \mathbf{x}^+} = \mathbf{Q}^{\mathrm{T}}. \tag{5.6}$$

Also, we can verify that

$$\left(\mathbf{A}^{\mathrm{T}}\right)^+ = \left(\mathbf{A}^+\right)^{\mathrm{T}}, \qquad \left(\mathbf{A}^{-1}\right)^+ = \left(\mathbf{A}^+\right)^{-1}, \qquad \left(\dot{\mathbf{A}}\right)^+ = \dot{\overline{\mathbf{A}^+}} \tag{5.7}$$

for any tensor $\mathbf{A}$.

PROBLEM 5.1
Prove that motions $\mathbf{x} = \chi(\mathbf{X}, t)$ and $\mathbf{x}^+ = \chi^+(\mathbf{X}, t^+)$ differ by a SRBM if and only if

$$\mathbf{x}^+ = \mathbf{Q}(t)\,\mathbf{x} + \mathbf{c}(t) \qquad \text{and} \qquad t^+ = t + c.$$

Solution
To prove *necessity*, we must show that if two motions $\mathbf{x} = \chi(\mathbf{X}, t)$ and $\mathbf{x}^+ = \chi^+(\mathbf{X}, t^+)$ differ by a SRBM, then

$$\mathbf{x}^+ = \mathbf{Q}(t)\,\mathbf{x} + \mathbf{c}(t) \quad \text{and} \quad t^+ = t + c$$

for all proper orthogonal $\mathbf{Q}(t)$. In other words, we must show that requirements (1)-(3) discussed in Section 5.2.1.1 imply (5.4a) and (5.4b).

Requirements (1) and (2) demand that $t^+ = t + c$, i.e., the two time scales differ only by their choice of an origin, so we have shown (5.4b). We now show that requirement (3) implies (5.4a).

At time t, a particle in the one motion occupies position $\mathbf{x}$, given by the function

$$\mathbf{x} = \chi(\mathbf{X}, t).$$

At corresponding instant $t^+ = t + c$ in the other motion, the same particle occupies position $\mathbf{x}^+$, given by the function

$$\mathbf{x}^+ = \chi^+(\mathbf{X}, t^+). \tag{a}$$

Another particle at time t in the first motion occupies position $\mathbf{y}$, given by

$$\mathbf{y} = \chi(\mathbf{Y}, t),$$

and at time t^+ in the second motion occupies position $\mathbf{y}^+$, given by

$$\mathbf{y}^+ = \chi^+(\mathbf{Y}, t^+). \tag{b}$$

To summarize, the motion χ takes the particles occupying reference positions $\mathbf{X}$ and $\mathbf{Y}$ to present positions $\mathbf{x}$ and $\mathbf{y}$ at time t, while the motion χ^+ takes those same particles to present positions $\mathbf{x}^+$ and $\mathbf{y}^+$ at time t^+. Recalling the inverse relation $\mathbf{X} = \chi^{-1}(\mathbf{x}, t)$ and using $t^+ = t + c$, we may rewrite (a) as

$$\mathbf{x}^+ = \chi^+(\mathbf{X}, t^+) = \chi^+(\chi^{-1}(\mathbf{x}, t),\ t + c) \equiv \tilde{\chi}^+(\mathbf{x}, t).$$

Similarly, (b) may be rewritten as

$$\mathbf{y}^+ = \tilde{\chi}^+(\mathbf{y}, t).$$

Requirement (3) demands that the distance $|\mathbf{x} - \mathbf{y}|$ between $\mathbf{x}$ and $\mathbf{y}$ at time t is the same as the distance $|\mathbf{x}^+ - \mathbf{y}^+|$ between $\mathbf{x}^+$ and $\mathbf{y}^+$ at time $t^+ = t + c$. This is equivalent to

$$(\mathbf{x}^+ - \mathbf{y}^+) \cdot (\mathbf{x}^+ - \mathbf{y}^+) = (\mathbf{x} - \mathbf{y}) \cdot (\mathbf{x} - \mathbf{y}),$$

or

$$\left[\tilde{\chi}^+(\mathbf{x}, t) - \tilde{\chi}^+(\mathbf{y}, t)\right] \cdot \left[\tilde{\chi}^+(\mathbf{x}, t) - \tilde{\chi}^+(\mathbf{y}, t)\right] = (\mathbf{x} - \mathbf{y}) \cdot (\mathbf{x} - \mathbf{y}), \tag{c}$$

for all $\mathbf{x}$, $\mathbf{y}$, and t. It follows from $(2.99)_3$ that differentiation of (c) with respect to $\mathbf{x}$ gives

$$2\left[\frac{\partial \tilde{\chi}^+(\mathbf{x}, t)}{\partial \mathbf{x}}\right]^{\mathrm{T}} \left[\tilde{\chi}^+(\mathbf{x}, t) - \tilde{\chi}^+(\mathbf{y}, t)\right] = 2(\mathbf{x} - \mathbf{y}). \tag{d}$$

Then, differentiation of (d) with respect to $\mathbf{y}$ gives

$$-2\left[\frac{\partial \tilde{\chi}^+(\mathbf{x}, t)}{\partial \mathbf{x}}\right]^{\mathrm{T}} \left[\frac{\partial \tilde{\chi}^+(\mathbf{y}, t)}{\partial \mathbf{y}}\right] = -2\mathbf{I},$$

from which we deduce that

$$\left[\frac{\partial\tilde{\chi}^+(\mathbf{x},t)}{\partial\mathbf{x}}\right]^{\mathrm{T}} = \left[\frac{\partial\tilde{\chi}^+(\mathbf{y},t)}{\partial\mathbf{y}}\right]^{-1}. \tag{e}$$

In (e) we have that a tensor-valued function of $(\mathbf{y},t)$ is equal to a tensor-valued function of $(\mathbf{x},t)$. Since $\mathbf{x}$ and $\mathbf{y}$ are independent, we deduce that both tensor-valued functions depend on t alone, i.e.,

$$\frac{\partial\tilde{\chi}^+(\mathbf{x},t)}{\partial\mathbf{x}} = \mathbf{Q}(t), \qquad \frac{\partial\tilde{\chi}^+(\mathbf{y},t)}{\partial\mathbf{y}} = \mathbf{Q}(t), \tag{f}$$

and thus, by (e),

$$\mathbf{Q}^{\mathrm{T}}(t) = \mathbf{Q}^{-1}(t). \tag{g}$$

Equation (g) implies that $\mathbf{Q}(t)$ is orthogonal (refer to (2.51)), so $\det\mathbf{Q} = \pm 1$. However, the SRBM under consideration must include as a special case $\tilde{\chi}^+(\mathbf{x},t) = \mathbf{x}$. For this particular case, $\mathbf{Q} = \mathbf{I}$ and $\det\mathbf{Q} = 1$. Since motions are continuous, we must have $\det\mathbf{Q} = 1$ always, so $\mathbf{Q}(t)$ is proper orthogonal. The first-order PDE $(\mathrm{f})_1$ (or system of first-order PDEs in component form) can then be integrated to give

$$\mathbf{x}^+ = \tilde{\chi}^+(\mathbf{x},t) = \mathbf{Q}(t)\,\mathbf{x} + \mathbf{c}(t),$$

which is (5.4a). We have thus completed the *necessity* proof.

Now, to prove *sufficiency*, we must show that if

$$\mathbf{x}^+ = \mathbf{Q}(t)\,\mathbf{x} + \mathbf{c}(t) \quad \text{and} \quad t^+ = t + c,$$

with $\mathbf{Q}(t)$ proper orthogonal, then motions $\mathbf{x} = \chi(\mathbf{X},t)$ and $\mathbf{x}^+ = \chi^+(\mathbf{X},t^+)$ differ by a SRBM. In other words, we must show that (5.4a) and (5.4b) imply conditions (1)-(3) discussed in Section 5.2.1.1.

Equation (5.4b) obviously implies requirements (1) and (2). From (5.4a), it follows that for any pair of particles,

$$\begin{aligned}
|\mathbf{x}^+ - \mathbf{y}^+|^2 &= (\mathbf{x}^+ - \mathbf{y}^+)\cdot(\mathbf{x}^+ - \mathbf{y}^+) \\
&= (\mathbf{Qx} + \mathbf{c} - \mathbf{Qy} - \mathbf{c})\cdot(\mathbf{Qx} + \mathbf{c} - \mathbf{Qy} - \mathbf{c}) \\
&= (\mathbf{Qx} - \mathbf{Qy})\cdot(\mathbf{Qx} - \mathbf{Qy}) \\
&= \mathbf{Q}(\mathbf{x} - \mathbf{y})\cdot\mathbf{Q}(\mathbf{x} - \mathbf{y}) \\
&= (\mathbf{x} - \mathbf{y})\cdot(\mathbf{x} - \mathbf{y}) \\
&= |\mathbf{x} - \mathbf{y}|^2,
\end{aligned}$$

so the distance $|\mathbf{x}^+ - \mathbf{y}^+|$ between $\mathbf{x}^+$ and $\mathbf{y}^+$ is the same as the distance $|\mathbf{x} - \mathbf{y}|$ between $\mathbf{x}$ and $\mathbf{y}$. Thus, requirement (3) is satisfied.

EXERCISES

1. Verify that for any tensor $\mathbf{A}$,
   (a) $\left(\mathbf{A}^{\mathrm{T}}\right)^{+} = \left(\mathbf{A}^{+}\right)^{\mathrm{T}}$,
   (b) $\left(\mathbf{A}^{-1}\right)^{+} = \left(\mathbf{A}^{+}\right)^{-1}$,
   (c) $\left(\dot{\mathbf{A}}\right)^{+} = \overline{\mathbf{A}^{+}}^{\cdot}$.

5.2.1.2 Relationships between geometric and kinematic quantities under a SRBM

In addition to preserving the distance between a pair of particles (refer to Problem 5.1), it can be shown that the transformation (5.4a) also preserves the angle θ between any two nonzero vectors, so

$$\cos\theta^{+} = \cos\theta; \tag{5.8}$$

refer to Problem 5.2. Now consider a surface in the present configuration that contains two points $\mathbf{x}$ and $\mathbf{y}$, and has outward unit normal $\mathbf{n}$. Since the SRBM (5.4a) is angle preserving, the angle between $\mathbf{n}$ and $\mathbf{x}-\mathbf{y}$ is the same as the angle between $\mathbf{n}^{+}$ and $\mathbf{x}^{+}-\mathbf{y}^{+}$. As a consequence, it can be shown that

$$\mathbf{n}^{+} = \mathbf{Q}\mathbf{n}; \tag{5.9}$$

refer to Problem 5.3. Similarly, it can be shown that the infinitesimal line elements $d\mathbf{x}^{+}$ and $d\mathbf{x}$ (refer to Section 3.3.1) differ only by a rotation, i.e.,

$$d\mathbf{x}^{+} = \mathbf{Q}\,d\mathbf{x}. \tag{5.10}$$

Recall that many kinematical quantities, such as $\mathbf{v}$, $\mathbf{F}$, $\mathbf{C}$, $\mathbf{L}$, $\mathbf{D}$, and $\mathbf{W}$, are derived from the motion χ. The SRBM χ^{+} will likewise be accompanied by $\mathbf{v}^{+}$, $\mathbf{F}^{+}$, $\mathbf{C}^{+}$, $\mathbf{L}^{+}$, $\mathbf{D}^{+}$, and $\mathbf{W}^{+}$. The transformation (5.4a) imposes relationships between these two sets of quantities. For instance,

$$\mathbf{v}^{+} = \dot{\mathbf{x}}^{+} = \overline{\mathbf{Q}(t)\,\mathbf{x} + \mathbf{c}(t)}^{\cdot} = \dot{\mathbf{Q}}(t)\,\mathbf{x} + \mathbf{Q}(t)\mathbf{v} + \dot{\mathbf{c}}(t). \tag{5.11}$$

Defining

$$\boldsymbol{\Omega}(t) = \dot{\mathbf{Q}}(t)\,\mathbf{Q}^{\mathrm{T}}(t) \tag{5.12}$$

allows us to rewrite (5.11) as

$$\mathbf{v}^{+} = \boldsymbol{\Omega}\mathbf{Q}\mathbf{x} + \mathbf{Q}\mathbf{v} + \dot{\mathbf{c}}. \tag{5.13}$$

It can be shown (refer to Problem 5.4) that the tensor $\boldsymbol{\Omega}(t)$ is skew. Another fundamental kinematic relationship is

$$\mathbf{F}^{+} = \frac{\partial \mathbf{x}^{+}}{\partial \mathbf{X}} = \frac{\partial \mathbf{x}^{+}}{\partial \mathbf{x}}\frac{\partial \mathbf{x}}{\partial \mathbf{X}} = \frac{\partial}{\partial \mathbf{x}}\left[\mathbf{Q}(t)\mathbf{x} + \mathbf{c}(t)\right]\frac{\partial \mathbf{x}}{\partial \mathbf{X}} = \mathbf{Q}\mathbf{F}. \tag{5.14}$$

Using (5.13) and (5.14), we can verify that

$$\mathbf{C}^{+} = \mathbf{C}, \qquad \mathbf{B}^{+} = \mathbf{Q}\mathbf{B}\mathbf{Q}^{\mathrm{T}}, \qquad \mathbf{U}^{+} = \mathbf{U}, \qquad \mathbf{V}^{+} = \mathbf{Q}\mathbf{V}\mathbf{Q}^{\mathrm{T}}, \qquad \mathbf{R}^{+} = \mathbf{Q}\mathbf{R},$$

$$\mathbf{E}^+ = \mathbf{E}, \qquad \mathbf{L}^+ = \mathbf{QLQ}^{\mathrm{T}} + \boldsymbol{\Omega}, \qquad \mathbf{D}^+ = \mathbf{QDQ}^{\mathrm{T}}, \qquad \mathbf{W}^+ = \mathbf{QWQ}^{\mathrm{T}} + \boldsymbol{\Omega}, \tag{5.15}$$

$$J^+ = J, \qquad \rho^+ = \rho, \qquad da^+ = da, \qquad dv^+ = dv,$$

where $\mathbf{Q}(t)$ is proper orthogonal and $\boldsymbol{\Omega}(t)$ is skew; refer to Problems 5.5–5.10.

PROBLEM 5.2

Show that $\cos\theta^+ = \cos\theta$.

Solution

Consider three different particles: one at reference position $\mathbf{X}$, another at reference position $\mathbf{Y}$, and a third at reference position $\mathbf{Z}$. The motions χ and χ^+, which differ by a SRBM, map these particles to positions $\mathbf{x}, \mathbf{y}, \mathbf{z}$ at time t, and $\mathbf{x}^+, \mathbf{y}^+, \mathbf{z}^+$ at time t^+, respectively. Let θ be the angle between vectors $\mathbf{x}-\mathbf{y}$ and $\mathbf{x}-\mathbf{z}$, and θ^+ be the angle between vectors $\mathbf{x}^+ - \mathbf{y}^+$ and $\mathbf{x}^+ - \mathbf{z}^+$. It follows from definition (2.54) that

$$\cos\theta^+ = \frac{\mathbf{x}^+ - \mathbf{y}^+}{|\mathbf{x}^+ - \mathbf{y}^+|} \cdot \frac{\mathbf{x}^+ - \mathbf{z}^+}{|\mathbf{x}^+ - \mathbf{z}^+|}.$$

The relations

$$\mathbf{x}^+ = \mathbf{Qx} + \mathbf{c}, \qquad \mathbf{y}^+ = \mathbf{Qy} + \mathbf{c}, \qquad \mathbf{z}^+ = \mathbf{Qz} + \mathbf{c},$$

and the preservation of length under a SRBM, then imply that

$$\cos\theta^+ = \frac{\mathbf{Q}(\mathbf{x}-\mathbf{y}) \cdot \mathbf{Q}(\mathbf{x}-\mathbf{z})}{|\mathbf{x}-\mathbf{y}||\mathbf{x}-\mathbf{z}|}.$$

Orthogonality of $\mathbf{Q}$ (refer to definition (2.50)) demands that

$$\cos\theta^+ = \frac{\mathbf{Q}(\mathbf{x}-\mathbf{y}) \cdot \mathbf{Q}(\mathbf{x}-\mathbf{z})}{|\mathbf{x}-\mathbf{y}||\mathbf{x}-\mathbf{z}|} = \frac{(\mathbf{x}-\mathbf{y}) \cdot (\mathbf{x}-\mathbf{z})}{|\mathbf{x}-\mathbf{y}||\mathbf{x}-\mathbf{z}|} = \cos\theta.$$

PROBLEM 5.3

Prove that $\mathbf{n}^+ = \mathbf{Qn}$.

Solution

Consider a surface in the present configuration that contains two points $\mathbf{x}$ and $\mathbf{y}$, and has outward unit normal $\mathbf{n}$. Since the SRBM (5.4a) is angle preserving (refer, for instance, to Problem 5.2), the angle between $\mathbf{n}$ and $\mathbf{x}-\mathbf{y}$ is the same as the angle between $\mathbf{n}^+$ and $\mathbf{x}^+ - \mathbf{y}^+$, i.e.,

$$\frac{\mathbf{n} \cdot (\mathbf{x}-\mathbf{y})}{|\mathbf{n}|\,|\mathbf{x}-\mathbf{y}|} = \frac{\mathbf{n}^+ \cdot (\mathbf{x}^+ - \mathbf{y}^+)}{|\mathbf{n}^+|\,|\mathbf{x}^+ - \mathbf{y}^+|}.$$

The vectors $\mathbf{n}$ and $\mathbf{n}^+$ are unit vectors, so $|\mathbf{n}| = |\mathbf{n}^+| = 1$. Since SRBMs are length preserving, $|\mathbf{x}^+ - \mathbf{y}^+| = |\mathbf{x} - \mathbf{y}|$. Also,

$$\mathbf{x}^+ - \mathbf{y}^+ = (\mathbf{Q}\mathbf{x} + \mathbf{c}) - (\mathbf{Q}\mathbf{y} + \mathbf{c}) = \mathbf{Q}\mathbf{x} - \mathbf{Q}\mathbf{y} = \mathbf{Q}(\mathbf{x} - \mathbf{y}).$$

These results imply that the original expression simplifies to

$$\mathbf{n} \cdot (\mathbf{x} - \mathbf{y}) = \mathbf{n}^+ \cdot \mathbf{Q}(\mathbf{x} - \mathbf{y}),$$

or

$$(\mathbf{n} - \mathbf{Q}^T\mathbf{n}^+) \cdot (\mathbf{x} - \mathbf{y}) = 0.$$

Since $\mathbf{n} - \mathbf{Q}^T\mathbf{n}^+$ is independent of $\mathbf{x} - \mathbf{y}$, we have

$$\mathbf{n} = \mathbf{Q}^T\mathbf{n}^+,$$

or, premultiplying both sides by $\mathbf{Q}$,

$$\mathbf{n}^+ = \mathbf{Q}\mathbf{n}.$$

PROBLEM 5.4

Prove that the tensor $\mathbf{\Omega} = \dot{\mathbf{Q}}\mathbf{Q}^T$ is skew.

Solution

Since $\mathbf{Q}$ is orthogonal,

$$\mathbf{Q}^T\mathbf{Q} = \mathbf{I}.$$

The material derivative of this expression is

$$\dot{\overline{\mathbf{Q}^T\mathbf{Q}}} = \dot{\mathbf{I}}.$$

It follows from (3.22) and (3.23) that

$$\dot{\mathbf{Q}}^T\mathbf{Q} + \mathbf{Q}^T\dot{\mathbf{Q}} = \mathbf{0}.$$

Use of definition (5.12) leads to

$$\mathbf{Q}^T\mathbf{\Omega}^T\mathbf{Q} + \mathbf{Q}^T\mathbf{\Omega}\mathbf{Q} = \mathbf{0},$$

which implies that

$$\mathbf{\Omega}^T + \mathbf{\Omega} = \mathbf{0},$$

or

$$\mathbf{\Omega}^T = -\mathbf{\Omega}.$$

Hence, $\mathbf{\Omega}$ is skew (refer to definition (2.16)).

PROBLEM 5.5
Prove that $\mathbf{C}^+ = \mathbf{C}$.

Solution

$$\mathbf{C}^+ = (\mathbf{F}^\mathrm{T})^+\mathbf{F}^+ = (\mathbf{F}^+)^\mathrm{T}\mathbf{F}^+ = (\mathbf{QF})^\mathrm{T}\mathbf{QF} = \mathbf{F}^\mathrm{T}\mathbf{Q}^\mathrm{T}\mathbf{QF} = \mathbf{F}^\mathrm{T}\mathbf{IF} = \mathbf{F}^\mathrm{T}\mathbf{F} = \mathbf{C}.$$

PROBLEM 5.6
Prove that $\mathbf{U}^+ = \mathbf{U}$.

Solution

$$\mathbf{U}^+ = \sqrt{\mathbf{C}^+} = \sqrt{\mathbf{C}} = \mathbf{U}.$$

PROBLEM 5.7
Prove that $\mathbf{L}^+ = \mathbf{QLQ}^\mathrm{T} + \mathbf{\Omega}$.

Solution

$$\mathbf{L}^+ = \frac{\partial \mathbf{v}^+}{\partial \mathbf{x}^+} = \frac{\partial \mathbf{v}^+}{\partial \mathbf{x}}\frac{\partial \mathbf{x}}{\partial \mathbf{x}^+} = \frac{\partial}{\partial \mathbf{x}}(\mathbf{\Omega Qx} + \mathbf{Qv} + \dot{\mathbf{c}})\frac{\partial \mathbf{x}}{\partial \mathbf{x}^+} = (\mathbf{\Omega Q} + \mathbf{QL})\mathbf{Q}^\mathrm{T} = \mathbf{\Omega} + \mathbf{QLQ}^\mathrm{T}.$$

Alternatively,

$$\mathbf{L}^+ = (\dot{\mathbf{F}})^+(\mathbf{F}^{-1})^+ = \dot{\overline{\mathbf{F}^+}}(\mathbf{F}^+)^{-1} = \dot{\overline{\mathbf{QF}}}(\mathbf{QF})^{-1} = (\dot{\mathbf{Q}}\mathbf{F} + \mathbf{Q}\dot{\mathbf{F}})\mathbf{F}^{-1}\mathbf{Q}^{-1}$$

$$= \dot{\mathbf{Q}}\mathbf{FF}^{-1}\mathbf{Q}^{-1} + \mathbf{Q}\dot{\mathbf{F}}\mathbf{F}^{-1}\mathbf{Q}^{-1} = \dot{\mathbf{Q}}\mathbf{Q}^{-1} + \mathbf{QLQ}^{-1} = \dot{\mathbf{Q}}\mathbf{Q}^\mathrm{T} + \mathbf{QLQ}^\mathrm{T} = \mathbf{\Omega} + \mathbf{QLQ}^\mathrm{T}.$$

PROBLEM 5.8
Prove that $\mathbf{D}^+ = \mathbf{QDQ}^\mathrm{T}$.

Solution

$$\mathbf{D}^+ = \frac{1}{2}\left[\mathbf{L}^+ + (\mathbf{L}^\mathrm{T})^+\right] \qquad \text{(definition (3.58)}_1\text{)}$$

$$= \frac{1}{2}\left[\mathbf{L}^+ + (\mathbf{L}^+)^\mathrm{T}\right] \qquad \text{(result (5.7)}_1\text{)}$$

$$= \frac{1}{2}\left[\mathbf{QLQ}^\mathrm{T} + \mathbf{\Omega} + (\mathbf{QLQ}^\mathrm{T} + \mathbf{\Omega})^\mathrm{T}\right] \qquad \text{(result (5.15)}_7\text{)}$$

$$= \frac{1}{2}\left[\mathbf{QLQ}^{\mathrm{T}} + \boldsymbol{\Omega} + \left(\mathbf{QLQ}^{\mathrm{T}}\right)^{\mathrm{T}} + \boldsymbol{\Omega}^{\mathrm{T}}\right] \qquad \text{(result (2.14)}_1\text{)}$$

$$= \frac{1}{2}\left[\mathbf{QLQ}^{\mathrm{T}} + \boldsymbol{\Omega} + \mathbf{QL}^{\mathrm{T}}\mathbf{Q}^{\mathrm{T}} + \boldsymbol{\Omega}^{\mathrm{T}}\right] \qquad \text{(results (2.14)}_2\text{ and (2.14)}_3\text{)}$$

$$= \frac{1}{2}\left[\mathbf{QLQ}^{\mathrm{T}} + \mathbf{QL}^{\mathrm{T}}\mathbf{Q}^{\mathrm{T}}\right] \qquad (\boldsymbol{\Omega}\text{ is skew, definition (2.16))}$$

$$= \mathbf{Q}\left[\frac{1}{2}\left(\mathbf{L} + \mathbf{L}^{\mathrm{T}}\right)\right]\mathbf{Q}^{\mathrm{T}} \qquad \text{(result (2.12))}$$

$$= \mathbf{QDQ}^{\mathrm{T}} \qquad \text{(definition (3.58)}_1\text{).}$$

PROBLEM 5.9

Prove that $J^+ = J$.

Solution

$$J^+ = \det \mathbf{F}^+ = \det(\mathbf{QF}) = (\det \mathbf{Q})(\det \mathbf{F}) = \det \mathbf{F} = J.$$

PROBLEM 5.10

Prove that $da^+ = da$.

Solution

It follows from (3.75) that

$$da^+ = \mathbf{n}^+ \cdot \left(\overset{1+}{d\mathbf{x}} \times \overset{2+}{d\mathbf{x}}\right).$$

Then, using (2.62), (5.9), and (5.10), we have

$$da^+ = \mathbf{Qn} \cdot \left(\mathbf{Q}\overset{1}{d\mathbf{x}} \times \mathbf{Q}\overset{2}{d\mathbf{x}}\right) = \left[\mathbf{Qn} \quad \mathbf{Q}\overset{1}{d\mathbf{x}} \quad \mathbf{Q}\overset{2}{d\mathbf{x}}\right].$$

Definition (2.64) and proper orthogonality of $\mathbf{Q}$ then imply that

$$da^+ = (\det \mathbf{Q})\left[\mathbf{n} \quad \overset{1}{d\mathbf{x}} \quad \overset{2}{d\mathbf{x}}\right] = \left[\mathbf{n} \quad \overset{1}{d\mathbf{x}} \quad \overset{2}{d\mathbf{x}}\right] = \mathbf{n} \cdot \left(\overset{1}{d\mathbf{x}} \times \overset{2}{d\mathbf{x}}\right) = da.$$

This result—that the element of area remains unaltered under a SRBM—is consistent with the length-preserving and angle-preserving properties of transformation (5.4a).

EXERCISES

1. Prove that $d\mathbf{x}^+ = \mathbf{Q}\,d\mathbf{x}$.
2. Prove that $\mathbf{B}^+ = \mathbf{QBQ}^T$.
3. Prove that $\mathbf{V}^+ = \mathbf{QVQ}^T$.
4. Show that $\mathbf{R}^+ = \mathbf{QR}$.
5. Verify that $\mathbf{E}^+ = \mathbf{E}$.
6. Show that $\mathbf{W}^+ = \mathbf{QWQ}^T + \boldsymbol{\Omega}$.
7. Verify that $\rho^+ = \rho$.
8. Prove that $dv^+ = dv$.

5.2.1.3 Relationships between kinetic quantities under a SRBM

Consider now the tractions $\mathbf{t}$ and $\mathbf{t}^+$ associated with the motions χ and χ^+, respectively, i.e.,

$$\mathbf{t} = \mathbf{t}(\mathbf{x}, t, \mathbf{n}), \qquad \mathbf{t}^+ = \mathbf{t}^+(\mathbf{x}^+, t^+, \mathbf{n}^+),$$

where

$$\mathbf{x}^+ = \mathbf{Qx} + \mathbf{c}, \qquad t^+ = t + c, \qquad \mathbf{n}^+ = \mathbf{Qn}.$$

It is physical to *assume* that $\mathbf{t}$ is unaltered apart from orientation by a SRBM, i.e.,

(1) the magnitude of $\mathbf{t}(\mathbf{x}, t, \mathbf{n})$ is equal to the magnitude of $\mathbf{t}^+(\mathbf{x}^+, t^+, \mathbf{n}^+)$, and
(2) the angle θ between $\mathbf{t}(\mathbf{x}, t, \mathbf{n})$ and $\mathbf{x} - \mathbf{y}$, for any $\mathbf{y}$, is the same as the angle θ^+ between $\mathbf{t}^+(\mathbf{x}^+, t^+, \mathbf{n}^+)$ and $\mathbf{x}^+ - \mathbf{y}^+$.

In mathematical language, these **invariance assumptions** become

$$|\mathbf{t}| = |\mathbf{t}^+| \tag{5.16a}$$

and

$$\frac{\mathbf{t}\cdot(\mathbf{x}-\mathbf{y})}{|\mathbf{t}||\mathbf{x}-\mathbf{y}|} = \frac{\mathbf{t}^+\cdot(\mathbf{x}^+-\mathbf{y}^+)}{|\mathbf{t}^+||\mathbf{x}^+-\mathbf{y}^+|}. \tag{5.16b}$$

It can be shown that as a consequence of (5.16a) and (5.16b),

$$\mathbf{t}^+ = \mathbf{Qt}, \tag{5.17}$$

which, in turn, implies that

$$\mathbf{T}^+ = \mathbf{QTQ}^T; \tag{5.18}$$

refer to Problems 5.11 and 5.12. Also,

$$\mathbf{P}^+ = \mathbf{QP}, \qquad \mathbf{S}^+ = \mathbf{S}. \tag{5.19}$$

PROBLEM 5.11
Prove that $\mathbf{t}^{+} = \mathbf{Qt}$.

Solution
Recall invariance assumption (5.16b), i.e.,

$$\frac{\mathbf{t}\cdot(\mathbf{x}-\mathbf{y})}{|\mathbf{t}||\mathbf{x}-\mathbf{y}|} = \frac{\mathbf{t}^{+}\cdot(\mathbf{x}^{+}-\mathbf{y}^{+})}{|\mathbf{t}^{+}||\mathbf{x}^{+}-\mathbf{y}^{+}|}.$$

Since the distance between two particles is unaltered by a SRBM,

$$|\mathbf{x}^{+}-\mathbf{y}^{+}| = |\mathbf{x}-\mathbf{y}|.$$

This observation, together with invariance assumption (5.16a), implies that (5.16b) becomes

$$\mathbf{t}\cdot(\mathbf{x}-\mathbf{y}) = \mathbf{t}^{+}\cdot(\mathbf{x}^{+}-\mathbf{y}^{+}).$$

Then, use of

$$\mathbf{x}^{+}-\mathbf{y}^{+} = (\mathbf{Qx}+\mathbf{c})-(\mathbf{Qy}+\mathbf{c}) = \mathbf{Qx}-\mathbf{Qy} = \mathbf{Q}(\mathbf{x}-\mathbf{y})$$

leads us to conclude that

$$\mathbf{t}\cdot(\mathbf{x}-\mathbf{y}) = \mathbf{t}^{+}\cdot\mathbf{Q}(\mathbf{x}-\mathbf{y}),$$

or

$$(\mathbf{t}-\mathbf{Q}^{\mathrm{T}}\mathbf{t}^{+})\cdot(\mathbf{x}-\mathbf{y}) = 0.$$

Since $\mathbf{t}-\mathbf{Q}^{\mathrm{T}}\mathbf{t}^{+}$ and $\mathbf{x}$ are independent of $\mathbf{y}$, it follows from $(2.99)_3$ that differentiation of this scalar expression with respect to vector $\mathbf{y}$ gives

$$\mathbf{Q}^{\mathrm{T}}\mathbf{t}^{+}-\mathbf{t} = \mathbf{0},$$

or

$$\mathbf{t}^{+} = \mathbf{Qt}.$$

PROBLEM 5.12
Prove that $\mathbf{T}^{+} = \mathbf{QTQ}^{\mathrm{T}}$.

Solution
Recall from Problem 5.11 that $\mathbf{t}^{+} = \mathbf{Qt}$. Also recall from Section 4.6 that $\mathbf{t} = \mathbf{Tn}$. Together, these results imply that

$$\mathbf{T}^{+}\mathbf{n}^{+} = \mathbf{QTn}.$$

Use of $\mathbf{n}^+ = \mathbf{Qn}$ leads to

$$(\mathbf{T}^+\mathbf{Q} - \mathbf{QT})\mathbf{n} = \mathbf{0}.$$

Since the coefficient of the unit normal $\mathbf{n}$ is independent of $\mathbf{n}$ itself, and $\mathbf{n}$ is arbitrary, we conclude that the coefficient vanishes:

$$\mathbf{T}^+\mathbf{Q} - \mathbf{QT} = \mathbf{0}.$$

Thus,

$$\mathbf{T}^+ = \mathbf{QTQ}^{\mathrm{T}}.$$

EXERCISES

1. Show that $\mathbf{P}^+ = \mathbf{QP}$.
2. Verify that $\mathbf{S}^+ = \mathbf{S}$.

5.2.1.4 Invariance requirements

In summary, for the Cauchy stress $\mathbf{T}$ to be invariant under all possible SRBMs of the body, any proposed constitutive equation for $\mathbf{T}$ must satisfy

$$\mathbf{T}^+ = \mathbf{QTQ}^{\mathrm{T}}$$

when

$$\mathbf{x}^+ = \mathbf{Qx} + \mathbf{c}, \qquad \mathbf{v}^+ = \mathbf{\Omega Qx} + \mathbf{Qv} + \dot{\mathbf{c}}, \qquad \mathbf{F}^+ = \mathbf{QF},$$

$$\mathbf{D}^+ = \mathbf{QDQ}^{\mathrm{T}}, \qquad \mathbf{W}^+ = \mathbf{QWQ}^{\mathrm{T}} + \mathbf{\Omega}, \qquad \text{etc.,}$$

for all proper orthogonal $\mathbf{Q}(t)$, skew $\mathbf{\Omega}(t)$, and vector functions $\mathbf{c}(t)$ and $\dot{\mathbf{c}}(t)$. Collectively, these conditions are referred to as **invariance requirements**. As will become evident in Chapters 6 and 7, invariance requirements allow us to reduce postulated forms of the constitutive equation for the Cauchy stress $\mathbf{T}$ for different materials. Such reductions are welcomed since they allow the characterization of a material from fewer laboratory experiments.

5.2.2 MATERIAL SYMMETRY

In addition to angular momentum and invariance requirements, the constitutive equation for the Cauchy stress $\mathbf{T}$ generally must satisfy material symmetry conditions. The motion of a body $\mathcal{B}$ may be referred to any number of reference configurations (refer to Section 3.1). Let ${}_0\kappa$ and ${}_1\kappa$ be two such reference configurations. Consider a typical particle Y of $\mathcal{B}$ located at position ${}_0\mathbf{X}$ in reference configuration ${}_0\kappa$, position ${}_1\mathbf{X}$ in reference configuration ${}_1\kappa$, and position $\mathbf{x}$ in the present configuration at time t (see Figure 5.2). In the notation of Section 3.1,

$${}_0\mathbf{X} = {}_0\kappa(Y), \qquad \mathbf{x} = \bar{\chi}(Y,t) = \bar{\chi}\left({}_0\kappa^{-1}({}_0\mathbf{X}),t\right) = {}_0\chi({}_0\mathbf{X},t),$$

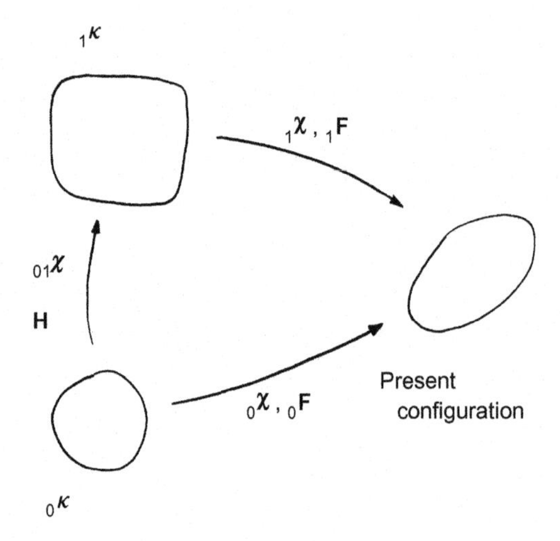

FIGURE 5.2

A motion referred to two reference configurations.

$$_1\mathbf{X} = {}_1\kappa(Y), \qquad \mathbf{x} = \bar{\chi}(Y,t) = \bar{\chi}\left({}_1\kappa^{-1}({}_1\mathbf{X}),t\right) = {}_1\chi({}_1\mathbf{X},t).$$

It follows that

$$_1\mathbf{X} = {}_1\chi^{-1}(\mathbf{x},t) = {}_1\chi^{-1}\left({}_0\chi({}_0\mathbf{X},t),t\right) = {}_{01}\chi({}_0\mathbf{X},t).$$

We denote

$$_0\mathbf{F} = \frac{\partial\, {}_0\chi}{\partial\, {}_0\mathbf{X}}, \qquad {}_1\mathbf{F} = \frac{\partial\, {}_1\chi}{\partial\, {}_1\mathbf{X}}, \qquad \mathbf{H} = \frac{\partial\, {}_1\mathbf{X}}{\partial\, {}_0\mathbf{X}}. \tag{5.20}$$

From (5.20) we have

$$\mathbf{H} = {}_1\mathbf{F}^{-1}{}_0\mathbf{F} \qquad \text{or} \qquad {}_1\mathbf{F} = {}_0\mathbf{F}\mathbf{H}^{-1}.$$

Suppose the constitutive equation for the Cauchy stress **T** is assumed to depend on the deformation gradient with respect to a specific reference configuration, say, $_0\kappa$. (This assumption is appropriate for an elastic solid, as will soon become apparent in Chapter 6.) Then,

$$\mathbf{T} = {}_0\check{\mathbf{T}}({}_0\mathbf{F}).$$

The choice of reference configuration is arbitrary. Hence, the constitutive equation may just as well be assumed to depend on the deformation gradient with respect to another reference configuration, say, $_1\kappa$:

$$\mathbf{T} = {}_1\check{\mathbf{T}}({}_1\mathbf{F}).$$

Hence, it follows that

$$_0\check{\mathbf{T}}({}_0\mathbf{F}) = {}_1\check{\mathbf{T}}({}_0\mathbf{F}\mathbf{H}^{-1}). \tag{5.21}$$

In general, the constitutive equation for **T** depends on the choice of reference configuration. For instance, ${}_0\kappa$ may be stress-free, and ${}_1\kappa$ may be stressed. We would not expect the addition of the same amount of stretch to each configuration to illicit the same response. It may happen, however, that the two reference configurations ${}_0\kappa$ and ${}_1\kappa$ are *equivalent* in the sense that constitutive functions referred to ${}_0\kappa$ are identical to those referred to ${}_1\kappa$. If ${}_0\kappa$ and ${}_1\kappa$ are equivalent, then it follows from (5.21) that

$$ {}_0\check{\mathbf{T}}\,({}_0\mathbf{F}) = {}_0\check{\mathbf{T}}\,({}_0\mathbf{F}\mathbf{H}^{-1}), \tag{5.22}$$

and the tensor **H** is called a **symmetry transformation** with respect to the reference configuration ${}_0\kappa$.

Consider now a particular reference configuration κ, and the set of all configurations ${}_i\kappa$, $i = 1, 2, \ldots$, equivalent to κ. Denote by ${}_i\mathbf{H}$ the gradient of the reference map ${}_i\kappa$ with respect to the reference map κ. Each of ${}_i\mathbf{H}$ is then a symmetry transformation with respect to the reference configuration κ; see Figure 5.3. The set of all such symmetry transformations forms a group with respect to tensor multiplication, and is called the **symmetry group** $\mathcal{G}$ of the material relative to the reference configuration κ.

One of the properties of a material is the amount of symmetry it possesses, which is reflected in its particular symmetry group $\mathcal{G}$ for a given κ. Relations such as (5.22) must hold for all **H** in $\mathcal{G}$. *The restrictions thus placed on the constitutive functions can therefore tell us much about the form of these functions.* Hence, material symmetry considerations play a role in reducing the functional dependence of the response functions for a particular material.

On physical grounds, it can be argued that if the density of a material is altered, then the material will exhibit different behavior, i.e., the response function will be altered. Therefore, for **H** to be an element of the symmetry group $\mathcal{G}$, it must be

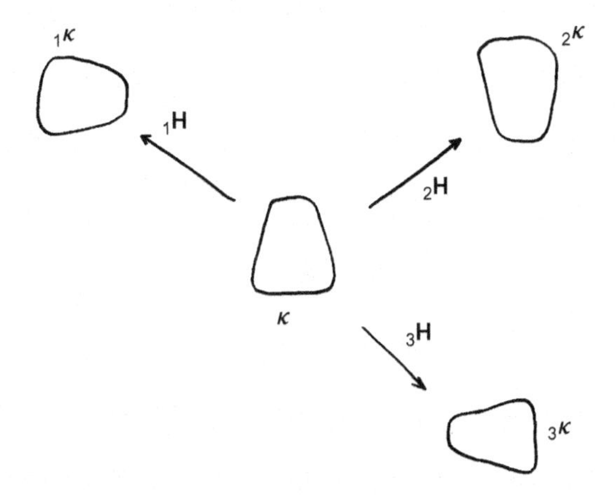

FIGURE 5.3

Equivalent reference configurations.

unimodular (i.e., $\det \mathbf{H} = \pm 1$), so density is preserved between the two reference configurations.

The most symmetry a material can possess, relative to a particular reference configuration κ, is $\mathcal{G} = \mu$, where μ is the full unimodular group. For such a material, there is no difference in the response from any two configurations having the same density. This has been taken as one definition of a **fluid.** By this definition, a fluid is a material for which $\mathcal{G} = \mu$, i.e., the symmetry group is the full unimodular group.

The least symmetry that a material can possess in a particular reference configuration is $\mathcal{G} = \{\mathbf{I}\}$. In this case the response functions in the reference configuration are different from those in any other reference configuration.

A material is said to be **isotropic** if there exists a reference configuration, called the undistorted or stress-free configuration, for which the symmetry group $\mathcal{G}$ contains the full orthogonal group $\mathcal{O}$. Hence, for the isotropic material, there is a stress-free state in which there is no preferred orientation.

A **solid** can be defined as a material that has different responses from two reference configurations if the change from one configuration to the other involves a deformation (as opposed to a rigid rotation or translation). Therefore, under this definition, a solid possesses a reference configuration for which the symmetry group $\mathcal{G}$ is a subgroup of the full orthogonal group $\mathcal{O}$.

Reviewing our definitions, we have the following:

fluid (maximal symmetry): $\mathcal{G} = \mu$;

isotropic material: $\mathcal{O} \subseteq \mathcal{G}$;

solid: $\mathcal{G} \subseteq \mathcal{O}$.

Since all orthogonal tensors are unimodular (i.e., $\mathcal{O} \subset \mu$), we note that fluids are isotropic.

It can be shown that there exists no group $\mathcal{A}$ such that $\mathcal{O} \subset \mathcal{A} \subset \mu$, so $\mathcal{O}$ is maximal in μ. This implies that all isotropic materials are either solids ($\mathcal{G} = \mathcal{O}$) or fluids ($\mathcal{G} = \mu$). However, not all solids are isotropic, since there are proper subgroups of the full orthogonal group (e.g., $\{\mathbf{I}\} \subset \mathcal{O}$). Therefore,

isotropic solid: $\mathcal{G} = \mathcal{O}$ (for the stress-free configuration);

anisotropic solid: $\mathcal{G} \subset \mathcal{O}$ (for the stress-free configuration).

PART II: THERMOMECHANICS

In the thermomechanical theory, the mechanical equations of motion are supplemented by the first and second laws of thermodynamics, and an appropriate set of constitutive equations for the thermal and mechanical quantities.

5.3 FUNDAMENTAL LAWS, CONSTITUTIVE EQUATIONS, THERMOMECHANICAL PROCESSES

Recall from Chapter 4 that the field equations and second law inequality in the thermomechanical theory are

$$\dot{\rho} + \rho \operatorname{div} \dot{\mathbf{x}} = 0, \tag{5.23}$$

$$\operatorname{div} \mathbf{T} + \rho \mathbf{b} = \rho \ddot{\mathbf{x}}, \tag{5.24}$$

$$\mathbf{T} = \mathbf{T}^{\mathrm{T}}, \tag{5.25}$$

$$\rho \dot{\varepsilon} = \mathbf{T} \cdot \mathbf{D} + \rho r - \operatorname{div} \mathbf{q}, \tag{5.26}$$

$$-\rho \dot{\varepsilon} + \mathbf{T} \cdot \mathbf{D} + \rho \Theta \dot{\eta} - \frac{1}{\Theta} \mathbf{q} \cdot \operatorname{grad} \Theta \geq 0. \tag{5.27}$$

Equations (5.23)–(5.26) are the local Eulerian forms of conservation of mass, conservation of linear momentum, conservation of angular momentum, and the first law of thermodynamics (conservation of energy). Inequality (5.27) is the Clausius-Duhem inequality, our particular statement of the second law of thermodynamics. Involved in (5.23)–(5.27) are the quantities

$$\{\mathbf{x}, \eta\}, \tag{5.28}$$

$$\{\mathbf{T}, \mathbf{q}, \varepsilon, \Theta\}, \tag{5.29}$$

$$\{\rho, \mathbf{b}, r\}. \tag{5.30}$$

These quantities have been conceptually divided into three sets. *This division is not for the purposes of solving initial-value boundary-value problems, but rather for the purposes of placing restrictions on the constitutive equations (or response functions).*

The elements of set (5.28) are the **independent variables.** Other quantities will be determined by constitutive equations that depend on these independent variables. For the purposes of restricting the forms of the constitutive equations, we will be able to specify the motion $\bar{\chi}(Y, t)$ and entropy $\bar{\eta}(Y, t)$ as any function of material particle Y and time t, with the confidence that (5.23), (5.24), and (5.26) can be satisfied.

The elements of set (5.29) are the **dependent variables.** The stress $\mathbf{T}$, heat flux vector $\mathbf{q}$, internal energy ε, and temperature Θ at a particular particle Y and time t are determined through constitutive equations from $\bar{\chi}(Y, \tau)$ and $\bar{\eta}(Y, \tau)$, including time and space derivatives, with $\tau \leq t$ (called the **history** of χ and η). Therefore, the specification of the motion χ and entropy η throughout the body and time interval of interest also fixes $\mathbf{T}$, $\mathbf{q}$, ε, and Θ.

The elements set of (5.30) are the **balancing terms.** With χ and η arbitrarily specified and $\mathbf{T}$, $\mathbf{q}$, ε, and Θ determined from χ and η through the response functions, the density ρ, body force $\mathbf{b}$, and heat source r are assigned the values necessary to ensure that (5.23), (5.24), and (5.26) are satisfied. Note that the balancing term ρ

depends only on the motion $\boldsymbol{\chi}$, and is independent of the choice of the constitutive functions, whereas the balancing terms $\mathbf{b}$ and r can be determined only after the constitutive assumptions have been made.

Therefore, the field equations (5.23)–(5.26) and inequality (5.27) are employed as follows: The conservation laws for mass (5.23), linear momentum (5.24), and energy (5.26) are employed to determine ρ, $\mathbf{b}$, and r. Then, the conservation law for angular momentum (5.25) and the Clausius-Duhem inequality (5.27) place restrictions on the response functions for $\mathbf{T}$, $\mathbf{q}$, ε, and Θ. A group of quantities $\boldsymbol{\chi}$, η, $\mathbf{T}$, $\mathbf{q}$, ε, Θ, ρ, $\mathbf{b}$, and r, each a function of material particle Y and time t, which satisfies the conservation laws (5.23)–(5.26) and Clausius-Duhem inequality (5.27) for all space and time in the domain of interest is called a **thermomechanical process**.

From the above discussion, we see that we can specify the motion $\boldsymbol{\chi}$ and entropy η as functions of space and time arbitrarily and still have them part of a thermomechanical process (i.e., a process that satisfies all the equations and inequality of motion), as long as the constitutive equations for $\mathbf{T}$, $\mathbf{q}$, ε, and Θ satisfy the restrictions imposed by conservation of angular momentum and the Clausius-Duhem inequality, and ρ, $\mathbf{b}$, and r are the functions of space and time demanded by conservation of mass (to determine ρ), conservation of linear momentum (to determine $\mathbf{b}$), and conservation of energy (to determine r).

As an example, suppose we specify

$$\boldsymbol{\chi}(\mathbf{X}, t) = \mathbf{X}, \qquad \hat{\eta}(\mathbf{X}, t) = 10t, \tag{5.31}$$

for all $\mathbf{X}$, t, so each particle remains at its reference location, and the entropy of each particle increases linearly in time. We can then explicitly calculate

$$\dot{\mathbf{x}} = \ddot{\mathbf{x}} = \mathbf{0}, \qquad \mathbf{F} = \mathbf{I}, \qquad J = 1, \qquad \dot{\eta} = 10, \qquad \text{etc.} \tag{5.32}$$

Constitutive equations have been provided for $\mathbf{T}$, $\mathbf{q}$, ε, and Θ, so they are determined as

$$\mathbf{T} = \tilde{\mathbf{T}}(\boldsymbol{\chi}, \eta) = \tilde{\mathbf{T}}(\mathbf{X}, 10t),$$

$$\mathbf{q} = \tilde{\mathbf{q}}(\boldsymbol{\chi}, \eta) = \tilde{\mathbf{q}}(\mathbf{X}, 10t),$$

$$\varepsilon = \tilde{\varepsilon}(\boldsymbol{\chi}, \eta) = \tilde{\varepsilon}(\mathbf{X}, 10t),$$

$$\Theta = \tilde{\Theta}(\boldsymbol{\chi}, \eta) = \tilde{\Theta}(\mathbf{X}, 10t).$$

With the functions $\tilde{\mathbf{T}}$, $\tilde{\mathbf{q}}$, $\tilde{\varepsilon}$, and $\tilde{\Theta}$ so prescribed for all $\mathbf{X}$ and t, the derivatives $\operatorname{div} \mathbf{T}$, $\dot{\varepsilon}$, $\operatorname{div} \mathbf{q}$, and $\operatorname{grad} \Theta$ appearing in (5.23)–(5.27) are also prescribed. Then from (5.23), (5.24), and (5.26) we can solve for

$$\hat{\rho}(\mathbf{X}, t) = \rho_{\mathrm{R}}, \qquad \hat{\mathbf{b}}(\mathbf{X}, t) = -\frac{1}{\rho_{\mathrm{R}}} \operatorname{div} \mathbf{T}, \qquad \hat{r}(\mathbf{X}, t) = \frac{\operatorname{div} \mathbf{q}}{\rho_{\mathrm{R}}} + \dot{\varepsilon},$$

which are now uniquely determined, from conservation of mass, linear momentum, and energy, in terms of the known functions ρ_{R}, $\operatorname{div} \mathbf{T}$, $\operatorname{div} \mathbf{q}$, and $\dot{\varepsilon}$. We know that if we could create these density, body force, and heat supply fields, and produce a material characterized by our chosen constitutive equations, we would observe the thermomechanical process with motion $\mathbf{x} = \mathbf{X}$ and entropy $\eta = 10t$.

The conceptual division given in (5.28)–(5.30) is arbitrary, and was chosen on the basis of the form of the Clausius-Duhem inequality (5.27). Its drawback is the choice of entropy η as an independent variable. Any physical understanding that most people have of entropy is as a vague, conceptual measure of "disorder in the universe." This understanding is then often exceeded when one is asked to propose a constitutive equation, say, for stress $\mathbf{T}$, as a function of entropy.

A much more comfortable choice as an independent variable is absolute temperature Θ rather than entropy η, since the notion of stress depending on temperature (which can easily be measured with a thermometer or thermocouple) is more familiar. To accomplish the change of independent variable from η to Θ, we must replace the dependent variable ε with ψ through the **Legendre transformation**

$$\psi = \varepsilon - \Theta\eta, \tag{5.33}$$

where ψ is the **Helmholtz free energy** and ε is the internal energy. Under the Legendre transformation (5.33), the fundamental laws (5.23)–(5.26) and the Clausius-Duhem inequality (5.27) become

$$\dot{\rho} + \rho \operatorname{div} \dot{\mathbf{x}} = 0, \tag{5.23}$$

$$\operatorname{div} \mathbf{T} + \rho \mathbf{b} = \rho \ddot{\mathbf{x}}, \tag{5.24}$$

$$\mathbf{T} = \mathbf{T}^{\mathrm{T}}, \tag{5.25}$$

$$\rho(\dot{\psi} + \Theta\dot{\eta} + \dot{\Theta}\eta) = \mathbf{T} \cdot \mathbf{D} + \rho r - \operatorname{div} \mathbf{q}, \tag{5.34}$$

$$-\rho\dot{\psi} + \mathbf{T} \cdot \mathbf{D} - \rho\eta\dot{\Theta} - \frac{1}{\Theta}\mathbf{q} \cdot \operatorname{grad} \Theta \geq 0. \tag{5.35}$$

The independent variables, dependent variables, and balancing terms are then, as an alternative to (5.28)–(5.30),

$$\{\mathbf{x}, \Theta\}, \tag{5.36}$$

$$\{\mathbf{T}, \mathbf{q}, \psi, \eta\}, \tag{5.37}$$

$$\{\rho, \mathbf{b}, r\}. \tag{5.38}$$

The conservation laws for mass (5.23), linear momentum (5.24), and energy (5.26) or (5.34) hold for *all* thermomechanical materials. To complete the continuum model for a *particular* material or class of materials, we must

- postulate constitutive relations for $\mathbf{T}$, $\mathbf{q}$, ε, and Θ at point Y and time t, in terms of $\bar{\chi}(Y, \tau)$ and $\bar{\eta}(Y, \tau)$ and their space and time derivatives, for $\tau \leq t$, or
- postulate constitutive relations for $\mathbf{T}$, $\mathbf{q}$, ψ, and η in terms of $\bar{\chi}(Y, \tau)$ and $\bar{\eta}(Y, \tau)$ and their space and time derivatives, for $\tau \leq t$.

Either way, the specific choice of constitutive equations characterizes the material for all motion and entropy histories in the first case, or all motion and temperature histories in the second case.

EXERCISES

1. Show that use of the Legendre transformation (5.33) in inequality (5.27) leads to inequality (5.35).

5.4 RESTRICTIONS ON THE CONSTITUTIVE EQUATIONS

In addition to being grounded on physical experiments, the constitutive equations proposed for a specific material or class of materials in the thermomechanical theory must satisfy

(1) conservation of angular momentum (5.25),
(2) invariance requirements under SRBMs (there is more on this in Section 5.4.1),
(3) the Clausius-Duhem inequality (5.27) or (5.35) for *all* thermomechanical processes, and perhaps
(4) some material symmetry requirements (refer to Section 5.2.2).

5.4.1 INVARIANCE UNDER SUPERPOSED RIGID BODY MOTIONS

In this section, we present invariance relations for the thermal quantities, which supplement the invariance relations developed in Section 5.2.1 for the geometric, kinematic, and kinetic quantities.

5.4.1.1 Relationships between thermal quantities under a SRBM

It is reasonable to *assume* that the specific internal energy ε, specific heat supply r, heat flux h, specific entropy η, and absolute temperature Θ are unaltered by a SRBM, i.e.,

$$\varepsilon^+ = \varepsilon, \qquad r^+ = r, \qquad h^+ = h, \qquad \eta^+ = \eta, \qquad \Theta^+ = \Theta. \tag{5.39}$$

As a consequence of these **invariance assumptions**, we can show that

$$\mathbf{g}^+ = \mathbf{Q}\mathbf{g}, \qquad \mathbf{g}_{\mathrm{R}}^+ = \mathbf{g}_{\mathrm{R}}, \qquad \psi^+ = \psi, \qquad \mathbf{q}^+ = \mathbf{Q}\mathbf{q}, \qquad \mathbf{q}_{\mathrm{R}}^+ = \mathbf{q}_{\mathrm{R}}; \tag{5.40}$$

refer to Problems 5.13–5.15. Recall that $\mathbf{g}$ and $\mathbf{g}_{\mathrm{R}}$ are the spatial and referential temperature gradients (refer, for instance, to (4.64) and (4.65)), ψ is the Helmholtz free energy (refer, for instance, to (5.33)), and $\mathbf{q}$ and $\mathbf{q}_{\mathrm{R}}$ are the spatial and referential heat flux vectors (refer, for instance, to (4.31), $(4.49)_2$, and (4.53)).

If, in addition, we *assume*

$$\mathbf{b}^+ - \ddot{\mathbf{x}}^+ = \mathbf{Q}(\mathbf{b} - \ddot{\mathbf{x}}), \qquad \mathcal{P}^+ = \mathcal{P}, \tag{5.41}$$

it can be shown that our invariance relationships together with the integral form of conservation of energy (4.19) can be used to obtain the integral forms of conservation of mass (4.11a), linear momentum (4.11b), and angular momentum (4.11c). This demonstrates that our invariance *assumptions* (5.16a), (5.16b), (5.39), and (5.41) are sensible in that they are consistent with the conservation laws.

5.4.1.2 Invariance requirements

In summary, we find that for formulation (5.36)–(5.38), the constitutive equations for the Cauchy stress $\mathbf{T}$, heat flux vector $\mathbf{q}$, Helmholtz free energy ψ, and entropy η must satisfy

$$\mathbf{T}^+ = \mathbf{Q}\mathbf{T}\mathbf{Q}^{\mathrm{T}}, \qquad \mathbf{q}^+ = \mathbf{Q}\mathbf{q}, \qquad \psi^+ = \psi, \qquad \eta^+ = \eta$$

when

$$\mathbf{x}^+ = \mathbf{Q}\mathbf{x} + \mathbf{c}, \qquad \mathbf{v}^+ = \boldsymbol{\Omega}\mathbf{Q}\mathbf{x} + \mathbf{Q}\mathbf{v} + \dot{\mathbf{c}}, \qquad \mathbf{F}^+ = \mathbf{Q}\,\mathbf{F},$$

$$\mathbf{D}^+ = \mathbf{Q}\mathbf{D}\mathbf{Q}^{\mathrm{T}}, \qquad \mathbf{W}^+ = \mathbf{Q}\mathbf{W}\mathbf{Q}^{\mathrm{T}} + \boldsymbol{\Omega}, \qquad \Theta^+ = \Theta, \qquad \text{etc.},$$

for all proper orthogonal $\mathbf{Q}(t)$, skew $\boldsymbol{\Omega}(t)$, and vector functions $\mathbf{c}(t)$ and $\dot{\mathbf{c}}(t)$.

PROBLEM 5.13

Prove that $\mathbf{g}^+ = \mathbf{Q}\mathbf{g}$.

Solution

From definition $(4.64)_1$ and use of the chain rule, we have

$$\mathbf{g}^+ = \operatorname{grad}\Theta^+ = \frac{\partial\Theta^+}{\partial\mathbf{x}^+} = \left(\frac{\partial\mathbf{x}}{\partial\mathbf{x}^+}\right)^{\mathrm{T}}\frac{\partial\Theta^+}{\partial\mathbf{x}}.$$

It then follows from (5.6) and $(5.39)_5$ that

$$\mathbf{g}^+ = (\mathbf{Q}^{\mathrm{T}})^{\mathrm{T}}\frac{\partial\Theta}{\partial\mathbf{x}} = \mathbf{Q}\mathbf{g}.$$

PROBLEM 5.14

Prove that $\mathbf{g}_{\mathrm{R}}^+ = \mathbf{g}_{\mathrm{R}}$.

Solution

We have

$$\mathbf{g}_{\mathrm{R}}^+ = \operatorname{Grad}\Theta^+ = \frac{\partial\Theta^+}{\partial\mathbf{X}} = \frac{\partial\Theta}{\partial\mathbf{X}} = \operatorname{Grad}\Theta = \mathbf{g}_{\mathrm{R}}.$$

Alternatively,

$$\mathbf{g}_{\mathrm{R}}^+ = (\mathbf{F}^{\mathrm{T}})^+\mathbf{g}^+ = (\mathbf{F}^+)^{\mathrm{T}}\mathbf{g}^+ = (\mathbf{Q}\mathbf{F})^{\mathrm{T}}\mathbf{Q}\mathbf{g} = \mathbf{F}^{\mathrm{T}}\mathbf{Q}^{\mathrm{T}}\mathbf{Q}\mathbf{g} = \mathbf{F}^{\mathrm{T}}\mathbf{I}\,\mathbf{g} = \mathbf{F}^{\mathrm{T}}\mathbf{g} = \mathbf{g}_{\mathrm{R}}.$$

PROBLEM 5.15
Prove that $\mathbf{q}^+ = \mathbf{Qq}$.

Solution
It follows from the invariance assumption $h^+ = h$ that

$$\mathbf{q}^+ \cdot \mathbf{n}^+ = \mathbf{q} \cdot \mathbf{n}.$$

Use of the result $\mathbf{n}^+ = \mathbf{Qn}$ and definition (2.13) of the transpose leads to

$$(\mathbf{Q}^T\mathbf{q}^+ - \mathbf{q}) \cdot \mathbf{n} = 0.$$

Since $\mathbf{Q}^T\mathbf{q}^+ - \mathbf{q}$ is independent of $\mathbf{n}$,

$$\mathbf{Q}^T\mathbf{q}^+ - \mathbf{q} = \mathbf{0},$$

or

$$\mathbf{q}^+ = \mathbf{Qq}.$$

EXERCISES

1. Prove that $\psi^+ = \psi$.

2. Verify that $\mathbf{q}_R^+ = \mathbf{q}_R$.

3. Show that the invariance relations developed in Sections 5.2.1 and 5.4.1, together with the integral form of conservation of energy (4.19), can be used to obtain the integral forms of conservation of mass (4.11a), conservation of linear momentum (4.11b), and conservation of angular momentum (4.11c). (Hint: Conservation of energy must be invariant under *all possible* SRBMs, i.e., *for any* proper orthogonal $\mathbf{Q}(t)$, skew $\mathbf{\Omega}(t)$, and vector functions $\mathbf{c}(t)$ and $\dot{\mathbf{c}}(t)$. To derive the conservation laws of mass, linear momentum, and angular momentum, consider invariance of conservation of energy under several *particular* SRBMs. For instance, to derive conservation of mass and conservation of linear momentum, consider the particular SRBM that has, at time t, $\mathbf{Q} = \mathbf{I}$, $\mathbf{\Omega} = \dot{\mathbf{\Omega}} = \mathbf{0}$, $\dot{\mathbf{c}} = \mathbf{c}_0 \equiv$ constant, and $\ddot{\mathbf{c}} = \mathbf{0}$. Then, to derive conservation of angular momentum, consider the particular SRBM that has, at time t, $\mathbf{Q} = \mathbf{I}$, $\mathbf{\Omega} = \mathbf{\Omega}_0 \equiv$ constant, $\dot{\mathbf{\Omega}} = \mathbf{0}$, and $\dot{\mathbf{c}} = \ddot{\mathbf{c}} = \mathbf{0}$.)

CHAPTER

Nonlinear Elasticity

6

This chapter presents constitutive equations appropriate for a broad class of engineering materials known as **nonlinear elastic solids**. Common examples include rubber, elastomers (rubberlike polymers), and soft biological tissues. Nonlinear elastic solids are characterized by their ability to undergo large deformations before yielding, and their highly nonlinear stress-strain response.[1,2] Hence, they exhibit *geometric nonlinearity* (or strain-displacement nonlinearity) owing to finite elastic deformations, and *material nonlinearity* owing to nonlinear constitutive response. Sections 6.1 and 6.2 discuss nonlinear elastic materials in the context of the mechanical (isothermal) theory and the thermomechanical theory, respectively. In the latter case, we explicitly illustrate how the constitutive equations must satisfy the second law of thermodynamics, invariance, conservation of angular momentum, and material symmetry (isotropy).

6.1 MECHANICAL THEORY

Recall from Section 5.1 that in the Eulerian formulation of the mechanical theory, a constitutive assumption is made on the Cauchy stress $\mathbf{T}$ as a function of the motion χ, and possibly its rates, gradients, and history, i.e.,

$$\mathbf{T} = \check{\mathbf{T}}\,(\text{motion}).$$

For an **elastic material**, which has a perfect memory of some reference configuration, the Cauchy stress $\mathbf{T}$ at position $\mathbf{x}$ and time t depends on the motion only through the strain at $\mathbf{x}$ and t with respect to the reference configuration. The deformation gradient $\mathbf{F} = \text{Grad}\,\chi$ (refer to Section 3.3.1) is a measure of this relative strain, so we may write

$$\mathbf{T} = \check{\mathbf{T}}(\mathbf{F}). \tag{6.1}$$

[1] An elastic material has a perfect memory of its reference configuration. So, if an elastic material deforms away from its reference configuration, it returns once the force responsible for the deformation is removed. Further, the response of an elastic material does not depend on the history or the rate of the deformation.

[2] *Large* deformations are also referred to as *finite* deformations, much like *small* deformations are referred to as *infinitesimal* deformations.

Fundamentals of Continuum Mechanics

We emphasize that the stress at position $\mathbf{x}$ and time t in an elastic material depends only on the strain at $\mathbf{x}$ and t, and neither the *history* nor the *rate* of the strain (refer to Appendix B).

Equation (6.1) implies that the stress $\mathbf{T}$ at position $\mathbf{x}$ and time t is an *explicit* function of the deformation gradient $\mathbf{F} = \text{Grad}\, \chi$ with respect to some reference configuration $\mathbf{X}$, evaluated at the same $\mathbf{x}$ and t. Therefore, *implicit* in (6.1) is a dependence of $\mathbf{T}$ on a reference configuration $\mathbf{X}$. Also *implicit* in (6.1) is a dependence of $\mathbf{T}$ on $\mathbf{x}$ and t, and a possible dependence of $\mathbf{T}$ on the present density ρ, since ρ is an algebraic function of the reference density ρ_R and the deformation gradient $\mathbf{F}$:

$$\rho = \frac{\rho_R}{\det \mathbf{F}}.$$

At this stage, it is useful to appeal to the *mechanical* energy balance

$$\rho \dot{\varepsilon} = \mathbf{T} \cdot \mathbf{D}, \tag{6.2}$$

which is the first law of thermodynamics (5.26) specialized to a mechanical setting. Essentially, (6.2) indicates that in the mechanical theory, the rate of change of the internal energy ε is due solely to the stress power $\mathbf{T} \cdot \mathbf{D}$; refer to Section 4.7. Arguing that the internal energy ε (like the Cauchy stress $\mathbf{T}$) is a function of the strain $\mathbf{F}$ alone, i.e.,

$$\varepsilon = \breve{\varepsilon}(\mathbf{F}),$$

and using the results of Problem 6.2 (which appear in the upcoming section), we have

$$\dot{\varepsilon} = \frac{d\breve{\varepsilon}}{d\mathbf{F}} \cdot \dot{\mathbf{F}} = \frac{d\breve{\varepsilon}}{d\mathbf{F}} \mathbf{F}^{\mathrm{T}} \cdot \mathbf{L}.$$

Hence, the mechanical energy balance (6.2) becomes

$$\left(\mathbf{T} - \rho \frac{d\breve{\varepsilon}}{d\mathbf{F}} \mathbf{F}^{\mathrm{T}} \right) \cdot \mathbf{L} = 0, \tag{6.3}$$

noting that $\mathbf{T} \cdot \mathbf{L} = \mathbf{T} \cdot \mathbf{D}$ (refer to Problem 6.1 in the upcoming section). Since the coefficient of $\mathbf{L}$ in (6.3) is independent of $\mathbf{L}$, and $\mathbf{L}$ is arbitrary, it follows that

$$\mathbf{T} = \rho \frac{d\breve{\varepsilon}}{d\mathbf{F}} \mathbf{F}^{\mathrm{T}}. \tag{6.4}$$

That is, the derivative of the internal energy ε with respect to the strain $\mathbf{F}$ gives the stress $\mathbf{T}$. Hence, the internal energy is a **strain energy**. The class of materials for which the stress-strain response is derived from a strain energy potential, or stored elastic energy, is called **hyperelastic**.

Recall from Section 5.2 that the constitutive equation (6.4) for the Cauchy stress $\mathbf{T}$ must satisfy conservation of angular momentum, invariance requirements, and material symmetry conditions. We postpone the exploration of these requirements until Section 6.2.

6.2 THERMOMECHANICAL THEORY

Recall from Section 5.3 that in formulation (5.36)–(5.38) of the thermomechanical theory, constitutive assumptions are made on the Cauchy stress $\mathbf{T}$, heat flux vector $\mathbf{q}$, Helmholtz free energy ψ, and entropy η as functions of the motion χ and temperature Θ, and possibly their rates, gradients, and histories, i.e.,

$$\mathbf{T} = \bar{\mathbf{T}} \text{ (motion and temperature)}, \qquad \mathbf{q} = \bar{\mathbf{q}} \text{ (motion and temperature)},$$

$$\psi = \bar{\psi} \text{ (motion and temperature)}, \qquad \eta = \bar{\eta} \text{ (motion and temperature)}.$$

For a **thermoelastic material**, which has a perfect memory of its reference configuration and temperature, the list of arguments in (6.1) is expanded from χ (implicit) and $\mathbf{F} = \operatorname{Grad} \chi$ (explicit) to include the analogous thermal quantities Θ and $\mathbf{g}_R = \operatorname{Grad}\Theta$. Therefore, for a thermoelastic material, we write

$$\mathbf{T} = \bar{\mathbf{T}}(\mathbf{F}, \Theta, \mathbf{g}_R), \quad \mathbf{q} = \bar{\mathbf{q}}(\mathbf{F}, \Theta, \mathbf{g}_R), \quad \psi = \bar{\psi}(\mathbf{F}, \Theta, \mathbf{g}_R), \quad \eta = \bar{\eta}(\mathbf{F}, \Theta, \mathbf{g}_R). \tag{6.5}$$

This notation means that $\mathbf{T}$, $\mathbf{q}$, ψ, and η at position $\mathbf{x}$ and time t are *explicit* functions of $\mathbf{F}$, Θ, and $\mathbf{g}_R$ evaluated at $\mathbf{x}$ and t (and, thus, *implicit* functions of $\mathbf{x}$ and t). There is an implicit dependence on the reference configuration $\mathbf{X}$ through $\mathbf{F} = \operatorname{Grad} \chi$ and $\mathbf{g}_R = \operatorname{Grad} \Theta$. Recall from (4.65) the relationship $\mathbf{g}_R = \mathbf{F}^{\mathrm{T}}\mathbf{g}$ between the referential and spatial temperature gradients, so $\mathbf{g}_R$ is a function of $\mathbf{g}$ and $\mathbf{F}$. We may therefore replace (6.5) with

$$\mathbf{T} = \breve{\mathbf{T}}(\mathbf{F}, \Theta, \mathbf{g}), \quad \mathbf{q} = \breve{\mathbf{q}}(\mathbf{F}, \Theta, \mathbf{g}), \quad \psi = \breve{\psi}(\mathbf{F}, \Theta, \mathbf{g}), \quad \eta = \breve{\eta}(\mathbf{F}, \Theta, \mathbf{g}). \tag{6.6}$$

Note that $\bar{\mathbf{T}}$ in (6.5) and $\breve{\mathbf{T}}$ in (6.6) denote two different response functions for $\mathbf{T}$. Also note that the constitutive functions for $\mathbf{T}$, $\mathbf{q}$, ψ, and η in (6.6) do not depend on the *history* or the *rate* of the strain or temperature (refer to Appendix B).

Now that a list of arguments has been specified for the constitutive functions that characterize a thermoelastic material, the number of independent constitutive functions, and the list of arguments itself, can be reduced. As described in Section 5.4 and illustrated in the following sections, this reduction is accomplished via the second law of thermodynamics, invariance requirements, conservation of angular momentum, and material symmetry considerations.

6.2.1 RESTRICTIONS IMPOSED BY THE SECOND LAW OF THERMODYNAMICS

Substitution of the constitutive assumptions (6.6) into the Clausius-Duhem inequality (5.35) gives

$$-\rho \frac{\partial \breve{\psi}}{\partial \mathbf{F}} \cdot \dot{\mathbf{F}} + \breve{\mathbf{T}} \cdot \mathbf{D} - \rho \left(\frac{\partial \breve{\psi}}{\partial \Theta} + \breve{\eta} \right) \dot{\Theta} - \rho \frac{\partial \breve{\psi}}{\partial \mathbf{g}} \cdot \dot{\mathbf{g}} - \frac{1}{\Theta} \breve{\mathbf{q}} \cdot \mathbf{g} \geq 0, \tag{6.7}$$

where we have used the chain rule (refer to Section 2.5.2):

$$\dot{\psi} = \frac{\partial \breve{\psi}}{\partial \mathbf{F}} \cdot \dot{\mathbf{F}} + \frac{\partial \breve{\psi}}{\partial \Theta} \dot{\Theta} + \frac{\partial \breve{\psi}}{\partial \mathbf{g}} \cdot \dot{\mathbf{g}}.$$

It can be shown (refer to Problems 6.1 and 6.2) that

$$\breve{\mathbf{T}} \cdot \mathbf{D} = \breve{\mathbf{T}} \cdot \mathbf{L}, \quad \frac{\partial \breve{\psi}}{\partial \mathbf{F}} \cdot \dot{\mathbf{F}} = \frac{\partial \breve{\psi}}{\partial \mathbf{F}} \mathbf{F}^{\mathrm{T}} \cdot \mathbf{L},$$

so inequality (6.7) becomes

$$\left(\breve{\mathbf{T}} - \rho \frac{\partial \breve{\psi}}{\partial \mathbf{F}} \mathbf{F}^{\mathrm{T}}\right) \cdot \mathbf{L} - \rho \left(\frac{\partial \breve{\psi}}{\partial \Theta} + \breve{\eta}\right) \dot{\Theta} - \rho \frac{\partial \breve{\psi}}{\partial \mathbf{g}} \cdot \dot{\mathbf{g}} - \frac{1}{\Theta} \breve{\mathbf{q}} \cdot \mathbf{g} \geq 0. \tag{6.8}$$

Inequality (6.8) must hold for *all* processes. In particular, it must hold for the family of processes with $\mathbf{g} = \mathbf{0}$, $\dot{\mathbf{g}} = \mathbf{0}$, and $\mathbf{L} = \mathbf{0}$, but $\dot{\Theta}$ arbitrary, at particular position $\mathbf{x}$ and time t. (An example is the family of processes (D.1) in Appendix D with $\mathbf{A} = \mathbf{0}$, $\mathbf{a} = \mathbf{0}$, and $\mathbf{g}_0 = \mathbf{0}$, but a any real number.) For members of this family of processes, (6.8) simplifies to

$$-\rho \left(\frac{\partial \breve{\psi}}{\partial \Theta} + \breve{\eta}\right) \dot{\Theta} \geq 0.$$

The density ρ is positive, which implies

$$\left(\frac{\partial \breve{\psi}}{\partial \Theta} + \breve{\eta}\right) \dot{\Theta} \leq 0. \tag{6.9}$$

The coefficient of $\dot{\Theta}$ *in (6.9) is independent of* $\dot{\Theta}$. For (6.9) to hold for *all* members of this family, the coefficient of $\dot{\Theta}$ must vanish, i.e.,

$$\eta = -\frac{\partial \breve{\psi}}{\partial \Theta}. \tag{6.10}$$

Thus, the partial derivative of the response function for the Helmholtz free energy $\breve{\psi}$ with respect to temperature Θ gives the value of the entropy η, so $\breve{\eta}$ is *not an independent response function*. Result (6.10) holds for all processes, not just the special family considered above. Hence, inequality (6.8) reduces to

$$\left(\breve{\mathbf{T}} - \rho \frac{\partial \breve{\psi}}{\partial \mathbf{F}} \mathbf{F}^{\mathrm{T}}\right) \cdot \mathbf{L} - \rho \frac{\partial \breve{\psi}}{\partial \mathbf{g}} \cdot \dot{\mathbf{g}} - \frac{1}{\Theta} \breve{\mathbf{q}} \cdot \mathbf{g} \geq 0. \tag{6.11}$$

Inequality (6.11) must hold for all processes, in particular the family of processes with $\mathbf{g} = \mathbf{0}$ and $\mathbf{L} = \mathbf{0}$, but $\dot{\mathbf{g}}$ arbitrary, at $\mathbf{x}$ and t (e.g., the family (D.1) in Appendix D with $\mathbf{A} = \mathbf{0}$ and $\mathbf{g}_0 = \mathbf{0}$, but $\mathbf{a}$ any vector). For members of this family, (6.11) simplifies to

$$-\rho \frac{\partial \breve{\psi}}{\partial \mathbf{g}} \cdot \dot{\mathbf{g}} \geq 0.$$

The density is positive; hence,

$$\frac{\partial \breve{\psi}}{\partial \mathbf{g}} \cdot \dot{\mathbf{g}} \leq 0. \tag{6.12}$$

The coefficient of $\dot{\mathbf{g}}$ *in (6.12) is independent of* $\dot{\mathbf{g}}$. Hence, the rate $\dot{\mathbf{g}}$ may be chosen to violate the inequality (6.12) unless its coefficient is the zero vector. Therefore,

$$\frac{\partial \breve{\psi}}{\partial \mathbf{g}} = \mathbf{0}.$$

Thus, if the response function for ψ depended on the temperature gradient $\mathbf{g}$, the second law would be violated for some processes. Since the second law must be obeyed for *all* processes, we conclude that a thermoelastic response function for ψ must be independent of $\mathbf{g}$. Inequality (6.11) thus reduces to

$$\left(\breve{\mathbf{T}} - \rho \frac{\partial \breve{\psi}}{\partial \mathbf{F}} \mathbf{F}^{\mathrm{T}}\right) \cdot \mathbf{L} - \frac{1}{\Theta} \breve{\mathbf{q}} \cdot \mathbf{g} \geq 0, \tag{6.13}$$

which must hold for all processes.

Consider those processes with $\mathbf{g} = \mathbf{0}$ but $\mathbf{L}$ arbitrary at $\mathbf{x}$ and t (e.g., the family (D.1) in Appendix D with $\mathbf{g}_0 = \mathbf{0}$, $\mathbf{F}_0$ fixed and invertible, and $\mathbf{A}$ any tensor). For these processes, (6.13) simplifies to

$$\left(\breve{\mathbf{T}} - \rho \frac{\partial \breve{\psi}}{\partial \mathbf{F}} \mathbf{F}^{\mathrm{T}}\right) \cdot \mathbf{L} \geq 0. \tag{6.14}$$

The coefficient of the rate $\mathbf{L}$ in (6.14) is independent of rates. Hence, we obtain

$$\mathbf{T} = \rho \frac{\partial \breve{\psi}}{\partial \mathbf{F}} \mathbf{F}^{\mathrm{T}}.$$

Thus, the partial derivative of the response function for the Helmholtz free energy ψ with respect to strain $\mathbf{F}$ gives the value of the stress $\mathbf{T}$, i.e., the Helmholtz free energy is a **strain energy**. Recall that the class of materials for which the stress-strain response is derived from a strain energy potential is called **hyperelastic**. The inequality (6.13) thus reduces to

$$-\frac{1}{\Theta} \breve{\mathbf{q}} \cdot \mathbf{g} \geq 0$$

or

$$\breve{\mathbf{q}} \cdot \mathbf{g} \leq 0, \tag{6.15}$$

since the absolute temperature Θ is strictly positive. Inequality (6.15) cannot be simplified any further since $\breve{\mathbf{q}}$ depends on $\mathbf{g}$. Thus, the Clausius-Duhem inequality has been reduced for thermoelastic materials to an intuitive statement of the second law: heat flows against the temperature gradient, from hot to cold.

We now pause and take stock of our accomplishments. By demanding that the Clausius-Duhem inequality holds for all thermoelastic processes, we have reduced the constitutive assumption (6.6) to

$$\psi = \breve{\psi}(\mathbf{F}, \Theta), \quad \mathbf{q} = \breve{\mathbf{q}}(\mathbf{F}, \Theta, \mathbf{g}) \quad \text{with} \quad \breve{\mathbf{q}} \cdot \mathbf{g} \leq 0, \tag{6.16a}$$

and

$$\mathbf{T} = \rho \frac{\partial \breve{\psi}}{\partial \mathbf{F}} \mathbf{F}^{\mathrm{T}}, \quad \eta = -\frac{\partial \breve{\psi}}{\partial \Theta}. \tag{6.16b}$$

Thus, as a consequence of the second law, we have found that:

(1) Only two response functions ($\breve{\psi}$ and $\breve{\mathbf{q}}$) are needed to characterize a thermoelastic material, rather than the four initially supposed. (Nice! Fewer response functions means fewer unknowns in the characterization problem.)

(2) The response function for ψ is independent of $\mathbf{g}$. (Again, fewer experiments to perform.)

(3) The response function for $\mathbf{q}$ is restricted by the second law through the inequality $\breve{\mathbf{q}} \cdot \mathbf{g} \leq 0$, i.e., *heat flows against the temperature gradient.*

It can be shown (refer to Problem 6.3) that if the heat flux vector $\breve{\mathbf{q}}$ in (6.6) is a continuous function of temperature gradient $\mathbf{g}$ at $\mathbf{g} = \mathbf{0}$, restriction (3) above has an additional implication: $\breve{\mathbf{q}}$ evaluated at $\mathbf{g} = \mathbf{0}$ is zero, i.e., *heat does not flow in the absence of a temperature gradient.*

PROBLEM 6.1

In direct notation, show that $\mathbf{T} \cdot \mathbf{L} = \mathbf{T} \cdot \mathbf{D}$.

Solution

$$\begin{aligned} \mathbf{T} \cdot \mathbf{L} &= \mathbf{T} \cdot (\mathbf{D} + \mathbf{W}) && \text{(decomposition (3.57))} \\ &= \mathbf{T} \cdot \mathbf{D} + \mathbf{T} \cdot \mathbf{W} && \text{(property (2.6)}_3\text{)} \\ &= \mathbf{T} \cdot \mathbf{D} && \text{(result (2.44)).} \end{aligned}$$

In the second step, we were able to employ the distributive property $(2.6)_3$ of the inner product since it was demonstrated (in Problem 2.34) that the set of all second-order tensors is an inner product space. In the third step, we have exploited that $\mathbf{T} \cdot \mathbf{W} = 0$, since $\mathbf{T}$ is symmetric (by conservation of angular momentum; see (5.25)) and $\mathbf{W}$ is skew (by construction; see $(3.58)_2$).

PROBLEM 6.2

Prove in direct notation that $\dfrac{\partial \psi}{\partial \mathbf{F}} \cdot \dot{\mathbf{F}} = \dfrac{\partial \psi}{\partial \mathbf{F}} \mathbf{F}^{\mathrm{T}} \cdot \mathbf{L}$.

Solution

$$\begin{aligned} \frac{\partial \psi}{\partial \mathbf{F}} \cdot \dot{\mathbf{F}} &= \frac{\partial \psi}{\partial \mathbf{F}} \cdot (\mathbf{LF}) && \text{(result (3.60)}_1\text{)} \\ &= \operatorname{tr}\left[\frac{\partial \psi}{\partial \mathbf{F}} (\mathbf{LF})^{\mathrm{T}}\right] && \text{(definition (2.41))} \\ &= \operatorname{tr}\left[\frac{\partial \psi}{\partial \mathbf{F}} (\mathbf{F}^{\mathrm{T}} \mathbf{L}^{\mathrm{T}})\right] && \text{(result (2.14)}_2\text{)} \end{aligned}$$

$$= \mathrm{tr}\left[\left(\frac{\partial \psi}{\partial \mathbf{F}}\mathbf{F}^{\mathrm{T}}\right)\mathbf{L}^{\mathrm{T}}\right] \qquad \text{(associativity of tensor multiplication)}$$

$$= \frac{\partial \psi}{\partial \mathbf{F}}\mathbf{F}^{\mathrm{T}} \cdot \mathbf{L} \qquad \text{(definition (2.41))}.$$

PROBLEM 6.3

Show that if the heat flux vector $\check{\mathbf{q}}$ in (6.6) is a continuous function of the temperature gradient $\mathbf{g}$ at $\mathbf{g} = \mathbf{0}$, then it follows from $\check{\mathbf{q}} \cdot \mathbf{g} \leq 0$ that $\check{\mathbf{q}}$ evaluated at $\mathbf{g} = \mathbf{0}$ is zero. That is, heat does not flow in the absence of a temperature gradient.

Solution

If $\check{\mathbf{q}}$ is a continuous function of $\mathbf{g}$ at $\mathbf{g} = \mathbf{0}$, then it follows that

$$\mathbf{q} = \check{\mathbf{q}}(\mathbf{F}, \Theta, \mathbf{g}) = \mathbf{a}(\mathbf{F}, \Theta) + \mathbf{b}(\mathbf{F}, \Theta, \mathbf{g}),$$

with

$$\lim_{\alpha \to 0} \mathbf{b}(\mathbf{F}, \Theta, \alpha \mathbf{g}) = \mathbf{0},$$

so

$$\check{\mathbf{q}}(\mathbf{F}, \Theta, \mathbf{g})\Big|_{\mathbf{g}=\mathbf{0}} = \lim_{\alpha \to 0} \check{\mathbf{q}}(\mathbf{F}, \Theta, \alpha \mathbf{g}) = \mathbf{a}(\mathbf{F}, \Theta).$$

Then $\check{\mathbf{q}} \cdot \mathbf{g} \leq 0$ implies

$$\mathbf{a}(\mathbf{F}, \Theta) \cdot \mathbf{g} + \mathbf{b}(\mathbf{F}, \Theta, \mathbf{g}) \cdot \mathbf{g} \leq 0.$$

Replacing $\mathbf{g}$ by $\alpha\mathbf{g}$, taking $\alpha > 0$, and dividing by α leads to

$$\mathbf{a}(\mathbf{F}, \Theta) \cdot \mathbf{g} + \mathbf{b}(\mathbf{F}, \Theta, \alpha \mathbf{g}) \cdot \mathbf{g} \leq 0.$$

Taking the limit as $\alpha \to 0$ gives

$$\mathbf{a}(\mathbf{F}, \Theta) \cdot \mathbf{g} \leq 0,$$

which must hold for *all* processes. Since $\mathbf{a}(\mathbf{F}, \Theta)$ is independent of $\mathbf{g}$, this implies that $\mathbf{a}(\mathbf{F}, \Theta) = \mathbf{0}$, or

$$\check{\mathbf{q}}(\mathbf{F}, \Theta, \mathbf{g})\Big|_{\mathbf{g}=\mathbf{0}} = \mathbf{0}.$$

6.2.2 RESTRICTIONS IMPOSED BY INVARIANCE UNDER SUPERPOSED RIGID BODY MOTIONS AND CONSERVATION OF ANGULAR MOMENTUM

Our appeal to the second law has proven fruitful. We now appeal to invariance requirements under superposed rigid body motions (refer to Sections 5.2.1 and 5.4.1) and obtain further restrictions. (Note that any restrictions on the response functions are very helpful, as they reduce the number of experiments needed to characterize a particular material.) In particular, we must have

$$\psi^{+} = \psi, \quad \mathbf{q}^{+} = \mathbf{Q}\mathbf{q}$$

when

$$\mathbf{F}^{+} = \mathbf{Q}\mathbf{F}, \quad \Theta^{+} = \Theta, \quad \mathbf{g}^{+} = \mathbf{Q}\mathbf{g}$$

for all proper orthogonal tensors $\mathbf{Q}(t)$. It can be shown (refer to Problems 6.4 and 6.5) that these invariance requirements demand

$$\psi = \breve{\psi}(\mathbf{F}, \Theta) = \grave{\psi}(\mathbf{C}, \Theta) \tag{6.17}$$

and

$$\mathbf{q} = \breve{\mathbf{q}}(\mathbf{F}, \Theta, \mathbf{g}) = \mathbf{F}\grave{\mathbf{q}}(\mathbf{C}, \Theta, \mathbf{g}_{\mathrm{R}}). \tag{6.18}$$

Note that $\breve{\psi}$ and $\grave{\psi}$ denote two different response functions for ψ. Equation (6.17) implies that the response function for ψ can depend on $\mathbf{F}$ only through the combination $\mathbf{C} = \mathbf{F}^{\mathrm{T}}\mathbf{F}$, where $\mathbf{C}$ is Green's deformation tensor; the dependence on Θ is unrestricted. Equation (6.18) implies that the dependence of the response function for $\mathbf{q}$ on $\mathbf{g}$ must be through $\mathbf{g}_{\mathrm{R}} = \mathbf{F}^{\mathrm{T}}\mathbf{g}$ (where $\mathbf{g}_{\mathrm{R}}$ is the referential temperature gradient), there must be a linear dependence of $\mathbf{q}$ on $\mathbf{F}$, and any further dependence on $\mathbf{F}$ must be through the combination $\mathbf{C} = \mathbf{F}^{\mathrm{T}}\mathbf{F}$.

Now that we have established $\psi = \grave{\psi}(\mathbf{C}, \Theta)$, it can be shown, through a change of independent variable from $\mathbf{F}$ to $\mathbf{C}$, that $(6.16\mathrm{b})_1$ becomes

$$\mathbf{T} = \rho \frac{\partial \breve{\psi}(\mathbf{F}, \Theta)}{\partial \mathbf{F}} \mathbf{F}^{\mathrm{T}} = \rho \mathbf{F} \left(\frac{\partial \grave{\psi}(\mathbf{C}, \Theta)}{\partial \mathbf{C}} + \frac{\partial \grave{\psi}(\mathbf{C}, \Theta)}{\partial \mathbf{C}^{\mathrm{T}}} \right) \mathbf{F}^{\mathrm{T}}. \tag{6.19}$$

If $\grave{\psi}(\mathbf{C}, \Theta)$ is a symmetric function of $\mathbf{C}$, then (6.19) becomes

$$\mathbf{T} = 2\rho \mathbf{F} \frac{\partial \grave{\psi}(\mathbf{C}, \Theta)}{\partial \mathbf{C}} \mathbf{F}^{\mathrm{T}}. \tag{6.20}$$

It can be shown that (6.19) and (6.20) ensure that the Cauchy stress $\mathbf{T}$ is symmetric, so conservation of angular momentum (5.25) is satisfied.

To summarize the results of this section thus far: A thermoelastic material has a perfect memory of its reference state. On physical grounds, it is assumed that such a material is characterized by at most four response functions, each of which could conceivably depend on the deformation gradient $\mathbf{F}$ (with respect to some reference configuration), temperature Θ, and temperature gradient $\mathbf{g}$, all evaluated at present time t (i.e., the response at present time t depends on the present values of $\mathbf{F}$, Θ, and $\mathbf{g}$). The four response functions and their lists of arguments are

$$\mathbf{T} = \breve{\mathbf{T}}(\mathbf{F}, \Theta, \mathbf{g}), \quad \mathbf{q} = \breve{\mathbf{q}}(\mathbf{F}, \Theta, \mathbf{g}), \quad \psi = \breve{\psi}(\mathbf{F}, \Theta, \mathbf{g}), \quad \eta = \breve{\eta}(\mathbf{F}, \Theta, \mathbf{g}).$$

Note that the dependence on deformation does not involve its history, but rather just the present value of the deformation gradient from some reference configuration.

Because these response functions must be consistent with conservation of angular momentum, the second law of thermodynamics for all thermomechanical processes, and invariance requirements under any superposed rigid body motion, *we find that only two of the response functions are independent,* i.e.,

$$\psi = \grave{\psi}(\mathbf{C}, \Theta), \quad \mathbf{q} = \mathbf{F}\grave{\mathbf{q}}(\mathbf{C}, \Theta, \mathbf{g}_R), \tag{6.21a}$$

from which

$$\mathbf{T} = 2\rho\mathbf{F}\frac{\partial\grave{\psi}(\mathbf{C}, \Theta)}{\partial\mathbf{C}}\mathbf{F}^T, \quad \eta = -\frac{\partial\grave{\psi}(\mathbf{C}, \Theta)}{\partial\Theta} \tag{6.21b}$$

can be deduced, with the restrictions

$$\grave{\mathbf{q}}(\mathbf{C}, \Theta, \mathbf{0}) = \mathbf{0}, \quad \mathbf{q}\cdot\mathbf{g} \leq 0. \tag{6.21c}$$

Note that the Helmholtz free energy, Cauchy stress, and entropy are independent of temperature gradient. Also note that the Clausius-Duhem inequality reduces to the intuitive second law statements $(6.21c)_1$ and $(6.21c)_2$, i.e., heat does not flow in the absence of a temperature gradient, and in the presence of a temperature gradient, heat flows opposite the temperature gradient, from hot to cold.

PROBLEM 6.4

Prove that the invariance requirement $\psi^+ = \psi$ is satisfied *if and only if*

$$\psi = \check{\psi}(\mathbf{F}, \Theta) = \grave{\psi}(\mathbf{C}, \Theta).$$

Solution

The phrase "if and only if" requires that we demonstrate both *necessity* and *sufficiency* in our proof. To demonstrate necessity, we must show that the invariance requirement $\psi^+ = \psi$ implies that $\psi = \grave{\psi}(\mathbf{C}, \Theta)$. To demonstrate sufficiency, we must show that $\psi = \grave{\psi}(\mathbf{C}, \Theta)$ satisfies the invariance requirement $\psi^+ = \psi$.

Necessity: First, note that

$$\psi = \check{\psi}(\mathbf{F}, \Theta), \quad \psi^+ = \check{\psi}(\mathbf{F}^+, \Theta^+).$$

The invariance requirement

$$\psi^+ = \psi \quad \text{when} \quad \mathbf{F}^+ = \mathbf{QF}, \quad \Theta^+ = \Theta$$

for all proper orthogonal $\mathbf{Q}$ implies that

$$\check{\psi}(\mathbf{QF}, \Theta) = \check{\psi}(\mathbf{F}, \Theta).$$

Since this must hold for *all* proper orthogonal $\mathbf{Q}$, it must hold for the particular case $\mathbf{Q} = \mathbf{R}^T$, so

$$\check{\psi}(\mathbf{F}, \Theta) = \check{\psi}(\mathbf{R}^T\mathbf{F}, \Theta).$$

(Recall from Section 3.3.3 that $\mathbf{R}$ is the proper orthogonal rotation tensor in the polar decomposition of $\mathbf{F}$.) Since

$$\mathbf{R}^{\mathrm{T}}\mathbf{F} = \mathbf{R}^{\mathrm{T}}\mathbf{R}\mathbf{U} = \mathbf{I}\mathbf{U} = \mathbf{U} = \mathbf{C}^{1/2},$$

it follows that

$$\psi = \check{\psi}(\mathbf{F}, \Theta) = \check{\psi}(\mathbf{R}^{\mathrm{T}}\mathbf{F}, \Theta) = \check{\psi}(\mathbf{U}, \Theta) = \check{\psi}(\mathbf{C}^{1/2}, \Theta) = \grave{\psi}(\mathbf{C}, \Theta).$$

Sufficiency: It follows from $\psi = \grave{\psi}(\mathbf{C}, \Theta)$ that

$$\psi^{+} = \grave{\psi}(\mathbf{C}^{+}, \Theta^{+}) = \grave{\psi}(\mathbf{C}, \Theta) = \psi,$$

where we have used the results $\mathbf{C}^{+} = \mathbf{C}$ and $\Theta^{+} = \Theta$ from Sections 5.2.1.2 and 5.4.1.1.

PROBLEM 6.5

Prove that the invariance requirement $\mathbf{q}^{+} = \mathbf{Q}\mathbf{q}$ is satisfied *if and only if*

$$\mathbf{q} = \check{\mathbf{q}}(\mathbf{F}, \Theta, \mathbf{g}) = \mathbf{F}\grave{\mathbf{q}}(\mathbf{C}, \Theta, \mathbf{F}^{\mathrm{T}}\mathbf{g}).$$

Solution

As was the case in Problem 6.4, the phrase "if and only if" requires that we demonstrate both *necessity* and *sufficiency* in our proof. To demonstrate necessity, we must show that the invariance requirement $\mathbf{q}^{+} = \mathbf{Q}\mathbf{q}$ implies that $\mathbf{q} = \mathbf{F}\grave{\mathbf{q}}(\mathbf{C}, \Theta, \mathbf{F}^{\mathrm{T}}\mathbf{g})$. To demonstrate sufficiency, we must show that $\mathbf{q} = \mathbf{F}\grave{\mathbf{q}}(\mathbf{C}, \Theta, \mathbf{F}^{\mathrm{T}}\mathbf{g})$ satisfies the invariance requirement $\mathbf{q}^{+} = \mathbf{Q}\mathbf{q}$.

Necessity: First, note that

$$\mathbf{q} = \check{\mathbf{q}}(\mathbf{F}, \Theta, \mathbf{g}), \quad \mathbf{q}^{+} = \check{\mathbf{q}}(\mathbf{F}^{+}, \Theta^{+}, \mathbf{g}^{+}).$$

The invariance requirement

$$\mathbf{q}^{+} = \mathbf{Q}\mathbf{q} \quad \text{when} \quad \mathbf{F}^{+} = \mathbf{Q}\mathbf{F}, \quad \Theta^{+} = \Theta, \quad \mathbf{g}^{+} = \mathbf{Q}\mathbf{g}$$

implies that

$$\check{\mathbf{q}}(\mathbf{Q}\mathbf{F}, \Theta, \mathbf{Q}\mathbf{g}) = \mathbf{Q}\check{\mathbf{q}}(\mathbf{F}, \Theta, \mathbf{g}),$$

which must hold for *all* proper orthogonal $\mathbf{Q}$, and, in particular, $\mathbf{Q} = \mathbf{R}^{\mathrm{T}}$. Since

$$\mathbf{R}^{\mathrm{T}}\mathbf{F} = \mathbf{R}^{\mathrm{T}}\mathbf{R}\mathbf{U} = \mathbf{I}\mathbf{U} = \mathbf{U},$$

it follows that

$$\check{\mathbf{q}}(\mathbf{F}, \Theta, \mathbf{g}) = \mathbf{R}\check{\mathbf{q}}(\mathbf{U}, \Theta, \mathbf{R}^{\mathrm{T}}\mathbf{g}).$$

We define a new function $\bar{\mathbf{q}}$ by

$$\check{\mathbf{q}}(\mathbf{U}, \Theta, \mathbf{R}^{\mathrm{T}}\mathbf{g}) = \mathbf{U}\bar{\mathbf{q}}(\mathbf{U}, \Theta, \mathbf{U}\mathbf{R}^{\mathrm{T}}\mathbf{g}),$$

which gives

$$\breve{\mathbf{q}}(\mathbf{F}, \Theta, \mathbf{g}) = \mathbf{R}\mathbf{U}\bar{\mathbf{q}}(\mathbf{U}, \Theta, \mathbf{U}\mathbf{R}^{\mathrm{T}}\mathbf{g}).$$

We have

$$\mathbf{F} = \mathbf{R}\mathbf{U}, \quad \mathbf{F}^{\mathrm{T}} = (\mathbf{R}\mathbf{U})^{\mathrm{T}} = \mathbf{U}^{\mathrm{T}}\mathbf{R}^{\mathrm{T}} = \mathbf{U}\mathbf{R}^{\mathrm{T}}, \quad \mathbf{U} = \mathbf{C}^{1/2},$$

which imply that

$$\breve{\mathbf{q}}(\mathbf{F}, \Theta, \mathbf{g}) = \mathbf{F}\bar{\mathbf{q}}(\mathbf{C}^{1/2}, \Theta, \mathbf{F}^{\mathrm{T}}\mathbf{g}) = \mathbf{F}\grave{\mathbf{q}}(\mathbf{C}, \Theta, \mathbf{F}^{\mathrm{T}}\mathbf{g}).$$

Sufficiency: It follows from $\mathbf{q} = \mathbf{F}\grave{\mathbf{q}}(\mathbf{C}, \Theta, \mathbf{F}^{\mathrm{T}}\mathbf{g})$ that

$$\begin{aligned}
\mathbf{q}^{+} &= \mathbf{F}^{+}\grave{\mathbf{q}}(\mathbf{C}^{+}, \Theta^{+}, (\mathbf{F}^{+})^{\mathrm{T}}\mathbf{g}^{+}) \\
&= \mathbf{Q}\mathbf{F}\grave{\mathbf{q}}(\mathbf{C}, \Theta, \mathbf{F}^{\mathrm{T}}\mathbf{Q}^{\mathrm{T}}\mathbf{Q}\mathbf{g}) \\
&= \mathbf{Q}\mathbf{F}\grave{\mathbf{q}}(\mathbf{C}, \Theta, \mathbf{F}^{\mathrm{T}}\mathbf{g}) \\
&= \mathbf{Q}\mathbf{q},
\end{aligned}$$

where we have used the results $\mathbf{F}^{+} = \mathbf{Q}\mathbf{F}$, $\mathbf{C}^{+} = \mathbf{C}$, $\mathbf{g}^{+} = \mathbf{Q}\mathbf{g}$, and $\Theta^{+} = \Theta$ from Sections 5.2.1.2 and 5.4.1.1.

EXERCISES

1. Using indicial notation, verify through a change of independent variable from $\mathbf{F}$ to $\mathbf{C}$ that

$$\mathbf{T} = \rho \frac{\partial \breve{\psi}(\mathbf{F}, \Theta)}{\partial \mathbf{F}} \mathbf{F}^{\mathrm{T}} = \rho \mathbf{F} \left(\frac{\partial \grave{\psi}(\mathbf{C}, \Theta)}{\partial \mathbf{C}} + \frac{\partial \grave{\psi}(\mathbf{C}, \Theta)}{\partial \mathbf{C}^{\mathrm{T}}} \right) \mathbf{F}^{\mathrm{T}} = 2\rho \mathbf{F} \frac{\partial \grave{\psi}(\mathbf{C}, \Theta)}{\partial \mathbf{C}} \mathbf{F}^{\mathrm{T}},$$

   where the last equality holds if $\grave{\psi}(\mathbf{C}, \Theta)$ is a symmetric function of $\mathbf{C}$, i.e.,

$$\frac{\partial \grave{\psi}(\mathbf{C}, \Theta)}{\partial \mathbf{C}} = \frac{\partial \grave{\psi}(\mathbf{C}, \Theta)}{\partial \mathbf{C}^{\mathrm{T}}}.$$

2. Verify that (6.19) and (6.20) ensure that the Cauchy stress $\mathbf{T}$ is symmetric, so conservation of angular momentum is satisfied.

6.2.3 RESTRICTIONS IMPOSED BY MATERIAL SYMMETRY: ISOTROPY

In this section, we apply material symmetry considerations (refer to Section 5.2.2) to further reduce and simplify the form of the response functions. In particular, we consider thermoelastic materials that are **isotropic**. Loosely speaking, isotropic materials have properties that are identical in all directions.

Recall from Section 6.2.1 that by demanding the Clausius-Duhem inequality hold for all thermoelastic processes, we found that only two response functions are necessary to characterize a thermoelastic material, i.e.,

$$\psi = \breve{\psi}(\mathbf{F}, \Theta), \quad \mathbf{q} = \breve{\mathbf{q}}(\mathbf{F}, \Theta, \mathbf{g}),$$

where the dependence of the deformation gradient $\mathbf{F}$ on a particular reference configuration κ is understood. Here, κ is chosen to be the stress-free configuration. By analogy to (5.22), we have

$$\breve{\psi}(\mathbf{F}, \Theta) = \breve{\psi}(\mathbf{F}\mathbf{H}^{-1}, \Theta) \quad \text{for all } \mathbf{H} \in \mathcal{G}, \tag{6.22}$$

where $\mathcal{G}$ is the **symmetry group** of the material relative to the stress-free reference configuration κ, and $\mathbf{H}$ is a **symmetry transformation** with respect to κ. Since a thermoelastic material is a solid, we have $\mathcal{G} \subseteq \mathcal{O}$, where $\mathcal{O}$ is the full orthogonal group. Then (6.22) becomes

$$\breve{\psi}(\mathbf{F}, \Theta) = \breve{\psi}(\mathbf{F}\mathbf{H}^{\mathrm{T}}, \Theta) \quad \text{for all } \mathbf{H} \in \mathcal{G}. \tag{6.23}$$

Recall from Section 6.2.2 that in order to satisfy invariance under superposed rigid body motions and conservation of angular momentum, we must specify ψ to be a function of $\mathbf{F}$ in the particular combination $\mathbf{C} = \mathbf{F}^{\mathrm{T}}\mathbf{F}$, i.e.,

$$\psi = \grave{\psi}(\mathbf{C}, \Theta),$$

where $\mathbf{C}$ is Green's deformation tensor. The condition on $\grave{\psi}$ corresponding to (6.23) is

$$\grave{\psi}(\mathbf{C}, \Theta) = \grave{\psi}(\mathbf{H}\mathbf{C}\mathbf{H}^{\mathrm{T}}, \Theta) \quad \text{for all } \mathbf{H} \in \mathcal{G}. \tag{6.24}$$

For *isotropic* thermoelastic solids, the symmetry group $\mathcal{G}$ relative to the stress-free reference configuration κ is the full orthogonal group $\mathcal{O}$. Therefore, (6.24) becomes

$$\grave{\psi}(\mathbf{C}, \Theta) = \grave{\psi}(\mathbf{H}\mathbf{C}\mathbf{H}^{\mathrm{T}}, \Theta) \quad \text{for all } \mathbf{H} \in \mathcal{O}. \tag{6.25}$$

The requirement (6.25) on the functional form of $\grave{\psi}$ is equivalent to the condition that $\grave{\psi}$ depend on $\mathbf{C}$ only through the principal scalar invariants of $\mathbf{C}$, i.e.,

$$\psi = \bar{\psi}(I_1, I_2, I_3, \Theta), \tag{6.26a}$$

where

$$I_1 = \operatorname{tr}\mathbf{C}, \quad I_2 = \frac{1}{2}\left[(\operatorname{tr}\mathbf{C})^2 - \operatorname{tr}\left(\mathbf{C}^2\right)\right], \quad I_3 = \det\mathbf{C}. \tag{6.26b}$$

Recalling that the Cauchy stress $\mathbf{T}$ is obtained from the Helmholtz free energy ψ by

$$\mathbf{T} = 2\rho\mathbf{F}\frac{\partial\grave{\psi}(\mathbf{C}, \Theta)}{\partial\mathbf{C}}\mathbf{F}^{\mathrm{T}},$$

we have, through a change of independent variable,

$$\mathbf{T} = 2\rho\mathbf{F}\left(\frac{\partial\bar{\psi}}{\partial I_1}\frac{dI_1}{d\mathbf{C}} + \frac{\partial\bar{\psi}}{\partial I_2}\frac{dI_2}{d\mathbf{C}} + \frac{\partial\bar{\psi}}{\partial I_3}\frac{dI_3}{d\mathbf{C}}\right)\mathbf{F}^{\mathrm{T}}. \tag{6.27}$$

Recall from Section 2.5.1 the results

$$\frac{dI_1}{d\mathbf{C}} = \mathbf{I}, \quad \frac{dI_2}{d\mathbf{C}} = I_1\mathbf{I} - \mathbf{C}, \quad \frac{dI_3}{d\mathbf{C}} = I_3\mathbf{C}^{-1}. \tag{6.28}$$

Substitution of (6.28) into (6.27) gives

$$\mathbf{T} = \alpha_0\mathbf{I} + \alpha_1\mathbf{B} + \alpha_2\mathbf{B}^2, \tag{6.29a}$$

where $\mathbf{B} = \mathbf{F}\mathbf{F}^{\mathrm{T}}$ is the Finger deformation tensor and

$$\alpha_0 = 2\rho I_3\frac{\partial\bar{\psi}}{\partial I_3}, \quad \alpha_1 = 2\rho\left(\frac{\partial\bar{\psi}}{\partial I_1} + I_1\frac{\partial\bar{\psi}}{\partial I_2}\right), \quad \alpha_2 = -2\rho\frac{\partial\bar{\psi}}{\partial I_2}. \tag{6.29b}$$

Note that since $\mathbf{B}$ and $\mathbf{C}$ have the same eigenvalues, they also have the same principal invariants (refer to Sections 2.3 and 3.3.5). Use of the Cayley-Hamilton theorem (2.75) gives the alternative form

$$\mathbf{T} = \beta_0\mathbf{I} + \beta_1\mathbf{B} + \beta_{-1}\mathbf{B}^{-1}, \tag{6.30a}$$

where

$$\beta_0 = 2\rho\left(I_2\frac{\partial\bar{\psi}}{\partial I_2} + I_3\frac{\partial\bar{\psi}}{\partial I_3}\right), \quad \beta_1 = 2\rho\frac{\partial\bar{\psi}}{\partial I_1}, \quad \beta_{-1} = -2\rho I_3\frac{\partial\bar{\psi}}{\partial I_2}. \tag{6.30b}$$

In nonlinear elasticity it is common to speak of the strain energy W per unit reference volume, i.e., the **strain energy density**, defined by

$$W = \rho_{\mathrm{R}}\psi,$$

rather than the Helmholtz free energy. For isotropic thermoelastic materials, we have

$$W = \bar{W}(I_1, I_2, I_3, \Theta).$$

It follows that (6.30b) becomes

$$\beta_0 = 2I_3^{-\frac{1}{2}}\left(I_2\frac{\partial\bar{W}}{\partial I_2} + I_3\frac{\partial\bar{W}}{\partial I_3}\right), \quad \beta_1 = 2I_3^{-1/2}\frac{\partial\bar{W}}{\partial I_1}, \quad \beta_{-1} = -2I_3^{1/2}\frac{\partial\bar{W}}{\partial I_2}, \tag{6.31}$$

where we have used

$$\frac{\rho}{\rho_{\mathrm{R}}} = \frac{1}{J} = \frac{1}{\det\mathbf{F}} = I_3^{-1/2},$$

which follows from conservation of mass (4.56a). Recall that ρ is the density in the present configuration, ρ_{R} is the density in the reference configuration, and J is the determinant of the deformation gradient $\mathbf{F}$.

EXERCISES

1. Verify that substitution of (6.28) into (6.27) gives (6.29a) and (6.29b).
2. Verify that (6.30a) and (6.30b) follow from (6.29a) and (6.29b) and use of the Cayley-Hamilton theorem.
3. Verify (6.31).

6.3 STRAIN ENERGY MODELS

We found in Section 6.2 that the Cauchy stress for an isotropic nonlinear elastic material can be expressed as

$$\mathbf{T} = \beta_0 \mathbf{I} + \beta_1 \mathbf{B} + \beta_{-1} \mathbf{B}^{-1},$$

where

$$\beta_0 = 2I_3^{-1/2}\left(I_2 \frac{\partial \bar{W}}{\partial I_2} + I_3 \frac{\partial \bar{W}}{\partial I_3}\right), \quad \beta_1 = 2I_3^{-1/2} \frac{\partial \bar{W}}{\partial I_1}, \quad \beta_{-1} = -2I_3^{1/2} \frac{\partial \bar{W}}{\partial I_2},$$

$\mathbf{B} = \mathbf{F}\mathbf{F}^{\mathrm{T}}$ is the Finger deformation tensor, $W = \bar{W}(I_1, I_2, I_3, \Theta)$ is the strain energy density, and I_1, I_2, and I_3 are the principal invariants of $\mathbf{B}$. We now specialize this constitutive model to the mechanical (isothermal) theory by eliminating the temperature dependence of W, so $W = \bar{W}(I_1, I_2, I_3)$. It remains only to specify the dependence of the strain energy $\bar{W}$ on the invariants I_1, I_2, and I_3.

Several examples of invariant-based strain energy models for compressible rubberlike materials are the *Blatz-Ko model* [10],

$$W = \frac{\mu}{2}\left\{f\left[(I_1 - 3) + \frac{1}{\gamma}(I_3^{-\gamma} - 1)\right] + (1-f)\left[\left(\frac{I_2}{I_3} - 3\right) + \frac{1}{\gamma}(I_3^{\gamma} - 1)\right]\right\}, \tag{6.32}$$

the *compressible neo-Hookean model* [11, p. 247],

$$W = \frac{\mu}{2}\left[(I_1 - 3) + \frac{1}{\gamma}(I_3^{-\gamma} - 1)\right], \tag{6.33}$$

the *compressible Mooney-Rivlin model* [11, p. 247],

$$W = c_1(I_1 - 3) + c_2(I_2 - 3) + c_3\left(I_3^{(1/2)} - 1\right)^2 - (c_1 + 2c_2)\ln I_3, \tag{6.34}$$

and the *Levinson-Burgess (polynomial) model* [12],

$$W = \frac{\mu}{2}\Bigg[f(I_1 - 3) + (1-f)\left(\frac{I_2}{I_3} - 3\right) + 2(1-2f)(I_3^{(1/2)} - 1) + \left(2f + \frac{4\nu - 1}{1 - 2\nu}\right)(I_3^{(1/2)} - 1)^2\Bigg]. \tag{6.35}$$

In (6.32)–(6.35), $\gamma = \nu/(1-2\nu)$; μ and ν are the shear modulus and Poisson's ratio evaluated at small strains; and f, c_1, c_2, and c_3 are parameters that can be adjusted to fit experimental data for a particular rubbery material. Note that (6.33) can be obtained as a special case of (6.32) by setting $f = 1$.

In contrast to the strain energy models (6.32)–(6.35) that are based on the principal invariants I_1, I_2, I_3 of $\mathbf{B}$, the *compressible Ogden model* [13] is based on the principal stretches $\lambda_1, \lambda_2, \lambda_3$:

$$W = \sum_n \mu_n \left[\frac{\lambda_1^{\alpha_n} + \lambda_2^{\alpha_n} + \lambda_3^{\alpha_n} - 3}{\alpha_n} - \ln J\right] + \lambda\beta^{-2}\left(\beta \ln J + J^{-\beta} - 1\right), \tag{6.36}$$

where J is the determinant of the deformation gradient $\mathbf{F}$, λ is the second Lamé constant evaluated at small strains, and μ_n, α_n, β, and n are adjustable parameters. Recall from Sections 2.3 and 3.3.5 that the principal stretches λ_1, λ_2, λ_3 are related to the principal invariants I_1, I_2, I_3 of $\mathbf{B}$ through

$$I_1 = \lambda_1^2 + \lambda_2^2 + \lambda_3^2, \quad I_2 = \lambda_1^2\lambda_2^2 + \lambda_1^2\lambda_3^2 + \lambda_2^2\lambda_3^2, \quad I_3 = \lambda_1^2\lambda_2^2\lambda_3^2. \tag{6.37}$$

Other models, such as those developed by Anand [14] and Bischoff et al. [15], are based on statistical mechanics, and thus account for the underlying deformation physics of the polymer chains.

Many other invariant-based, stretch-based, and statistical-mechanics-based strain energy models for compressible rubbery materials—beyond the representative few presented here—can be found, for instance, in books by Holzapfel [11], Treloar [16], and Ogden [17], and review articles by Ogden [18] and Boyce and Arruda [19].

CHAPTER

Fluid Mechanics

7

This chapter presents constitutive equations appropriate for modeling the flow of several technologically important classes of fluids, namely, viscous and inviscid fluids.[1] **Viscous fluids** are sensitive to the rate at which they are deformed, whereas **inviscid fluids** are insensitive to the rate at which they are deformed. Of course, almost no fluids of practical importance are truly "inviscid" (i.e., zero viscosity). Nevertheless, from a modeling perspective, the notion of an inviscid fluid is quite useful. In particular, it can be used as an *idealization* for modeling flows where viscous effects only weakly influence the flow physics. For instance, in most applications, water is modeled as a viscous fluid. However, in modeling the high-speed flow of water far from a bounding surface, it may be more appropriate to model water as an inviscid fluid. Hence, it is the fluid and physical application *together* that dictate if neglecting viscous effects is an appropriate simplification. In this chapter, we discuss viscous and inviscid fluids in the context of the mechanical theory (Section 7.1) and the thermomechanical theory (Section 7.2).

7.1 MECHANICAL THEORY

Recall from Section 5.1 that in the mechanical theory, we only consider processes that are independent of temperature. Also recall from Section 5.1 that a constitutive assumption is made on the Cauchy stress $\mathbf{T}$ as a function of the motion $\boldsymbol{\chi}$, and possibly its rates, gradients, and history, i.e.,

$$\mathbf{T} = \bar{\mathbf{T}}(\text{motion}).$$

7.1.1 VISCOUS FLUIDS

A **viscous fluid** has no memory of any reference configuration, so the Cauchy stress $\mathbf{T}$ depends on the kinematics of the fluid *in its present configuration alone*. Accordingly, in the mechanical theory, we assume that the stress $\mathbf{T}$ at present position $\mathbf{x}$ and time t

[1] Fluids, unlike solids, do not have a preferred shape, and deform continuously under the application of a shear stress, regardless of how small the shear stress is.

depends on the motion $\boldsymbol{\chi}$ through the present density ρ, velocity $\mathbf{v} = \dot{\boldsymbol{\chi}}$, and Eulerian velocity gradient $\mathbf{L} = \operatorname{grad} \mathbf{v}$ evaluated at $\mathbf{x}$ and t, i.e.,

$$\mathbf{T} = \bar{\mathbf{T}}(\rho, \mathbf{v}, \mathbf{L}). \tag{7.1}$$

Refer to Appendix B. Note that none of the arguments in (7.1) involve the concept of a reference configuration. Since the velocity gradient can be additively decomposed into $\mathbf{L} = \mathbf{D} + \mathbf{W}$, where $\mathbf{D}$ is the rate of deformation tensor and $\mathbf{W}$ is the vorticity tensor (refer to (3.57) and (3.58)), we may write

$$\mathbf{T} = \check{\mathbf{T}}(\rho, \mathbf{v}, \mathbf{D}, \mathbf{W}) \tag{7.2}$$

in place of (7.1). Notation (7.2) implies that $\mathbf{T}$ is an *explicit* function of ρ, $\mathbf{v}$, $\mathbf{D}$, and $\mathbf{W}$ evaluated at $\mathbf{x}$ and t, and an *implicit* function of $\mathbf{x}$ and t. Note that the dependence of stress $\mathbf{T}$ on density ρ must be *explicitly* included, since ρ is not a function of the other arguments in the list.[2] *Recall from Section 5.2 that, as with all materials in the mechanical theory, the response function (7.2) for a viscous fluid must satisfy invariance requirements and conservation of angular momentum.*

7.1.1.1 Restrictions imposed by invariance under superposed rigid body motions

Now that a list of arguments has been specified for the constitutive function for $\mathbf{T}$, this list can be reduced using invariance under superposed rigid body motions (SRBMs). In particular, recall from Section 5.2.1.4 the invariance requirement

$$\mathbf{T}^+ = \mathbf{Q}\mathbf{T}\mathbf{Q}^{\mathrm{T}}$$

when

$$\rho^+ = \rho, \quad \mathbf{v}^+ = \boldsymbol{\Omega}\mathbf{Q}\mathbf{x} + \mathbf{Q}\mathbf{v} + \dot{\mathbf{c}}, \quad \mathbf{D}^+ = \mathbf{Q}\mathbf{D}\mathbf{Q}^{\mathrm{T}}, \quad \mathbf{W}^+ = \mathbf{Q}\mathbf{W}\mathbf{Q}^{\mathrm{T}} + \boldsymbol{\Omega}$$

for all proper orthogonal $\mathbf{Q}(t)$, skew $\boldsymbol{\Omega}(t)$, and vector-valued functions $\dot{\mathbf{c}}(t)$. Noting that

$$\mathbf{T} = \check{\mathbf{T}}(\rho, \mathbf{v}, \mathbf{D}, \mathbf{W}), \quad \mathbf{T}^+ = \check{\mathbf{T}}(\rho^+, \mathbf{v}^+, \mathbf{D}^+, \mathbf{W}^+),$$

we find the invariance requirement on $\mathbf{T}$ demands that

$$\check{\mathbf{T}}(\rho, \boldsymbol{\Omega}\mathbf{Q}\mathbf{x} + \mathbf{Q}\mathbf{v} + \dot{\mathbf{c}}, \mathbf{Q}\mathbf{D}\mathbf{Q}^{\mathrm{T}}, \mathbf{Q}\mathbf{W}\mathbf{Q}^{\mathrm{T}} + \boldsymbol{\Omega}) = \mathbf{Q}\check{\mathbf{T}}(\rho, \mathbf{v}, \mathbf{D}, \mathbf{W})\mathbf{Q}^{\mathrm{T}}. \tag{7.3}$$

Result (7.3) must hold for *all possible SRBMs*, i.e., for *all* proper orthogonal $\mathbf{Q}(t)$, skew $\boldsymbol{\Omega}(t)$, and vector-valued functions $\dot{\mathbf{c}}(t)$. In what follows, we show that enforcing (7.3) for *special families of SRBMs* puts restrictions on the manner in which the tensor-valued function $\check{\mathbf{T}}$ depends on its arguments.

We first consider the family of SRBMs, which all have, at a particular time $t = \tau$, $\mathbf{Q} = \mathbf{I}$, $\dot{\mathbf{Q}} = \mathbf{0}$ ($\Rightarrow \boldsymbol{\Omega} = \dot{\mathbf{Q}}\mathbf{Q}^{\mathrm{T}} = \mathbf{0}$), but $\dot{\mathbf{c}}$ arbitrary. This is the family of rigid body translations. For this family, (7.3) demands

---

[2] Conversely, recall that for an elastic material, ρ does not appear *explicitly* in the argument list, but rather appears *implicitly*, since ρ is a function of $\mathbf{F}$ (refer to (4.56a) and (6.1)).

$$\check{\mathbf{T}}(\rho, \mathbf{v} + \dot{\mathbf{c}}, \mathbf{D}, \mathbf{W}) = \check{\mathbf{T}}(\rho, \mathbf{v}, \mathbf{D}, \mathbf{W}) \tag{7.4}$$

for arbitrary vector $\dot{\mathbf{c}}$. Since (7.4) implies that the value (output) of the function $\check{\mathbf{T}}$ is unaltered by the addition of an arbitrary vector $\dot{\mathbf{c}}$ to the argument $\mathbf{v}$, it follows that the response function $\check{\mathbf{T}}$ cannot depend on $\mathbf{v}$ (i.e., our guess that $\check{\mathbf{T}}$ depends on $\mathbf{v}$ is inconsistent with invariance requirements).

We now consider the family of SRBMs, which have, at a particular time $t = \tau$, $\mathbf{Q} = \mathbf{I}$ but $\dot{\mathbf{Q}}$ arbitrary ($\Rightarrow \boldsymbol{\Omega}$ arbitrary). (At first thought, it might seem impossible to hold $\mathbf{Q} = \mathbf{I}$ fixed, but vary $\dot{\mathbf{Q}}$. "Isn't the derivative of a constant always zero?" is a likely first response. However, it is important to realize that $\mathbf{Q} = \mathbf{I}$ not for *all* time t, but just at a *particular* time $t = \tau$; the SRBMs are passing through $\mathbf{I}$. It is analogous to the vertical motion of a projectile at the top of its flight, at which instant $\mathbf{v} = \mathbf{0}$ but $\mathbf{a} = \dot{\mathbf{v}} \neq \mathbf{0}$.) Some of the SRBMs in this family are those with components, as functions of time t,

$$\left[\mathbf{Q}(t)\right] = \begin{bmatrix} \cos\left[\omega(t-\tau)\right] & \sin\left[\omega(t-\tau)\right] & 0 \\ -\sin\left[\omega(t-\tau)\right] & \cos\left[\omega(t-\tau)\right] & 0 \\ 0 & 0 & 1 \end{bmatrix},$$

$$\left[\dot{\mathbf{Q}}(t)\right] = \omega \begin{bmatrix} -\sin\left[\omega(t-\tau)\right] & \cos\left[\omega(t-\tau)\right] & 0 \\ -\cos\left[\omega(t-\tau)\right] & -\sin\left[\omega(t-\tau)\right] & 0 \\ 0 & 0 & 0 \end{bmatrix},$$

and

$$\left[\boldsymbol{\Omega}(t)\right] = \left[\dot{\mathbf{Q}}(t)\mathbf{Q}^{\mathrm{T}}(t)\right] = \omega \begin{bmatrix} 0 & 1 & 0 \\ -1 & 0 & 0 \\ 0 & 0 & 0 \end{bmatrix}.$$

Then, at particular time $t = \tau$, we have $\mathbf{Q} = \mathbf{I}$ and $\boldsymbol{\Omega}$ arbitrary (the arbitrariness in $\boldsymbol{\Omega}$ lies in the choice of the constant ω). Thus, for this family of SRBMs with $\mathbf{Q} = \mathbf{I}$ and $\boldsymbol{\Omega}$ arbitrary, (7.3) produces the restriction

$$\check{\mathbf{T}}(\rho, \mathbf{D}, \mathbf{W} + \boldsymbol{\Omega}) = \check{\mathbf{T}}(\rho, \mathbf{D}, \mathbf{W}). \tag{7.5}$$

The invariance requirement (7.3) must hold for *all* SRBMs, and, *in particular*, those SRBMs that give condition (7.5) for arbitrary tensor $\boldsymbol{\Omega}$. Since (7.5) implies that the value of the function $\check{\mathbf{T}}$ is unaltered by the addition of an arbitrary skew tensor $\boldsymbol{\Omega}$ to the argument $\mathbf{W}$, it follows that the response function $\check{\mathbf{T}}$ cannot depend on $\mathbf{W}$.

We have therefore reduced the constitutive assumption (7.2) by appealing to the invariance requirement (7.3) in the forms (7.4) and (7.5) produced by certain subclasses of SRBMs. In particular, (7.2) has been reduced to

$$\mathbf{T} = \check{\mathbf{T}}(\rho, \mathbf{D}), \tag{7.6}$$

since the proposed dependence on $\mathbf{v}$ and $\mathbf{W}$ was shown to be inconsistent with invariance under SRBMs. Again, appealing to the invariance requirement (7.3), we see that the response function $\check{\mathbf{T}}(\rho, \mathbf{D})$ must satisfy

$$\check{\mathbf{T}}(\rho, \mathbf{Q}\mathbf{D}\mathbf{Q}^{\mathrm{T}}) = \mathbf{Q}\check{\mathbf{T}}(\rho, \mathbf{D})\mathbf{Q}^{\mathrm{T}} \tag{7.7}$$

for all *proper orthogonal* $\mathbf{Q}$. We further observe that condition (7.7) is unaffected if $\mathbf{Q}$ is replaced by the *improper orthogonal* tensor $-\mathbf{Q}$; hence, (7.7) must hold for all *orthogonal* tensors $\mathbf{Q}$. It follows, then, that $\check{\mathbf{T}}$ must be an *isotropic* tensor function of $\mathbf{D}$.

We now pause and review our accomplishments up to this point: We assumed at the outset that the Cauchy stress $\mathbf{T}$ in a viscous fluid depends on the present density ρ, velocity $\mathbf{v}$, and Eulerian velocity gradient $\mathbf{L}$. Appeal to invariance requirements revealed that the stress $\mathbf{T}$: (1) cannot depend on velocity $\mathbf{v}$, (2) depends on the velocity gradient $\mathbf{L}$ only through its symmetric part $\mathbf{D}$, and (3) is an isotropic function of $\mathbf{D}$. Restrictions such as these are welcomed by the experimentalist. Rather than being forced to sift through all possible functions of ρ, $\mathbf{v}$, and $\mathbf{L}$ to find one that acceptably matches physical experiments, one sees from the above thought experiments that only functions of ρ and isotropic functions of $\mathbf{D}$ need be considered. This enables one to characterize a particular viscous fluid from fewer laboratory experiments.

7.1.1.2 Linear viscous (Newtonian) fluids

A special case of the viscous fluid (7.6) is the **linear viscous** (or **Newtonian**) **fluid**:

$$\mathbf{T} = \mathbf{A}^{(4)}\mathbf{D} + \mathbf{A}, \tag{7.8a}$$

or, in Cartesian components,

$$T_{ij} = A_{ijkl}D_{kl} + A_{ij}. \tag{7.8b}$$

Note that $\mathbf{A}^{(4)}$ is a fourth-order tensor function of density ρ, $\mathbf{A}$ is a second-order tensor function of density ρ, and the Cauchy stress $\mathbf{T}$ is linearly dependent on the rate of deformation $\mathbf{D}$. The constitutive equation (7.8a) must satisfy conservation of angular momentum (i.e., $\mathbf{T} = \mathbf{T}^{\mathrm{T}}$) and the invariance requirement (7.7). It can be shown (refer to Problem 7.1) that these conditions demand that

$$A_{ijkl} = \lambda\delta_{ij}\delta_{kl} + \mu\left(\delta_{ik}\delta_{jl} + \delta_{il}\delta_{jk}\right), \quad A_{ij} = -p\delta_{ij}, \tag{7.9}$$

i.e., $\mathbf{A}^{(4)}$ is an *isotropic* fourth-order tensor and $\mathbf{A}$ is an *isotropic* second-order tensor; refer to Appendix C. In (7.9), δ_{ij} is the Kronecker delta, and λ, μ, and p are scalar functions of density ρ. It can be verified (refer to Problem 7.2) that substitution of (7.9) into (7.8b) gives

$$T_{ij} = -p\delta_{ij} + \lambda D_{kk}\delta_{ij} + 2\mu D_{ij}, \tag{7.10a}$$

or, in direct notation,

$$\mathbf{T} = -p\mathbf{I} + \lambda(\mathrm{tr}\mathbf{D})\mathbf{I} + 2\mu\mathbf{D}. \tag{7.10b}$$

The density-dependent coefficients λ, μ, and p are identified as the **dilatational viscosity**, **shear viscosity**, and **thermodynamic pressure**, respectively. The contribution $\lambda(\mathrm{tr}\mathbf{D})\mathbf{I}$ to the stress $\mathbf{T}$ is called the **viscous pressure.**

PROBLEM 7.1

In the mechanical theory, the constitutive equation for the stress in a Newtonian fluid is

$$T_{ij} = A_{ijkl}D_{kl} + A_{ij},$$

where A_{ij} and A_{ijkl} are density dependent. Confirm that invariance requirements and conservation of angular momentum demand that

$$A_{ijkl} = \lambda\delta_{ij}\delta_{kl} + \mu\left(\delta_{ik}\delta_{jl} + \delta_{il}\delta_{jk}\right), \quad A_{ij} = -p\delta_{ij}.$$

Solution

Use of the constitutive equation (7.8a) in the invariance requirement (7.7) gives the condition

$$\mathbf{A}^{(4)}\mathbf{QDQ}^{\mathrm{T}} + \mathbf{A} = \mathbf{Q}\left(\mathbf{A}^{(4)}\mathbf{D} + \mathbf{A}\right)\mathbf{Q}^{\mathrm{T}}.$$

This condition must hold for any choice of $\mathbf{D}$, in particular $\mathbf{D} = \mathbf{0}$ (i.e., a fluid at rest), which implies that

$$\mathbf{A} = \mathbf{QAQ}^{\mathrm{T}} \tag{a}$$

and, consequently,

$$\mathbf{A}^{(4)}\mathbf{QDQ}^{\mathrm{T}} = \mathbf{Q}\left(\mathbf{A}^{(4)}\mathbf{D}\right)\mathbf{Q}^{\mathrm{T}} \tag{b}$$

for all proper orthogonal $\mathbf{Q}$. We immediately conclude (refer to Appendix C) that result (a) demands that $\mathbf{A}$ is isotropic, i.e.,

$$\mathbf{A} = -p\mathbf{I},$$

or, in indicial notation,

$$A_{ij} = -p\delta_{ij},$$

where p is a scalar function of ρ. We proceed by considering the indicial form of (b):

$$A_{ijmn}Q_{mp}D_{pq}Q_{nq} = Q_{im}A_{mnpq}D_{pq}Q_{jn},$$

which can be rewritten as

$$\left(A_{ijmn}Q_{mp}Q_{nq} - Q_{im}Q_{jn}A_{mnpq}\right)D_{pq} = 0.$$

The coefficient of D_{pq} is independent of D_{pq}, and D_{pq} is arbitrary, which implies that the coefficient vanishes, i.e.,

$$A_{ijmn}Q_{mp}Q_{nq} = Q_{im}Q_{jn}A_{mnpq}.$$

Multiplying both sides by $Q_{kp}Q_{lq}$ gives

$$A_{ijmn}Q_{mp}Q_{kp}Q_{nq}Q_{lq} = Q_{im}Q_{jn}Q_{kp}Q_{lq}A_{mnpq}.$$

Use of

$$Q_{mp}Q_{kp} = \delta_{mk}, \quad Q_{nq}Q_{lq} = \delta_{nl},$$

which are both indicial forms of $\mathbf{QQ}^{\mathrm{T}} = \mathbf{I}$, along with (2.30), leads to

$$A_{ijkl} = Q_{im}Q_{jn}Q_{kp}Q_{lq}A_{mnpq}. \qquad \text{(c)}$$

We conclude (see Appendix C) that result (c) demands that $\mathbf{A}^{(4)}$ is isotropic, i.e.,

$$A_{ijkl} = \lambda\delta_{ij}\delta_{kl} + \mu\delta_{ik}\delta_{jl} + \gamma\delta_{il}\delta_{jk},$$

where λ, μ, and γ are functions of ρ. Lastly, conservation of angular momentum ($T_{ij} = T_{ji}$) requires that

$$A_{ijkl} = A_{jikl}, \quad A_{ij} = A_{ji},$$

i.e., A_{ij} and A_{ijkl} must be symmetric in i and j. This required symmetry, in turn, demands that $\gamma = \mu$ (refer to Appendix C), so

$$A_{ijkl} = \lambda\delta_{ij}\delta_{kl} + \mu\left(\delta_{ik}\delta_{jl} + \delta_{il}\delta_{jk}\right).$$

PROBLEM 7.2

Show that

$$A_{ijkl} = \lambda\delta_{ij}\delta_{kl} + \mu\left(\delta_{ik}\delta_{jl} + \delta_{il}\delta_{jk}\right), \quad A_{ij} = -p\delta_{ij}$$

imply that the constitutive equation (7.8b) for a Newtonian fluid becomes

$$T_{ij} = -p\delta_{ij} + \lambda D_{kk}\delta_{ij} + 2\mu D_{ij},$$

with p the thermodynamic pressure, λ the dilatational viscosity, and μ the shear viscosity.

Solution

$$\begin{aligned}
T_{ij} &= A_{ijkl}D_{kl} + A_{ij} && \text{(definition (7.8b))}\\
&= \left[\lambda\delta_{ij}\delta_{kl} + \mu\left(\delta_{ik}\delta_{jl} + \delta_{il}\delta_{jk}\right)\right]D_{kl} - p\delta_{ij} && \text{(result (7.9))}\\
&= \lambda\delta_{ij}\delta_{kl}D_{kl} + \mu\delta_{ik}\delta_{jl}D_{kl} + \mu\delta_{il}\delta_{jk}D_{kl} - p\delta_{ij} && \text{(property (2.1)}_{12}\text{)}\\
&= \lambda D_{kk}\delta_{ij} + \mu D_{ij} + \mu D_{ji} - p\delta_{ij} && \text{(result 2.30)}\\
&= \lambda D_{kk}\delta_{ij} + 2\mu D_{ij} - p\delta_{ij} && \text{(}\mathbf{D}\text{ is symmetric).}
\end{aligned}$$

7.1.1.3 The Navier-Stokes equations

In the mechanical theory, the field equations for a Newtonian fluid are

$$\dot{\rho} + \rho \operatorname{div} \mathbf{v} = 0, \quad \rho\dot{\mathbf{v}} = \operatorname{div}\mathbf{T} + \rho\mathbf{b}, \quad \mathbf{T} = -p\mathbf{I} + \lambda(\operatorname{tr}\mathbf{D})\mathbf{I} + 2\mu\mathbf{D}, \qquad (7.11)$$

which correspond to Eulerian statements of conservation of mass, conservation of linear momentum, and the linear viscous constitutive equation. It can be shown (refer to Problem 7.3) that the constitutive equation $(7.11)_3$ can be combined with conservation of linear momentum $(7.11)_2$ to obtain the **Navier-Stokes equations**

$$\rho\dot{\mathbf{v}} = -\operatorname{grad} p + \mu \operatorname{div}\left(\operatorname{grad}\mathbf{v}\right) + (\lambda+\mu)\operatorname{grad}\left(\operatorname{div}\mathbf{v}\right) + \rho\mathbf{b}, \tag{7.12a}$$

whose Cartesian component (indicial) form is

$$\rho\dot{v}_i = -p_{,i} + \mu v_{i,jj} + (\lambda+\mu)v_{j,ji} + \rho b_i. \tag{7.12b}$$

Note that the dilatational viscosity λ and shear viscosity μ are assumed to be constant, a customary assumption in the derivation of the Navier-Stokes equations. It can be verified (refer to Problem 7.4) that **Archimedes' principle** follows from (7.12) for the special case of a fluid at rest.

PROBLEM 7.3

Demonstrate that the Newtonian constitutive equation can be combined with conservation of linear momentum to obtain the Navier-Stokes equations.

Solution

The Cartesian component form of the Newtonian constitutive equation $(7.11)_3$ is

$$T_{ij} = -p\delta_{ij} + \lambda D_{kk}\delta_{ij} + 2\mu D_{ij}.$$

Use of the definition

$$D_{ij} = \frac{1}{2}\left(v_{i,j} + v_{j,i}\right)$$

implies that

$$T_{ij} = -p\delta_{ij} + \lambda v_{k,k}\delta_{ij} + \mu\left(v_{i,j} + v_{j,i}\right).$$

Then, we have

$$\begin{aligned}
T_{ij,j} &= -p_{,j}\delta_{ij} + \lambda v_{k,kj}\delta_{ij} + \mu\left(v_{i,jj} + v_{j,ij}\right) \\
&= -p_{,i} + \lambda v_{k,ki} + \mu v_{i,jj} + \mu v_{j,ij} && \text{(result (2.30))} \\
&= -p_{,i} + \lambda v_{k,ki} + \mu v_{i,jj} + \mu v_{j,ji} && \text{(continuity of velocity } \mathbf{v}\text{)} \\
&= -p_{,i} + \mu v_{i,jj} + (\lambda+\mu)v_{j,ji} && \text{(change of repeated subscript from } k \text{ to } j\text{).}
\end{aligned}$$

Subsequent use of this result in the Cartesian component form of conservation of linear momentum $(7.11)_2$

$$\rho\dot{v}_i = T_{ij,j} + \rho b_i$$

leads to

$$\rho\dot{v}_i = -p_{,i} + \mu v_{i,jj} + (\lambda+\mu)v_{j,ji} + \rho b_i,$$

the Cartesian component form of the Navier-Stokes equations.

PROBLEM 7.4

Prove Archimedes' principle: The resultant contact force acting on the surface of a body submerged in a fluid at rest is equal in magnitude to the weight of the displaced fluid and directed upwards.

Solution

For a fluid at rest, $\mathbf{v} = \mathbf{0}$. In this special case, the Navier-Stokes equations (7.12a) reduce to

$$\operatorname{grad} p = -\rho g \mathbf{e}_3, \tag{a}$$

where we have taken the gravitational body force per unit mass (or gravitational acceleration) to be coincident with the negative $\mathbf{e}_3$ axis, i.e., $\mathbf{b} = -g\mathbf{e}_3$. (Note that (a) implies that if the density ρ of the fluid is constant, then the pressure p in the fluid varies linearly in the direction of the gravitational acceleration, a well-known result in fluid statics.) Integration of (a) over the volume $\mathcal{R}$ occupied by the submerged body (or, equivalently, the volume occupied by the displaced fluid), and subsequent use of the divergence theorem $(2.104)_1$, leads to

$$\int_{\partial \mathcal{R}} p\mathbf{n}\, da = -\int_{\mathcal{R}} \rho g \mathbf{e}_3\, dv. \tag{b}$$

With the aid of definition (4.2), we recognize that the right-hand side of (b) is nothing but the weight W of the displaced fluid, i.e.,

$$\int_{\partial \mathcal{R}} p\mathbf{n}\, da = -W\mathbf{e}_3. \tag{c}$$

For a fluid at rest, the Newtonian constitutive equation $(7.11)_3$ simplifies to

$$\mathbf{T} = -p\mathbf{I},$$

which implies that the traction is

$$\mathbf{t} = -p\mathbf{n}. \tag{d}$$

The resultant contact force acting on the surface $\partial\mathcal{R}$ of the submerged body is obtained by integrating the traction (d) over $\partial\mathcal{R}$ and using result (c):

$$\int_{\partial \mathcal{R}} \mathbf{t}\, da = -\int_{\partial \mathcal{R}} p\mathbf{n}\, da = W\mathbf{e}_3.$$

Thus, the resultant of the tractions acting on the surface of a body submerged in a fluid at rest is equal in magnitude to the weight of the displaced fluid and directed upwards.

EXERCISES

1. Verify that (7.12b) is the Cartesian component form of (7.12a).

2. Starting with the Navier-Stokes equations in Cartesian component notation,

$$\rho \dot{v}_i = -p_{,i} + \mu v_{i,jj} + (\lambda + \mu) v_{j,ji} + \rho b_i:$$

(a) Confirm that it follows that

$$\rho\left(v_i' + v_{i,j} v_j\right) = -p_{,i} + \mu v_{i,jj} + (\lambda + \mu) v_{j,ji} + \rho b_i.$$

(b) Fully expand the above expression (i.e., no summation convention). Together with conservation of mass $(7.11)_1$, this yields a system of four nonlinear partial differential equations for the density ρ and the three velocity components v_1, v_2, v_3.

7.1.2 INVISCID FLUIDS

In contrast to a viscous fluid, an **inviscid fluid** is insensitive to the rate at which it is deformed. (We refer to an inviscid fluid in the mechanical theory as an **ideal fluid**.) Hence, for an ideal fluid, we assume that the Cauchy stress $\mathbf{T}$ at present position $\mathbf{x}$ and time t depends on the motion χ only through the present density ρ evaluated at $\mathbf{x}$ and t, i.e.,

$$\mathbf{T} = \check{\mathbf{T}}(\rho). \tag{7.13}$$

Compare (7.13) with (7.6), note that it is independent of the rate of deformation $\mathbf{D}$, and refer to Appendix B. Notation (7.13) implies that $\mathbf{T}$ is an *explicit* function of ρ evaluated at $\mathbf{x}$ and t, and an *implicit* function of $\mathbf{x}$ and t. It can be shown that invariance requirements demand that the dependence (7.13) is of the form

$$\mathbf{T} = -p\mathbf{I}, \tag{7.14}$$

where the thermodynamic pressure p is a scalar function of density ρ. Note that an ideal fluid is a special case of a linear viscous fluid, with the stress $\mathbf{T}$ independent of the rate of deformation $\mathbf{D}$, the dilatational viscosity λ, and the shear viscosity μ (compare (7.14) with (7.10b)). Also note that it can be shown (using the relationship $\rho = \rho_{\mathrm{R}}/\det \mathbf{F}$; refer to Problem 7.5) that an ideal fluid is a special case of an elastic solid, and can thus be referred to as an **elastic fluid**.

PROBLEM 7.5

Demonstrate that an ideal fluid is a special case of an elastic solid.

Solution

The constitutive equation for an ideal fluid is of the form

$$\mathbf{T} = \check{\mathbf{T}}(\rho).$$

Recall from conservation of mass (refer to (4.56a)) that

$$\rho = \frac{\rho_{\mathrm{R}}}{J} = \frac{\rho_{\mathrm{R}}}{\det \mathbf{F}}.$$

It follows that

$$\mathbf{T} = \check{\mathbf{T}}(\rho) = \check{\mathbf{T}}\left(\frac{\rho_R}{\det \mathbf{F}}\right) = \grave{\mathbf{T}}(\mathbf{F}),$$

so an ideal fluid is a special case of an elastic solid (refer to (6.1)) when the dependence of stress $\mathbf{T}$ on the deformation gradient $\mathbf{F}$ is through its determinant $\det \mathbf{F}$.

EXERCISES

1. The constitutive equation for the stress in an ideal fluid is of the form

$$\mathbf{T} = \check{\mathbf{T}}(\rho).$$

Prove that invariance requirements demand that

$$\mathbf{T} = -p\mathbf{I},$$

where p is a scalar function of ρ.

7.2 THERMOMECHANICAL THEORY

Recall from Section 5.3 that in formulation (5.36)–(5.38) of the thermomechanical theory, constitutive assumptions are made on the Cauchy stress $\mathbf{T}$, heat flux vector $\mathbf{q}$, Helmholtz free energy ψ, and entropy η as functions of the motion χ and temperature Θ, and possibly their rates, gradients, and histories, i.e.,

$$\mathbf{T} = \bar{\mathbf{T}}\ \text{(motion and temperature)}, \qquad \mathbf{q} = \bar{\mathbf{q}}\ \text{(motion and temperature)},$$

$$\psi = \bar{\psi}\ \text{(motion and temperature)}, \qquad \eta = \bar{\eta}\ \text{(motion and temperature)}.$$

7.2.1 VISCOUS FLUIDS

For a **viscous fluid** in the thermomechanical theory (which we call a **thermoviscous fluid**), we begin with the constitutive assumption that the Cauchy stress $\mathbf{T}$, heat flux $\mathbf{q}$, Helmholtz free energy ψ, and entropy η at present position $\mathbf{x}$ and time t depend on the motion χ and temperature Θ through the present density ρ, velocity $\mathbf{v} = \dot{\chi}$, Eulerian velocity gradient $\mathbf{L} = \operatorname{grad} \mathbf{v}$, temperature Θ, and Eulerian temperature gradient $\mathbf{g} = \operatorname{grad} \Theta$, all evaluated at $\mathbf{x}$ and t, i.e.,

$$\mathbf{T} = \bar{\mathbf{T}}(\rho, \mathbf{v}, \mathbf{L}, \Theta, \mathbf{g}), \qquad \mathbf{q} = \bar{\mathbf{q}}(\rho, \mathbf{v}, \mathbf{L}, \Theta, \mathbf{g}),$$

$$\psi = \bar{\psi}(\rho, \mathbf{v}, \mathbf{L}, \Theta, \mathbf{g}), \qquad \eta = \bar{\eta}(\rho, \mathbf{v}, \mathbf{L}, \Theta, \mathbf{g}). \qquad (7.15)$$

Compare (7.15) with (7.1), and refer to Appendix B. This notation means that $\mathbf{T}$, $\mathbf{q}$, ψ, and η at position $\mathbf{x}$ and time t are *explicit* functions of ρ, $\mathbf{v}$, $\mathbf{L}$, Θ, and $\mathbf{g}$ evaluated at $\mathbf{x}$ and t, and *implicit* functions of $\mathbf{x}$ and t. We emphasize that there is no notion of a reference configuration in constitutive assumption (7.15). Note that the density ρ is not an algebraic function of $\mathbf{v}$, $\mathbf{L}$, Θ, or $\mathbf{g}$, and therefore must be explicitly included in the argument list of (7.15). *Recall from Section 5.4 that, as with all materials in the thermomechanical theory, the response functions (7.15) for a thermoviscous fluid must satisfy invariance requirements, the second law of thermodynamics, and conservation of angular momentum.*

7.2.1.1 Restrictions imposed by invariance under SRBMs and the second law of thermodynamics

Similar to what was done in Section 7.1.1.1, it can be shown that to satisfy invariance under SRBMs, the response functions (7.15) must be independent of $\mathbf{v}$, and can depend on the velocity gradient $\mathbf{L}$ only through its symmetric part $\mathbf{D}$, i.e.,

$$\mathbf{T} = \breve{\mathbf{T}}(\rho, \mathbf{D}, \Theta, \mathbf{g}), \qquad \mathbf{q} = \breve{\mathbf{q}}(\rho, \mathbf{D}, \Theta, \mathbf{g}),$$
$$\psi = \breve{\psi}(\rho, \mathbf{D}, \Theta, \mathbf{g}), \qquad \eta = \breve{\eta}(\rho, \mathbf{D}, \Theta, \mathbf{g}). \tag{7.16}$$

Substitution of the simplified constitutive assumptions (7.16) into the Clausius-Duhem inequality (5.35) gives

$$\left(\breve{\mathbf{T}} + \rho^2 \frac{\partial \breve{\psi}}{\partial \rho} \mathbf{I}\right) \cdot \mathbf{D} - \rho \left(\frac{\partial \breve{\psi}}{\partial \Theta} + \breve{\eta}\right) \dot{\Theta} - \rho \frac{\partial \breve{\psi}}{\partial \mathbf{D}} \cdot \dot{\mathbf{D}}$$
$$- \rho \frac{\partial \breve{\psi}}{\partial \mathbf{g}} \cdot \dot{\mathbf{g}} - \frac{1}{\Theta} \breve{\mathbf{q}} \cdot \mathbf{g} \geq 0, \tag{7.17}$$

where we have used the chain rule

$$\dot{\psi} = \frac{\partial \breve{\psi}}{\partial \rho} \dot{\rho} + \frac{\partial \breve{\psi}}{\partial \mathbf{D}} \cdot \dot{\mathbf{D}} + \frac{\partial \breve{\psi}}{\partial \Theta} \dot{\Theta} + \frac{\partial \breve{\psi}}{\partial \mathbf{g}} \cdot \dot{\mathbf{g}}$$

and the result (refer to Problem 7.6)

$$\dot{\rho} = -\rho \mathbf{I} \cdot \mathbf{D}.$$

Inequality (7.17) must hold for all processes; with use of arguments similar to those employed in Section 6.2.1, it can be shown that

$$\psi = \breve{\psi}(\rho, \Theta), \quad \eta = -\frac{\partial \breve{\psi}}{\partial \Theta} \tag{7.18a}$$

and

$$\left(\breve{\mathbf{T}} + \rho^2 \frac{\partial \breve{\psi}}{\partial \rho} \mathbf{I}\right) \cdot \mathbf{D} \geq 0, \quad \breve{\mathbf{q}} \cdot \mathbf{g} \leq 0. \tag{7.18b}$$

Comparing (7.18) with (7.16), we see that as a consequence of the second law of thermodynamics:

(1) Only three response functions ($\breve{\psi}$, $\breve{\mathbf{T}}$, and $\breve{\mathbf{q}}$) are needed to characterize a thermoviscous fluid, rather than the four initially supposed. (Nice! Fewer response functions means fewer unknowns in the characterization problem.)

(2) The response function for ψ is independent of $\mathbf{D}$ and $\mathbf{g}$. (Again, fewer experiments to perform.)

(3) The response function for $\mathbf{q}$ is restricted by the second law through the inequality $\breve{\mathbf{q}} \cdot \mathbf{g} \leq 0$, i.e., *heat flows against the temperature gradient.*

We can demonstrate (using a proof similar to the one employed in Problem 6.3) that if the heat flux vector $\breve{\mathbf{q}}$ in (7.16) is a continuous function of the temperature gradient $\mathbf{g}$ at $\mathbf{g} = \mathbf{0}$, then it follows from $\breve{\mathbf{q}} \cdot \mathbf{g} \leq 0$ that $\breve{\mathbf{q}}$ evaluated at $\mathbf{g} = \mathbf{0}$ is zero. That is, *heat does not flow in the absence of a temperature gradient.*

PROBLEM 7.6

Prove that $\dot{\rho} = -\rho \mathbf{I} \cdot \mathbf{D}$.

Solution

This result follows from conservation of mass (4.36a) and the additive decomposition (3.57) of the velocity gradient. In particular, we have:

$$\dot{\rho} = -\rho \operatorname{div} \mathbf{v} = -\rho \operatorname{tr}(\operatorname{grad} \mathbf{v}) = -\rho \operatorname{tr} \mathbf{L} = -\rho (\operatorname{tr} \mathbf{D} + \operatorname{tr} \mathbf{W}) = -\rho \operatorname{tr} \mathbf{D} = -\rho \mathbf{I} \cdot \mathbf{D}.$$

EXERCISES

1. Starting with the constitutive assumptions

$$\mathbf{T} = \bar{\mathbf{T}}(\rho, \mathbf{v}, \mathbf{L}, \Theta, \mathbf{g}), \qquad \mathbf{q} = \bar{\mathbf{q}}(\rho, \mathbf{v}, \mathbf{L}, \Theta, \mathbf{g}),$$

$$\psi = \bar{\psi}(\rho, \mathbf{v}, \mathbf{L}, \Theta, \mathbf{g}), \qquad \eta = \bar{\eta}(\rho, \mathbf{v}, \mathbf{L}, \Theta, \mathbf{g})$$

for a thermoviscous fluid, use invariance requirements to show that that these response functions must be independent of $\mathbf{v}$, and can depend on the velocity gradient $\mathbf{L}$ only through its symmetric part $\mathbf{D}$, i.e.,

$$\mathbf{T} = \breve{\mathbf{T}}(\rho, \mathbf{D}, \Theta, \mathbf{g}), \qquad \mathbf{q} = \breve{\mathbf{q}}(\rho, \mathbf{D}, \Theta, \mathbf{g}),$$

$$\psi = \breve{\psi}(\rho, \mathbf{D}, \Theta, \mathbf{g}), \qquad \eta = \breve{\eta}(\rho, \mathbf{D}, \Theta, \mathbf{g}).$$

2. Prove that as a consequence of the second law of thermodynamics, the response functions in the previous exercise further simplify to

$$\psi = \breve{\psi}(\rho, \Theta), \quad \eta = -\frac{\partial \breve{\psi}}{\partial \Theta}$$

and

$$\left(\breve{\mathbf{T}} + \rho^2 \frac{\partial \breve{\psi}}{\partial \rho} \mathbf{I}\right) \cdot \mathbf{D} \geq 0, \quad \breve{\mathbf{q}} \cdot \mathbf{g} \leq 0.$$

3. Show that if the heat flux vector $\breve{\mathbf{q}}$ in (7.16) is a continuous function of the temperature gradient $\mathbf{g}$ at $\mathbf{g} = \mathbf{0}$, then it follows from $\breve{\mathbf{q}} \cdot \mathbf{g} \leq 0$ that $\breve{\mathbf{q}}$ evaluated at $\mathbf{g} = \mathbf{0}$ is zero. That is, heat does not flow in the absence of a temperature gradient.

7.2.1.2 Linear thermoviscous (Newtonian) fluids

In this section, we specialize to **linear thermoviscous** (or **Newtonian**) **fluids**. Hence, we assume that the response functions for stress $\mathbf{T}$ and heat flux $\mathbf{q}$ are linear in temperature gradient $\mathbf{g}$ and rate of deformation $\mathbf{D}$, i.e.,

$$\mathbf{T} = \mathbf{A}^{(2)} + \mathbf{A}^{(3)}\mathbf{g} + \mathbf{A}^{(4)}\mathbf{D}, \quad \mathbf{q} = \mathbf{K}^{(1)} + \mathbf{K}^{(2)}\mathbf{g} + \mathbf{K}^{(3)}\mathbf{D}, \tag{7.19a}$$

or, in Cartesian component form,

$$T_{ij} = A_{ij} + A_{ijk}g_k + A_{ijkl}D_{kl}, \quad q_i = K_i + K_{ij}g_j + K_{ijk}D_{jk}. \tag{7.19b}$$

Note that $\mathbf{K}^{(1)}$, $\mathbf{K}^{(2)}$ and $\mathbf{A}^{(2)}$, $\mathbf{K}^{(3)}$ and $\mathbf{A}^{(3)}$, and $\mathbf{A}^{(4)}$ are vector, tensor, third-order tensor, and fourth-order tensor functions, respectively, of density ρ and temperature Θ. It can be shown that the invariance requirements

$$\mathbf{T}^+ = \mathbf{Q}\mathbf{T}\mathbf{Q}^{\mathrm{T}}, \quad \mathbf{q}^+ = \mathbf{Q}\mathbf{q}$$

when

$$\rho^+ = \rho, \quad \mathbf{D}^+ = \mathbf{Q}\mathbf{D}\mathbf{Q}^{\mathrm{T}}, \quad \Theta^+ = \Theta, \quad \mathbf{g}^+ = \mathbf{Q}\mathbf{g}$$

for all proper orthogonal $\mathbf{Q}$ demand

$$A_{ij} = -p\delta_{ij}, \quad A_{ijk} = a\epsilon_{ijk}, \quad A_{ijkl} = \lambda\delta_{ij}\delta_{kl} + \mu\delta_{ik}\delta_{jl} + \gamma\delta_{il}\delta_{jk}, \tag{7.20a}$$

and

$$K_i = 0, \quad K_{ij} = -k\delta_{ij}, \quad K_{ijk} = b\epsilon_{ijk}, \tag{7.20b}$$

where δ_{ij} is the Kronecker delta, ϵ_{ijk} is the permutation symbol, and p, λ, μ, γ, k, a, and b are scalar functions of density ρ and temperature Θ; refer to Appendix C. It can be verified that substitution of (7.20a) and (7.20b) into (7.19b) gives

$$T_{ij} = -p\delta_{ij} + a\epsilon_{ijk}g_k + \lambda\delta_{ij}D_{kk} + (\mu + \gamma)D_{ij}, \quad q_i = -kg_i. \tag{7.21}$$

We can then show that conservation of angular momentum (which demands that $T_{ij} = T_{ji}$) implies that $a = 0$ and $\gamma = \mu$, so

$$T_{ij} = -p\delta_{ij} + \lambda D_{kk}\delta_{ij} + 2\mu D_{ij}, \quad q_i = -kg_i, \tag{7.22a}$$

or, in direct notation,

$$\mathbf{T} = -p\mathbf{I} + \lambda(\mathrm{tr}\,\mathbf{D})\mathbf{I} + 2\mu\mathbf{D}, \quad \mathbf{q} = -k\mathbf{g}. \tag{7.22b}$$

In (7.22b), we identify p as the **thermodynamic pressure**, μ as the **shear viscosity**, λ as the **dilatational viscosity**, and k as the **thermal conductivity**; all are functions of density ρ and temperature Θ.

We now return to second law considerations. Recall from (7.18) the inequality

$$\breve{\mathbf{q}} \cdot \mathbf{g} \leq 0.$$

With $(7.22b)_2$ this implies that

$$k(\mathbf{g} \cdot \mathbf{g}) = k|\mathbf{g}|^2 \geq 0,$$

so the thermal conductivity k must be nonnegative, i.e.,

$$k \geq 0. \tag{7.23}$$

Also recall from (7.18) the inequality

$$\left(\breve{\mathbf{T}} + \rho^2 \frac{\partial \breve{\psi}}{\partial \rho} \mathbf{I}\right) \cdot \mathbf{D} \geq 0,$$

which, with use of use of $(7.22b)_1$, becomes

$$\left(-p\mathbf{I} + \lambda(\operatorname{tr}\mathbf{D})\mathbf{I} + 2\mu\mathbf{D} + \rho^2 \frac{\partial \breve{\psi}}{\partial \rho} \mathbf{I}\right) \cdot \mathbf{D} \geq 0. \tag{7.24}$$

This inequality must hold for all $\mathbf{D}$, in particular $\alpha\mathbf{D}$, with α a positive scalar, so

$$\left(-p\mathbf{I} + \alpha\lambda(\operatorname{tr}\mathbf{D})\mathbf{I} + 2\alpha\mu\mathbf{D} + \rho^2 \frac{\partial \breve{\psi}}{\partial \rho} \mathbf{I}\right) \cdot \alpha\mathbf{D} \geq 0. \tag{7.25}$$

Dividing (7.25) by α and taking the limit as $\alpha \to 0$ gives

$$\left(-p\mathbf{I} + \rho^2 \frac{\partial \breve{\psi}}{\partial \rho} \mathbf{I}\right) \cdot \mathbf{D} \geq 0. \tag{7.26}$$

Since the coefficient of $\mathbf{D}$ in (7.26) is independent of $\mathbf{D}$, we conclude that

$$p = \rho^2 \frac{\partial \breve{\psi}}{\partial \rho}, \tag{7.27}$$

and there remains in (7.24)

$$\left(\lambda(\operatorname{tr}\mathbf{D})\mathbf{I} + 2\mu\mathbf{D}\right) \cdot \mathbf{D} \geq 0. \tag{7.28}$$

If we decompose $\mathbf{D}$ into its *spherical* and *deviatoric* parts, i.e.,

$$\mathbf{D} = \frac{1}{3}(\operatorname{tr}\mathbf{D})\mathbf{I} + \mathbf{D}_{\mathrm{d}}, \tag{7.29}$$

with $\operatorname{tr}\mathbf{D}_{\mathrm{d}} = 0$, it can be shown that use of (7.29) in (7.28) leads to

$$\left(\lambda + \frac{2}{3}\mu\right)(\operatorname{tr}\mathbf{D})^2 + 2\mu\mathbf{D}_{\mathrm{d}} \cdot \mathbf{D}_{\mathrm{d}} \geq 0, \tag{7.30}$$

where $\lambda + \frac{2}{3}\mu$ is the **bulk viscosity**. Since $\operatorname{tr}\mathbf{D}$ and $\mathbf{D}_{\mathrm{d}}$ are independent, $\lambda + \frac{2}{3}\mu$ is independent of $\operatorname{tr}\mathbf{D}$, and μ is independent of $\mathbf{D}_{\mathrm{d}}$, we conclude from (7.30) that

$$\lambda + \frac{2}{3}\mu \geq 0, \quad \mu \geq 0, \tag{7.31}$$

i.e., the bulk viscosity and the shear viscosity must be nonnegative.

In summary, we have found that the constitutive response of a linear thermoviscous fluid is given by

$$\psi = \breve{\psi}(\rho, \Theta), \quad p = \rho^2 \frac{\partial \breve{\psi}(\rho, \Theta)}{\partial \rho}, \quad \eta = -\frac{\partial \breve{\psi}(\rho, \Theta)}{\partial \Theta},$$

$$\mathbf{T} = -p\mathbf{I} + \lambda(\operatorname{tr}\mathbf{D})\mathbf{I} + 2\mu\mathbf{D}, \quad \mathbf{q} = -k\mathbf{g}, \tag{7.32}$$

subject to

$$k \geq 0, \quad \lambda + \frac{2}{3}\mu \geq 0, \quad \mu \geq 0,$$

where λ, μ, and k are functions of density ρ and temperature Θ. Thus, as a consequence of conservation of angular momentum, invariance requirements, and the second law of thermodynamics, a linear thermoviscous fluid is fully characterized by only four scalar functions of density ρ and temperature Θ (namely, the Helmholtz free energy $\breve{\psi}(\rho, \Theta)$, dilatational viscosity $\lambda(\rho, \Theta)$, shear viscosity $\mu(\rho, \Theta)$, and thermal conductivity $k(\rho, \Theta)$) subject to three inequalities.

We remark that Eqs. $(7.32)_2$ and $(7.32)_3$ are referred to as the **equations of state**. Formulations of the theory that use different independent variables than density ρ and temperature Θ, and hence different energy potentials than the Helmholtz free energy ψ, can be found in Appendix E, along with their corresponding equations of state.

EXERCISES

1. In the thermomechanical theory, the constitutive equations for the stress and heat flux in a Newtonian fluid are

$$T_{ij} = A_{ij} + A_{ijk}g_k + A_{ijkl}D_{kl}, \quad q_i = K_i + K_{ij}g_j + K_{ijk}D_{jk},$$

where A_{ij}, A_{ijk}, A_{ijkl}, K_i, K_{ij}, and K_{ijk} are density and temperature dependent.

(a) Confirm that invariance requirements demand that

$$A_{ij} = -p\delta_{ij}, \quad A_{ijk} = a\epsilon_{ijk}, \quad A_{ijkl} = \lambda\delta_{ij}\delta_{kl} + \mu\delta_{ik}\delta_{jl} + \gamma\delta_{il}\delta_{jk},$$

and

$$K_i = 0, \quad K_{ij} = -k\delta_{ij}, \quad K_{ijk} = b\epsilon_{ijk}.$$

(b) Show that the results obtained in part (a) imply that

$$T_{ij} = -p\delta_{ij} + a\epsilon_{ijk}g_k + \lambda\delta_{ij}D_{kk} + (\mu + \gamma)D_{ij}, \quad q_i = -kg_i.$$

(c) Demonstrate that as a consequence of conservation of angular momentum (which demands that $T_{ij} = T_{ji}$), we must have $a = 0$ and $\gamma = \mu$, so

$$T_{ij} = -p\delta_{ij} + \lambda D_{kk}\delta_{ij} + 2\mu D_{ij}, \quad q_i = -kg_i.$$

2. Show that use of (7.29) in (7.28) leads to (7.30).

7.2.2 INVISCID FLUIDS

Recall that an **inviscid fluid** is insensitive to the rate at which it is deformed. (We refer to an inviscid fluid in the thermomechanical theory as a **perfect fluid**.) Hence, for a perfect fluid, we begin with the constitutive assumption that the Cauchy stress $\mathbf{T}$ at present position $\mathbf{x}$ and time t depends on the motion χ and temperature Θ through the present density ρ, temperature Θ, and Eulerian temperature gradient $\mathbf{g} = \text{grad}\,\Theta$, all evaluated at $\mathbf{x}$ and t, i.e.,

$$\mathbf{T} = \check{\mathbf{T}}(\rho, \Theta, \mathbf{g}), \qquad \mathbf{q} = \check{\mathbf{q}}(\rho, \Theta, \mathbf{g}),$$
$$\psi = \check{\psi}(\rho, \Theta, \mathbf{g}), \qquad \eta = \check{\eta}(\rho, \Theta, \mathbf{g}). \tag{7.33}$$

Compare (7.33) with (7.16), note that the response functions are independent of the rate of deformation $\mathbf{D}$, and refer to Appendix B. Notation (7.33) implies that $\mathbf{T}$, $\mathbf{q}$, ψ, and η at position $\mathbf{x}$ and time t are *explicit* functions of ρ, Θ, and $\mathbf{g}$ evaluated at $\mathbf{x}$ and t, and *implicit* functions of $\mathbf{x}$ and t. It can be shown (refer to Problem 7.7) that second law considerations reduce (7.33) to

$$\psi = \check{\psi}(\rho, \Theta), \quad p = \rho^2 \frac{\partial \check{\psi}(\rho, \Theta)}{\partial \rho}, \quad \eta = -\frac{\partial \check{\psi}(\rho, \Theta)}{\partial \Theta},$$
$$\mathbf{T} = -p\mathbf{I}, \quad \check{\mathbf{q}} \cdot \mathbf{g} \leq 0, \tag{7.34}$$

where the thermodynamic pressure p is a scalar function of density ρ and temperature Θ. Note that a perfect fluid is a special case of a linear thermoviscous fluid, with the stress $\mathbf{T}$ independent of the rate of deformation $\mathbf{D}$, the dilatational viscosity λ, and the shear viscosity μ (compare (7.34) with (7.32)). Also note that it can be shown (using the relationship $\rho = \rho_{\mathrm{R}}/\det \mathbf{F}$; refer to Problem 7.8) that a perfect fluid is a special case of a thermoelastic solid, and can thus be referred to as a **thermoelastic fluid**.

PROBLEM 7.7

Prove that as a consequence of the second law of thermodynamics, the response functions for a perfect fluid

$$\mathbf{T} = \check{\mathbf{T}}(\rho, \Theta, \mathbf{g}), \quad \mathbf{q} = \check{\mathbf{q}}(\rho, \Theta, \mathbf{g}), \quad \psi = \check{\psi}(\rho, \Theta, \mathbf{g}), \quad \eta = \check{\eta}(\rho, \Theta, \mathbf{g})$$

simplify to

$$\psi = \check{\psi}(\rho, \Theta), \quad p = \rho^2 \frac{\partial \check{\psi}}{\partial \rho}, \quad \eta = -\frac{\partial \check{\psi}}{\partial \Theta}, \quad \mathbf{T} = -p\mathbf{I}, \quad \check{\mathbf{q}} \cdot \mathbf{g} \leq 0.$$

Solution

Substitution of the constitutive assumptions

$$\mathbf{T} = \check{\mathbf{T}}(\rho, \Theta, \mathbf{g}), \quad \mathbf{q} = \check{\mathbf{q}}(\rho, \Theta, \mathbf{g}), \quad \psi = \check{\psi}(\rho, \Theta, \mathbf{g}), \quad \eta = \check{\eta}(\rho, \Theta, \mathbf{g})$$

for a perfect fluid into the Clausius-Duhem inequality

$$-\rho\dot{\psi} + \mathbf{T} \cdot \mathbf{D} - \rho\eta\dot{\Theta} - \frac{1}{\Theta}\mathbf{q} \cdot \mathbf{g} \geq 0$$

leads to

$$\left(\breve{\mathbf{T}}+\rho^2\frac{\partial\breve{\psi}}{\partial\rho}\mathbf{I}\right)\cdot\mathbf{D}-\rho\left(\frac{\partial\breve{\psi}}{\partial\Theta}+\breve{\eta}\right)\dot{\Theta}-\rho\frac{\partial\breve{\psi}}{\partial\mathbf{g}}\cdot\dot{\mathbf{g}}-\frac{1}{\Theta}\breve{\mathbf{q}}\cdot\mathbf{g}\geq 0, \tag{a}$$

where we have used the chain rule

$$\dot{\breve{\psi}}=\frac{\partial\breve{\psi}}{\partial\rho}\dot{\rho}+\frac{\partial\breve{\psi}}{\partial\Theta}\dot{\Theta}+\frac{\partial\breve{\psi}}{\partial\mathbf{g}}\cdot\dot{\mathbf{g}}$$

and the result

$$\dot{\rho}=-\rho\mathbf{I}\cdot\mathbf{D}.$$

Inequality (a) must hold for *all* processes. *In particular*, it must hold for the family of processes with $\mathbf{g}=\mathbf{0}$, $\dot{\mathbf{g}}=\mathbf{0}$, and $\mathbf{D}=\mathbf{0}$, but $\dot{\Theta}$ arbitrary, at particular position $\mathbf{x}$ and time t. (An example is the family of processes (D.1) in Appendix D with $\mathbf{A}=\mathbf{0}$, $\mathbf{a}=\mathbf{0}$, and $\mathbf{g}_0=\mathbf{0}$, but a any real number.) For this family of processes, (a) simplifies to

$$-\rho\left(\frac{\partial\breve{\psi}}{\partial\Theta}+\breve{\eta}\right)\dot{\Theta}\geq 0,$$

or, since the density ρ is positive,

$$\left(\frac{\partial\breve{\psi}}{\partial\Theta}+\breve{\eta}\right)\dot{\Theta}\leq 0.$$

Since the coefficient of $\dot{\Theta}$ is independent of $\dot{\Theta}$, and $\dot{\Theta}$ is arbitrary, the coefficient must vanish, i.e.,

$$\eta=-\frac{\partial\breve{\psi}}{\partial\Theta}. \tag{b}$$

Result (b) reduces inequality (a) to

$$\left(\breve{\mathbf{T}}+\rho^2\frac{\partial\breve{\psi}}{\partial\rho}\mathbf{I}\right)\cdot\mathbf{D}-\rho\frac{\partial\breve{\psi}}{\partial\mathbf{g}}\cdot\dot{\mathbf{g}}-\frac{1}{\Theta}\breve{\mathbf{q}}\cdot\mathbf{g}\geq 0. \tag{c}$$

Inequality (c) must hold for all processes, in particular the family of processes with $\mathbf{g}=\mathbf{0}$ and $\mathbf{D}=\mathbf{0}$, but $\dot{\mathbf{g}}$ arbitrary, at $\mathbf{x}$ and t (e.g., the family (D.1) in Appendix D with $\mathbf{A}=\mathbf{0}$ and $\mathbf{g}_0=\mathbf{0}$, but $\mathbf{a}$ any vector). For members of this family, (c) simplifies to

$$-\rho\frac{\partial\breve{\psi}}{\partial\mathbf{g}}\cdot\dot{\mathbf{g}}\geq 0,$$

or

$$\frac{\partial\breve{\psi}}{\partial\mathbf{g}}\cdot\dot{\mathbf{g}}\leq 0.$$

Since the coefficient of $\dot{\mathbf{g}}$ is independent of $\dot{\mathbf{g}}$, and $\dot{\mathbf{g}}$ is arbitrary, we conclude that

$$\frac{\partial \check{\psi}}{\partial \mathbf{g}} = \mathbf{0}, \tag{d}$$

i.e., the response function for the Helmholtz free energy ψ is independent of the temperature gradient $\mathbf{g}$. Result (d) reduces inequality (c) to

$$\left(\check{\mathbf{T}} + \rho^2 \frac{\partial \check{\psi}}{\partial \rho} \mathbf{I}\right) \cdot \mathbf{D} - \frac{1}{\Theta} \check{\mathbf{q}} \cdot \mathbf{g} \geq 0.$$

A series of similar arguments allows us to deduce that

$$\mathbf{T} = -p\mathbf{I}, \quad p = \rho^2 \frac{\partial \check{\psi}}{\partial \rho}, \quad \check{\mathbf{q}} \cdot \mathbf{g} \leq 0. \tag{e}$$

PROBLEM 7.8

Verify that a perfect fluid is a special case of a thermoelastic solid.

Solution

For a perfect fluid, we may write

$$\psi = \check{\psi}(\rho, \Theta) = \check{\psi}\left(\frac{\rho_R}{\det \mathbf{F}}, \Theta\right) = \bar{\psi}(\mathbf{F}, \Theta),$$

where we have used the relationship $\rho = \rho_R / \det \mathbf{F}$. Thus, a perfect fluid is a special case of a thermoelastic solid (refer to $(6.16\text{a})_1$) when the dependence of the Helmholtz free energy ψ on the deformation gradient $\mathbf{F}$ is through its determinant $J = \det \mathbf{F}$.

Then, starting with the constitutive equation $(6.16\text{b})_1$ for stress in a thermoelastic solid,

$$\mathbf{T} = \rho \frac{\partial \bar{\psi}}{\partial \mathbf{F}} \mathbf{F}^{\mathrm{T}},$$

we find, using the chain rule, that the constitutive equation for stress in a perfect fluid falls out as a special case:

$$\mathbf{T} = \rho \frac{\partial \bar{\psi}}{\partial \mathbf{F}} \mathbf{F}^{\mathrm{T}} = \rho \frac{\partial \check{\psi}}{\partial \rho} \frac{\mathrm{d}\rho}{\mathrm{d}J} \frac{\mathrm{d}J}{\mathrm{d}\mathbf{F}} \mathbf{F}^{\mathrm{T}} = -\rho^2 \frac{\partial \check{\psi}}{\partial \rho} \mathbf{I} = -p\mathbf{I},$$

where we have used

$$\frac{\mathrm{d}\rho}{\mathrm{d}J} = \frac{\mathrm{d}}{\mathrm{d}J}\left(\frac{\rho_R}{J}\right) = -\frac{\rho_R}{J^2} = -\frac{\rho J}{J^2} = -\frac{\rho}{J}$$

and, from Problem 2.54,

$$\frac{\mathrm{d}J}{\mathrm{d}\mathbf{F}} = \frac{\mathrm{d}}{\mathrm{d}\mathbf{F}}(\det \mathbf{F}) = (\det \mathbf{F})\mathbf{F}^{-\mathrm{T}} = J\mathbf{F}^{-\mathrm{T}}.$$

CHAPTER

Incompressibility and Thermal Expansion

8

In this chapter, we develop thermomechanical models for (1) *incompressible* viscous fluids, (2) *incompressible* nonlinear elastic solids, and (3) viscous fluids that *thermally expand* and *contract*. A material is referred to as **incompressible** if, within a certain class of loadings, only **isochoric** or volume-preserving motions are possible.[1] In other words, the volume of the material cannot be changed appreciably by any means, be they thermal or mechanical. In the same vein, for certain classes of loadings, some materials are *mechanically* incompressible, yet experience significant **thermal expansion**, i.e., thermally induced volume change. For instance, polymer melts are relatively insensitive to changes in pressure, but can experience significant volume shrinkage (sometimes 30% or more) as they cool from their melt-processing temperature to room temperature.

8.1 INTRODUCTION

If we model incompressible viscous fluids, for instance, using the fully compressible theory presented in Chapter 7 and obtain a solution to the resulting mathematical problem, we will find that div $\mathbf{v}$ is essentially zero. Tremendous effort will be saved when solving the mathematical problem if we instead develop a new theory for these "incompressible viscous fluids" in which we a priori assume that only volume-preserving flows (i.e., flows with div $\mathbf{v}$ identically zero) are allowable. The resulting theory is referred to as a **constrained theory**, reflecting the fact that only a *subset* of all motions are possible (in this case, only those motions preserving incompressibility); this contrasts with the **unconstrained theory** presented in Chapters 5–7, where any smooth motion is possible. (We will soon see that constrained theories produce mathematical problems that are easier to solve than those from the corresponding unconstrained theories, although the actual development of a constrained theory is more tedious.)

[1]It is important to note that the title "incompressible material" is somewhat misleading. A material *in itself* is not incompressible; it is the material, class of loading, and physical application *together* that dictate if incompressibility is an appropriate simplification. For example, in most applications, water can be practically modeled as incompressible. However, in modeling the propagation of sound in water, one must allow for compressibility. Therefore, it is somewhat misleading to classify water itself as being either compressible or incompressible.

With incompressibility and thermal expansion as motivation, we therefore modify the unconstrained theory presented in Chapters 5–7. We now suppose that the continuum is subjected to one or more **internal constraints**, so that the class of all possible motions is reduced.

8.1.1 MOTION-TEMPERATURE CONSTRAINTS

The motion χ and temperature Θ, formerly *independent variables* in Chapters 5–7 (refer, in particular, to Section 5.3), are no longer *arbitrary* smooth functions of space and time, but are instead interrelated through a specified **constraint equation.** A constraint equation that captures a wide range of thermomechanical constraints (e.g., incompressibility and thermal expansion) has the general form

$$\mathbf{A} \cdot \mathbf{D} + \mathbf{b} \cdot \mathbf{g} + \alpha \dot{\Theta} = 0, \tag{8.1}$$

where $\mathbf{A}$, $\mathbf{b}$, and α are material-dependent quantities,

$$\mathbf{D} = \frac{1}{2}\left[\operatorname{grad} \mathbf{v} + (\operatorname{grad} \mathbf{v})^{\mathrm{T}}\right]$$

is the rate of deformation, and

$$\mathbf{g} = \operatorname{grad} \Theta$$

is the Eulerian temperature gradient. As invariance requirements, we must have

$$\mathbf{A}^{+} \cdot \mathbf{D}^{+} = \mathbf{A} \cdot \mathbf{D}, \quad \mathbf{b}^{+} \cdot \mathbf{g}^{+} = \mathbf{b} \cdot \mathbf{g}, \quad \alpha^{+}\dot{\Theta}^{+} = \alpha\dot{\Theta}. \tag{8.2}$$

The internal constraint (8.1) disrupts the arguments used to produce the theory presented in Chapters 5–7. In particular, the functions χ and Θ can no longer be considered as independent variables since they are related to each other by the constraint (8.1). Concomitantly, the *dependent variables* (i.e., the Cauchy stress $\mathbf{T}$, heat flux $\mathbf{q}$, Helmholtz free energy ψ, and entropy η) are no longer completely determined by constitutive equations for the motion χ and temperature Θ; parts of $\mathbf{T}$, $\mathbf{q}$, ψ, and η are necessary to maintain the constraint (8.1) and depend only on $\mathbf{A}$ and $\mathbf{b}$. Internal constraints, since they prevent certain types of thermomechanical processes, must be maintained by appropriate forces and heat flows. Since the constraints are specified a priori, for all processes, *the forces and heat flows maintaining them can depend on the constraint only and not on the particular process χ and Θ.*

For a constrained material, the dependent variables $\mathbf{T}$, $\mathbf{q}$, ψ, and η must include terms that allow for forces and heat flows necessary to maintain the constraint, regardless of the thermomechanical process. The part of $\mathbf{T}$, $\mathbf{q}$, ψ, and η involved in the maintenance of the constraint is called the **constraint response.** We therefore assume that the dependent quantities $\mathbf{T}$, $\mathbf{q}$, ψ, and η are not *completely* determined by **constitutive equations** evaluated on the particular process (as was the case previously), but rather are determined by constitutive equations *only up to* an additive constraint response, i.e.,

$$\mathbf{T} = \mathbf{T}_r + \mathbf{T}_c, \quad \mathbf{q} = \mathbf{q}_r + \mathbf{q}_c, \quad \psi = \psi_r + \psi_c, \quad \eta = \eta_r + \eta_c, \tag{8.3}$$

where $\mathbf{T}_r$, $\mathbf{q}_r$, ψ_r, and η_r are the specified constitutive equations, and $\mathbf{T}_c$, $\mathbf{q}_c$, ψ_c, and η_c are the additive constraint responses. These functions are defined only for those motions satisfying constraint (8.1). (As an example, for an incompressible material, the stresses $\mathbf{T}_r$ and $\mathbf{T}_c$ need be defined only on those motions satisfying $\operatorname{div} \mathbf{v} = 0$, since those are the only motions to which the material can be subjected.) We further assume that the constraint response $\mathbf{T}_c$, $\mathbf{q}_c$, ψ_c, and η_c maintains the constraint *without increasing entropy*, so from the Clausius-Duhem inequality (5.35) we have

$$-\rho\dot{\psi}_c + \mathbf{T}_c \cdot \mathbf{D} - \rho\eta_c\dot{\Theta} - \frac{1}{\Theta}\mathbf{q}_c \cdot \mathbf{g} = 0 \tag{8.4a}$$

and

$$-\rho\dot{\psi}_r + \mathbf{T}_r \cdot \mathbf{D} - \rho\eta_r\dot{\Theta} - \frac{1}{\Theta}\mathbf{q}_r \cdot \mathbf{g} \geq 0 \tag{8.4b}$$

for all processes satisfying the constraint

$$\mathbf{A} \cdot \mathbf{D} + \mathbf{b} \cdot \mathbf{g} + \alpha\dot{\Theta} = 0.$$

We emphasize that the constraint response $\mathbf{T}_c$, $\mathbf{q}_c$, ψ_c, η_c *is independent of the particular process* χ *and* Θ.

8.1.2 MOTION-ENTROPY CONSTRAINTS

We now recall the formulation of the unconstrained thermomechanical theory wherein motion χ and entropy η (instead of temperature Θ) are taken as the independent variables; refer again to Section 5.3. Similar to what was done in Section 8.1.1, we demand that χ and η satisfy a general constraint equation of the form

$$\mathbf{A} \cdot \mathbf{D} + \beta\dot{\eta} = 0, \tag{8.5}$$

so that χ and η are no longer independent. Note that β is a material-dependent quantity. As invariance requirements, we must have

$$\mathbf{A}^+ \cdot \mathbf{D}^+ = \mathbf{A} \cdot \mathbf{D}, \quad \beta^+\dot{\eta}^+ = \beta\dot{\eta}. \tag{8.6}$$

As was the case before, the dependent quantities (the Cauchy stress $\mathbf{T}$, heat flux $\mathbf{q}$, internal energy ε, and temperature Θ in this formulation) are determined by constitutive functions only up to an additive constraint response, i.e.,

$$\mathbf{T} = \mathbf{T}_r + \mathbf{T}_c, \quad \mathbf{q} = \mathbf{q}_r + \mathbf{q}_c, \quad \varepsilon = \varepsilon_r + \varepsilon_c, \quad \Theta = \Theta_r + \Theta_c. \tag{8.7}$$

Corresponding to this formulation of the constrained theory is the Clausius-Duhem inequality (5.27), which, if we use (8.7) and note that the constraint response generates no entropy, becomes

$$-\rho\dot{\varepsilon}_c + \mathbf{T}_c \cdot \mathbf{D} + \rho\Theta_c\dot{\eta} - \frac{1}{\Theta_c}\mathbf{q}_c \cdot \mathbf{g} = 0 \tag{8.8a}$$

and

$$-\rho\dot{\varepsilon}_r + \mathbf{T}_r \cdot \mathbf{D} + \rho\Theta_r\dot{\eta} - \frac{1}{\Theta_r}\mathbf{q}_r \cdot \mathbf{g} \geq 0 \tag{8.8b}$$

for all processes satisfying the constraint

$$\mathbf{A} \cdot \mathbf{D} + \beta\dot{\eta} = 0.$$

We emphasize that the constraint response $\mathbf{T}_c$, $\mathbf{q}_c$, ε_c, Θ_c *is independent of the particular process* χ *and* η.

Similar to the unconstrained theory presented in Chapters 5–7, the constitutive equations (8.3) corresponding to the motion-temperature formulation (or (8.7) corresponding to the motion-entropy formulation) must satisfy (1) the Clausius-Duhem inequality for *all* thermomechanical processes satisfying constraint (8.1) (or (8.5)), (2) invariance requirements under superposed rigid body motions, (3) conservation of angular momentum, and (4) perhaps some material symmetry requirements. Additionally, for constrained theories, **stability** of the thermodynamic equilibrium must be established; refer to Section E.2.

8.2 NEWTONIAN FLUIDS

This section is structured as follows: In Section 8.2.1, we review the *unconstrained* theory for *compressible* Newtonian fluids presented in Chapter 7 and Appendix E. In Sections 8.2.2 and 8.2.3, we present the *constrained* theory for *incompressible* Newtonian fluids and derive the incompressible Navier-Stokes equations. Our examination of constrained theories continues in Sections 8.2.4 and 8.2.5, where we present models for the *thermal expansion* of a Newtonian fluid and discuss the implications of the associated stability conditions.

8.2.1 THE COMPRESSIBLE THEORY: A BRIEF REVIEW

Recall that the fundamental laws for a continuum in the *compressible* (i.e., *unconstrained*) thermomechanical theory are

$$\dot{\rho} + \rho \operatorname{div} \mathbf{v} = 0, \quad \rho\dot{\mathbf{v}} = \operatorname{div} \mathbf{T} + \rho\mathbf{b}, \quad \rho\dot{\varepsilon} = \mathbf{T} \cdot \mathbf{D} + \rho r - \operatorname{div} \mathbf{q}, \tag{8.9}$$

which correspond to Eulerian statements of conservation of mass, linear momentum, and energy. In (8.9), ρ is the density, $\mathbf{v}$ is the velocity, $\mathbf{T}$ is the (symmetric) Cauchy stress tensor, $\mathbf{b}$ is the body force per unit mass, ε is the internal energy per unit mass, r is the rate of heat absorbed per unit mass, $\mathbf{q}$ is the heat flux vector, and $\mathbf{D}$ is the rate of deformation. A **compressible Newtonian fluid** is characterized by the constitutive equations

$$\mathbf{T} = -p\mathbf{I} + \lambda(\operatorname{tr} \mathbf{D})\mathbf{I} + 2\mu\mathbf{D}, \quad \mathbf{q} = -k \operatorname{grad} \Theta, \tag{8.10}$$

where p is the pressure, Θ is the absolute temperature, λ and μ are the dilatational and shear viscosities, respectively, and k is the thermal conductivity. As a consequence of the second law of thermodynamics,

$$\lambda + \frac{2}{3}\mu \geq 0, \quad \mu \geq 0, \quad k \geq 0.$$

The mathematical model for a compressible Newtonian fluid is completed by specifying the **equations of state**. To this end, from Appendix E we have four equivalent formulations of the equations of state, *each employing a different set of independent variables* (density ρ or pressure p for the independent mechanical variable, and entropy η or temperature Θ for the independent thermal variable):

(1) the *density-entropy formulation*

$$p = \rho^2 \frac{\partial \breve{\varepsilon}(\rho, \eta)}{\partial \rho}, \quad \Theta = \frac{\partial \breve{\varepsilon}(\rho, \eta)}{\partial \eta}; \tag{8.11a}$$

(2) the *density-temperature formulation*

$$p = \rho^2 \frac{\partial \breve{\psi}(\rho, \Theta)}{\partial \rho}, \quad \eta = -\frac{\partial \breve{\psi}(\rho, \Theta)}{\partial \Theta}; \tag{8.11b}$$

(3) the *pressure-entropy formulation*

$$\frac{1}{\rho} = \frac{\partial \breve{\chi}(p, \eta)}{\partial p}, \quad \Theta = \frac{\partial \breve{\chi}(p, \eta)}{\partial \eta}; \tag{8.11c}$$

(4) the *pressure-temperature formulation*

$$\frac{1}{\rho} = \frac{\partial \breve{\phi}(p, \Theta)}{\partial p}, \quad \eta = -\frac{\partial \breve{\phi}(p, \Theta)}{\partial \Theta}. \tag{8.11d}$$

Also from Appendix E we have that the **Helmholtz free energy** ψ, **enthalpy** χ, and **Gibbs free energy** ϕ are related to the internal energy ε through the **Legendre transformations**

$$\psi = \varepsilon - \Theta\eta, \quad \chi = \varepsilon + \frac{1}{\rho}p, \quad \phi = \varepsilon - \Theta\eta + \frac{1}{\rho}p. \tag{8.12}$$

EXERCISE

1. Write the Eulerian conservation laws (8.9) and constitutive equations (8.10) in indicial notation. Then fully expand them into their Cartesian component forms (i.e., no summation convention). Confirm that the resulting equations, together with the density-entropy formulation (8.11a) of the equations of state, constitute a closed system for the density ρ, entropy η, and velocity components v_1, v_2, v_3. Regard the internal energy function $\breve{\varepsilon}(\rho, \eta)$ as specified.

8.2.2 INCOMPRESSIBILITY

For an **incompressible Newtonian fluid**, we impose the restriction that the density remain constant, i.e.,

$$\rho = \bar{\rho} \equiv \text{constant}, \tag{8.13}$$

so that only *isochoric*, or volume-preserving, flows by the fluid are allowed. Use of restriction (8.13) in the Eulerian form of conservation of mass $(8.9)_1$ gives the **incompressibility constraint**

$$\operatorname{div} \mathbf{v} = \operatorname{tr} \mathbf{D} = \mathbf{I} \cdot \mathbf{D} = 0. \tag{8.14}$$

Note that (8.14) is a special case of the general thermomechanical constraint (8.1), i.e.,

$$\mathbf{A} \cdot \mathbf{D} + \mathbf{b} \cdot \mathbf{g} + \alpha \dot{\Theta} = 0,$$

with

$$\mathbf{A} = \mathbf{I}, \quad \mathbf{b} = \mathbf{0}, \quad \alpha = 0.$$

It can be verified that (8.14) satisfies the invariance requirements (8.2); refer to Problem 8.2.

The second law equality (8.4a), i.e.,

$$-\bar{\rho}\dot{\psi}_c + \mathbf{T}_c \cdot \mathbf{D} - \bar{\rho}\eta_c\dot{\Theta} - \frac{1}{\Theta}\mathbf{q}_c \cdot \mathbf{g} = 0, \tag{8.4a}$$

for the constraint response $\mathbf{T}_c$, $\mathbf{q}_c$, ψ_c, η_c must hold for all thermomechanical processes satisfying the incompressibility constraint (8.14), i.e.,

$$\mathbf{I} \cdot \mathbf{D} = 0. \tag{8.14}$$

In particular, (8.4a) must hold for the family of processes with $\mathbf{D} = \mathbf{0}$, $\mathbf{g} = \mathbf{0}$, and $\dot{\Theta} = 0$, which necessarily satisfy constraint (8.14). For this family, equality (8.4a) reduces to

$$\dot{\psi}_c = 0.$$

Without loss of generality, we take the constant of integration (the datum of the free energy) to be zero, which gives

$$\psi_c = 0. \tag{8.15}$$

Recalling that the constraint response is independent of the particular process, we find that (8.15) must hold for all processes satisfying (8.14), not just the above subset, so (8.4a) becomes

$$\mathbf{T}_c \cdot \mathbf{D} - \bar{\rho}\eta_c\dot{\Theta} - \frac{1}{\Theta}\mathbf{q}_c \cdot \mathbf{g} = 0. \tag{8.16}$$

Consider now those processes with $\mathbf{D} = \mathbf{0}$ and $\mathbf{g} = \mathbf{0}$, but $\dot{\Theta}$ arbitrary. This family of processes satisfies constraint (8.14) and reduces (8.16) to

$$-\bar{\rho}\eta_c\dot{\Theta} = 0.$$

The constraint response η_c is independent of the process and thus independent of $\dot{\Theta}$, and $\dot{\Theta}$ is arbitrary; together, they imply that

$$\eta_c = 0. \tag{8.17}$$

Again, since the constraint response is independent of the particular process, (8.17) must hold for all processes satisfying (8.14), not just the above subset, so (8.16) becomes

$$\mathbf{T}_c \cdot \mathbf{D} - \frac{1}{\Theta}\mathbf{q}_c \cdot \mathbf{g} = 0. \tag{8.18}$$

Consider now those processes with $\mathbf{D} = \mathbf{0}$, but $\mathbf{g}$ and $\dot{\Theta}$ arbitrary; this family of processes necessarily satisfies constraint (8.14). A series of arguments similar to the ones used previously enables us to conclude that

$$\mathbf{q}_c = \mathbf{0} \tag{8.19}$$

for all processes satisfying constraint (8.14). Thus, (8.18) reduces to

$$\mathbf{T}_c \cdot \mathbf{D} = 0. \tag{8.20}$$

Representing $\mathbf{D}$ as a vector in the six-dimensional inner product space $\mathcal{E}^6$, we find that admissible $\mathbf{D}$ (i.e., those satisfying the incompressibility constraint (8.14)) are those orthogonal to the vector $\mathbf{I}$. Condition (8.20) demands that $\mathbf{D}$ also be orthogonal to $\mathbf{T}_c$. Hence, we conclude that $\mathbf{T}_c$ is parallel to $\mathbf{I}$, i.e.,

$$\mathbf{T}_c = -\bar{p}\,\mathbf{I}, \tag{8.21}$$

where the **constraint pressure** $\bar{p}$, a scalar function of position and time, acts as a **Lagrange multiplier**. That is, in contrast to the thermodynamic pressure p, which is given by an equation of state (namely, $(8.11\text{b})_1$), the constraint pressure $\bar{p}$ is a *primitive unknown* and takes whatever value is necessary to maintain the incompressibility constraint (8.14).

We now assemble our results. For an incompressible Newtonian fluid, we have

$$\mathbf{T} = \mathbf{T}_r + \mathbf{T}_c, \quad \mathbf{q} = \mathbf{q}_r + \mathbf{q}_c, \quad \psi = \psi_r + \psi_c, \quad \eta = \eta_r + \eta_c.$$

The constitutive response is given by (8.10) and (8.11b), i.e.,

$$\mathbf{T}_r = -p\mathbf{I} + 2\mu\mathbf{D}, \quad \mathbf{q}_r = -k \operatorname{grad}\Theta, \quad \psi_r = \check{\psi}(\bar{\rho},\Theta) = \bar{\psi}(\Theta), \quad \eta_r = -\bar{\psi}'(\Theta),$$

where we have used (8.14) and (8.13), and the constraint response is given by (8.15), (8.17), (8.19), and (8.21), i.e.,

$$\mathbf{T}_c = -\bar{p}\,\mathbf{I}, \quad \mathbf{q}_c = \mathbf{0}, \quad \psi_c = 0, \quad \eta_c = 0,$$

so the total response (constitutive plus constraint) is

$$\mathbf{T} = -\bar{p}\,\mathbf{I} + 2\mu\mathbf{D}, \quad \mathbf{q} = -k \operatorname{grad}\Theta, \quad \psi = \bar{\psi}(\Theta), \quad \eta = -\bar{\psi}'(\Theta). \tag{8.22}$$

Note that the thermodynamic pressure p was absorbed into the Lagrange multiplier $\bar{p}$ that maintains the incompressibility constraint (8.14). Also note that a superscript prime denotes differentiation of a function of a single variable with respect to that variable.[2] The total response (8.22) satisfies the second law of thermodynamics, conservation of angular momentum, and the necessary invariance requirements.

We can show (refer to Problem 8.3) that the Eulerian conservation laws (8.9) in the unconstrained theory become, for the special case of an incompressible Newtonian fluid,

$$\operatorname{div} \mathbf{v} = 0, \quad \bar{\rho}\dot{\mathbf{v}} = \operatorname{div} \mathbf{T} + \bar{\rho}\mathbf{b}, \quad -\bar{\rho}\,\Theta\bar{\psi}''(\Theta)\dot{\Theta} = \mathbf{T} \cdot \mathbf{D} + \bar{\rho}r - \operatorname{div} \mathbf{q}. \tag{8.23}$$

With the Helmholtz potential $\bar{\psi}(\Theta)$ a given function of temperature, the total response $(8.22)_1$ and $(8.22)_2$ and conservation laws (8.23) form a closed set of equations for the constraint pressure $\bar{p}$, temperature Θ, and velocity $\mathbf{v}$. It can be verified that use of expression $(8.22)_1$ for the Cauchy stress in conservation of linear momentum $(8.23)_2$ leads to the **incompressible Navier-Stokes equations**

$$\bar{\rho}\dot{\mathbf{v}} = -\operatorname{grad}\bar{p} + \mu \operatorname{div}(\operatorname{grad} \mathbf{v}) + \bar{\rho}\mathbf{b}, \tag{8.24a}$$

whose indicial form is

$$\bar{\rho}\dot{v}_i = -\bar{p}_{,i} + \mu v_{i,jj} + \bar{\rho}b_i. \tag{8.24b}$$

Refer to Problem 8.4 for the fully expanded version of (8.24b). Contrast (8.24a) and (8.24b) with the **compressible Navier-Stokes equations** (7.12a) and (7.12b) derived in Chapter 7.

PROBLEM 8.1

Prove that $\operatorname{div} \mathbf{v} = \mathbf{I} \cdot \mathbf{D}$.

Solution

$$\operatorname{div} \mathbf{v} = \operatorname{tr}(\operatorname{grad} \mathbf{v}) = \operatorname{tr} \mathbf{L} = \operatorname{tr}(\mathbf{D} + \mathbf{W}) = \operatorname{tr} \mathbf{D} + \operatorname{tr} \mathbf{W} = \operatorname{tr} \mathbf{D} = \operatorname{tr}\left(\mathbf{I}\mathbf{D}^{\mathrm{T}}\right) = \mathbf{I} \cdot \mathbf{D}.$$

PROBLEM 8.2

Verify that the incompressibility constraint (8.14) satisfies the invariance requirements (8.2). In other words, confirm that

$$\mathbf{I}^{+} \cdot \mathbf{D}^{+} = \mathbf{I} \cdot \mathbf{D}.$$

[2]This departs from other chapters, where a superscript prime denotes the partial derivative of the Eulerian representation of a particular quantity with respect to time; refer, for instance, to (3.19).

Solution

$$\begin{aligned}
\mathbf{I}^{+}\cdot\mathbf{D}^{+} &= \mathbf{I}\cdot\mathbf{QDQ}^{\mathrm{T}} && \text{(result (5.15))}\\
&= \mathrm{tr}\Big[\mathbf{I}\big(\mathbf{QDQ}^{\mathrm{T}}\big)^{\mathrm{T}}\Big] && \text{(definition (2.41))}\\
&= \mathrm{tr}\Big(\mathbf{QD}^{\mathrm{T}}\mathbf{Q}^{\mathrm{T}}\Big) && \text{(results } (2.14)_{2,3})\\
&= \mathrm{tr}\Big(\mathbf{Q}^{\mathrm{T}}\mathbf{QD}^{\mathrm{T}}\Big) && \text{(result } (2.40)_{3})\\
&= \mathrm{tr}\Big(\mathbf{ID}^{\mathrm{T}}\Big) && \text{(result (2.51))}\\
&= \mathbf{I}\cdot\mathbf{D} && \text{(definition (2.41)).}
\end{aligned}$$

PROBLEM 8.3

Show that for the special case of an incompressible Newtonian fluid, the conservation laws (8.9) become (8.23).

Solution

Straightforward substitution of the incompressibility restriction (8.13), i.e.,

$$\rho = \bar{\rho} \equiv \text{constant},$$

into the conservation laws (8.9) gives

$$\mathrm{div}\,\mathbf{v} = 0, \quad \bar{\rho}\dot{\mathbf{v}} = \mathrm{div}\,\mathbf{T} + \bar{\rho}\mathbf{b}, \quad \bar{\rho}\dot{\varepsilon} = \mathbf{T}\cdot\mathbf{D} + \bar{\rho}r - \mathrm{div}\,\mathbf{q}.$$

We then have

$$\begin{aligned}
\dot{\varepsilon} &= \overline{\psi + \Theta\eta}^{\,\boldsymbol{\cdot}} && \text{(Legendre transform } (8.12)_{1})\\
&= \dot{\psi} + \Theta\dot{\eta} + \eta\dot{\Theta} && \text{(product rule)}\\
&= \bar{\psi}'(\Theta)\dot{\Theta} - \Theta\bar{\psi}''(\Theta)\dot{\Theta} - \bar{\psi}'(\Theta)\dot{\Theta} && \text{(results } (8.22)_{3,4} \text{ and the chain rule)}\\
&= -\Theta\bar{\psi}''(\Theta)\dot{\Theta},
\end{aligned}$$

so conservation of energy becomes

$$-\bar{\rho}\,\Theta\bar{\psi}''(\Theta)\dot{\Theta} = \mathbf{T}\cdot\mathbf{D} + \bar{\rho}r - \mathrm{div}\,\mathbf{q}.$$

PROBLEM 8.4

Using relationships (3.19) and (3.20), one can rewrite conservation of mass $(8.23)_1$ and the incompressible Navier-Stokes equations (8.24a) as

$$\operatorname{div}\mathbf{v} = 0, \quad \bar{\rho}\left[\frac{\partial \mathbf{v}}{\partial t} + (\operatorname{grad}\mathbf{v})\mathbf{v}\right] = -\operatorname{grad}\bar{p} + \mu\operatorname{div}(\operatorname{grad}\mathbf{v}) + \bar{\rho}\mathbf{b},$$

respectively. Re-express these equations in indicial notation. Then fully expand the resulting expressions in Cartesian component form (i.e., no summation convention).

Solution

With the aid of the results presented in Section 2.5.3, we can write

$$v_{i,i} = 0, \quad \bar{\rho}\left(\frac{\partial v_i}{\partial t} + v_j v_{i,j}\right) = -\bar{p}_{,i} + \mu v_{i,jj} + \bar{\rho} b_i.$$

Fully expanding these expressions gives

$$\frac{\partial v_1}{\partial x_1} + \frac{\partial v_2}{\partial x_2} + \frac{\partial v_3}{\partial x_3} = 0$$

and

$$\bar{\rho}\left(\frac{\partial v_1}{\partial t} + v_1\frac{\partial v_1}{\partial x_1} + v_2\frac{\partial v_1}{\partial x_2} + v_3\frac{\partial v_1}{\partial x_3}\right) = -\frac{\partial \bar{p}}{\partial x_1} + \mu\left(\frac{\partial^2 v_1}{\partial x_1^2} + \frac{\partial^2 v_1}{\partial x_2^2} + \frac{\partial^2 v_1}{\partial x_3^2}\right) + b_1,$$

$$\bar{\rho}\left(\frac{\partial v_2}{\partial t} + v_1\frac{\partial v_2}{\partial x_1} + v_2\frac{\partial v_2}{\partial x_2} + v_3\frac{\partial v_2}{\partial x_3}\right) = -\frac{\partial \bar{p}}{\partial x_2} + \mu\left(\frac{\partial^2 v_2}{\partial x_1^2} + \frac{\partial^2 v_2}{\partial x_2^2} + \frac{\partial^2 v_2}{\partial x_3^2}\right) + b_2,$$

$$\bar{\rho}\left(\frac{\partial v_3}{\partial t} + v_1\frac{\partial v_3}{\partial x_1} + v_2\frac{\partial v_3}{\partial x_2} + v_3\frac{\partial v_3}{\partial x_3}\right) = -\frac{\partial \bar{p}}{\partial x_3} + \mu\left(\frac{\partial^2 v_3}{\partial x_1^2} + \frac{\partial^2 v_3}{\partial x_2^2} + \frac{\partial^2 v_3}{\partial x_3^2}\right) + b_3.$$

PROBLEM 8.5

Consider the steady laminar flow of an incompressible Newtonian fluid between two infinite parallel plates (refer to Figure 8.1).The flow is unidirectional and fully developed. Thermal effects and gravity have a negligible effect on the flow. Derive the velocity field for two special cases:

(1) plane Poiseuille flow, where the flow is pressure driven ($p_l > p_r$) and the top plate is fixed ($V = 0$);

(2) plane Couette flow, where the flow is driven by the top plate (which moves with *constant* velocity V) and there is no pressure gradient ($p_l = p_r$).

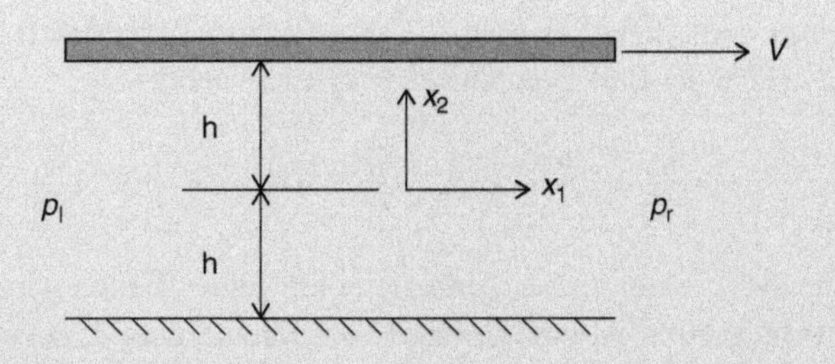

FIGURE 8.1

Steady laminar flow between two parallel plates. In plane Poiseuille flow, the flow is pressure driven ($p_l > p_r$) and the top plate is stationary ($V = 0$). In plane Couette flow, the flow is driven by the top plate (which moves with *constant* velocity V) and there is no pressure gradient ($p_l = p_r$).

Solution

In the absence of thermal effects and gravitational body forces, the flow of an incompressible Newtonian fluid is governed by conservation of mass $(8.23)_1$ and the Navier-Stokes equations (8.24a), which, in Cartesian component form, are

$$\frac{\partial v_1}{\partial x_1} + \frac{\partial v_2}{\partial x_2} + \frac{\partial v_3}{\partial x_3} = 0 \tag{a}$$

and

$$\bar{\rho}\left(\frac{\partial v_1}{\partial t} + v_1\frac{\partial v_1}{\partial x_1} + v_2\frac{\partial v_1}{\partial x_2} + v_3\frac{\partial v_1}{\partial x_3}\right) = -\frac{\partial \bar{p}}{\partial x_1} + \mu\left(\frac{\partial^2 v_1}{\partial x_1^2} + \frac{\partial^2 v_1}{\partial x_2^2} + \frac{\partial^2 v_1}{\partial x_3^2}\right), \tag{ba}$$

$$\bar{\rho}\left(\frac{\partial v_2}{\partial t} + v_1\frac{\partial v_2}{\partial x_1} + v_2\frac{\partial v_2}{\partial x_2} + v_3\frac{\partial v_2}{\partial x_3}\right) = -\frac{\partial \bar{p}}{\partial x_2} + \mu\left(\frac{\partial^2 v_2}{\partial x_1^2} + \frac{\partial^2 v_2}{\partial x_2^2} + \frac{\partial^2 v_2}{\partial x_3^2}\right), \tag{bb}$$

$$\bar{\rho}\left(\frac{\partial v_3}{\partial t} + v_1\frac{\partial v_3}{\partial x_1} + v_2\frac{\partial v_3}{\partial x_2} + v_3\frac{\partial v_3}{\partial x_3}\right) = -\frac{\partial \bar{p}}{\partial x_3} + \mu\left(\frac{\partial^2 v_3}{\partial x_1^2} + \frac{\partial^2 v_3}{\partial x_2^2} + \frac{\partial^2 v_3}{\partial x_3^2}\right), \tag{bc}$$

respectively; refer to Problem 8.4. The mechanical theory (a) and (ba)–(bc) represents a closed system of equations for the velocity components v_1, v_2, v_3 and pressure $\bar{p}$, all functions of present position x_1, x_2, x_3 and time t. We now explore our assumptions on the flow (refer to the problem statement) and their implications:

(1) Steady flow implies that the pressure and velocity (at a particular fixed point) do not vary with time, i.e., $\bar{p}$, v_1, v_2, and v_3 are independent of t.

(2) Unidirectional flow implies that the transverse velocity components vanish, i.e., $v_2 = v_3 = 0$.

(3) Fully developed flow implies that the velocity is independent of the axial coordinate, i.e., v_1 is independent of x_1.

(4) Infinite plates imply that the velocity does not vary along the width of the plates (into or out of the page), i.e., v_1 is independent of x_3.

As a consequence of these assumptions, conservation of mass (a) is trivially satisfied, and the Navier-Stokes equations (ba)–(bc) become

$$-\frac{\partial \bar{p}}{\partial x_1} + \mu \frac{d^2 v_1}{dx_2^2} = 0, \quad \frac{\partial \bar{p}}{\partial x_2} = 0, \quad \frac{\partial \bar{p}}{\partial x_3} = 0, \tag{c}$$

noting that the axial velocity v_1 is a function of x_2 alone. Equations $(c)_2$ and $(c)_3$ imply that the pressure $\bar{p}$ is independent of x_2 and x_3, i.e., $\bar{p}$ is a function of x_1 alone, so $(c)_1$ becomes

$$\frac{d\bar{p}}{dx_1} = \mu \frac{d^2 v_1}{dx_2^2}. \tag{d}$$

Equation (d) holds only if both sides are equal to a constant, i.e.,

$$\frac{d\bar{p}}{dx_1} = \mu \frac{d^2 v_1}{dx_2^2} = c_0 \equiv \text{constant}. \tag{e}$$

Integrating (e) twice gives

$$v_1 = \frac{1}{2\mu} \frac{d\bar{p}}{dx_1} x_2^2 + c_1 x_2 + c_2. \tag{f}$$

We emphasize that the pressure gradient $d\bar{p}/dx_1$ is a constant. In what follows, the constants of integration c_1 and c_2 are deduced using the no-slip boundary conditions at the top and bottom plates.

For *plane Poiseuille flow*, both plates are stationary; hence, the no-slip boundary conditions are $v_1 = 0$ at $x_2 = \pm h$. For (f) to satisfy these boundary conditions, we must have

$$c_1 = 0, \quad c_2 = -\frac{h^2}{2\mu} \frac{d\bar{p}}{dx_1},$$

and thus

$$v_1 = \frac{h^2}{2\mu} \frac{d\bar{p}}{dx_1} \left[\left(\frac{x_2}{h} \right)^2 - 1 \right].$$

Note that the axial velocity profile is parabolic in x_2; refer to Figure 8.2.

For *plane Couette flow*, the bottom plate is stationary and the top plate moves at a constant velocity V; hence, the no-slip boundary conditions are $v_1 = V$ at $x_2 = h$ and $v_1 = 0$ at $x_2 = -h$. Noting that the pressure gradient $d\bar{p}/dx_1 = 0$ in Couette flow, we find that (f) satisfies the no-slip boundary conditions if

$$c_1 = \frac{V}{2h}, \quad c_2 = \frac{V}{2}.$$

Thus, (f) becomes

$$v_1 = \frac{V}{2} \left(\frac{x_2}{h} + 1 \right).$$

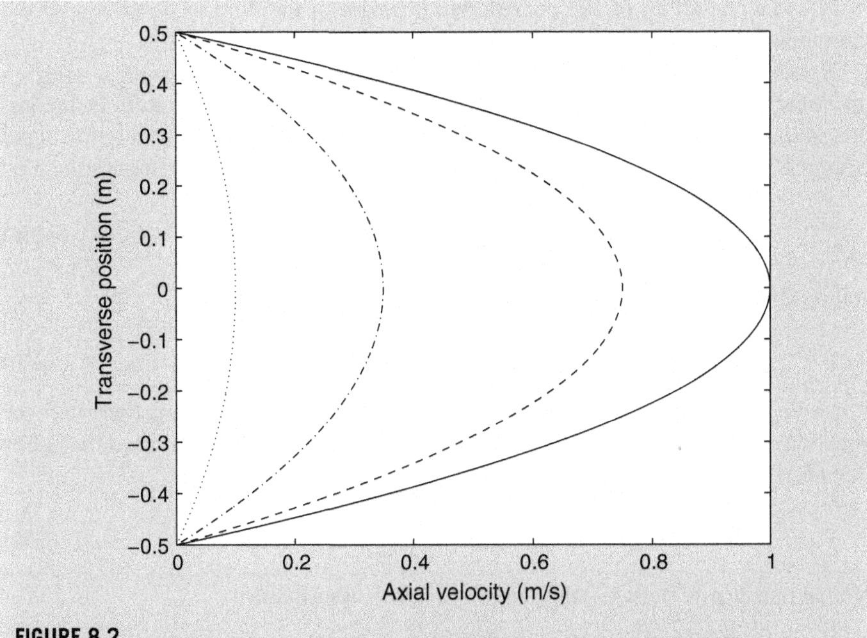

FIGURE 8.2

Axial velocity profile for the plane Poiseuille flow of crude oil ($\mu = 0.125$ Pa · s) driven by pressure gradients of −0.1 Pa/m (dotted line), −0.35 Pa/m (dotted-dashed line), −0.75 Pa/m (dashed line), and −1 Pa/m (solid line).

Note that the axial velocity profile is linear in x_2 and independent of the shear viscosity μ.

EXERCISES

1. Argue that (8.19) follows from (8.18), so (8.18) reduces to (8.20).
2. Using indicial notation, verify that use of the Cauchy stress $(8.22)_1$ in balance of linear momentum $(8.23)_2$ leads to the incompressible Navier-Stokes equations (8.24b). Assume that the shear viscosity μ is a constant.

8.2.3 INCOMPRESSIBILITY AS A CONSTITUTIVE LIMIT: AN ALTERNATIVE PERSPECTIVE

In this section, we present an alternative derivation of the equations (8.22) and (8.23) that govern an incompressible Newtonian fluid. This derivation follows from a fundamental shift in how incompressibility is regarded, from an a priori imposed restriction on the class of allowable processes (refer to the incompressibility constraint

(8.14)) to a restriction on the constitutive behavior of the fluid as described by the thermodynamic potential [20].

Recall from Section 8.2.1 the pressure-temperature formulation of the compressible theory, where the density ρ is a function of pressure p and temperature Θ. Incompressibility is modeled by imposing two **constitutive limits** on this formulation of the compressible theory: no pressure dependence of density at constant temperature, i.e.,

$$\frac{\partial \rho(p, \Theta)}{\partial p} \rightarrow 0, \tag{8.25a}$$

and no temperature dependence of density at constant pressure, i.e.,

$$\frac{\partial \rho(p, \Theta)}{\partial \Theta} \rightarrow 0. \tag{8.25b}$$

It can be shown that the constitutive limits (8.25a) and (8.25b) together with the equation of state $(8.11\text{d})_1$ produce the following restrictions on the form of the Gibbs free energy:

$$\frac{\partial^2 \phi}{\partial p^2} = 0, \quad \frac{\partial^2 \phi}{\partial p\, \partial \Theta} = 0. \tag{8.26}$$

Noting that $\phi(p, \Theta)$, if we integrate $(8.26)_1$ once, we obtain

$$\frac{\partial \phi}{\partial p} = a_1(\Theta). \tag{8.27}$$

Differentiating (8.27) with respect to temperature and using $(8.26)_2$ yields

$$a_1'(\Theta) = 0,$$

where the superscript prime denotes differentiation with respect to temperature Θ. Thus,

$$a_1 = \text{constant} \equiv \frac{1}{\bar{\rho}}.$$

Note that $\bar{\rho}$ is a specified positive constant. Integrating (8.27) produces

$$\phi(p, \Theta) = a_1 p + a_2(\Theta) = \frac{p}{\bar{\rho}} + a_2(\Theta), \tag{8.28}$$

where $a_2(\Theta)$ is a specified function of temperature. As a consequence of (8.28), the equations of state (8.11d) become

$$\rho = \left(\frac{\partial \phi}{\partial p}\right)^{-1} = \bar{\rho}, \quad \eta = -\frac{\partial \phi}{\partial \Theta} = -a_2'(\Theta), \tag{8.29}$$

i.e., density is constant and entropy is a function of temperature alone. It can be verified that the restricted form (8.28) of the Gibbs free energy together with results (8.29) reduce the conservation laws (8.9) and constitutive equations (8.10) from the compressible theory to

$$\operatorname{div} \mathbf{v} = 0, \quad \bar{\rho}\dot{\mathbf{v}} = \operatorname{div} \mathbf{T} + \bar{\rho}\mathbf{b}, \quad -\bar{\rho}\,\Theta a_2''(\Theta)\dot{\Theta} = \mathbf{T} \cdot \mathbf{D} + \bar{\rho} r - \operatorname{div} \mathbf{q} \tag{8.30}$$

and

$$\mathbf{T} = -p\mathbf{I} + 2\mu\mathbf{D}, \quad \mathbf{q} = -k \text{ grad } \Theta, \tag{8.31}$$

respectively, for the incompressible case. These equations are precisely the governing equations (8.22) and (8.23) for an incompressible Newtonian fluid obtained in the previous section, if we equate the primitive unknown $\bar{p}$ with p and the specified function $\bar{\psi}(\Theta)$ with $a_2(\Theta)$. It can be verified that the stability conditions (E.19) for the Gibbs free energy (refer to Appendix E) are satisfied as long as $a_2''(\Theta) \leq 0$.

EXERCISES

1. Show that the constitutive limits (8.25a) and (8.25b), together with the equation of state $(8.11\text{d})_1$, imply restrictions (8.26) on the Gibbs free energy.
2. Verify that the restricted form (8.28) of the Gibbs free energy, together with results (8.29), reduce the conservation laws (8.9) and constitutive equations (8.10) to (8.30) and (8.31), respectively.

8.2.4 THERMAL EXPANSION

Viscous fluids in many processes are mechanically incompressible (e.g., insensitive to pressure changes), but experience significant thermally induced volume change. A **thermal expansion constraint** enforces that the mechanical variable (density or pressure) is *specified completely* by the thermal variable (temperature or entropy). Thus, there are four candidate constraints: $\rho = \hat{\rho}(\Theta)$, $\rho = \hat{\rho}(\eta)$, $p = \hat{p}(\Theta)$, and $p = \hat{p}(\eta)$. In this section, we present two of the four theories that follow from these constraints; the remaining two are discussed in [21].

8.2.4.1 A density-temperature constraint (isothermal incompressibility)

We first investigate the customary model of thermal expansion, where the fluid is essentially incompressible under **isothermal** (or constant-temperature) conditions, but whose density changes in response to changes in temperature. We thus adopt the density-temperature formulation (8.11b) with independent mechanical and thermal variables ρ and Θ, and posit

$$\rho = \hat{\rho}(\Theta), \tag{8.32}$$

i.e., density is a prescribed function of temperature only. Taking the rate (material derivative) of (8.32) and using the chain rule, we obtain

$$\dot{\rho} = \hat{\rho}'(\Theta)\dot{\Theta},$$

where the superscript prime denotes differentiation with respect to Θ. Use of this result in conservation of mass $(8.9)_1$ leads to the thermal expansion constraint

$$\mathbf{I} \cdot \mathbf{D} + \frac{\hat{\rho}'(\Theta)}{\hat{\rho}(\Theta)} \dot{\Theta} = 0, \tag{8.33}$$

which is a special case of the general motion-temperature constraint (8.1) with

$$\mathbf{A} = \mathbf{I}, \quad \mathbf{b} = \mathbf{0}, \quad \alpha = \frac{\hat{\rho}'(\Theta)}{\hat{\rho}(\Theta)}.$$

Constraint (8.33) satisfies the invariance requirements (8.2).

We now investigate restrictions imposed by the second law of thermodynamics. Recall that the constraint response $\mathbf{T}_c$, $\mathbf{q}_c$, ψ_c, η_c must maintain the constraint (8.33) without generating entropy, i.e.,

$$-\hat{\rho}(\Theta)\dot{\psi}_c + \mathbf{T}_c \cdot \mathbf{D} - \hat{\rho}(\Theta)\eta_c\dot{\Theta} - \frac{1}{\Theta}\mathbf{q}_c \cdot \mathbf{g} = 0. \tag{8.34}$$

Condition (8.34) must hold for all processes satisfying the thermal expansion constraint (8.33). In particular, it must hold for the family of processes with $\mathbf{D} = \mathbf{0}$, $\mathbf{g} = \mathbf{0}$, and $\dot{\Theta} = 0$, which necessarily satisfies (8.33). For this family of processes, condition (8.34) becomes

$$\dot{\psi}_c = 0,$$

or, without loss of generality,

$$\psi_c = 0. \tag{8.35}$$

Result (8.35) holds for the particular case $\mathbf{D} = \mathbf{0}$, $\mathbf{g} = \mathbf{0}$, and $\dot{\Theta} = 0$. However, since the constraint response ψ_c is independent of the process $\mathbf{D}$, $\mathbf{g}$, $\dot{\Theta}$, we conclude that (8.35) holds for *all* $\mathbf{D}$, $\mathbf{g}$, and $\dot{\Theta}$ that satisfy the constraint (8.33). Hence, condition (8.34) becomes

$$\mathbf{T}_c \cdot \mathbf{D} - \hat{\rho}(\Theta)\eta_c\dot{\Theta} - \frac{1}{\Theta}\mathbf{q}_c \cdot \mathbf{g} = 0. \tag{8.36}$$

We now consider the family of processes with $\mathbf{D} = \mathbf{0}$ and $\dot{\Theta} = 0$, but $\mathbf{g}$ arbitrary, which necessarily satisfies constraint (8.33). For this family, condition (8.36) reduces to

$$-\frac{1}{\Theta}\mathbf{q}_c \cdot \mathbf{g} = 0.$$

The constraint response $\mathbf{q}_c$ is independent of the process $\mathbf{g}$, and $\mathbf{g}$ is arbitrary, which together imply that

$$\mathbf{q}_c = \mathbf{0} \tag{8.37}$$

for *all* processes satisfying constraint (8.33), not just the family with $\mathbf{D} = \mathbf{0}$ and $\dot{\Theta} = 0$. Thus, (8.36) becomes

$$\mathbf{T}_c \cdot \mathbf{D} - \hat{\rho}(\Theta)\eta_c\dot{\Theta} = 0. \tag{8.38}$$

Since $\mathbf{D}$ and $\dot{\Theta}$ are not independent (recall that they are interrelated through the constraint (8.33)), condition (8.38) cannot be reduced any further.

If we represent a thermomechanical process $(\mathbf{D}, \dot{\Theta})$ as a vector in the seven-dimensional inner product space $\mathcal{E}^7$, then the thermal expansion constraint (8.33) and entropy condition (8.38) can be rewritten as

$$\left(\mathbf{I}, \frac{\hat{\rho}\,'(\Theta)}{\hat{\rho}(\Theta)}\right) \cdot \left(\mathbf{D}, \dot{\Theta}\right) = 0 \tag{8.39a}$$

and

$$\left(\mathbf{T}_\mathrm{c}, -\hat{\rho}(\Theta)\eta_\mathrm{c}\right) \cdot \left(\mathbf{D}, \dot{\Theta}\right) = 0, \tag{8.39b}$$

respectively. It follows from (8.39a) and (8.39b) that $\mathbf{T}_\mathrm{c}$ is parallel to $\mathbf{I}$ and $-\hat{\rho}\eta_\mathrm{c}$ is parallel to $\hat{\rho}\,'/\hat{\rho}$, i.e.,

$$\mathbf{T}_\mathrm{c} = -\hat{p}\,\mathbf{I}, \quad \eta_\mathrm{c} = \hat{p}\,\frac{\hat{\rho}\,'(\Theta)}{\left(\hat{\rho}(\Theta)\right)^2}, \tag{8.40}$$

where, as with the incompressible theory, the *constraint pressure* $\hat{p}$ functions as a *Lagrange multiplier*, this time maintaining the thermal expansion constraint (8.33).

We now assemble our results. For a Newtonian fluid that thermally expands according to (8.32), we have

$$\mathbf{T} = \mathbf{T}_\mathrm{r} + \mathbf{T}_\mathrm{c}, \quad \mathbf{q} = \mathbf{q}_\mathrm{r} + \mathbf{q}_\mathrm{c}, \quad \psi = \psi_\mathrm{r} + \psi_\mathrm{c}, \quad \eta = \eta_\mathrm{r} + \eta_\mathrm{c}.$$

The constitutive response is given by (8.10) and (8.11b), i.e.,

$$\mathbf{T}_\mathrm{r} = -p\mathbf{I} + \lambda(\mathrm{tr}\,\mathbf{D})\mathbf{I} + 2\mu\mathbf{D}, \qquad \mathbf{q}_\mathrm{r} = -k\,\mathrm{grad}\,\Theta,$$

$$\psi_\mathrm{r} = \breve{\psi}\left(\hat{\rho}(\Theta), \Theta\right) = \hat{\psi}(\Theta), \qquad \eta_\mathrm{r} = -\,\hat{\psi}\,'(\Theta),$$

where we have used (8.32), and the constraint response is given by (8.35), (8.37), and (8.40), i.e.,

$$\mathbf{T}_\mathrm{c} = -\hat{p}\,\mathbf{I}, \quad \mathbf{q}_\mathrm{c} = \mathbf{0}, \quad \psi_\mathrm{c} = 0, \quad \eta_\mathrm{c} = \hat{p}\,\frac{\hat{\rho}\,'(\Theta)}{\left(\hat{\rho}(\Theta)\right)^2},$$

so the total response (constitutive plus constraint) is

$$\mathbf{T} = -\hat{p}\,\mathbf{I} + \lambda(\mathrm{tr}\,\mathbf{D})\mathbf{I} + 2\mu\mathbf{D}, \qquad \psi = \hat{\psi}(\Theta),$$

$$\eta = -\,\hat{\psi}\,'(\Theta) + \hat{p}\,\frac{\hat{\rho}\,'(\Theta)}{\left(\hat{\rho}(\Theta)\right)^2}, \qquad \mathbf{q} = -k\,\mathrm{grad}\,\Theta. \tag{8.41}$$

Note that the thermodynamic pressure p was absorbed into the Lagrange multiplier $\hat{p}$ in the equation for the Cauchy stress $\mathbf{T}$. The total response (8.41) satisfies the second law of thermodynamics, conservation of angular momentum, and the necessary invariance requirements.

It can be shown that the conservation laws of mass $(8.9)_1$, linear momentum $(8.9)_2$, and energy $(8.9)_3$ become, for the special case of a Newtonian fluid that thermally expands according to (8.32),

$$\frac{\hat{\rho}'(\Theta)}{\hat{\rho}(\Theta)}\dot{\Theta} + \operatorname{div}\mathbf{v} = 0, \quad \hat{\rho}(\Theta)\dot{\mathbf{v}} = \operatorname{div}\mathbf{T} + \hat{\rho}(\Theta)\mathbf{b}, \tag{8.42a}$$

and

$$-\hat{\rho}(\Theta)\Theta\,\hat{\psi}''(\Theta)\dot{\Theta} + \frac{\hat{\rho}'(\Theta)}{\hat{\rho}(\Theta)}\left(\Theta\dot{\hat{p}} + \hat{p}\,\dot{\Theta}\right) + \frac{\hat{p}\,\Theta\dot{\Theta}}{\hat{\rho}(\Theta)}\left(\hat{\rho}''(\Theta) - 2\frac{(\hat{\rho}'(\Theta))^2}{\hat{\rho}(\Theta)}\right)$$

$$= \mathbf{T}\cdot\mathbf{D} + \hat{\rho}(\Theta)r - \operatorname{div}\mathbf{q}, \tag{8.42b}$$

respectively. Augmented by the specified functions $\hat{\psi}(\Theta)$ and $\hat{\rho}(\Theta)$ for the Helmholtz potential and density, respectively, the constitutive equations $(8.41)_1$ and $(8.41)_4$ and the conservation laws (8.42a) and (8.42b) form a closed set for the fundamental unknowns pressure $\hat{p}$, temperature Θ, and velocity $\mathbf{v}$. Note that this set of governing equations reduces to the equations (8.22) and (8.23) governing incompressible Newtonian fluids if we restrict density to be constant.

EXERCISES

1. Prove that restriction (8.32) implies the motion-temperature constraint (8.33).
2. Show that for the special case of a Newtonian fluid that thermally expands according to (8.32), the conservation laws (8.9) become (8.42a) and (8.42b).

8.2.4.2 A density-entropy constraint (isentropic incompressibility)

We now pursue an alternative model of thermal expansion, where the fluid is essentially incompressible under **isentropic** (or constant entropy) conditions, but whose density changes in response to changes in entropy. The density-entropy formulation (8.11a) with density ρ and entropy η as independent variables is thus adopted, and we posit

$$\rho = \hat{\rho}(\eta), \tag{8.43}$$

i.e., the density is specified completely by the entropy. It can be shown that the thermal expansion constraint

$$\mathbf{I}\cdot\mathbf{D} + \frac{\hat{\rho}'(\eta)}{\hat{\rho}(\eta)}\dot{\eta} = 0 \tag{8.44}$$

follows from use of relation (8.43) in conservation of mass $(8.9)_1$. Constraint (8.44) is a special case of the general motion-entropy constraint (8.5) with

$$\mathbf{A} = \mathbf{I}, \quad \beta = \frac{\hat{\rho}'(\eta)}{\hat{\rho}(\eta)}.$$

Constraint (8.44) satisfies the invariance requirements (8.2).

We now consider restrictions imposed by the second law of thermodynamics. Starting with condition (8.8a), i.e.,

$$-\hat{\rho}(\eta)\dot{\varepsilon}_c + \mathbf{T}_c \cdot \mathbf{D} + \hat{\rho}(\eta)\Theta_c\dot{\eta} - \frac{1}{\Theta_c}\mathbf{q}_c \cdot \mathbf{g} = 0, \tag{8.45}$$

which demands that the constraint response generate no entropy, we can employ a series of arguments similar to the ones employed in Sections 8.2.2 and 8.2.4.1 to show that

$$\mathbf{T}_c = -\hat{p}\,\mathbf{I}, \quad \mathbf{q}_c = \mathbf{0}, \quad \varepsilon_c = 0, \quad \Theta_c = -\hat{p}\,\frac{\hat{\rho}\,'(\eta)}{\left(\hat{\rho}(\eta)\right)^2}, \tag{8.46}$$

where $\hat{p}$ is the constraint pressure that maintains (8.44). It then follows from (8.7), (8.10), and (8.11a) that the total response (constitutive plus constraint) is

$$\mathbf{T} = -\hat{p}\,\mathbf{I} + \lambda(\mathrm{tr}\,\mathbf{D})\mathbf{I} + 2\mu\mathbf{D}, \qquad \varepsilon = \breve{\varepsilon}\left(\hat{\rho}(\eta),\eta\right) = \hat{\varepsilon}(\eta),$$

$$\Theta = \hat{\varepsilon}\,'(\eta) - \hat{p}\,\frac{\hat{\rho}\,'(\eta)}{\left(\hat{\rho}(\eta)\right)^2}, \qquad \mathbf{q} = -k\,\mathrm{grad}\,\Theta. \tag{8.47}$$

The total response (8.47) satisfies the second law of thermodynamics, conservation of angular momentum, and the appropriate invariance requirements.

We can show that the conservation laws of mass $(8.9)_1$, linear momentum $(8.9)_2$, and energy $(8.9)_3$ become, for the special case of a Newtonian fluid that thermally expands according to (8.43),

$$\frac{\hat{\rho}\,'(\eta)}{\hat{\rho}(\eta)}\,\dot{\eta} + \mathrm{div}\,\mathbf{v} = 0, \quad \hat{\rho}(\eta)\dot{\mathbf{v}} = \mathrm{div}\,\mathbf{T} + \hat{\rho}(\eta)\mathbf{b}, \tag{8.48a}$$

and

$$\hat{\rho}(\eta)\hat{\varepsilon}\,'(\eta)\,\dot{\eta} = \mathbf{T}\cdot\mathbf{D} + \hat{\rho}(\eta)r - \mathrm{div}\,\mathbf{q}, \tag{8.48b}$$

respectively. Augmented by the specified functions $\hat{\varepsilon}(\eta)$ and $\hat{\rho}(\eta)$ for the internal energy and density, respectively, the constitutive equations (8.47) and conservation laws (8.48a) and (8.48b) form a closed set for the fundamental unknowns pressure $\hat{p}$, entropy η, and velocity $\mathbf{v}$.

EXERCISES

1. Prove that restriction (8.43) implies the motion-entropy constraint (8.44).

2. Argue that the constraint response (8.46) follows from the second law equality (8.45) and the constraint (8.44). Then show that the total response (constitutive plus constraint) is (8.47).

3. Verify that for the special case of a Newtonian fluid that thermally expands according to (8.43), the conservation laws (8.9) become (8.48a) and (8.48b).

8.2.5 THERMAL EXPANSION AS A CONSTITUTIVE LIMIT: AN ALTERNATIVE PERSPECTIVE

In this section, we present alternative derivations of the thermal expansion models developed in Sections 8.2.4.1 and 8.2.4.2. In particular, we show that these models can be regarded as constitutive limits of the fully compressible theory.

8.2.5.1 Isothermal incompressibility

To obtain the thermal expansion model of Section 8.2.4.1, we impose the **constitutive limit**

$$\frac{\partial \rho(p,\Theta)}{\partial p} \to 0 \tag{8.49}$$

on the pressure-temperature formulation (8.11d) of the compressible theory. The constitutive limit (8.49) demands no pressure dependence of density at constant temperature, i.e., *isothermal incompressibility*. It can be shown that the constitutive limit (8.49) together with the equation of state $(8.11\text{d})_1$ imposes the following restriction on the form of the Gibbs free energy:

$$\frac{\partial^2 \phi}{\partial p^2} = 0. \tag{8.50}$$

Noting that $\phi(p,\Theta)$, if we integrate (8.50) once, we obtain

$$\frac{\partial \phi}{\partial p} = b_1(\Theta). \tag{8.51}$$

It then follows from use of (8.51) in equation of state $(8.11\text{d})_1$ that

$$\rho = \left(\frac{\partial \phi}{\partial p}\right)^{-1} = \left(b_1(\Theta)\right)^{-1} = \hat{\rho}(\Theta), \tag{8.52}$$

i.e., density is a function of temperature alone. Note that in the internal constraint theory presented in Section 8.2.4.1, $\rho = \hat{\rho}(\Theta)$ was *posited* at the outset rather than *deduced* downstream. Integrating (8.50) a second time gives

$$\phi(p,\Theta) = b_1(\Theta)p + b_2(\Theta) = \frac{p}{\hat{\rho}(\Theta)} + b_2(\Theta). \tag{8.53}$$

It can be verified that use of the restricted form of the Gibbs free energy (8.53) in equation of state $(8.11\text{d})_2$ leads to

$$\eta = -\frac{\partial \phi}{\partial \Theta} = p\,\frac{\hat{\rho}\,'(\Theta)}{\left(\hat{\rho}(\Theta)\right)^2} - b_2'(\Theta). \tag{8.54}$$

We can show that as a consequence of results (8.52)–(8.54), the conservation laws (8.9) and constitutive equations (8.10) become

$$\frac{\hat{\rho}\,'(\Theta)}{\hat{\rho}(\Theta)}\,\dot{\Theta} + \operatorname{div}\mathbf{v} = 0, \quad \hat{\rho}(\Theta)\dot{\mathbf{v}} = \operatorname{div}\mathbf{T} + \hat{\rho}(\Theta)\mathbf{b}, \tag{8.55a}$$

$$-\hat{\rho}(\Theta)\Theta\, b_2''(\Theta)\dot{\Theta} + \frac{\hat{\rho}\,'(\Theta)}{\hat{\rho}(\Theta)}\left(\Theta\dot{p} + p\,\dot{\Theta}\right) + \frac{p\,\Theta\dot{\Theta}}{\hat{\rho}(\Theta)}\left(\hat{\rho}\,''(\Theta) - 2\,\frac{\left(\hat{\rho}\,'(\Theta)\right)^2}{\hat{\rho}(\Theta)}\right)$$

$$= \mathbf{T}\cdot\mathbf{D} + \hat{\rho}(\Theta)r - \operatorname{div}\mathbf{q}, \tag{8.55b}$$

and

$$\mathbf{T} = -p\mathbf{I} + \lambda(\operatorname{tr}\mathbf{D})\mathbf{I} + 2\mu\mathbf{D}, \quad \mathbf{q} = -k\,\operatorname{grad}\Theta, \tag{8.56}$$

respectively. These equations are precisely the governing equations (8.41), (8.42a), and (8.42b) obtained in Section 8.2.4.1 if we equate the primitive unknown $\hat{p}$ with p and the specified function $\hat{\psi}(\Theta)$ with $b_2(\Theta)$.

It is important to note that this set of governing equations violates the stability conditions (E.19) in Appendix E *absent any additional conditions placed on the flow*. The interested reader is encouraged to consult [21, 22] for additional details.

EXERCISES

1. Show that the constitutive limit (8.49) together with the equation of state $(8.11\text{d})_1$ impose restriction (8.50) on the Gibbs free energy.
2. Confirm result (8.54).
3. Verify that results (8.52)–(8.54) imply that the conservation laws (8.9) and constitutive equations (8.10) become (8.55a) and (8.55b), and (8.56), respectively.

8.2.5.2 Isentropic incompressibility

To obtain the thermal expansion model of Section 8.2.4.2, we impose the **constitutive limit**

$$\frac{\partial\rho(p,\eta)}{\partial p} \to 0 \tag{8.57}$$

on the pressure-entropy formulation (8.11c) of the compressible theory. Limit (8.57) demands no pressure dependence of density at constant entropy, i.e., *isentropic incompressibility*. Using arguments similar to those employed in Sections 8.2.3 and 8.2.5.1, it can be shown that the constitutive limit (8.57) demands that the density, enthalpy, and temperature take the following forms:

$$\rho = \hat{\rho}(\eta), \quad \chi(p,\eta) = \frac{p}{\hat{\rho}(\eta)} + c_2(\eta), \quad \Theta(p,\eta) = -p\,\frac{\hat{\rho}\,'(\eta)}{\left(\hat{\rho}(\eta)\right)^2} + c_2'(\eta). \tag{8.58}$$

We can confirm that use of results (8.58) in the conservation laws (8.9) and constitutive equations (8.10) leads to

$$\frac{\hat{\rho}\,'(\eta)}{\hat{\rho}(\eta)}\,\dot{\eta} + \operatorname{div}\mathbf{v} = 0, \quad \hat{\rho}(\eta)\dot{\mathbf{v}} = \operatorname{div}\mathbf{T} + \hat{\rho}(\eta)\mathbf{b}, \tag{8.59a}$$

$$\hat{p}(\eta)c_2'(\eta)\,\dot{\eta} = \mathbf{T}\cdot\mathbf{D} + \hat{\rho}(\eta)r - \operatorname{div}\mathbf{q}, \tag{8.59b}$$

and

$$\mathbf{T} = -p\mathbf{I} + \lambda(\operatorname{tr}\mathbf{D})\mathbf{I} + 2\mu\mathbf{D}, \quad \mathbf{q} = -k \operatorname{grad}\Theta, \tag{8.60}$$

respectively. Note that these equations are precisely the governing equations (8.47), (8.48a), and (8.48b) obtained in Section 8.2.4.2 using the formalism of internal constraints, provided we equate the primitive unknown $\hat{p}$ with p and the specified function $\hat{\varepsilon}(\eta)$ with $c_2(\eta)$. For this thermal expansion model, the stability conditions (E.18) in Appendix E are preserved; refer to [21, 22] for details.

EXERCISE

1. Argue that the constitutive limit (8.57) on the pressure-entropy formulation (8.11c) of the compressible theory leads to (8.58). Then show that (8.58) implies that the conservation laws (8.9) and constitutive equations (8.10) become (8.59a) and (8.59b), and (8.60), respectively.

8.3 NONLINEAR ELASTIC SOLIDS

This section is structured as follows: In Section 8.3.1, we review the *unconstrained* theory for *compressible*, isotropic, nonlinear elastic solids presented in Chapter 6. In Section 8.3.2, we present the *constrained* theory for *incompressible* nonlinear elastic solids. Finally, in Section 8.3.3, we discuss some representative strain energy models.

8.3.1 THE COMPRESSIBLE THEORY: A BRIEF REVIEW

For nonlinear elastic materials, it is customary to adopt the *Lagrangian* form of the fundamental laws, which, in the *compressible* (i.e., *unconstrained*) thermomechanical theory, are

$$\rho J = \rho_R, \quad \rho_R\,\ddot{\mathbf{x}} = \operatorname{Div}\mathbf{P} + \rho_R\mathbf{b}, \quad \rho_R\left(\dot{\psi} + \Theta\dot{\eta} + \eta\dot{\Theta}\right) = \mathbf{P}\cdot\dot{\mathbf{F}} + \rho_R r - \operatorname{Div}\mathbf{q}_R. \tag{8.61}$$

Equations $(8.61)_1$, $(8.61)_2$, and $(8.61)_3$ are Lagrangian statements of conservation of mass, balance of linear momentum, and the first law of thermodynamics, respectively. In (8.61), ρ is the density in the present configuration, ρ_R is the density in the reference configuration, $\mathbf{P}$ is the first Piola-Kirchhoff stress, $\mathbf{b}$ is the body force per unit mass, ψ is the Helmholtz free energy per unit mass, Θ is the absolute temperature, η is the entropy per unit mass, r is the rate of heat absorbed per unit mass, $\mathbf{q}_R$ is the referential heat flux vector, and

$$\mathbf{F} = \frac{\partial\mathbf{x}}{\partial\mathbf{X}}, \quad J = \det\mathbf{F}$$

are the deformation gradient and its determinant. Note that $\mathbf{x}$ and $\mathbf{X}$ are the positions of a continuum particle in the present and reference configurations, respectively.

The constitutive equations and equation of state for a compressible, isotropic, nonlinear elastic material are

$$\mathbf{T} = \beta_0 \mathbf{I} + \beta_1 \mathbf{B} + \beta_{-1} \mathbf{B}^{-1}, \quad \mathbf{q} = -k \operatorname{grad} \Theta, \quad \eta = -\frac{\partial \psi}{\partial \Theta}, \tag{8.62}$$

where $\mathbf{B} = \mathbf{F}\mathbf{F}^{\mathrm{T}}$ is the left Cauchy-Green deformation tensor and k is the thermal conductivity. Note that the Cauchy stress $\mathbf{T}$ in (8.62) is related to the first Piola-Kirchhoff stress $\mathbf{P}$ in (8.61) through

$$\mathbf{P} = J\mathbf{T}\mathbf{F}^{-\mathrm{T}},$$

and the spatial heat flux vector $\mathbf{q}$ in (8.62) is related to the referential heat flux vector $\mathbf{q}_{\mathrm{R}}$ in (8.61) through

$$\mathbf{q}_{\mathrm{R}} = J\mathbf{F}^{-1}\mathbf{q}.$$

The material-dependent and deformation-dependent coefficients β_0, β_1, and β_{-1} in (8.62) are given by

$$\beta_0 = 2I_3^{-\frac{1}{2}} \left(I_2 \frac{\partial \breve{W}}{\partial I_2} + I_3 \frac{\partial \breve{W}}{\partial I_3} \right), \quad \beta_1 = 2I_3^{-\frac{1}{2}} \frac{\partial \breve{W}}{\partial I_1}, \quad \beta_{-1} = -2I_3^{\frac{1}{2}} \frac{\partial \breve{W}}{\partial I_2},$$

where $W = \breve{W}(I_1, I_2, I_3, \Theta)$ is the strain energy density, related to the Helmholtz free energy by

$$W = \rho_{\mathrm{R}} \psi,$$

and I_1, I_2, and I_3 are the principal invariants of $\mathbf{B}$, i.e.,

$$I_1 = \operatorname{tr} \mathbf{B}, \quad I_2 = \frac{1}{2}\left[(\operatorname{tr} \mathbf{B})^2 - \operatorname{tr}(\mathbf{B}^2) \right], \quad I_3 = \det \mathbf{B}.$$

With the strain energy $W = \breve{W}(I_1, I_2, I_3, \Theta)$ specified, the fundamental laws (8.61) and constitutive equations (8.62) form a closed system for the present position $\mathbf{x}$, present density ρ, and temperature Θ, all functions of reference position $\mathbf{X}$ and time t.

8.3.2 INCOMPRESSIBILITY

For an **incompressible nonlinear elastic solid**, the density remains constant as the material deforms, i.e.,

$$\rho = \bar{\rho} \equiv \text{constant}. \tag{8.63}$$

Use of restriction (8.63) in the Lagrangian form of conservation of mass $(8.61)_1$ implies that

$$J = \det \mathbf{F} = 1, \tag{8.64}$$

i.e., only isochoric deformations are permitted by the material; refer to Section 3.6 and, in particular, (3.71). We can show that taking the material derivative of (8.64) leads to the familiar incompressibility constraint

$$\mathbf{I} \cdot \mathbf{D} = 0. \tag{8.65}$$

Recall that (8.65) is a special case of (8.1) with $\mathbf{A} = \mathbf{I}$, $\mathbf{b} = \mathbf{0}$, and $\alpha = 0$. Also recall that (8.65) satisfies the invariance requirements (8.2).

We now investigate restrictions on the elastic response imposed by the second law of thermodynamics. Recall that the *constitutive equations* $\mathbf{T}_\mathrm{r}$, $\mathbf{q}_\mathrm{r}$, ψ_r, η_r for elastic solids must satisfy inequality (8.4b), i.e.,

$$-\bar{\rho}\dot{\psi}_\mathrm{r} + \mathbf{T}_\mathrm{r} \cdot \mathbf{D} - \bar{\rho}\eta_\mathrm{r}\dot{\Theta} - \frac{1}{\Theta}\mathbf{q}_\mathrm{r} \cdot \mathbf{g} \geq 0,$$

which we successfully demonstrated in Chapter 6. Thus, it remains for the *constraint response* $\mathbf{T}_\mathrm{c}$, $\mathbf{q}_\mathrm{c}$, ψ_c, η_c to satisfy equality (8.4a), i.e.,

$$-\bar{\rho}\dot{\psi}_\mathrm{c} + \mathbf{T}_\mathrm{c} \cdot \mathbf{D} - \bar{\rho}\eta_\mathrm{c}\dot{\Theta} - \frac{1}{\Theta}\mathbf{q}_\mathrm{c} \cdot \mathbf{g} = 0. \tag{8.66}$$

Condition (8.66) must hold for all processes satisfying (8.65). In particular, it must hold for the family of processes with $\mathbf{D} = \mathbf{0}$, $\mathbf{g} = \mathbf{0}$, and $\dot{\Theta} = 0$, which necessarily satisfies (8.65). For this family of processes, condition (8.66) becomes

$$\dot{\psi}_\mathrm{c} = 0.$$

Without loss of generality, we take the constant of integration to be zero, which gives

$$\psi_\mathrm{c} = 0. \tag{8.67}$$

Result (8.67) holds for the particular case $\mathbf{D} = \mathbf{0}$, $\mathbf{g} = \mathbf{0}$, $\dot{\Theta} = 0$. However, since the constraint response ψ_c is independent of the process $\mathbf{D}$, $\mathbf{g}$, $\dot{\Theta}$, we conclude that (8.67) holds for *all* $\mathbf{D}$, $\mathbf{g}$, and $\dot{\Theta}$ that satisfy the constraint (8.65). Hence, condition (8.66) reduces to

$$\mathbf{T}_\mathrm{c} \cdot \mathbf{D} - \bar{\rho}\eta_\mathrm{c}\dot{\Theta} - \frac{1}{\Theta}\mathbf{q}_\mathrm{c} \cdot \mathbf{g} = 0. \tag{8.68}$$

Consider now the family of processes with $\mathbf{D} = \mathbf{0}$ and $\mathbf{g} = \mathbf{0}$, but $\dot{\Theta}$ arbitrary, which necessarily satisfies constraint (8.65). For this family, condition (8.68) becomes

$$-\bar{\rho}\eta_\mathrm{c}\dot{\Theta} = 0.$$

The constraint response η_c is independent of the process $\dot{\Theta}$, and $\dot{\Theta}$ is arbitrary, which together imply that

$$\eta_\mathrm{c} = 0 \tag{8.69}$$

for *all* processes satisfying constraint (8.65), not just the family with $\mathbf{D} = \mathbf{0}$ and $\mathbf{g} = \mathbf{0}$. Thus, condition (8.68) reduces to

$$\mathbf{T}_\mathrm{c} \cdot \mathbf{D} - \frac{1}{\Theta}\mathbf{q}_\mathrm{c} \cdot \mathbf{g} = 0. \tag{8.70}$$

Consider now the family of processes with $\mathbf{D} = \mathbf{0}$, but $\mathbf{g}$ and $\dot{\Theta}$ arbitrary; this family of processes necessarily satisfies the constraint (8.65). For this family, condition (8.70) reduces to

$$-\frac{1}{\Theta}\mathbf{q}_c \cdot \mathbf{g} = 0.$$

The constraint response $\mathbf{q}_c$ is independent of the process $\mathbf{g}$, and $\mathbf{g}$ is arbitrary, so we may argue

$$\mathbf{q}_c = \mathbf{0} \tag{8.71}$$

for *all* processes satisfying constraint (8.65), not just the family with $\mathbf{D} = \mathbf{0}$. Thus, condition (8.70) becomes

$$\mathbf{T}_c \cdot \mathbf{D} = 0. \tag{8.72}$$

Representing $\mathbf{D}$ as a vector in the six-dimensional inner product space $\mathcal{E}^6$, we find that admissible $\mathbf{D}$ (i.e., those satisfying the incompressibility constraint (8.65)) are those orthogonal to the vector $\mathbf{I}$. Condition (8.72) demands that $\mathbf{D}$ also be orthogonal to $\mathbf{T}_c$. Hence, we deduce that $\mathbf{T}_c$ is parallel to $\mathbf{I}$, i.e.,

$$\mathbf{T}_c = -\bar{p}\,\mathbf{I}, \tag{8.73}$$

where the **constraint pressure** $\bar{p}$, a scalar function of position and time, acts as a **Lagrange multiplier**. That is, the constraint pressure $\bar{p}$ is a *primitive unknown* and takes whatever value is necessary to maintain the incompressibility constraint (8.65).

We now assemble our results. For an incompressible, isotropic, nonlinear elastic solid, we have

$$\mathbf{T} = \mathbf{T}_r + \mathbf{T}_c, \quad \mathbf{q} = \mathbf{q}_r + \mathbf{q}_c, \quad W = W_r + W_c, \quad \eta = \eta_r + \eta_c,$$

recalling that the strain energy W is related to the Helmholtz free energy ψ by $W = \rho_R \psi$. The constitutive response is given by (8.62), i.e.,

$$\mathbf{T}_r = \beta_0 \mathbf{I} + \beta_1 \mathbf{B} + \beta_{-1}\mathbf{B}^{-1}, \qquad \mathbf{q}_r = -k \,\text{grad}\,\Theta,$$

$$W_r = \breve{W}(I_1, I_2, I_3 = 1, \Theta) = \bar{W}(I_1, I_2, \Theta), \qquad \eta_r = -\frac{1}{\rho_R}\frac{\partial \bar{W}}{\partial \Theta},$$

where we have used (8.64), and the constraint response is given by (8.67), (8.69), (8.71), and (8.73), i.e.,

$$\mathbf{T}_c = -\bar{p}\,\mathbf{I}, \quad \mathbf{q}_c = \mathbf{0}, \quad W_c = 0, \quad \eta_c = 0,$$

so the total response (constitutive plus constraint) is

$$\mathbf{T} = -\bar{p}\,\mathbf{I} + \beta_1 \mathbf{B} + \beta_{-1}\mathbf{B}^{-1}, \quad \mathbf{q} = -k \,\text{grad}\,\Theta, \quad \eta = -\frac{1}{\rho_R}\frac{\partial \bar{W}}{\partial \Theta}, \tag{8.74a}$$

with

$$W = \bar{W}(I_1, I_2, \Theta), \quad \beta_1 = 2\frac{\partial \bar{W}}{\partial I_1}, \quad \beta_{-1} = -2\frac{\partial \bar{W}}{\partial I_2}. \tag{8.74b}$$

Note that in $(8.74a)_1$, the term $\beta_0\mathbf{I}$ from the constitutive response was absorbed into the Lagrange multiplier $\bar{p}$ that maintains the incompressibility constraint. Also note that (8.74a) and (8.74b) satisfy the second law of thermodynamics, conservation of angular momentum, and the appropriate invariance requirements and material symmetry conditions.

We can show that for the special case of an incompressible, isotropic, nonlinear elastic solid, the Lagrangian conservation laws of mass $(8.61)_1$, linear momentum $(8.61)_2$, and energy $(8.61)_3$ become

$$\det \mathbf{F} = 1, \quad \bar{\rho}\,\ddot{\mathbf{x}} = \operatorname{Div}\mathbf{P} + \bar{\rho}\,\mathbf{b}, \tag{8.75a}$$

and

$$\left(\frac{\partial \bar{W}}{\partial I_1} - \Theta\frac{\partial^2 \bar{W}}{\partial \Theta \partial I_1}\right)\dot{I}_1 + \left(\frac{\partial \bar{W}}{\partial I_2} - \Theta\frac{\partial^2 \bar{W}}{\partial \Theta \partial I_2}\right)\dot{I}_2 - \Theta\dot{\Theta}\frac{\partial^2 \bar{W}}{\partial \Theta^2} = \mathbf{P}\cdot\dot{\mathbf{F}} + \bar{\rho}r - \operatorname{Div}\mathbf{q}_{\mathrm{R}}, \tag{8.75b}$$

respectively. The fundamental laws (8.75a) and (8.75b), together with the material response (8.74a) and (8.74b) and the relationships

$$\mathbf{F} = \frac{\partial \mathbf{x}}{\partial \mathbf{X}}, \quad \mathbf{B} = \mathbf{F}\mathbf{F}^{\mathrm{T}}, \quad \mathbf{P} = \mathbf{T}\mathbf{F}^{-\mathrm{T}}, \quad \mathbf{q}_{\mathrm{R}} = \mathbf{F}^{-1}\mathbf{q},$$

and

$$I_1 = \operatorname{tr}\mathbf{B}, \quad I_2 = \frac{1}{2}\left[(\operatorname{tr}\mathbf{B})^2 - \operatorname{tr}(\mathbf{B}^2)\right],$$

form a closed system of equations for the constraint pressure $\bar{p}$, present position $\mathbf{x}$, and temperature Θ, all functions of reference position $\mathbf{X}$ and time t. Note that the strain energy $W = \bar{W}(I_1, I_2, \Theta)$ is a specified function, of which some representative examples are presented in the following section.

PROBLEM 8.6

Verify that restriction (8.63) implies the incompressibility constraint (8.65).

Solution

Use of restriction (8.63) in the Lagrangian form of conservation of mass $(8.61)_1$ implies that

$$J = \det \mathbf{F} = 1.$$

Taking the material derivative of this equation, we obtain

$$\overline{\det \mathbf{F}}^{\,\cdot} = 0.$$

Then,

$$\overline{\det \mathbf{F}}^{\,\cdot} = \frac{\mathrm{d}(\det \mathbf{F})}{\mathrm{d}\mathbf{F}}\cdot\dot{\mathbf{F}} \qquad \text{(chain rule } (2.95)_1\text{)}$$

$$
\begin{aligned}
&= (\det \mathbf{F})\mathbf{F}^{-\mathrm{T}} \cdot \dot{\mathbf{F}} && \text{(Problem 2.48)} \\
&= \mathbf{F}^{-\mathrm{T}} \cdot \dot{\mathbf{F}} && \text{(restriction (8.64))} \\
&= \mathbf{F}^{-\mathrm{T}} \cdot \mathbf{L}\mathbf{F} && \text{(result (3.60)}_1\text{)} \\
&= \operatorname{tr}\left[\mathbf{F}^{-\mathrm{T}}(\mathbf{L}\mathbf{F})^{\mathrm{T}}\right] && \text{(definition (2.41))} \\
&= \operatorname{tr}\left(\mathbf{F}^{-\mathrm{T}}\mathbf{F}^{\mathrm{T}}\mathbf{L}^{\mathrm{T}}\right) && \text{(result (2.14)}_2\text{)} \\
&= \operatorname{tr}\mathbf{L} && \text{(result (2.40)}_2\text{)} \\
&= \mathbf{I} \cdot \mathbf{D} && \text{(Problem 8.1).}
\end{aligned}
$$

Thus, we conclude that

$$\mathbf{I} \cdot \mathbf{D} = 0.$$

EXERCISES

1. Prove that as a consequence of restriction (8.64), $I_3 = \det \mathbf{B} = 1$. (Hint: Start with $\mathbf{B} = \mathbf{F}\mathbf{F}^{\mathrm{T}}$ and use the properties (2.47) of the determinant.)

2. Show that for the special case of an incompressible, isotropic, nonlinear elastic solid, the Lagrangian conservation laws (8.61) become (8.75a) and (8.75b).

8.3.3 INCOMPRESSIBLE STRAIN ENERGY MODELS

In this section, we present some representative strain energy models for incompressible, isotropic, rubbery materials. We do so in the context of the *mechanical* (or isothermal) variant of the theory presented in Section 8.3.2, which is

$$\det \mathbf{F} = 1, \quad \bar{\rho}\,\ddot{\mathbf{x}} = \operatorname{Div} \mathbf{P} + \bar{\rho}\,\mathbf{b},$$

and

$$\mathbf{T} = -\bar{p}\,\mathbf{I} + 2\frac{\partial \bar{W}}{\partial I_1}\mathbf{B} - 2\frac{\partial \bar{W}}{\partial I_2}\mathbf{B}^{-1},$$

with

$$\mathbf{P} = \mathbf{T}\mathbf{F}^{-\mathrm{T}}, \quad I_1 = \operatorname{tr}\mathbf{B}, \quad I_2 = \frac{1}{2}\left[(\operatorname{tr}\mathbf{B})^2 - \operatorname{tr}(\mathbf{B}^2)\right], \quad \mathbf{B} = \mathbf{F}\mathbf{F}^{\mathrm{T}}, \quad \mathbf{F} = \frac{\partial \mathbf{x}}{\partial \mathbf{X}}.$$

Together, these mechanical governing equations constitute a closed system for the present position $\mathbf{x}$ and constraint pressure $\bar{p}$, provided the material-specific strain energy density function $W = \bar{W}(I_1, I_2)$ is specified. Two common strain energy models for rubbery materials are the **incompressible Mooney-Rivlin model** [23, 24],

$$W = c_1\left(I_1 - 3\right) + c_2\left(I_2 - 3\right), \tag{8.76}$$

and the **incompressible neo-Hookean model** [24, 25],

$$W = \frac{\mu}{2}\left(I_1 - 3\right). \tag{8.77}$$

In (8.76) and (8.77), c_1 and c_2 are parameters that can be adjusted to fit experimental data for a particular rubbery material, and μ is the shear modulus evaluated at small strains. The strain energy models (8.76) and (8.77) are referred to as *invariant-based models* since they are formulated in terms of the principal invariants I_1, I_2, and I_3.

In contrast, the **incompressible Ogden model** [26] is a *stretch-based model;* i.e., it is formulated in terms of powers of the principal stretches λ_1, λ_2, λ_3,

$$W = \sum_n \frac{\mu_n}{\alpha_n}\left(\lambda_1^{\alpha_n} + \lambda_2^{\alpha_n} + \lambda_3^{\alpha_n} - 3\right), \tag{8.78}$$

as is the **Varga model** [11, p. 238],

$$W = 2\mu\left(\lambda_1 + \lambda_2 + \lambda_3 - 3\right), \tag{8.79}$$

which is a special case of the Ogden model. In (8.78) and (8.79), μ_n, α_n, and n are parameters that can be adjusted to fit experimental data, and μ is the shear modulus evaluated at small strains.

Other models, such as the one developed by Arruda and Boyce [27], are based on statistical mechanics, and thus account for the underlying deformation physics of the polymer chains. It should be noted that a strain energy function that is formally identical to the neo-Hookean model (8.77) of Rivlin was developed several years prior by Treloar [28] using statistical mechanics.

Many other invariant-based, stretch-based, and statistical-mechanics-based strain energy models for incompressible rubbery materials—beyond the representative few presented here—can be found, for instance, in the books by Holzapfel [11], Treloar [16], and Ogden [17], and the review articles by Ogden [18] and Boyce and Arruda [19].

PROBLEM 8.7

Consider the inflation of a balloon, described by the motion

$$r^3 = C + R^3, \quad \theta = \Theta, \quad \phi = \Phi,$$

where (r, θ, ϕ) and (R, Θ, Φ) are the spherical coordinates of a representative continuum particle in the present (deformed) and reference (undeformed) configurations, respectively, and the inflation C is positive. We model the balloon as a thin, hollow, spherical shell composed of an incompressible, isotropic, rubbery material. The balloon is subjected to an internal pressure p_i on its inner surface $r = a$, while its outer surface $r = b$ is traction-free. The deformation is essentially static and isothermal, and the effects of gravity are negligible.

(a) Verify that the inflation pressure p_i can be written in terms of the radial stretch $\lambda = a/a_o$ as

$$p_i(\lambda) = \frac{2t_o}{a_o\lambda}\left(1 - \frac{1}{\lambda^6}\right)\left(\beta_1 - \lambda^2\beta_{-1}\right),$$

where $t_o = b_o - a_o$ is the thickness of the undeformed shell, a_o and b_o are the inner and outer radii of the undeformed shell, a and b are the inner and outer radii of the deformed shell, and

$$W = \bar{W}(I_1, I_2), \quad \beta_1 = 2\frac{\partial \bar{W}}{\partial I_1}, \quad \beta_{-1} = -2\frac{\partial \bar{W}}{\partial I_2}.$$

(b) Determine the inflation pressure p_i as a function of the radial stretch $\lambda = a/a_o$ for the special cases of a neo-Hookean material and a Mooney-Rivlin material. Using the properties of vulcanized natural rubber, plot p_i versus λ for both cases.

Solution

We briefly recall from Section 8.3.3 the *mechanical* equations governing an incompressible, isotropic, nonlinear elastic material. Specializing to static deformations and in the absence of body forces, we have

$$\det \mathbf{F} = 1, \quad \operatorname{div} \mathbf{T} = \mathbf{0}, \quad \mathbf{T} = -\bar{p}\mathbf{I} + \beta_1\mathbf{B} + \beta_{-1}\mathbf{B}^{-1}, \tag{a}$$

with

$$W = \bar{W}(I_1, I_2), \quad \beta_1 = 2\frac{\partial \bar{W}}{\partial I_1}, \quad \beta_{-1} = -2\frac{\partial \bar{W}}{\partial I_2}.$$

Equations $(a)_1$, $(a)_2$, and $(a)_3$ are statements of conservation of mass (incompressibility), balance of linear momentum (static equilibrium), and the nonlinear elastic constitutive response, respectively. From the motion

$$r^3 = C + R^3, \quad \theta = \Theta, \quad \phi = \Phi,$$

we deduce the spherical components of the deformation gradient:

$$[\mathbf{F}] = \begin{bmatrix} \dfrac{\partial r}{\partial R} & \dfrac{1}{R}\dfrac{\partial r}{\partial \Theta} & \dfrac{1}{R\sin\Theta}\dfrac{\partial r}{\partial \Phi} \\ r\dfrac{\partial \theta}{\partial R} & \dfrac{r}{R}\dfrac{\partial \theta}{\partial \Theta} & \dfrac{r}{R\sin\Theta}\dfrac{\partial \theta}{\partial \Phi} \\ r\sin\theta\dfrac{\partial \phi}{\partial R} & \dfrac{r\sin\theta}{R}\dfrac{\partial \phi}{\partial \Theta} & \dfrac{r\sin\theta}{R\sin\Theta}\dfrac{\partial \phi}{\partial \Phi} \end{bmatrix} = \begin{bmatrix} \alpha & 0 & 0 \\ 0 & \dfrac{1}{\sqrt{\alpha}} & 0 \\ 0 & 0 & \dfrac{1}{\sqrt{\alpha}} \end{bmatrix}, \tag{b}$$

where

$$\alpha = \frac{R^2}{r^2} = \left(1 - \frac{C}{r^3}\right)^{\frac{2}{3}} = \left(1 + \frac{C}{R^3}\right)^{-\frac{2}{3}}.$$

Note that since the balloon is being inflated ($C > 0$), the radial fibers are shortening ($\alpha < 1$). Hence, consistent with physical observation, the thickness of the balloon decreases as it inflates. Also note that $\det \mathbf{F} = 1$ (i.e., the deformation is isochoric), so the incompressibility constraint (a)$_1$ is satisfied. The left Cauchy-Green deformation $\mathbf{B} = \mathbf{F}\mathbf{F}^{\mathrm{T}}$ and its inverse follow from (b):

$$[\mathbf{B}] = \begin{bmatrix} \alpha^2 & 0 & 0 \\ 0 & \dfrac{1}{\alpha} & 0 \\ 0 & 0 & \dfrac{1}{\alpha} \end{bmatrix}, \qquad [\mathbf{B}^{-1}] = \begin{bmatrix} \dfrac{1}{\alpha^2} & 0 & 0 \\ 0 & \alpha & 0 \\ 0 & 0 & \alpha \end{bmatrix}.$$

The principal invariants of $\mathbf{B}$ are

$$I_1 = \operatorname{tr}\mathbf{B} = \alpha^2 + \frac{2}{\alpha}, \quad I_2 = \frac{1}{2}\left[(\operatorname{tr}\mathbf{B})^2 - \operatorname{tr}(\mathbf{B}^2)\right] = 2\alpha + \frac{1}{\alpha^2}.$$

With the deformation in hand, we now calculate the spherical components of the (symmetric) Cauchy stress tensor $\mathbf{T}$ using the constitutive equation (a)$_3$:

$$T_{rr} = -\bar{p} + \alpha^2\beta_1 + \frac{1}{\alpha^2}\beta_{-1}, \quad T_{\theta\theta} = T_{\phi\phi} = -\bar{p} + \frac{1}{\alpha}\beta_1 + \alpha\beta_{-1},$$

$$T_{r\theta} = T_{r\phi} = T_{\theta\phi} = 0. \tag{c}$$

The stresses in (c) must satisfy the equilibrium equation (a)$_2$, which can be written in spherical components as (refer to Section 2.7.3.2)

$$\frac{\partial T_{rr}}{\partial r} + \frac{1}{r}\frac{\partial T_{r\theta}}{\partial \theta} + \frac{1}{r\sin\theta}\frac{\partial T_{r\phi}}{\partial \phi} + \frac{1}{r}\left(2T_{rr} - T_{\theta\theta} - T_{\phi\phi}\right) + \frac{\cot\theta}{r}T_{r\theta} = 0, \tag{da}$$

$$\frac{\partial T_{r\theta}}{\partial r} + \frac{1}{r}\frac{\partial T_{\theta\theta}}{\partial \theta} + \frac{1}{r\sin\theta}\frac{\partial T_{\theta\phi}}{\partial \phi} + \frac{3}{r}T_{r\theta} + \frac{\cot\theta}{r}\left(T_{\theta\theta} - T_{\phi\phi}\right) = 0, \tag{db}$$

$$\frac{\partial T_{r\phi}}{\partial r} + \frac{1}{r}\frac{\partial T_{\theta\phi}}{\partial \theta} + \frac{1}{r\sin\theta}\frac{\partial T_{\phi\phi}}{\partial \phi} + \frac{3}{r}T_{r\phi} + \frac{2\cot\theta}{r}T_{\theta\phi} = 0. \tag{dc}$$

Use of (c) in (da)–(dc) leads to

$$\frac{\partial T_{rr}}{\partial r} + \frac{2}{r}\left(\alpha^2 - \frac{1}{\alpha}\right)\left(\beta_1 - \frac{1}{\alpha}\beta_{-1}\right) = 0, \quad \frac{\partial T_{\theta\theta}}{\partial \theta} = 0, \quad \frac{\partial T_{\phi\phi}}{\partial \phi} = 0. \tag{e}$$

Arguing that the radial stress T_{rr} is a function of r alone, we integrate $(\text{e})_1$ and obtain

$$T_{rr}(r) = -2\int_a^r \left(\alpha^2 - \frac{1}{\alpha}\right)\left(\beta_1 - \frac{1}{\alpha}\beta_{-1}\right)\frac{dr}{r} - p_i, \tag{f}$$

where we have used the boundary condition $T_{rr} = -p_i$ at the inner surface $r = a$. Evaluating (f) on $r = b$ and using the boundary condition $T_{rr}(b) = 0$ yields

$$p_i = -2\int_a^b \left(\alpha^2 - \frac{1}{\alpha}\right)\left(\beta_1 - \frac{1}{\alpha}\beta_{-1}\right)\frac{dr}{r}. \tag{g}$$

Noting that the shell is thin and taking $\alpha = a_o^2/a^2$, we can argue that (g) can be approximated as

$$p_i = -2\left(\alpha^2 - \frac{1}{\alpha}\right)\left(\beta_1 - \frac{1}{\alpha}\beta_{-1}\right)\frac{t}{a}, \tag{h}$$

recalling that a, b, and $t = b - a$ are the inner radius, outer radius, and thickness, respectively, of the deformed shell. By defining the radial stretch $\lambda = a/a_o$ and noting that the volume of the thin spherical shell remains constant as it deforms, i.e.,

$$4\pi a_o^2 t_o = 4\pi a^2 t,$$

we can rewrite (h) in terms of the stretch λ and the geometric properties of the undeformed shell, i.e.,

$$p_i(\lambda) = \frac{2t_o}{a_o\lambda}\left(1 - \frac{1}{\lambda^6}\right)\left(\beta_1 - \lambda^2\beta_{-1}\right),$$

recalling that a_o, b_o, and $t_o = b_o - a_o$, are the inner radius, outer radius, and thickness, respectively, of the undeformed shell.

For an *incompressible neo-Hookean material*, we have

$$W = \frac{\mu}{2}\left(I_1 - 3\right), \quad \beta_1 = 2\frac{\partial W}{\partial I_1} = \mu, \quad \beta_{-1} = -2\frac{\partial W}{\partial I_2} = 0,$$

so

$$p_i(\lambda) = \frac{2\mu t_o}{a_o\lambda}\left(1 - \frac{1}{\lambda^6}\right).$$

Similarly, for an *incompressible Mooney-Rivlin material*, we have

$$W = c_1\left(I_1 - 3\right) + c_2\left(I_2 - 3\right), \quad \beta_1 = 2\frac{\partial W}{\partial I_1} = 2c_1, \quad \beta_{-1} = -2\frac{\partial W}{\partial I_2} = -2c_2,$$

so

$$p_{\mathrm{i}}(\lambda) = \frac{4t_{\mathrm{o}}}{a_{\mathrm{o}}\lambda}\left(1 - \frac{1}{\lambda^6}\right)\left(c_1 + \lambda^2 c_2\right).$$

Note that in both cases, as the radial stretch $\lambda \to 1$, the inflation pressure $p_{\mathrm{i}} \to 0$, as expected (see Figure 8.3).

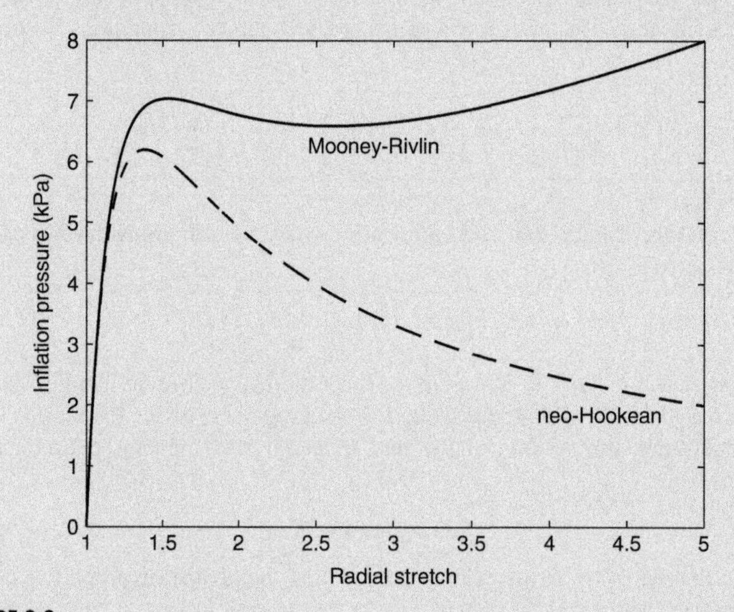

FIGURE 8.3

Variation of the inflation pressure p_{i} with increasing radial stretch λ in a rubber balloon ($\mu = 500\,\mathrm{kPa}$, $c_1 = 0.4375\,\mu$, and $c_2 = 0.0625\,\mu$ [29]) using neo-Hookean (dashed line) and Mooney-Rivlin (solid line) material models. The inner radius and thickness of the undeformed balloon are $a_{\mathrm{o}} = 5\,\mathrm{cm}$ and $t_{\mathrm{o}} = 0.5\,\mathrm{mm}$, respectively. Note that the two material models show excellent agreement at small stretches, but diverge at larger stretches.

PART

IV

Beyond Mechanics and Thermomechanics

CHAPTER

9 Modeling of Thermo-Electro-Magneto-Mechanical Behavior, with Application to Smart Materials

Sushma Santapuri, Robert L. Lowe, Stephen E. Bechtel
Department of Mechanical and Aerospace Engineering, The Ohio State University
Columbus, OH

In this chapter, we present a continuum approach to modeling **smart materials**. A key feature of smart materials is their ability to convert energy from one form into another. For instance, *piezoelectric materials* deform when exposed to an electric field, thus converting electrical energy to mechanical energy. Other common smart materials include *magnetostrictives*, *magnetorheological fluids*, *shape memory alloys*, and *electroactive polymers*. These materials couple different physical effects, e.g., thermal, electrical, magnetic, and/or mechanical. Owing to their unique properties, smart materials are implemented in a wide variety of automotive, aerospace, and biomedical applications, to name but a few.

Compared with classical elastic solids and viscous fluids (refer to Chapters 4–8), additional balance laws (e.g., Maxwell's equations) are required to model the complex multiphysics behavior of smart materials. Also, the thermomechanical balance laws (i.e., linear momentum, angular momentum, and energy) must be modified to account for contributions from electromagnetic fields. Finally, the constitutive equations that describe the response of smart materials should highlight the coupling of different physical effects. For example, stress in a piezoelectric material is a function of both strain and electric field.

In what follows, we present a unified continuum framework that lays the groundwork for modeling a broad range of smart materials exhibiting coupled thermal, electrical, magnetic, and/or mechanical behavior. This framework has the breadth to accommodate dynamic electromagnetic fields, large deformations (i.e., geometric nonlinearities), anisotropy, and nonlinear constitutive response (i.e., material nonlinearity). We devote special attention to developing the fundamental laws of continuum electrodynamics and presenting key aspects of thermodynamic constitutive modeling. For instance, we show how the thermodynamic formalism that produced constitutive models for classical elastic solids (refer to Chapter 6) and viscous fluids (refer to Chapter 7) can also be used to facilitate the constitutive modeling of smart materials with thermo-electro-magneto-mechanical (TEMM) behavior. Finally, we illustrate that our modeling approach provides an overarching framework that encompasses many well-known types of smart material behavior. For instance, we explicitly demonstrate that the linear theory of piezoelectricity falls out as a special case of our more general finite-deformation TEMM framework, in much the same way linear elasticity falls out of finite-deformation elasticity.

Although some of the concepts presented in this chapter are introduced in earlier parts of the book, they are restated here so that this chapter is essentially self-contained.

9.1 THE FUNDAMENTAL LAWS OF CONTINUUM ELECTRODYNAMICS: INTEGRAL FORMS

In this section, we present the fundamental laws of continuum electrodynamics, i.e., the first principles for a deformable, polarizable, magnetizable, conductive thermo-electro-magneto-mechanical (TEMM) material. As was done in Chapter 4, the first principles are postulated at four different levels: primitive, material, integral, and pointwise. In particular, we begin by explicitly stating the first principles in their most primitive or fundamental form. These primitive statements are then expressed mathematically in material form. The material form is global, i.e., valid on the body as a whole and all subsets.[1] Specializing to a continuum leads to a corresponding set of integral equations. Boundedness and continuity then allow these integral equations to be localized, i.e., expressed in a pointwise fashion. Note that we carefully progress from primitive statements to pointwise equations, rather than starting directly with a set of pointwise equations, since (1) the pointwise equations must be derivable from an integral set and (2) the assumptions for continuum models are customarily imposed on the integral form [30, 31].

9.1.1 NOTATION AND NOMENCLATURE

We now present the notation necessary to describe the geometry of the deformable TEMM body. We label the body $\mathbb{B}$ and two arbitrary subsets $\mathcal{S}_1$ and $\mathcal{S}_2$.[2] Subset $\mathcal{S}_1$ is bounded by a *closed material surface,* while subset $\mathcal{S}_2$ is bounded by a *closed material curve*; see Figure 9.1. As will soon be evident, both closed material surfaces *and* closed material curves are required to formulate integral statements of the fundamental laws of continuum electrodynamics.

In the reference configuration, body $\mathbb{B}$ occupies open volume $\mathcal{R}_{\mathrm{R}}$ of Euclidean 3-space, bounded by closed surface $\partial\mathcal{R}_{\mathrm{R}}$. Subset $\mathcal{S}_1$ occupies open volume $\mathcal{P}_{\mathrm{R}} \subset$

[1] A familiar example of this notion is a truss in rigid body statics: not only is the entire structure in equilibrium, but also each joint and each member.

[2] In order to avoid notational conflicts later in the chapter, we change the notation for the body from $\mathcal{B}$ (as was used in previous chapters) to $\mathbb{B}$.

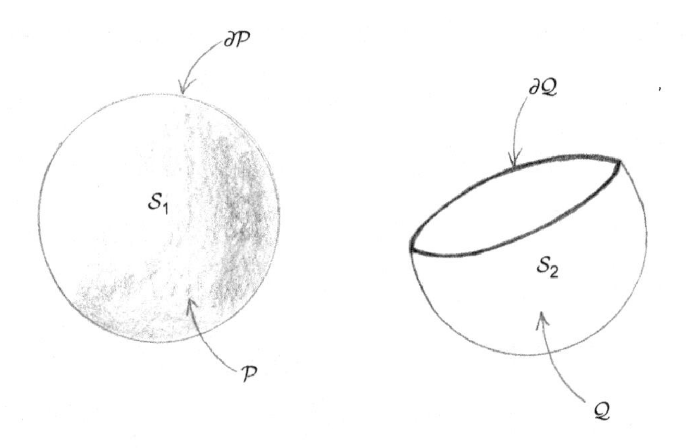

FIGURE 9.1

Subsets $\mathcal{S}_1$ and $\mathcal{S}_2$ as seen in the present configuration. Subset $\mathcal{S}_1$ is an open material volume $\mathcal{P}$ bounded by a closed material surface $\partial\mathcal{P}$, while subset $\mathcal{S}_2$ is an open material surface $\mathcal{Q}$ bounded by a closed material curve $\partial\mathcal{Q}$.

$\mathcal{R}_{\text{R}}$, bounded by closed surface $\partial\mathcal{P}_{\text{R}}$, and subset $\mathcal{S}_2$ occupies open surface $\mathcal{Q}_{\text{R}} \subset \mathcal{R}_{\text{R}}$, bounded by closed curve $\partial\mathcal{Q}_{\text{R}}$.

In the present configuration at time t, body $\mathbb{B}$ occupies open material volume $\mathcal{R}$, bounded by closed material surface $\partial\mathcal{R}$. Subset $\mathcal{S}_1$ occupies open material volume $\mathcal{P} \subset \mathcal{R}$, bounded by closed material surface $\partial\mathcal{P}$, and subset $\mathcal{S}_2$ occupies open material surface $\mathcal{Q} \subset \mathcal{R}$, bounded by closed material curve $\partial\mathcal{Q}$.[3]

9.1.2 CONSERVATION OF MASS

Primitively, conservation of mass postulates that the **mass** $\mathcal{M}$ of every subset of the body $\mathbb{B}$ is constant throughout its motion, or, equivalently, the time rate of change of the mass of every subset is zero. Applying this primitive statement to arbitrary subset $\mathcal{S}_1$ allows us to express conservation of mass mathematically in material form:

$$\frac{\mathrm{d}}{\mathrm{d}t}\mathcal{M}(\mathcal{S}_1, t) = 0 \quad \text{or} \quad \mathcal{M}(\mathcal{S}_1) \equiv \text{independent of } t. \tag{9.1}$$

Specializing to a continuum, the approach taken heretofore in this book, allows us to express the mass $\mathcal{M}$ of subset $\mathcal{S}_1$ in Eulerian and Lagrangian integral forms, i.e.,

$$\mathcal{M}(\mathcal{S}_1) = \int_{\mathcal{S}_1} \mathrm{d}m = \begin{cases} \displaystyle\int_{\mathcal{P}} \rho\, \mathrm{d}v, \\ \displaystyle\int_{\mathcal{P}_{\text{R}}} \rho_{\text{R}}\, \mathrm{d}V. \end{cases} \tag{9.2}$$

[3]Refer to Section 3.5 for a discussion of material curves, material surfaces, and material volumes.

Thus, specializing to a continuum is tantamount to a *smoothness assumption* on the mass $\mathcal{M}$; refer to Section 4.2. The Eulerian integral representation (top of (9.2)) corresponds to subset $\mathcal{S}_1$ as seen in its present configuration, while the Lagrangian integral representation (bottom of (9.2)) corresponds to $\mathcal{S}_1$ as seen in its reference configuration.[4] In (9.2), $\mathrm{d}v$ and $\mathrm{d}V$ are volume elements in the present and reference configurations (refer to Section 3.6), and ρ and ρ_{R} are the mass densities in the present and reference configurations (refer to Section 4.1). Note that ρ has units of mass per present volume, while ρ_{R} has units of mass per reference volume (see Table 9.1). Both ρ and ρ_{R} are bounded, continuous functions of space and time.

To perform the integrations in (9.2), it is natural to consider ρ in its Eulerian description, i.e., as a function of $\mathbf{x}$ and t, and to consider ρ_{R} in its Lagrangian description, i.e., as a function of $\mathbf{X}$ and t, although both ρ and ρ_{R} can be expressed using either an Eulerian or a Lagrangian description. Recall that $\mathbf{X}$ and $\mathbf{x}$ are the reference and present positions of a continuum particle, related through the motion $\mathbf{x} = \boldsymbol{\chi}(\mathbf{X}, t)$ (refer to Section 3.1).

Use of (9.2) in (9.1) leads to Eulerian and Lagrangian integral representations of conservation of mass:

$$\frac{\mathrm{d}}{\mathrm{d}t}\int_{\mathcal{P}} \rho \,\mathrm{d}v = 0, \tag{9.3a}$$

$$\int_{\mathcal{P}} \rho \,\mathrm{d}v = \int_{\mathcal{P}_{\mathrm{R}}} \rho_{\mathrm{R}} \,\mathrm{d}V. \tag{9.3b}$$

These integral statements are valid for any open volume $\mathcal{P}$ in the present configuration and corresponding open volume $\mathcal{P}_{\mathrm{R}}$ in the reference configuration. Note that (9.3a) and (9.3b) are identical to their counterparts in the thermomechanical theory (refer to Chapter 4).

9.1.3 BALANCE OF LINEAR MOMENTUM

Balance of linear momentum postulates that the time rate of change of the **linear momentum $\mathcal{L}$** of any subset of the body is equal to the **resultant external force $\mathbf{f}$** acting on that subset. Applying this primitive statement to subset $\mathcal{S}_1$ gives the material form:

$$\frac{\mathrm{d}}{\mathrm{d}t}\mathcal{L}(\mathcal{S}_1, t) = \mathbf{f}(\mathcal{S}_1, t). \tag{9.4}$$

Assuming smoothness of $\mathcal{L}$, we write its Eulerian and Lagrangian integral representations:

$$\mathcal{L}(\mathcal{S}_1, t) = \begin{cases} \displaystyle\int_{\mathcal{P}} \mathbf{v}\rho \,\mathrm{d}v, \\ \displaystyle\int_{\mathcal{P}_{\mathrm{R}}} \mathbf{v}\rho_{\mathrm{R}} \,\mathrm{d}V, \end{cases} \tag{9.5}$$

[4] Recall from Section 4.9 that we are free to label the volume occupied by subset $\mathcal{S}_1$ by its present volume $\mathcal{P}$ or its reference volume $\mathcal{P}_{\mathrm{R}}$.

Table 9.1 Units for Thermal, Electrical, Magnetic, and Mechanical Quantities

Type	Quantity	Representation	Symbol	Fundamental Units	Derived Units	SI Units
Electrical	Vacuum permittivity		ϵ_0	$\frac{C^2T^2}{ML_P^3}$	$\frac{\text{Capacitance}}{\text{Present length}}$	$\frac{\text{Farad}}{\text{Meter}}$
	Electric field	Referential	$\mathbf{e}_R$	$\frac{ML_P^2}{L_RT^2C}$	$\frac{\text{Force}}{\text{Charge}} \cdot \frac{\text{Present length}}{\text{Reference length}}$	$\frac{\text{Newton}}{\text{Coulomb}} \equiv \frac{\text{Volt}}{\text{Meter}}$
		Spatial	$\mathbf{e}^*$	$\frac{ML_P}{CT^2}$	$\frac{\text{Force}}{\text{Charge}}$	$\frac{\text{Newton}}{\text{Coulomb}} \equiv \frac{\text{Volt}}{\text{Meter}}$
	Electric polarization	Referential	$\mathbf{p}_R$	$\frac{C}{L_R^2}$	$\frac{\text{Charge}}{\text{Reference area}}$	$\frac{\text{Coulomb}}{\text{Meter}^2}$
		Spatial	$\mathbf{p}^*$	$\frac{C}{L_P^2}$	$\frac{\text{Charge}}{\text{Present area}}$	$\frac{\text{Coulomb}}{\text{Meter}^2}$
	Electric displacement	Referential	$\mathbf{d}_R$	$\frac{C}{L_R^2}$	$\frac{\text{Charge}}{\text{Reference area}}$	$\frac{\text{Coulomb}}{\text{Meter}^2}$
		Spatial	$\mathbf{d}^*$	$\frac{C}{L_P^2}$	$\frac{\text{Charge}}{\text{Present area}}$	$\frac{\text{Coulomb}}{\text{Meter}^2}$
	Free charge density	Referential	σ_R	$\frac{C}{L_R^3}$	$\frac{\text{Charge}}{\text{Reference volume}}$	$\frac{\text{Coulomb}}{\text{Meter}^3}$
		Spatial	σ^*	$\frac{C}{L_P^3}$	$\frac{\text{Charge}}{\text{Present volume}}$	$\frac{\text{Coulomb}}{\text{Meter}^3}$
	Conductive current density	Referential	$\mathbf{j}_R$	$\frac{C}{L_R^2T}$	$\frac{\text{Current}}{\text{Reference area}}$	$\frac{\text{Ampere}}{\text{Meter}^2}$
		Spatial	$\mathbf{j}^*$	$\frac{C}{L_P^2T}$	$\frac{\text{Current}}{\text{Present area}}$	$\frac{\text{Ampere}}{\text{Meter}^2}$

Table 9.1 Units for Thermal, Electrical, Magnetic, and Mechanical Quantities *Continued*

Type	Quantity	Representation	Symbol	Fundamental Units	Derived Units	SI Units
Magnetic	Vacuum permeability		μ_o	$\frac{ML_P}{C^2}$	$\frac{\text{Inductance}}{\text{Present length}}$	$\frac{\text{Henry}}{\text{Meter}}$
	Magnetic field	Referential	$\mathbf{h}_R$	$\frac{C}{L_R T}$	$\frac{\text{Current}}{\text{Reference length}}$	$\frac{\text{Ampere}}{\text{Meter}}$
		Spatial	$\mathbf{h}^*$	$\frac{C}{L_P T}$	$\frac{\text{Current}}{\text{Present length}}$	$\frac{\text{Ampere}}{\text{Meter}}$
	Magnetization	Referential	$\mathbf{m}_R$	$\frac{C}{L_R T}$	$\frac{\text{Current}}{\text{Reference length}}$	$\frac{\text{Ampere}}{\text{Meter}}$
		Spatial	$\mathbf{m}^*$	$\frac{C}{L_P T}$	$\frac{\text{Current}}{\text{Present length}}$	$\frac{\text{Ampere}}{\text{Meter}}$
	Magnetic flux density	Referential	$\mathbf{b}_R$	$\frac{ML_P^2}{L_R^2 TC}$	$\frac{\text{Magnetic flux}}{\text{Reference area}}$	$\frac{\text{Weber}}{\text{Meter}^2} \equiv \text{Tesla}$
		Spatial	$\mathbf{b}^*$	$\frac{M}{TC}$	$\frac{\text{Magnetic flux}}{\text{Present area}}$	$\frac{\text{Weber}}{\text{Meter}^2} \equiv \text{Tesla}$
Mechanical	Mass density	Referential	ρ_R	$\frac{M}{L_R^3}$	$\frac{\text{Mass}}{\text{Reference volume}}$	$\frac{\text{Kilogram}}{\text{Meter}^3}$
		Spatial	ρ	$\frac{M}{L_P^3}$	$\frac{\text{Mass}}{\text{Present volume}}$	$\frac{\text{Kilogram}}{\text{Meter}^3}$
	Traction	Referential	$\mathbf{t}_R$	$\frac{ML_P}{L_R^2 T^2}$	$\frac{\text{Force}}{\text{Reference area}}$	$\frac{\text{Newton}}{\text{Meter}^2} \equiv \text{Pascal}$
		Spatial	$\mathbf{t}$	$\frac{M}{L_P T^2}$	$\frac{\text{Force}}{\text{Present area}}$	$\frac{\text{Newton}}{\text{Meter}^2} \equiv \text{Pascal}$
	Stress	Referential	$\mathbf{P}$	$\frac{ML_P}{L_R^2 T^2}$	$\frac{\text{Force}}{\text{Reference area}}$	$\frac{\text{Newton}}{\text{Meter}^2} \equiv \text{Pascal}$
		Spatial	$\mathbf{T}$	$\frac{M}{L_P T^2}$	$\frac{\text{Force}}{\text{Present area}}$	$\frac{\text{Newton}}{\text{Meter}^2} \equiv \text{Pascal}$

Table 9.1 Units for Thermal, Electrical, Magnetic, and Mechanical Quantities *Continued*

Type	Quantity	Representation	Symbol	Fundamental Units	Derived Units	SI Units
Mechanical	Velocity		$\mathbf{v}$	$\frac{L_P}{T}$		$\frac{\text{Meter}}{\text{Second}}$
	Deformation gradient		$\mathbf{F}$	$\frac{L_P}{L_R}$		
	Green's deformation		$\mathbf{C}$	$\frac{L_P^2}{L_R^2}$		
	Volumetric deformation		J	$\frac{L_P^3}{L_R^3}$		
Thermal	Heat flux rate	Referential	h_R	$\frac{ML_P^2}{L_R^2T^3}$	$\frac{\text{Energy}}{\text{Reference area} \cdot \text{time}}$	$\frac{\text{Watt}}{\text{Meter}^2}$
		Spatial	h	$\frac{M}{T^3}$	$\frac{\text{Energy}}{\text{Present area} \cdot \text{time}}$	$\frac{\text{Watt}}{\text{Meter}^2}$
	Heat flux vector	Referential	$\mathbf{q}_R$	$\frac{ML_P^2}{L_R^2T^3}$	$\frac{\text{Energy}}{\text{Reference area} \cdot \text{time}}$	$\frac{\text{Watt}}{\text{Meter}^2}$
		Spatial	$\mathbf{q}$	$\frac{M}{T^3}$	$\frac{\text{Energy}}{\text{Present area} \cdot \text{time}}$	$\frac{\text{Watt}}{\text{Meter}^2}$
	Temperature		Θ	θ		Kelvin
	Specific internal energy		ε	$\frac{L_P^2}{T^2}$	$\frac{\text{Energy}}{\text{Mass}}$	$\frac{\text{Joule}}{\text{Kilogram}}$
	Specific entropy		η	$\frac{L_P^2}{T^2\theta}$	$\frac{\text{Energy}}{\text{Mass} \cdot \text{temperature}}$	$\frac{\text{Joule}}{\text{Kilogram} \cdot \text{kelvin}}$

Note that M *is mass,* L_P *is present length,* L_R *is reference length,* T *is time,* θ *is temperature, and* C *is charge.*

where $\mathbf{v}$ is the **velocity** of a continuum particle at the present time t, and the integrands are continuous, bounded functions of space and time. As was done in Section 4.2, it is assumed that the resultant external force $\mathbf{f}$ can be additively decomposed into a **body force** and a **contact force**. The effects of electromagnetism are modeled through an electromagnetic contribution to the body force [32, 33] so that

$$\mathbf{f}(\mathcal{S}_1, t) = \begin{cases} \displaystyle\int_{\mathcal{P}} (\mathbf{f}^{\mathrm{m}} + \mathbf{f}^{\mathrm{em}})\rho \,\mathrm{d}v + \int_{\partial\mathcal{P}} \mathbf{t} \,\mathrm{d}a, \\ \displaystyle\int_{\mathcal{P}_{\mathrm{R}}} (\mathbf{f}^{\mathrm{m}} + \mathbf{f}^{\mathrm{em}})\rho_{\mathrm{R}} \,\mathrm{d}V + \int_{\partial\mathcal{P}_{\mathrm{R}}} \mathbf{t}_{\mathrm{R}} \,\mathrm{d}A, \end{cases} \tag{9.6}$$

where $\mathbf{f}^{\mathrm{m}}$ is the **mechanically induced body force** per unit mass, $\mathbf{f}^{\mathrm{em}}$ is the **electromagnetically induced body force** per unit mass, $\mathbf{t}$ and $\mathbf{t}_{\mathrm{R}}$ are the **spatial** and **referential tractions** (refer to Section 4.9), and $\mathrm{d}a$ and $\mathrm{d}A$ are area elements in the present and reference configurations (refer to Section 3.6).[5] Elaborating, $\mathbf{t}$ is the traction acting on surface $\partial\mathcal{P}$ in the present configuration measured per unit area of $\partial\mathcal{P}$, whereas $\mathbf{t}_{\mathrm{R}}$ is the traction acting on surface $\partial\mathcal{P}$ in the present configuration but measured per unit area of the corresponding surface $\partial\mathcal{P}_{\mathrm{R}}$ in the reference configuration. Thus, $\mathbf{t}$ has units of force per present area, while $\mathbf{t}_{\mathrm{R}}$ has units of force per reference area (see Table 9.1).

Use of (9.5) and (9.6) in (9.4) gives Eulerian and Lagrangian integral representations of balance of linear momentum:

$$\frac{\mathrm{d}}{\mathrm{d}t}\int_{\mathcal{P}} \mathbf{v}\rho \,\mathrm{d}v = \int_{\mathcal{P}} (\mathbf{f}^{\mathrm{m}} + \mathbf{f}^{\mathrm{em}})\rho \,\mathrm{d}v + \int_{\partial\mathcal{P}} \mathbf{t} \,\mathrm{d}a, \tag{9.7a}$$

$$\frac{\mathrm{d}}{\mathrm{d}t}\int_{\mathcal{P}_{\mathrm{R}}} \mathbf{v}\rho_{\mathrm{R}} \,\mathrm{d}V = \int_{\mathcal{P}_{\mathrm{R}}} (\mathbf{f}^{\mathrm{m}} + \mathbf{f}^{\mathrm{em}})\rho_{\mathrm{R}} \,\mathrm{d}V + \int_{\partial\mathcal{P}_{\mathrm{R}}} \mathbf{t}_{\mathrm{R}} \,\mathrm{d}A. \tag{9.7b}$$

Comparing (9.7a) and (9.7b) with their counterparts in the thermomechanical theory (refer to Chapter 4), we see that (9.7a) and (9.7b) contain an additional body force $\mathbf{f}^{\mathrm{em}}$.

9.1.4 BALANCE OF ANGULAR MOMENTUM

Balance of angular momentum postulates that the time rate of change of the **angular momentum $\mathbf{H}_0$** of any subset of the body about the origin $\mathbf{0}$ is equal to the **resultant external moment $\mathbf{M}_0$** acting on that subset about the origin $\mathbf{0}$. In material form:

$$\frac{\mathrm{d}}{\mathrm{d}t}\mathbf{H}_0(\mathcal{S}_1, t, \mathbf{0}) = \mathbf{M}_0(\mathcal{S}_1, t, \mathbf{0}). \tag{9.8}$$

---

[5] In order to avoid notational conflicts later in the chapter, we change the notation for the referential traction from $\mathbf{p}$ (as was used in previous chapters) to $\mathbf{t}_{\mathrm{R}}$, and the notation for the mechanical body force from $\mathbf{b}$ (as was used in previous chapters) to $\mathbf{f}^{\mathrm{m}}$.

Assuming that $\mathbf{H}_0$ is smooth leads to Eulerian and Lagrangian integral representations of the angular momentum about $\mathbf{0}$, i.e.,

$$\mathbf{H}_0(\mathcal{S}_1, t, \mathbf{0}) = \begin{cases} \displaystyle\int_{\mathcal{P}} \mathbf{x} \times \mathbf{v}\rho \, \mathrm{d}v, \\ \displaystyle\int_{\mathcal{P}_\mathrm{R}} \mathbf{x} \times \mathbf{v}\rho_\mathrm{R} \, \mathrm{d}V. \end{cases} \tag{9.9}$$

The integrands in (9.9) are continuous, bounded functions of space and time. Similarly, smoothness of $\mathbf{M}_0$ implies that

$$\mathbf{M}_0(\mathcal{S}_1, t, \mathbf{0}) = \begin{cases} \displaystyle\int_{\mathcal{P}} \mathbf{x} \times (\mathbf{f}^\mathrm{m} + \mathbf{f}^\mathrm{em})\rho \, \mathrm{d}v + \int_{\partial\mathcal{P}} \mathbf{x} \times \mathbf{t} \, \mathrm{d}a + \int_{\mathcal{P}} \mathbf{c}^\mathrm{em} \rho \, \mathrm{d}v, \\ \displaystyle\int_{\mathcal{P}_\mathrm{R}} \mathbf{x} \times (\mathbf{f}^\mathrm{m} + \mathbf{f}^\mathrm{em})\rho_\mathrm{R} \, \mathrm{d}V + \int_{\partial\mathcal{P}_\mathrm{R}} \mathbf{x} \times \mathbf{t}_\mathrm{R} \, \mathrm{d}A + \int_{\mathcal{P}_\mathrm{R}} \mathbf{c}^\mathrm{em} \rho_\mathrm{R} \, \mathrm{d}V. \end{cases} \tag{9.10}$$

Following [32, 33], an **electromagnetically induced body couple** per unit mass $\mathbf{c}^\mathrm{em}$ is included in (9.10) to model the effects of electromagnetism. Note that the first two terms in $(9.10)_1$ and $(9.10)_2$ represent the moment about $\mathbf{0}$ due to the resultant external force $\mathbf{f}$, the first term being a contribution from the body force and the second term being a contribution from the contact force.

Use of (9.9) and (9.10) in (9.8) gives Eulerian and Lagrangian integral representations of balance of angular momentum:

$$\frac{\mathrm{d}}{\mathrm{d}t} \int_{\mathcal{P}} \mathbf{x} \times \mathbf{v}\rho \, \mathrm{d}v = \int_{\mathcal{P}} \mathbf{x} \times (\mathbf{f}^\mathrm{m} + \mathbf{f}^\mathrm{em})\rho \, \mathrm{d}v + \int_{\partial\mathcal{P}} \mathbf{x} \times \mathbf{t} \, \mathrm{d}a + \int_{\mathcal{P}} \mathbf{c}^\mathrm{em} \rho \, \mathrm{d}v, \tag{9.11a}$$

$$\frac{\mathrm{d}}{\mathrm{d}t} \int_{\mathcal{P}_\mathrm{R}} \mathbf{x} \times \mathbf{v}\rho_\mathrm{R} \, \mathrm{d}V = \int_{\mathcal{P}_\mathrm{R}} \mathbf{x} \times (\mathbf{f}^\mathrm{m} + \mathbf{f}^\mathrm{em})\rho_\mathrm{R} \, \mathrm{d}V + \int_{\partial\mathcal{P}_\mathrm{R}} \mathbf{x} \times \mathbf{t}_\mathrm{R} \, \mathrm{d}A + \int_{\mathcal{P}_\mathrm{R}} \mathbf{c}^\mathrm{em} \rho_\mathrm{R} \, \mathrm{d}V. \tag{9.11b}$$

Comparing (9.11a) and (9.11b) with their counterparts in the thermomechanical theory (refer to Chapter 4), we see that (9.11a) and (9.11b) contain additional moments due to (1) the electromagnetic body force $\mathbf{f}^\mathrm{em}$ and (2) the electromagnetic body couple $\mathbf{c}^\mathrm{em}$.

9.1.5 FIRST LAW OF THERMODYNAMICS

The first law of thermodynamics (or conservation of energy) postulates that the time rate of change of the **total energy** (i.e., **kinetic energy** K plus **internal energy** E) of any subset of the body is equal to the rate of work R generated by the resultant external force acting on that subset plus the rate of all other energies $\mathcal{A}$ (e.g., heat, electromagnetic, chemical) entering or exiting that subset. In material form,

$$\frac{\mathrm{d}}{\mathrm{d}t}\Big(K(\mathcal{S}_1, t) + E(\mathcal{S}_1, t)\Big) = R(\mathcal{S}_1, t) + \mathcal{A}(\mathcal{S}_1, t). \tag{9.12}$$

Assuming that the kinetic and internal energies of part $\mathcal{S}_1$ are smooth implies that

$$K(\mathcal{S}_1,t) = \begin{cases} \displaystyle\int_{\mathcal{P}} \frac{1}{2}\mathbf{v}\cdot\mathbf{v}\rho\,\mathrm{d}v, \\ \displaystyle\int_{\mathcal{P}_\mathrm{R}} \frac{1}{2}\mathbf{v}\cdot\mathbf{v}\rho_\mathrm{R}\,\mathrm{d}V, \end{cases} \qquad E(\mathcal{S}_1,t) = \begin{cases} \displaystyle\int_{\mathcal{P}} \varepsilon\rho\,\mathrm{d}v, \\ \displaystyle\int_{\mathcal{P}_\mathrm{R}} \varepsilon\rho_\mathrm{R}\,\mathrm{d}V, \end{cases} \tag{9.13}$$

where ε is the **specific internal energy** (or internal energy per unit mass), and the integrands are bounded, continuous functions of space and time. The **rate of work** R generated by the resultant external force $\mathbf{f}$ can be additively decomposed into contributions from the body force and the contact force, i.e.,

$$R(\mathcal{S}_1,t) = \begin{cases} \displaystyle\int_{\mathcal{P}} (\mathbf{f}^\mathrm{m}+\mathbf{f}^\mathrm{em})\cdot\mathbf{v}\rho\,\mathrm{d}v + \int_{\partial\mathcal{P}} \mathbf{t}\cdot\mathbf{v}\,\mathrm{d}a, \\ \displaystyle\int_{\mathcal{P}_\mathrm{R}} (\mathbf{f}^\mathrm{m}+\mathbf{f}^\mathrm{em})\cdot\mathbf{v}\rho_\mathrm{R}\,\mathrm{d}V + \int_{\partial\mathcal{P}_\mathrm{R}} \mathbf{t}_\mathrm{R}\cdot\mathbf{v}\,\mathrm{d}A. \end{cases} \tag{9.14}$$

Following [32, 33], the **auxiliary energy rate** $\mathcal{A}$ is additively decomposed into three contributions, two from radiation and one from conduction: the rate of heat absorption throughout the volume, the rate of electromagnetic energy absorption throughout the volume, and the rate of heat entering through the boundary, i.e.,

$$\mathcal{A}(\mathcal{S}_1,t) = \begin{cases} \displaystyle\int_{\mathcal{P}} r^\mathrm{t}\rho\,\mathrm{d}v + \int_{\mathcal{P}} r^\mathrm{em}\rho\,\mathrm{d}v - \int_{\partial\mathcal{P}} h\,\mathrm{d}a, \\ \displaystyle\int_{\mathcal{P}_\mathrm{R}} r^\mathrm{t}\rho_\mathrm{R}\,\mathrm{d}V + \int_{\mathcal{P}_\mathrm{R}} r^\mathrm{em}\rho_\mathrm{R}\,\mathrm{d}V - \int_{\partial\mathcal{P}_\mathrm{R}} h_\mathrm{R}\,\mathrm{d}A, \end{cases} \tag{9.15}$$

where r^t is the **specific heat supply rate**, r^em is the **specific electromagnetic energy supply rate**, and h_R and h are the **referential** and **spatial heat flux rates** (refer to Section 4.9.3).[6] Elaborating, h is the rate of heat flow *out* of the present boundary $\partial\mathcal{P}$ measured per unit area of the present boundary $\partial\mathcal{P}$. Conversely, h_R is the rate of heat flow *out* of the present boundary $\partial\mathcal{P}$, but measured per unit area of the corresponding boundary $\partial\mathcal{P}_\mathrm{R}$ in the reference configuration. Thus, h has units of energy per time per present area, while h_R has units of energy per time per reference area (see Table 9.1).

Use of (9.13)–(9.15) in (9.12) yields Eulerian and Lagrangian integral representations of the first law of thermodynamics:

$$\frac{\mathrm{d}}{\mathrm{d}t}\int_{\mathcal{P}} \frac{1}{2}\mathbf{v}\cdot\mathbf{v}\rho\,\mathrm{d}v + \frac{\mathrm{d}}{\mathrm{d}t}\int_{\mathcal{P}} \varepsilon\rho\,\mathrm{d}v = \int_{\mathcal{P}} (\mathbf{f}^\mathrm{m}+\mathbf{f}^\mathrm{em})\cdot\mathbf{v}\rho\,\mathrm{d}v + \int_{\partial\mathcal{P}} \mathbf{t}\cdot\mathbf{v}\,\mathrm{d}a + \int_{\mathcal{P}} (r^\mathrm{t}+r^\mathrm{em})\rho\,\mathrm{d}v - \int_{\partial\mathcal{P}} h\,\mathrm{d}a, \tag{9.16a}$$

[6] In this chapter, we change the notation for the heat supply rate from r (as was used in previous chapters) to r^t.

$$\frac{\mathrm{d}}{\mathrm{d}t}\int_{\mathcal{P}_{\mathrm{R}}}\frac{1}{2}\mathbf{v}\cdot\mathbf{v}\rho_{\mathrm{R}}\,\mathrm{d}V+\frac{\mathrm{d}}{\mathrm{d}t}\int_{\mathcal{P}_{\mathrm{R}}}\varepsilon\rho_{\mathrm{R}}\,\mathrm{d}V=\int_{\mathcal{P}_{\mathrm{R}}}(\mathbf{f}^{\mathrm{m}}+\mathbf{f}^{\mathrm{em}})\cdot\mathbf{v}\rho_{\mathrm{R}}\,\mathrm{d}V+\int_{\partial\mathcal{P}_{\mathrm{R}}}\mathbf{t}_{\mathrm{R}}\cdot\mathbf{v}\,\mathrm{d}A$$
$$+\int_{\mathcal{P}_{\mathrm{R}}}(r^{\mathrm{t}}+r^{\mathrm{em}})\rho_{\mathrm{R}}\,\mathrm{d}V-\int_{\partial\mathcal{P}_{\mathrm{R}}}h_{\mathrm{R}}\,\mathrm{d}A. \tag{9.16b}$$

Comparing (9.16a) and (9.16b) with their counterparts in the thermomechanical theory (refer to Chapter 4), we see that (9.16a) and (9.16b) contain additional energy contributions from (1) the work due to the electromagnetic body force $\mathbf{f}^{\mathrm{em}}$ and (2) the electromagnetic energy supply r^{em}.

9.1.6 SECOND LAW OF THERMODYNAMICS

In this chapter—as was done in the thermomechanical theory of Chapter 4—we adopt the Clausius-Duhem inequality as our particular statement of the second law of thermodynamics. The Clausius-Duhem inequality postulates that the rate of change of the **entropy** $\mathcal{N}$ of any subset of the body is greater than or equal to the rate of entropy generation $\mathcal{R}$ *due to the radiative heat supply* minus the rate of entropy loss $\mathcal{H}$ *due to the outward heat flux*. Applying this primitive statement of the second law to subset $\mathcal{S}_1$ yields the material form:

$$\frac{\mathrm{d}}{\mathrm{d}t}\mathcal{N}(\mathcal{S}_1,t)\geq\mathcal{R}(\mathcal{S}_1,t)-\mathcal{H}(\mathcal{S}_1,t). \tag{9.17}$$

Specializing to a continuum and assuming smoothness of $\mathcal{N}(\mathcal{S}_1,t)$, $\mathcal{R}(\mathcal{S}_1,t)$, and $\mathcal{H}(\mathcal{S}_1,t)$, we can write

$$\mathcal{N}(\mathcal{S}_1,t)=\begin{cases}\displaystyle\int_{\mathcal{P}}\eta\rho\,\mathrm{d}v,\\ \displaystyle\int_{\mathcal{P}_{\mathrm{R}}}\eta\rho_{\mathrm{R}}\,\mathrm{d}V,\end{cases}\qquad \mathcal{R}(\mathcal{S}_1,t)=\begin{cases}\displaystyle\int_{\mathcal{P}}\frac{r^{\mathrm{t}}}{\Theta}\rho\,\mathrm{d}v,\\ \displaystyle\int_{\mathcal{P}_{\mathrm{R}}}\frac{r^{\mathrm{t}}}{\Theta}\rho_{\mathrm{R}}\,\mathrm{d}V,\end{cases} \tag{9.18}$$

and

$$\mathcal{H}(\mathcal{S}_1,t)=\begin{cases}\displaystyle\int_{\partial\mathcal{P}}\frac{h}{\Theta}\,\mathrm{d}a,\\ \displaystyle\int_{\partial\mathcal{P}_{\mathrm{R}}}\frac{h_{\mathrm{R}}}{\Theta}\,\mathrm{d}A,\end{cases} \tag{9.19}$$

where η is the **specific entropy** (or entropy per unit mass) and Θ is the **absolute temperature**. The Eulerian integral representations of $\mathcal{N}(\mathcal{S}_1,t)$, $\mathcal{R}(\mathcal{S}_1,t)$, and $\mathcal{H}(\mathcal{S}_1,t)$ (top of (9.18) and (9.19)) correspond to subset $\mathcal{S}_1$ as seen in its present configuration, while the Lagrangian integral representations of these quantities (bottom of (9.18) and (9.19)) correspond to subset $\mathcal{S}_1$ as seen in its reference configuration.

Use of (9.18) and (9.19) in (9.17) leads to Eulerian and Lagrangian integral representations of the Clausius-Duhem inequality:

$$\frac{\mathrm{d}}{\mathrm{d}t}\int_{\mathcal{P}} \eta\rho \, \mathrm{d}v \geq \int_{\mathcal{P}} \frac{r^{t}}{\Theta}\rho \, \mathrm{d}v - \int_{\partial\mathcal{P}} \frac{h}{\Theta} \, \mathrm{d}a, \tag{9.20a}$$

$$\frac{\mathrm{d}}{\mathrm{d}t}\int_{\mathcal{P}_{\mathrm{R}}} \eta\rho_{\mathrm{R}} \, \mathrm{d}V \geq \int_{\mathcal{P}_{\mathrm{R}}} \frac{r^{t}}{\Theta}\rho_{\mathrm{R}} \, \mathrm{d}V - \int_{\partial\mathcal{P}_{\mathrm{R}}} \frac{h_{\mathrm{R}}}{\Theta} \, \mathrm{d}A. \tag{9.20b}$$

Note that these integral statements are valid for any open volume $\mathcal{P}$ bounded by closed surface $\partial\mathcal{P}$ in the present configuration, or corresponding open volume $\mathcal{P}_{\mathrm{R}}$ bounded by closed surface $\partial\mathcal{P}_{\mathrm{R}}$ in the reference configuration. Also note that (9.20a) and (9.20b) are identical to their counterparts in the thermomechanical theory (refer to Chapter 4).

9.1.7 CONSERVATION OF ELECTRIC CHARGE

Conservation of charge postulates that the time rate of change of the **total electric charge** (i.e., **free charge** Σ plus **bound charge** Σ_{b}) *within* any closed material surface is equal to the sum of the **free** (or **conductive**) **current** $\mathcal{J}$ and the **polarization current** $\mathcal{J}_{\mathrm{p}}$ *entering* that surface. Applying this primitive statement of the law to subset $\mathcal{S}_1$ (an open material volume bounded by a closed material surface) allows us to express conservation of charge mathematically in material form:

$$\frac{\mathrm{d}}{\mathrm{d}t}\Big(\Sigma(\mathcal{S}_1,t) + \Sigma_{\mathrm{b}}(\mathcal{S}_1,t)\Big) = \mathcal{J}(\mathcal{S}_1,t) + \mathcal{J}_{\mathrm{p}}(\mathcal{S}_1,t). \tag{9.21}$$

Loosely, free charges are unpaired and "free" to move; this motion gives rise to the conductive current. Conversely, bound charges are paired, and are thus "bound" to a particular atom. When a material experiences a spatially varying polarization, the bound charges realign; if this polarization is also time varying, a polarization current arises.

Physically, the conductive current $\mathcal{J}(\mathcal{S}_1,t)$ always enters and exits subset $\mathcal{S}_1$ at time t through its present surface $\partial\mathcal{P}$, but we are free to label this surface by its reference location $\partial\mathcal{P}_{\mathrm{R}}$ instead. Similarly, the free charge $\Sigma(\mathcal{S}_1,t)$ always resides in present volume $\mathcal{P}$, but we are free to label this volume by its reference location $\mathcal{P}_{\mathrm{R}}$ instead. Exploiting this freedom in how the geometry of $\mathcal{S}_1$ is labeled allows us to write Eulerian and Lagrangian integral representations of the free charge, bound charge, conductive current, and polarization current, i.e.,

$$\Sigma(\mathcal{S}_1,t) = \begin{cases} \displaystyle\int_{\mathcal{P}} \sigma^{*} \, \mathrm{d}v, \\ \displaystyle\int_{\mathcal{P}_{\mathrm{R}}} \sigma_{\mathrm{R}} \, \mathrm{d}V, \end{cases} \qquad \Sigma_{\mathrm{b}}(\mathcal{S}_1,t) = \begin{cases} \displaystyle -\int_{\mathcal{P}} \operatorname{div}\mathbf{p}^{*} \, \mathrm{d}v, \\ \displaystyle -\int_{\mathcal{P}_{\mathrm{R}}} \operatorname{Div}\mathbf{p}_{\mathrm{R}} \, \mathrm{d}V, \end{cases} \tag{9.22}$$

$$\mathcal{J}(\mathcal{S}_1,t)=\begin{cases}-\displaystyle\int_{\partial\mathcal{P}}\mathbf{j}^*\cdot\mathbf{n}\,\mathrm{d}a,\\-\displaystyle\int_{\partial\mathcal{P}_\mathrm{R}}\mathbf{j}_\mathrm{R}\cdot\mathbf{N}\,\mathrm{d}A,\end{cases}\qquad \mathcal{J}_\mathrm{p}(\mathcal{S}_1,t)=\begin{cases}-\dfrac{\mathrm{d}}{\mathrm{d}t}\displaystyle\int_{\partial\mathcal{P}}\mathbf{p}^*\cdot\mathbf{n}\,\mathrm{d}a,\\-\dfrac{\mathrm{d}}{\mathrm{d}t}\displaystyle\int_{\partial\mathcal{P}_\mathrm{R}}\mathbf{p}_\mathrm{R}\cdot\mathbf{N}\,\mathrm{d}A,\end{cases}\tag{9.23}$$

where $\mathbf{n}$ and $\mathbf{N}$ are outward unit normals in the present and reference configurations (refer to Section 3.6), "div" denotes the Eulerian divergence (i.e., the divergence calculated with respect to the present configuration), and "Div" denotes the Lagrangian divergence (i.e., the divergence calculated with respect to the reference configuration). Note that the minus signs in (9.23) are required to maintain consistency in our sign convention: positive $\mathcal{J}$ denotes current flowing *into* the boundary (see (9.21)), whereas positive $\mathbf{j}^*\cdot\mathbf{n}$ implies that current is flowing *out of* the boundary ($\mathbf{n}$ is an *outward* unit normal).

In (9.22) and (9.23), σ^* and σ_R are denoted the **spatial free charge density** and **referential free charge density**, $\mathbf{j}^*$ and $\mathbf{j}_\mathrm{R}$ the **spatial conductive current density** and **referential conductive current density**, and $\mathbf{p}^*$ and $\mathbf{p}_\mathrm{R}$ the **spatial electric polarization** and **referential electric polarization**. All are bounded, continuous functions of space and time. *Recall that the spatial and referential representations of a particular quantity are different since they are associated with different labels for the geometry of the subset:* σ^* has units of charge per present volume, while σ_R has units of charge per reference volume; $\mathbf{j}^*$ has units of current per present area, while $\mathbf{j}_\mathrm{R}$ has units of current per reference area; and $\mathbf{p}^*$ has units of charge per present area, while $\mathbf{p}_\mathrm{R}$ has units of charge per reference area (see Table 9.1).

σ^*, $\mathbf{j}^*$, and $\mathbf{p}^*$ are often called ***effective*** **electromagnetic fields** in the literature [33] to signify that they are measured with respect to a co-moving frame, or **rest frame**, i.e., one affixed to but not deforming with the continuum. In this book, an *effective* electromagnetic field is denoted by a superscript asterisk. In Section 9.3, we present transformations that relate the effective electromagnetic fields to the ***standard*** **electromagnetic fields**, the latter being measured with respect to a stationary frame, or **laboratory frame**, rather than a co-moving frame.

Use of (9.22) and (9.23) in (9.21) leads to Eulerian and Lagrangian integral representations of conservation of charge:

$$\frac{\mathrm{d}}{\mathrm{d}t}\int_{\mathcal{P}}\sigma^*\,\mathrm{d}v=-\int_{\partial\mathcal{P}}\mathbf{j}^*\cdot\mathbf{n}\,\mathrm{d}a,\tag{9.24a}$$

$$\frac{\mathrm{d}}{\mathrm{d}t}\int_{\mathcal{P}_\mathrm{R}}\sigma_\mathrm{R}\,\mathrm{d}V=-\int_{\partial\mathcal{P}_\mathrm{R}}\mathbf{j}_\mathrm{R}\cdot\mathbf{N}\,\mathrm{d}A.\tag{9.24b}$$

These integral statements are valid for any open volume $\mathcal{P}$ bounded by closed surface $\partial\mathcal{P}$ in the present configuration, or corresponding open volume $\mathcal{P}_\mathrm{R}$ bounded by closed surface $\partial\mathcal{P}_\mathrm{R}$ in the reference configuration. Note that (9.24a) and (9.24b) involve only free charge and free current. Also note that to perform the integrations in (9.24a) and

(9.24b), it is more natural to consider σ^* and $\mathbf{j}^*$ in their Eulerian descriptions, i.e., as functions of $\mathbf{x}$ and t, and σ_R and $\mathbf{j}_R$ in their Lagrangian descriptions, i.e., as functions of $\mathbf{X}$ and t.

9.1.8 FARADAY'S LAW

Faraday's law postulates that the time rate of change of the **magnetic flux** $\mathcal{B}$ through any open material surface is equal to and opposite the **electromotive force** $\mathcal{E}$ induced in the closed material curve bounding that surface. Mathematically, in material form for subset $\mathcal{S}_2$ (an open material surface bounded by a closed material curve), this amounts to

$$\frac{\mathrm{d}}{\mathrm{d}t}\mathcal{B}(\mathcal{S}_2,t) = -\mathcal{E}(\mathcal{S}_2,t). \tag{9.25}$$

Smoothness of $\mathcal{B}$ and of $\mathcal{E}$ imply that

$$\mathcal{B}(\mathcal{S}_2,t) = \begin{cases} \int_{\mathcal{Q}} \mathbf{b}^* \cdot \mathbf{n}\,\mathrm{d}a, \\ \int_{\mathcal{Q}_R} \mathbf{b}_R \cdot \mathbf{N}\,\mathrm{d}A, \end{cases} \qquad \mathcal{E}(\mathcal{S}_2,t) = \begin{cases} \int_{\partial\mathcal{Q}} \mathbf{e}^* \cdot \mathbf{l}\,\mathrm{d}l, \\ \int_{\partial\mathcal{Q}_R} \mathbf{e}_R \cdot \mathbf{l}_R\,\mathrm{d}L, \end{cases} \tag{9.26}$$

where dl and dL are line elements in the present and reference configurations, $\mathbf{l}$ and $\mathbf{l}_R$ are unit tangents in the present and reference configurations, $\mathbf{b}^*$ and $\mathbf{b}_R$ are the **spatial magnetic flux density** and **referential magnetic flux density** (or **magnetic induction**), and $\mathbf{e}^*$ and $\mathbf{e}_R$ are the **spatial electric field** and **referential electric field**. Refer to Table 9.1 for their respective units. Use of (9.26) in (9.25) leads to Eulerian and Lagrangian representations of Faraday's law:

$$\frac{\mathrm{d}}{\mathrm{d}t}\int_{\mathcal{Q}} \mathbf{b}^* \cdot \mathbf{n}\,\mathrm{d}a = -\int_{\partial\mathcal{Q}} \mathbf{e}^* \cdot \mathbf{l}\,\mathrm{d}l, \tag{9.27a}$$

$$\frac{\mathrm{d}}{\mathrm{d}t}\int_{\mathcal{Q}_R} \mathbf{b}_R \cdot \mathbf{N}\,\mathrm{d}A = -\int_{\partial\mathcal{Q}_R} \mathbf{e}_R \cdot \mathbf{l}_R\,\mathrm{d}L. \tag{9.27b}$$

Note that these integral statements are valid for any open surface $\mathcal{Q}$ bounded by a closed curve $\partial\mathcal{Q}$ in the present configuration, or corresponding open surface $\mathcal{Q}_R$ bounded by a closed curve $\partial\mathcal{Q}_R$ in the reference configuration.

9.1.9 GAUSS'S LAW FOR MAGNETISM

Gauss's law for magnetism (a statement of conservation of magnetic flux, or, alternatively, the absence of magnetic monopoles) postulates that the **magnetic flux** $\mathcal{B}$ through any closed material surface is zero, i.e.,

$$\mathcal{B}(\mathcal{S}_1,t) = 0. \tag{9.28}$$

Assuming that the magnetic flux $\mathcal{B}(\mathcal{S}_1, t)$ through the surface of $\mathcal{S}_1$ (recall that $\mathcal{S}_1$ consists of an open volume bounded by a closed surface) at time t is smooth, we can write

$$\mathcal{B}(\mathcal{S}_1, t) = \begin{cases} \displaystyle\int_{\partial\mathcal{P}} \mathbf{b}^* \cdot \mathbf{n}\, da, \\ \displaystyle\int_{\partial\mathcal{P}_R} \mathbf{b}_R \cdot \mathbf{N}\, dA. \end{cases} \tag{9.29}$$

Use of (9.29) in (9.28) gives Eulerian and Lagrangian representations of Gauss's law for magnetism:

$$\int_{\partial\mathcal{P}} \mathbf{b}^* \cdot \mathbf{n}\, da = 0, \tag{9.30a}$$

$$\int_{\partial\mathcal{P}_R} \mathbf{b}_R \cdot \mathbf{N}\, dA = 0. \tag{9.30b}$$

9.1.10 GAUSS'S LAW FOR ELECTRICITY

Gauss's law for electricity postulates that the **electric flux** $\mathcal{F}$ through any closed material surface is proportional to the **total electric charge** (i.e., **free charge** Σ plus **bound charge** Σ_b) *enclosed within* that surface, i.e.,

$$\mathcal{F}(\mathcal{S}_1, t) = \frac{\Sigma(\mathcal{S}_1, t) + \Sigma_b(\mathcal{S}_1, t)}{\epsilon_o}, \tag{9.31}$$

where ϵ_o is the **electric permittivity** in vacuo. Assuming that the electric flux $\mathcal{F}$ is smooth implies that

$$\mathcal{F}(\mathcal{S}_1, t) = \begin{cases} \displaystyle\int_{\partial\mathcal{P}} \mathbf{e}^* \cdot \mathbf{n}\, da, \\ \displaystyle\int_{\partial\mathcal{P}_R} J\mathbf{C}^{-1}\mathbf{e}_R \cdot \mathbf{N}\, dA, \end{cases} \tag{9.32}$$

where J is the determinant of the deformation gradient $\mathbf{F}$, and $\mathbf{C}^{-1}$ is the inverse of the right Cauchy-Green deformation tensor $\mathbf{C} = \mathbf{F}^T\mathbf{F}$. Recall that the free charge Σ and the bound charge Σ_b associated with subset $\mathcal{S}_1$ at time t are given in (9.22). Subsequent use of (9.22) and (9.32) in (9.31) leads to Eulerian and Lagrangian representations of Gauss's law for electricity:

$$\int_{\partial\mathcal{P}} \mathbf{d}^* \cdot \mathbf{n}\, da = \int_{\mathcal{P}} \sigma^*\, dv, \tag{9.33a}$$

$$\int_{\partial\mathcal{P}_R} \mathbf{d}_R \cdot \mathbf{N}\, dA = \int_{\mathcal{P}_R} \sigma_R\, dV, \tag{9.33b}$$

where the **spatial electric displacement** $\mathbf{d}^*$ and **referential electric displacement** $\mathbf{d}_{\mathrm{R}}$ are introduced through the algebraic relationships (see, for instance, [37])

$$\mathbf{d}^* = \mathbf{p}^* + \epsilon_{\mathrm{o}}\, \mathbf{e}^*, \quad \mathbf{d}_{\mathrm{R}} = \mathbf{p}_{\mathrm{R}} + \epsilon_{\mathrm{o}} J \mathbf{C}^{-1} \mathbf{e}_{\mathrm{R}}. \tag{9.34}$$

9.1.11 AMPÈRE-MAXWELL LAW

The Ampère-Maxwell law postulates that the time rate of change of the **electric flux** $\mathcal{F}$ through any open material surface plus the **conductive current** $\mathcal{J}$, **polarization current** $\mathcal{J}_{\mathrm{p}}$, and **magnetization current** $\mathcal{J}_{\mathrm{m}}$ passing through that surface is proportional to the **magnetic field** $\mathcal{T}$ around the closed material curve bounding that surface. Application of this primitive statement of the law to subset $\mathcal{S}_2$ (an open material surface bounded by a closed material curve) allows us to express the Ampère-Maxwell law mathematically in material form:

$$\mu_{\mathrm{o}} \left(\epsilon_{\mathrm{o}} \frac{\mathrm{d}}{\mathrm{d}t} \mathcal{F}(\mathcal{S}_2, t) + \mathcal{J}(\mathcal{S}_2, t) + \mathcal{J}_{\mathrm{p}}\,(\mathcal{S}_2, t) + \mathcal{J}_{\mathrm{m}}\,(\mathcal{S}_2, t) \right) = \mathcal{T}(\mathcal{S}_2, t), \tag{9.35}$$

where μ_{o} is the **magnetic permeability** in vacuo. Smoothness allows us to write

$$\mathcal{F}(\mathcal{S}_2, t) = \begin{cases} \displaystyle\int_{\mathcal{Q}} \mathbf{e}^* \cdot \mathbf{n}\, \mathrm{d}a, \\ \displaystyle\int_{\mathcal{Q}_{\mathrm{R}}} J \mathbf{C}^{-1} \mathbf{e}_{\mathrm{R}} \cdot \mathbf{N}\, \mathrm{d}A, \end{cases} \qquad \mathcal{J}(\mathcal{S}_2, t) = \begin{cases} \displaystyle\int_{\mathcal{Q}} \mathbf{j}^* \cdot \mathbf{n}\, \mathrm{d}a, \\ \displaystyle\int_{\mathcal{Q}_{\mathrm{R}}} \mathbf{j}_{\mathrm{R}} \cdot \mathbf{N}\, \mathrm{d}A, \end{cases} \tag{9.36}$$

$$\mathcal{J}_{\mathrm{p}}\,(\mathcal{S}_2, t) = \begin{cases} \displaystyle\frac{\mathrm{d}}{\mathrm{d}t} \int_{\mathcal{Q}} \mathbf{p}^* \cdot \mathbf{n}\, \mathrm{d}a, \\ \displaystyle\frac{\mathrm{d}}{\mathrm{d}t} \int_{\mathcal{Q}_{\mathrm{R}}} \mathbf{p}_{\mathrm{R}} \cdot \mathbf{N}\, \mathrm{d}A, \end{cases} \qquad \mathcal{J}_{\mathrm{m}}\,(\mathcal{S}_2, t) = \begin{cases} \displaystyle\int_{\mathcal{Q}} (\mathrm{curl}\, \mathbf{m}^*) \cdot \mathbf{n}\, \mathrm{d}a, \\ \displaystyle\int_{\mathcal{Q}_{\mathrm{R}}} (\mathrm{Curl}\, \mathbf{m}_{\mathrm{R}}) \cdot \mathbf{N}\, \mathrm{d}A, \end{cases} \tag{9.37}$$

$$\mathcal{T}(\mathcal{S}_2, t) = \begin{cases} \displaystyle\int_{\partial\mathcal{Q}} \mathbf{b}^* \cdot \mathbf{l}\, \mathrm{d}l, \\ \displaystyle\int_{\partial\mathcal{Q}_{\mathrm{R}}} \frac{1}{J} \mathbf{C} \mathbf{b}_{\mathrm{R}} \cdot \mathbf{l}_{\mathrm{R}}\, \mathrm{d}L, \end{cases} \tag{9.38}$$

where $\mathbf{m}^*$ and $\mathbf{m}_{\mathrm{R}}$ are the **spatial magnetization** and **referential magnetization** (or **magnetic polarization**), "curl" is the Eulerian curl (i.e., the curl calculated with respect to the present configuration), and "Curl" is the Lagrangian curl (i.e., the curl calculated with respect to the reference configuration). Use of (9.36)–(9.38) in (9.35) leads to

$$\frac{\mathrm{d}}{\mathrm{d}t} \int_{\mathcal{Q}} \mathbf{d}^* \cdot \mathbf{n}\, \mathrm{d}a + \int_{\mathcal{Q}} \mathbf{j}^* \cdot \mathbf{n}\, \mathrm{d}a = \int_{\partial\mathcal{Q}} \mathbf{h}^* \cdot \mathbf{l}\, \mathrm{d}l, \tag{9.39a}$$

$$\frac{\mathrm{d}}{\mathrm{d}t} \int_{\mathcal{Q}_{\mathrm{R}}} \mathbf{d}_{\mathrm{R}} \cdot \mathbf{N}\, \mathrm{d}A + \int_{\mathcal{Q}_{\mathrm{R}}} \mathbf{j}_{\mathrm{R}} \cdot \mathbf{N}\, \mathrm{d}A = \int_{\partial\mathcal{Q}_{\mathrm{R}}} \mathbf{h}_{\mathrm{R}} \cdot \mathbf{l}_{\mathrm{R}}\, \mathrm{d}L, \tag{9.39b}$$

where we have used (9.34) to introduce $\mathbf{d}^*$ and $\mathbf{d}_R$, and the algebraic relationships (see, for instance, [36])

$$\mathbf{h}^* = \frac{1}{\mu_0}\mathbf{b}^* - \mathbf{m}^*, \quad \mathbf{h}_R = \frac{1}{\mu_0 J}\mathbf{C}\mathbf{b}_R - \mathbf{m}_R \tag{9.40}$$

to introduce the **spatial magnetic field $\mathbf{h}^*$** and **referential magnetic field $\mathbf{h}_R$**.

9.1.12 TRANSFORMATIONS BETWEEN SPATIAL AND REFERENTIAL TEMM QUANTITIES

The spatial (Eulerian) and referential (Lagrangian) TEMM quantities appearing in Sections 9.1.2–9.1.11, along with their corresponding units, are listed in Table 9.1. (In Table 9.1, we employ the following notation for the fundamental units: M is mass, L_P is present length, L_R is reference length, T is time, θ is temperature, and C is charge.)[7] It can be shown that these spatial and referential quantities are related through the following linear algebraic transformations:

$$\begin{aligned} &\mathbf{e}_R = \mathbf{F}^T\mathbf{e}^*, && \mathbf{p}_R = J\mathbf{F}^{-1}\mathbf{p}^*, && \mathbf{d}_R = J\mathbf{F}^{-1}\mathbf{d}^*, && \sigma_R = J\sigma^*, \\ &\mathbf{h}_R = \mathbf{F}^T\mathbf{h}^*, && \mathbf{m}_R = \mathbf{F}^T\mathbf{m}^*, && \mathbf{b}_R = J\mathbf{F}^{-1}\mathbf{b}^*, && \rho_R = J\rho, \\ &\mathbf{j}_R = J\mathbf{F}^{-1}\mathbf{j}^*, && \mathbf{P} = J\mathbf{T}\mathbf{F}^{-T}, && \mathbf{q}_R = J\mathbf{F}^{-1}\mathbf{q}. \end{aligned} \tag{9.41}$$

Recall that J is the determinant of the deformation gradient $\mathbf{F}$, and is a measure of dilatation or volume change (refer to Section 3.6). Several of these results—namely, $(9.41)_8$, $(9.41)_{10}$, and $(9.41)_{11}$—were obtained in Sections 4.10 and 4.12. The other transformations can be obtained in a similar manner; refer to Problems 9.1–9.3.

PROBLEM 9.1

Verify that $\mathbf{e}_R = \mathbf{F}^T\mathbf{e}^*$.

Solution

Recall from $(9.26)_2$ that depending on whether we label the closed curve enclosing subset $\mathcal{S}_2$ by its present location $\partial\mathcal{Q}$ or its reference location $\partial\mathcal{Q}_R$, the electromotive force $\mathcal{E}$ induced in the boundary of $\mathcal{S}_2$ has the following Eulerian and Lagrangian integral representations:

$$\mathcal{E}(\mathcal{S}_2, t) = \begin{cases} \displaystyle\int_{\partial\mathcal{Q}} \mathbf{e}^* \cdot \mathbf{l}\, dl, \\ \displaystyle\int_{\partial\mathcal{Q}_R} \mathbf{e}_R \cdot \mathbf{l}_R\, dL. \end{cases}$$

[7] When validating the dimensional homogeneity of an equation in the finite-deformation theory, one finds it useful to differentiate the fundamental units of length from one another, with L_P corresponding to the present configuration and L_R corresponding to the reference configuration.

It follows that

$$\int_{\partial\mathcal{Q}} \mathbf{e}^* \cdot \mathbf{l}\, dl = \int_{\partial\mathcal{Q}_R} \mathbf{e}_R \cdot \mathbf{l}_R\, dL. \tag{a}$$

Upon a change of independent variable from $\mathbf{x}$ to $\mathbf{X}$, the left-hand side of (a) becomes

$$\int_{\partial\mathcal{Q}} \mathbf{e}^* \cdot \mathbf{l}\, dl = \int_{\partial\mathcal{Q}_R} \mathbf{e}^* \cdot \mathbf{F}\mathbf{l}_R\, dL, \tag{b}$$

where we have used

$$d\mathbf{l} = \mathbf{F} d\mathbf{l}_R,$$

i.e., the deformation gradient $\mathbf{F}$ linearly maps each line element $d\mathbf{l}_R = \mathbf{l}_R\, dL$ in the reference configuration into a line element $d\mathbf{l} = \mathbf{l}\, dl$ in the present configuration. (Refer to (3.28), and recall that $\mathbf{l}_R$ and $\mathbf{l}$ are unit tangents in the reference and present configurations, and dL and dl are the infinitesimal lengths of the line elements.) Substitution of (b) into (a), and subsequent use of the definition (2.13) of the transpose of a tensor, leads to

$$\int_{\partial\mathcal{Q}_R} \left(\mathbf{F}^T\mathbf{e}^* - \mathbf{e}_R\right) \cdot \mathbf{l}_R\, dL = 0.$$

Since the integrand is continuous and $\partial\mathcal{Q}_R$ is arbitrary, the localization theorem in Section 4.5.2 implies that

$$\left(\mathbf{F}^T\mathbf{e}^* - \mathbf{e}_R\right) \cdot \mathbf{l}_R = 0.$$

Since the coefficient of $\mathbf{l}_R$ is independent of $\mathbf{l}_R$, and $\mathbf{l}_R$ is arbitrary, it follows that

$$\mathbf{e}_R = \mathbf{F}^T\mathbf{e}^*.$$

An alternative proof that starts with the electric flux $\mathcal{F}$ (refer to (9.32)) instead of the electromotive force $\mathcal{E}$ (refer to $(9.26)_2$) is left as an exercise for the reader.

PROBLEM 9.2

Verify that $\sigma_R = J\sigma^*$.

Solution

Recall from $(9.22)_1$ that depending on whether we label the volume occupied by subset $\mathcal{S}_1$ by its present location $\mathcal{P}$ or its reference location $\mathcal{P}_R$, the free charge Σ within $\mathcal{S}_1$ has the following Eulerian and Lagrangian integral representations:

$$\Sigma(\mathcal{S}_1, t) = \begin{cases} \int_{\mathcal{P}} \sigma^* \, dv, \\ \int_{\mathcal{P}_R} \sigma_R \, dV. \end{cases}$$

It follows that

$$\int_{\mathcal{P}} \sigma^* \, dv = \int_{\mathcal{P}_R} \sigma_R \, dV. \tag{a}$$

The left-hand side of (a), after a change of independent variable from $\mathbf{x}$ to $\mathbf{X}$ and use of the relationship $dv = J \, dV$ (refer to (3.71)), becomes

$$\int_{\mathcal{P}} \sigma^* \, dv = \int_{\mathcal{P}_R} \sigma^* J \, dV. \tag{b}$$

Substitution of (b) into (a) and use of the localization theorem leads to

$$\sigma_R = J\sigma^*.$$

PROBLEM 9.3

Verify that $\mathbf{b}_R = J\mathbf{F}^{-1}\mathbf{b}^*$.

Solution

Recall from (9.29) that depending on whether we label the closed surface bounding subset $\mathcal{S}_1$ by its present location $\partial\mathcal{P}$ or its reference location $\partial\mathcal{P}_R$, the magnetic flux $\mathcal{B}$ through the boundary of $\mathcal{S}_1$ has the following Eulerian and Lagrangian integral representations:

$$\mathcal{B}(\mathcal{S}_1, t) = \begin{cases} \int_{\partial\mathcal{P}} \mathbf{b}^* \cdot \mathbf{n} \, da, \\ \int_{\partial\mathcal{P}_R} \mathbf{b}_R \cdot \mathbf{N} \, dA. \end{cases}$$

It follows that

$$\int_{\partial\mathcal{P}} \mathbf{b}^* \cdot \mathbf{n} \, da = \int_{\partial\mathcal{P}_R} \mathbf{b}_R \cdot \mathbf{N} \, dA. \tag{a}$$

Upon a change of independent variable from $\mathbf{X}$ to $\mathbf{x}$, the right-hand side of (a) becomes

$$\int_{\partial\mathcal{P}_R} \mathbf{b}_R \cdot \mathbf{N} \, dA = \int_{\partial\mathcal{P}} \frac{1}{J} \mathbf{b}_R \cdot \mathbf{F}^{\mathrm{T}} \mathbf{n} \, da, \tag{b}$$

where we have used the relationship

$$J\mathbf{N}\,\mathrm{d}A = \mathbf{F}^{\mathrm{T}}\mathbf{n}\,\mathrm{d}a$$

from Section 3.6. Substitution of (b) into (a), and subsequent use of the definition (2.13) of the transpose of a tensor, leads to

$$\int_{\partial\mathcal{P}} \left(\mathbf{b}^* - \frac{1}{J}\mathbf{F}\mathbf{b}_{\mathrm{R}}\right)\cdot\mathbf{n}\,\mathrm{d}a = 0.$$

The localization theorem then implies that

$$\left(\mathbf{b}^* - \frac{1}{J}\mathbf{F}\mathbf{b}_{\mathrm{R}}\right)\cdot\mathbf{n} = 0.$$

Since the coefficient of **n** is independent of **n**, and **n** is arbitrary, it follows that

$$\mathbf{b}_{\mathrm{R}} = J\mathbf{F}^{-1}\mathbf{b}^*.$$

Note that this relationship can also be obtained starting from the Eulerian and Lagrangian integral representations of the magnetic field $\mathcal{T}$ in (9.38), an exercise that we leave to the reader.

EXERCISES

1. Confirm that use of the free charge $(9.22)_1$, bound charge $(9.22)_2$, conductive current $(9.23)_1$, and polarization current $(9.23)_2$ in the material form of conservation of charge (9.21) leads to the integral forms (9.24a) and (9.24b).

2. Verify that use of the free charge $(9.22)_1$, bound charge $(9.22)_2$, and electric flux (9.32) in the material form of Gauss's law for electricity (9.31) leads to the integral forms (9.33a) and (9.33b).

3. Demonstrate that use of (9.36)–(9.38) in the material form of the Ampère-Maxwell law (9.35) leads to the integral forms (9.39a) and (9.39b). (Hint: You will need to use Stokes's theorem (9.44) to convert the magnetization current $(9.37)_2$ from a surface integral to a line integral.)

4. Prove all transformations in (9.41) that remain unverified.

5. Using the *fundamental units* provided in Table 9.1, verify that the following equations are dimensionally homogeneous (i.e., all terms in the equation have the same units):
   (a) The Eulerian and Lagrangian integral forms of the electromagnetic balance laws.
   (b) The algebraic relationships (9.34) and (9.40).
   (c) The transformations (9.41).

9.2 THE FUNDAMENTAL LAWS OF CONTINUUM ELECTRODYNAMICS: POINTWISE FORMS

In this section, we derive pointwise versions of the Eulerian and Lagrangian integral balance laws presented in Section 9.1.

9.2.1 EULERIAN FUNDAMENTAL LAWS

We begin by recalling the Eulerian integral forms of the first principles developed in Section 9.1:

Conservation of mass

$$\frac{\mathrm{d}}{\mathrm{d}t}\int_{\mathcal{P}} \rho \,\mathrm{d}v = 0, \tag{9.42a}$$

Balance of linear momentum

$$\frac{\mathrm{d}}{\mathrm{d}t}\int_{\mathcal{P}} \mathbf{v}\rho \,\mathrm{d}v = \int_{\mathcal{P}} (\mathbf{f}^{\mathrm{m}} + \mathbf{f}^{\mathrm{em}})\rho \,\mathrm{d}v + \int_{\partial\mathcal{P}} \mathbf{t} \,\mathrm{d}a, \tag{9.42b}$$

Balance of angular momentum

$$\frac{\mathrm{d}}{\mathrm{d}t}\int_{\mathcal{P}} \mathbf{x}\times\mathbf{v}\rho \,\mathrm{d}v = \int_{\mathcal{P}} \mathbf{x}\times(\mathbf{f}^{\mathrm{m}} + \mathbf{f}^{\mathrm{em}})\rho \,\mathrm{d}v + \int_{\partial\mathcal{P}} \mathbf{x}\times\mathbf{t} \,\mathrm{d}a + \int_{\mathcal{P}} \mathbf{c}^{\mathrm{em}}\rho \,\mathrm{d}v, \tag{9.42c}$$

First law of thermodynamics

$$\begin{aligned}\frac{\mathrm{d}}{\mathrm{d}t}\int_{\mathcal{P}} \frac{1}{2}\mathbf{v}\cdot\mathbf{v}\rho \,\mathrm{d}v + \frac{\mathrm{d}}{\mathrm{d}t}\int_{\mathcal{P}} \varepsilon\rho \,\mathrm{d}v &= \int_{\mathcal{P}} (\mathbf{f}^{\mathrm{m}} + \mathbf{f}^{\mathrm{em}})\cdot\mathbf{v}\rho \,\mathrm{d}v + \int_{\partial\mathcal{P}} \mathbf{t}\cdot\mathbf{v} \,\mathrm{d}a \\ &\quad + \int_{\mathcal{P}} (r^{\mathrm{t}} + r^{\mathrm{em}})\rho \,\mathrm{d}v - \int_{\partial\mathcal{P}} h \,\mathrm{d}a,\end{aligned} \tag{9.42d}$$

Second law of thermodynamics

$$\frac{\mathrm{d}}{\mathrm{d}t}\int_{\mathcal{P}} \eta\rho \,\mathrm{d}v \geq \int_{\mathcal{P}} \frac{r^{\mathrm{t}}}{\Theta}\rho \,\mathrm{d}v - \int_{\partial\mathcal{P}} \frac{h}{\Theta} \,\mathrm{d}a, \tag{9.42e}$$

Conservation of electric charge

$$\frac{\mathrm{d}}{\mathrm{d}t}\int_{\mathcal{P}} \sigma^{*} \,\mathrm{d}v = -\int_{\partial\mathcal{P}} \mathbf{j}^{*}\cdot\mathbf{n} \,\mathrm{d}a, \tag{9.42f}$$

Gauss's law for magnetism

$$\int_{\partial\mathcal{P}} \mathbf{b}^{*}\cdot\mathbf{n} \,\mathrm{d}a = 0, \tag{9.42g}$$

Faraday's law

$$\frac{\mathrm{d}}{\mathrm{d}t}\int_{\mathcal{Q}} \mathbf{b}^{*}\cdot\mathbf{n} \,\mathrm{d}a = -\int_{\partial\mathcal{Q}} \mathbf{e}^{*}\cdot\mathbf{l} \,\mathrm{d}l, \tag{9.42h}$$

Gauss's law for electricity

$$\int_{\partial\mathcal{P}} \mathbf{d}^* \cdot \mathbf{n}\, da = \int_{\mathcal{P}} \sigma^*\, dv, \tag{9.42i}$$

Ampère-Maxwell law

$$\frac{d}{dt}\int_{\mathcal{Q}} \mathbf{d}^* \cdot \mathbf{n}\, da + \int_{\mathcal{Q}} \mathbf{j}^* \cdot \mathbf{n}\, da = \int_{\partial\mathcal{Q}} \mathbf{h}^* \cdot \mathbf{l}\, dl. \tag{9.42j}$$

To obtain pointwise versions of the integral equations (9.42a)–(9.42j), we use tools similar to those employed in Section 4.8, including the transport theorem for *volume integrals* (refer to Section 4.5.1), the divergence theorem (refer to (2.104)), and the localization theorem (refer to Section 4.5.2). Several additional tools that will prove useful in this section include the **transport theorem** for *surface integrals* (refer to Problem 9.4)

$$\frac{d}{dt}\int_{\mathcal{Q}} \mathbf{a} \cdot \mathbf{n}\, da = \int_{\mathcal{Q}} \left[\mathbf{a}' + \mathrm{curl}\,(\mathbf{a} \times \mathbf{v}) + \mathbf{v}(\mathrm{div}\,\mathbf{a})\right] \cdot \mathbf{n}\, da \tag{9.43}$$

and **Stokes's theorem**

$$\int_{\partial\mathcal{Q}} \mathbf{a} \cdot \mathbf{l}\, dl = \int_{\mathcal{Q}} (\mathrm{curl}\,\mathbf{a}) \cdot \mathbf{n}\, da, \tag{9.44}$$

where $\mathbf{a} = \tilde{\mathbf{a}}(\mathbf{x}, t)$ is an arbitrary vector-valued function of present position $\mathbf{x}$ and time t, $\mathbf{v}$ is the velocity, "div" denotes the Eulerian divergence (i.e., the divergence calculated with respect to the present configuration), "curl" denotes the Eulerian curl, and

$$\mathbf{a}' = \frac{\partial}{\partial t}\tilde{\mathbf{a}}(\mathbf{x}, t)$$

denotes the Eulerian time derivative, i.e., the partial derivative of the spatial description of $\mathbf{a}$ with respect to time t. Also useful are the relations

$$\mathbf{t} = \mathbf{T}\mathbf{n}, \quad h = \mathbf{q} \cdot \mathbf{n}, \tag{9.45}$$

where $\mathbf{T}$ is the **Cauchy stress** and $\mathbf{q}$ is the **spatial heat flux vector**. Note that the proofs of $(9.45)_1$ and $(9.45)_2$ in a thermo-electro-magneto-mechanical setting are essentially identical to the corresponding proofs in a thermomechanical setting (refer to Section 4.6). With these tools in hand, it can be shown (refer, for instance, to Problem 9.5) that the pointwise variants of the Eulerian integral equations (9.42a)–(9.42j) are

$$\dot{\rho} + \rho\, \mathrm{div}\,\mathbf{v} = 0, \tag{9.46a}$$

$$\rho\dot{\mathbf{v}} = \rho\left(\mathbf{f}^{\mathrm{m}} + \mathbf{f}^{\mathrm{em}}\right) + \mathrm{div}\mathbf{T}, \tag{9.46b}$$

$$\rho\boldsymbol{\Gamma}^{\mathrm{em}} + \mathbf{T} - \mathbf{T}^{\mathrm{T}} = \mathbf{0}, \tag{9.46c}$$

$$\rho\dot{\varepsilon} = \mathbf{T} \cdot \mathbf{L} + \rho\left(r^{\mathrm{t}} + r^{\mathrm{em}}\right) - \mathrm{div}\,\mathbf{q}, \tag{9.46d}$$

$$\rho\dot{\eta} \geq \rho\frac{r^{\mathrm{t}}}{\Theta} - \mathrm{div}\left(\frac{\mathbf{q}}{\Theta}\right), \tag{9.46e}$$

$$\dot{\sigma}^* + \sigma^* \,\mathrm{div}\,\mathbf{v} + \mathrm{div}\,\mathbf{j}^* = 0, \tag{9.46f}$$

$$\mathrm{div}\,\mathbf{b}^* = 0, \tag{9.46g}$$

$$\mathrm{curl}\,\mathbf{e}^* = -\left(\mathbf{b}^*\right)' - \mathrm{curl}\,(\mathbf{b}^* \times \mathbf{v}), \tag{9.46h}$$

$$\mathrm{div}\,\mathbf{d}^* = \sigma^*, \tag{9.46i}$$

$$\mathrm{curl}\,\mathbf{h}^* = \left(\mathbf{d}^*\right)' + \mathrm{curl}\,(\mathbf{d}^* \times \mathbf{v}) + \sigma^*\mathbf{v} + \mathbf{j}^*, \tag{9.46j}$$

where $\mathbf{L} = \mathrm{grad}\,\mathbf{v}$ is the Eulerian velocity gradient, $\mathbf{\Gamma}^{\mathrm{em}}$ is a skew tensor whose corresponding axial vector is $\mathbf{c}^{\mathrm{em}}$, i.e., $\mathbf{\Gamma}^{\mathrm{em}}\mathbf{a} = \mathbf{c}^{\mathrm{em}} \times \mathbf{a}$ for any vector $\mathbf{a}$, and

$$\dot{\mathbf{a}} = \mathbf{a}' + (\mathbf{v} \cdot \mathrm{grad})\,\mathbf{a}$$

denotes the material time derivative of an arbitrary vector $\mathbf{a} = \tilde{\mathbf{a}}(\mathbf{x}, t)$. Note that in deriving the pointwise version of the first law of thermodynamics (9.46d) from its integral counterpart (9.42d), we have made use of the Eulerian form of the **energy theorem** for continuum electrodynamics:

$$\int_{\mathcal{P}} (\mathbf{f}^{\mathrm{m}} + \mathbf{f}^{\mathrm{em}}) \cdot \mathbf{v}\rho\,\mathrm{d}v + \int_{\partial\mathcal{P}} \mathbf{t} \cdot \mathbf{v}\,\mathrm{d}a - \frac{\mathrm{d}}{\mathrm{d}t}\int_{\mathcal{P}} \frac{1}{2}\mathbf{v} \cdot \mathbf{v}\rho\,\mathrm{d}v = \int_{\mathcal{P}} \mathbf{T} \cdot \mathbf{L}\,\mathrm{d}v. \tag{9.47}$$

Compare (9.47) with (4.33), the Eulerian form of the energy theorem for mechanics.

It is important to note that the electromagnetic equations (9.46f)–(9.46j) are not independent. For instance, it can be shown that conservation of charge (9.46f) is implicitly contained in **Maxwell's equations** (9.46g)–(9.46j); refer to Problem 9.6.

PROBLEM 9.4

Prove the transport theorem for surface integrals. That is, show that

$$\frac{\mathrm{d}}{\mathrm{d}t}\int_{\mathcal{Q}} \mathbf{a} \cdot \mathbf{n}\,\mathrm{d}a = \int_{\mathcal{Q}} \left[\mathbf{a}' + \mathrm{curl}\,(\mathbf{a} \times \mathbf{v}) + \mathbf{v}(\mathrm{div}\,\mathbf{a})\right] \cdot \mathbf{n}\,\mathrm{d}a.$$

Solution

We begin by using relationship (3.76) between $\mathbf{n}\,\mathrm{d}a$ and $\mathbf{N}\,\mathrm{d}A$, together with a change of independent variable from $\mathbf{x}$ to $\mathbf{X}$, to convert the Eulerian integration to a Lagrangian integration, i.e.,

$$\frac{\mathrm{d}}{\mathrm{d}t}\int_{\mathcal{Q}} \mathbf{a} \cdot \mathbf{n}\,\mathrm{d}a = \frac{\mathrm{d}}{\mathrm{d}t}\int_{\mathcal{Q}_{\mathrm{R}}} \mathbf{a} \cdot J\mathbf{F}^{-\mathrm{T}}\mathbf{N}\,\mathrm{d}A.$$

We emphasize that the integrand on the left-hand side is a function of $\mathbf{x}$ and t, while the integrand on the right-hand side is a function of $\mathbf{X}$ and t. Then, it follows that

$$\frac{\mathrm{d}}{\mathrm{d}t}\int_{\mathcal{Q}} \mathbf{a}\cdot\mathbf{n}\,\mathrm{d}a = \frac{\mathrm{d}}{\mathrm{d}t}\int_{\mathcal{Q}_\mathrm{R}} J\mathbf{F}^{-1}\mathbf{a}\cdot\mathbf{N}\,\mathrm{d}A \qquad \text{(definition (2.13))}$$

$$= \int_{\mathcal{Q}_\mathrm{R}} \overline{J\mathbf{F}^{-1}\mathbf{a}}^{\,\boldsymbol{\cdot}}\cdot\mathbf{N}\,\mathrm{d}A \qquad \text{(fixed region of integration } \mathcal{Q}_\mathrm{R})$$

$$= \int_{\mathcal{Q}_\mathrm{R}} \left(\dot{J}\mathbf{F}^{-1}\mathbf{a} + J\overline{\mathbf{F}^{-1}}^{\,\boldsymbol{\cdot}}\mathbf{a} + J\mathbf{F}^{-1}\dot{\mathbf{a}}\right)\cdot\mathbf{N}\,\mathrm{d}A \qquad \text{(product rule (3.22)).}$$

We now individually examine the three terms on the right-hand side of the above equation. For the first term, we have

$$\int_{\mathcal{Q}_\mathrm{R}} \dot{J}\mathbf{F}^{-1}\mathbf{a}\cdot\mathbf{N}\,\mathrm{d}A = \int_{\mathcal{Q}_\mathrm{R}} J(\mathrm{div}\,\mathbf{v})\mathbf{F}^{-1}\mathbf{a}\cdot\mathbf{N}\,\mathrm{d}A \qquad \text{(result (3.60)}_3)$$

$$= \int_{\mathcal{Q}_\mathrm{R}} \mathbf{a}(\mathrm{div}\,\mathbf{v})\cdot J\mathbf{F}^{-\mathrm{T}}\mathbf{N}\mathrm{d}A \qquad \text{(definition (2.13)).}$$

For the second term, we have

$$\int_{\mathcal{Q}_\mathrm{R}} J\overline{\mathbf{F}^{-1}}^{\,\boldsymbol{\cdot}}\mathbf{a}\cdot\mathbf{N}\,\mathrm{d}A = -\int_{\mathcal{Q}_\mathrm{R}} J\mathbf{F}^{-1}\mathbf{L}\mathbf{a}\cdot\mathbf{N}\,\mathrm{d}A \qquad \text{(result (3.61)}_2)$$

$$= -\int_{\mathcal{Q}_\mathrm{R}} \mathbf{L}\mathbf{a}\cdot J\mathbf{F}^{-\mathrm{T}}\mathbf{N}\,\mathrm{d}A \qquad \text{(definition (2.13))}$$

$$= -\int_{\mathcal{Q}_\mathrm{R}} (\mathrm{grad}\,\mathbf{v})\mathbf{a}\cdot J\mathbf{F}^{-\mathrm{T}}\mathbf{N}\,\mathrm{d}A \qquad \text{(definition (3.56)).}$$

For the third term, we have

$$\int_{\mathcal{Q}_\mathrm{R}} J\mathbf{F}^{-1}\dot{\mathbf{a}}\cdot\mathbf{N}\,\mathrm{d}A = \int_{\mathcal{Q}_\mathrm{R}} \dot{\mathbf{a}}\cdot J\mathbf{F}^{-\mathrm{T}}\mathbf{N}\,\mathrm{d}A \qquad \text{(definition (2.13)).}$$

Assembling the preceding results yields

$$\frac{\mathrm{d}}{\mathrm{d}t}\int_{\mathcal{Q}} \mathbf{a}\cdot\mathbf{n}\,\mathrm{d}a = \int_{\mathcal{Q}_\mathrm{R}} \left(\dot{\mathbf{a}} + \mathbf{a}(\mathrm{div}\,\mathbf{v}) - (\mathrm{grad}\,\mathbf{v})\mathbf{a}\right)\cdot J\mathbf{F}^{-\mathrm{T}}\mathbf{N}\,\mathrm{d}A.$$

We convert the Lagrangian integration to an Eulerian integration with a change of independent variable from $\mathbf{X}$ to $\mathbf{x}$ and use of the relationship (3.76) between $\mathbf{N}\,\mathrm{d}A$ and $\mathbf{n}\,\mathrm{d}a$:

$$\frac{\mathrm{d}}{\mathrm{d}t}\int_{Q} \mathbf{a}\cdot\mathbf{n}\,\mathrm{d}a = \int_{Q} \left(\dot{\mathbf{a}} + \mathbf{a}(\operatorname{div}\mathbf{v}) - (\operatorname{grad}\mathbf{v})\mathbf{a}\right)\cdot\mathbf{n}\,\mathrm{d}a.$$

Use of results (2.102) and $(3.20)_2$ leads to

$$\frac{\mathrm{d}}{\mathrm{d}t}\int_{Q} \mathbf{a}\cdot\mathbf{n}\,\mathrm{d}a = \int_{Q} \left[\mathbf{a}' + \operatorname{curl}(\mathbf{a}\times\mathbf{v}) + \mathbf{v}(\operatorname{div}\mathbf{a})\right]\cdot\mathbf{n}\,\mathrm{d}a.$$

PROBLEM 9.5

Starting with the Eulerian (or spatial) statement of the Ampère-Maxwell law in integral form,

$$\frac{\mathrm{d}}{\mathrm{d}t}\int_{Q} \mathbf{d}^*\cdot\mathbf{n}\,\mathrm{d}a + \int_{Q} \mathbf{j}^*\cdot\mathbf{n}\,\mathrm{d}a = \int_{\partial Q} \mathbf{h}^*\cdot\mathbf{l}\,\mathrm{d}l,$$

derive the corresponding pointwise form

$$\operatorname{curl}\mathbf{h}^* = (\mathbf{d}^*)' + \operatorname{curl}(\mathbf{d}^*\times\mathbf{v}) + \sigma^*\mathbf{v} + \mathbf{j}^*.$$

Solution

We begin with the Eulerian integral form of the Ampère-Maxwell law, i.e.,

$$\frac{\mathrm{d}}{\mathrm{d}t}\int_{Q} \mathbf{d}^*\cdot\mathbf{n}\,\mathrm{d}a + \int_{Q} \mathbf{j}^*\cdot\mathbf{n}\,\mathrm{d}a = \int_{\partial Q} \mathbf{h}^*\cdot\mathbf{l}\,\mathrm{d}l. \tag{a}$$

We consider the first term on the left-hand side of (a):

$$\begin{aligned}\frac{\mathrm{d}}{\mathrm{d}t}\int_{Q} \mathbf{d}^*\cdot\mathbf{n}\,\mathrm{d}a &= \int_{Q}\left[(\mathbf{d}^*)' + \operatorname{curl}(\mathbf{d}^*\times\mathbf{v}) + \mathbf{v}(\operatorname{div}\mathbf{d}^*)\right]\cdot\mathbf{n}\,\mathrm{d}a \quad \text{(transport theorem (9.43))}\\ &= \int_{Q}\left[(\mathbf{d}^*)' + \operatorname{curl}(\mathbf{d}^*\times\mathbf{v}) + \sigma^*\mathbf{v}\right]\cdot\mathbf{n}\,\mathrm{d}a \quad \text{(Gauss's law (9.46i)).}\end{aligned}$$

Next, we consider the right-hand side of (a), which, after use of Stokes's theorem (9.44), becomes

$$\int_{\partial Q} \mathbf{h}^*\cdot\mathbf{l}\mathrm{d}l = \int_{Q} (\operatorname{curl}\mathbf{h}^*)\cdot\mathbf{n}\,\mathrm{d}a.$$

Substitution of the preceding results into (a) leads to

$$\int_{Q}\left[(\mathbf{d}^*)' + \operatorname{curl}(\mathbf{d}^*\times\mathbf{v}) + \sigma^*\mathbf{v} + \mathbf{j}^* - \operatorname{curl}\mathbf{h}^*\right]\cdot\mathbf{n}\,\mathrm{d}a = 0.$$

Since the integrand is continuous and $\mathcal{Q}$ is arbitrary, the localization theorem implies that

$$\left[(\mathbf{d}^*)' + \text{curl}\,(\mathbf{d}^* \times \mathbf{v}) + \sigma^*\mathbf{v} + \mathbf{j}^* - \text{curl}\,\mathbf{h}^*\right] \cdot \mathbf{n} = 0.$$

Furthermore, since the coefficient of $\mathbf{n}$ is independent of $\mathbf{n}$, and $\mathbf{n}$ is arbitrary, it follows that the coefficient must vanish, i.e.,

$$\text{curl}\,\mathbf{h}^* = \left(\mathbf{d}^*\right)' + \text{curl}\,(\mathbf{d}^* \times \mathbf{v}) + \sigma^*\mathbf{v} + \mathbf{j}^*.$$

PROBLEM 9.6

Show that conservation of charge (9.46f) is a consequence of Maxwell's equations (9.46g)–(9.46j).

Solution

Taking the divergence of the Ampère-Maxwell law (9.46j), and subsequently using result (2.101), gives

$$\text{div}\left[(\mathbf{d}^*)'\right] + \text{div}\,(\sigma^*\mathbf{v}) + \text{div}\,\mathbf{j}^* = 0.$$

Continuity of $\mathbf{d}^*$ enables the order of partial differentiation to be exchanged, so

$$\text{div}\left[(\mathbf{d}^*)'\right] = (\text{div}\,\mathbf{d}^*)'.$$

This, together with result $(2.99)_2$, implies that

$$(\text{div}\,\mathbf{d}^*)' + \mathbf{v} \cdot \text{grad}\,\sigma^* + \sigma^*\text{div}\,\mathbf{v} + \text{div}\,\mathbf{j}^* = 0.$$

We then invoke Gauss's law for electricity (9.46i) and relationship $(3.20)_1$ to recover

$$\dot{\sigma}^* + \sigma^*\text{div}\,\mathbf{v} + \text{div}\,\mathbf{j}^* = 0,$$

the pointwise Eulerian statement of conservation of charge.

EXERCISES

1. Verify that $\mathbf{t} = \mathbf{T}\mathbf{n}$.
2. Confirm that $h = \mathbf{q} \cdot \mathbf{n}$.
3. Prove that (9.46a)–(9.46i) are the pointwise versions of the Eulerian integral forms of the fundamental laws (9.42a)–(9.42i).

9.2.2 LAGRANGIAN FUNDAMENTAL LAWS

Recall from Section 9.1 the Lagrangian integral forms of the first principles:

Conservation of mass

$$\int_{\mathcal{P}} \rho \, \mathrm{d}v = \int_{\mathcal{P}_\mathrm{R}} \rho_\mathrm{R} \mathrm{d}V, \tag{9.48a}$$

Balance of linear momentum

$$\frac{\mathrm{d}}{\mathrm{d}t} \int_{\mathcal{P}_\mathrm{R}} \mathbf{v} \rho_\mathrm{R} \, \mathrm{d}V = \int_{\mathcal{P}_\mathrm{R}} (\mathbf{f}^\mathrm{m} + \mathbf{f}^\mathrm{em}) \rho_\mathrm{R} \, \mathrm{d}V + \int_{\partial \mathcal{P}_\mathrm{R}} \mathbf{t}_\mathrm{R} \, \mathrm{d}A, \tag{9.48b}$$

Balance of angular momentum

$$\frac{\mathrm{d}}{\mathrm{d}t} \int_{\mathcal{P}_\mathrm{R}} \mathbf{x} \times \mathbf{v} \rho_\mathrm{R} \, \mathrm{d}V = \int_{\mathcal{P}_\mathrm{R}} \mathbf{x} \times (\mathbf{f}^\mathrm{m} + \mathbf{f}^\mathrm{em}) \rho_\mathrm{R} \, \mathrm{d}V + \int_{\partial \mathcal{P}_\mathrm{R}} \mathbf{x} \times \mathbf{t}_\mathrm{R} \, \mathrm{d}A + \int_{\mathcal{P}_\mathrm{R}} \mathbf{c}^\mathrm{em} \rho_\mathrm{R} \, \mathrm{d}V, \tag{9.48c}$$

First law of thermodynamics

$$\begin{aligned} \frac{\mathrm{d}}{\mathrm{d}t} \int_{\mathcal{P}_\mathrm{R}} \frac{1}{2} \mathbf{v} \cdot \mathbf{v} \rho_\mathrm{R} \, \mathrm{d}V + \frac{\mathrm{d}}{\mathrm{d}t} \int_{\mathcal{P}_\mathrm{R}} \varepsilon \rho_\mathrm{R} \, \mathrm{d}V &= \int_{\mathcal{P}_\mathrm{R}} (\mathbf{f}^\mathrm{m} + \mathbf{f}^\mathrm{em}) \cdot \mathbf{v} \rho_\mathrm{R} \, \mathrm{d}V + \int_{\partial \mathcal{P}_\mathrm{R}} \mathbf{t}_\mathrm{R} \cdot \mathbf{v} \, \mathrm{d}A \\ &+ \int_{\mathcal{P}_\mathrm{R}} (r^\mathrm{t} + r^\mathrm{em}) \rho_\mathrm{R} \, \mathrm{d}V - \int_{\partial \mathcal{P}_\mathrm{R}} h_\mathrm{R} \, \mathrm{d}A, \end{aligned} \tag{9.48d}$$

Second law of thermodynamics

$$\frac{d}{dt} \int_{\mathcal{P}_\mathrm{R}} \eta \rho_\mathrm{R} \, \mathrm{d}V \geq \int_{\mathcal{P}_\mathrm{R}} \frac{r^\mathrm{t}}{\Theta} \rho_\mathrm{R} \, \mathrm{d}V - \int_{\partial \mathcal{P}_\mathrm{R}} \frac{h_\mathrm{R}}{\Theta} \, \mathrm{d}A, \tag{9.48e}$$

Conservation of electric charge

$$\frac{\mathrm{d}}{\mathrm{d}t} \int_{\mathcal{P}_\mathrm{R}} \sigma_\mathrm{R} \, \mathrm{d}V = - \int_{\partial \mathcal{P}_\mathrm{R}} \mathbf{j}_\mathrm{R} \cdot \mathbf{N} \, \mathrm{d}A, \tag{9.48f}$$

Gauss's law for magnetism

$$\int_{\partial \mathcal{P}_\mathrm{R}} \mathbf{b}_\mathrm{R} \cdot \mathbf{N} \, \mathrm{d}A = 0, \tag{9.48g}$$

Faraday's law

$$\frac{\mathrm{d}}{\mathrm{d}t} \int_{\mathcal{Q}_\mathrm{R}} \mathbf{b}_\mathrm{R} \cdot \mathbf{N} \, \mathrm{d}A = - \int_{\partial \mathcal{Q}_\mathrm{R}} \mathbf{e}_\mathrm{R} \cdot \mathbf{l}_\mathrm{R} \, \mathrm{d}L, \tag{9.48h}$$

Gauss's law for electricity

$$\int_{\partial \mathcal{P}_\mathrm{R}} \mathbf{d}_\mathrm{R} \cdot \mathbf{N} \, \mathrm{d}A = \int_{\mathcal{P}_\mathrm{R}} \sigma_\mathrm{R} dV, \tag{9.48i}$$

Ampère-Maxwell law

$$\frac{\mathrm{d}}{\mathrm{d}t}\int_{\mathcal{Q}_R} \mathbf{d}_R \cdot \mathbf{N}\,\mathrm{d}A + \int_{\mathcal{Q}_R} \mathbf{j}_R \cdot \mathbf{N}\,\mathrm{d}A = \int_{\partial\mathcal{Q}_R} \mathbf{h}_R \cdot \mathbf{l}_R\,dL. \tag{9.48j}$$

An important feature of the Lagrangian integral conservation laws (9.48a)–(9.48j) is that the regions $\mathcal{P}_R$ and $\mathcal{Q}_R$ occupied by subsets $\mathcal{S}_1$ and $\mathcal{S}_2$ in the reference configuration are fixed. Hence, the regions of integration $\mathcal{P}_R$ and $\mathcal{Q}_R$ do not change with time, so time derivatives of Lagrangian surface and volume integrals can be passed directly inside the integrals. Conversely, with the Eulerian integral conservation laws (9.42a)–(9.42j), the regions of integration $\mathcal{P}$ and $\mathcal{Q}$ change with time. Thus, to take time derivatives of these time-varying integrals, the transport theorem for surface integrals (9.43) and the transport theorem for volume integrals (4.20) must be employed. The transport theorems are analogous to Leibniz's rule in multivariable calculus.

The continuous, bounded nature of the integrands in (9.48a)–(9.48j) enables the divergence theorem and Stokes's theorem, and the requirement that (9.48a)–(9.48j) be continuous and global, i.e., true for the entire body and all subsets, enables the localization theorem. With the traction $\mathbf{t}_R$ and heat flux h_R dependent on surface geometry only through the outward unit normal $\mathbf{N}$, so $\mathbf{t}_R = \mathbf{PN}$ and $h_R = \mathbf{q}_R \cdot \mathbf{N}$, it can be shown that application of the divergence theorem, Stokes's theorem, and the localization theorem to the Lagrangian integral equations (9.48a)–(9.48j) leads to the Lagrangian pointwise equations

$$\rho J = \rho_R, \tag{9.49a}$$

$$\rho_R \dot{\mathbf{v}} = \rho_R \left(\mathbf{f}^{m} + \mathbf{f}^{em}\right) + \mathrm{Div}\,\mathbf{P}, \tag{9.49b}$$

$$\rho_R \boldsymbol{\Gamma}^{em} + \mathbf{P}\mathbf{F}^{\mathrm{T}} - \mathbf{F}\mathbf{P}^{\mathrm{T}} = \mathbf{0}, \tag{9.49c}$$

$$\rho_R \dot{\varepsilon} = \mathbf{P} \cdot \mathrm{Grad}\,\mathbf{v} + \rho_R \left(r^{t} + r^{em}\right) - \mathrm{Div}\,\mathbf{q}_R, \tag{9.49d}$$

$$\rho_R \dot{\eta} \geq \rho_R \frac{r^{t}}{\Theta} - \mathrm{Div}\left(\frac{\mathbf{q}_R}{\Theta}\right), \tag{9.49e}$$

$$\dot{\sigma}_R + \mathrm{Div}\,\mathbf{j}_R = 0, \tag{9.49f}$$

$$\mathrm{Div}\,\mathbf{b}_R = 0, \tag{9.49g}$$

$$\mathrm{Curl}\,\mathbf{e}_R = -\dot{\mathbf{b}}_R, \tag{9.49h}$$

$$\mathrm{Div}\,\mathbf{d}_R = \sigma_R, \tag{9.49i}$$

$$\mathrm{Curl}\,\mathbf{h}_R = \dot{\mathbf{d}}_R + \mathbf{j}_R. \tag{9.49j}$$

Note that it can be shown that conservation of charge (9.49f) is not independent of **Maxwell's equations** (9.49g)–(9.49j), but rather is implicitly contained in them. Recall that $\mathbf{P}$ is the first Piola-Kirchhoff stress, $\mathbf{q}_R$ is the referential heat flux vector,

"Div" is the Lagrangian divergence (i.e., the divergence calculated with respect to the reference configuration), "Curl" is the Lagrangian curl, and

$$\dot{\mathbf{a}} = \frac{\partial}{\partial t}\hat{\mathbf{a}}(\mathbf{X}, t)$$

is the material derivative of an arbitrary vector $\mathbf{a} = \hat{\mathbf{a}}(\mathbf{X}, t)$, i.e., the partial derivative of the referential description of $\mathbf{a}$ with respect to time t. Also,

$$\boldsymbol{\Gamma}^{\mathrm{em}}\mathbf{a} = \mathbf{c}^{\mathrm{em}} \times \mathbf{a}$$

for any vector $\mathbf{a}$, i.e., $\boldsymbol{\Gamma}^{\mathrm{em}}$ is a skew tensor whose corresponding axial vector is $\mathbf{c}^{\mathrm{em}}$. Note that in deriving the pointwise version of the first law of thermodynamics (9.49d) from its integral counterpart (9.48d), we have made use of the Lagrangian form of the **energy theorem** for continuum electrodynamics:

$$\int_{\mathcal{P}_{\mathrm{R}}} (\mathbf{f}^{\mathrm{m}} + \mathbf{f}^{\mathrm{em}}) \cdot \mathbf{v}\rho_{\mathrm{R}}\, \mathrm{d}V + \int_{\partial\mathcal{P}_{\mathrm{R}}} \mathbf{t}_{\mathrm{R}} \cdot \mathbf{v}\, \mathrm{d}A - \frac{\mathrm{d}}{\mathrm{d}t}\int_{\mathcal{P}_{\mathrm{R}}} \frac{1}{2}\mathbf{v} \cdot \mathbf{v}\rho_{\mathrm{R}}\, \mathrm{d}V = \int_{\mathcal{P}_{\mathrm{R}}} \mathbf{P} \cdot \mathrm{Grad}\,\mathbf{v}\, \mathrm{d}V. \tag{9.50}$$

Compare (9.50) with (4.54), the Lagrangian form of the energy theorem for mechanics.

EXERCISES

1. Verify that $\mathbf{t}_{\mathrm{R}} = \mathbf{P}\mathbf{N}$.
2. Confirm that $h_{\mathrm{R}} = \mathbf{q}_{\mathrm{R}} \cdot \mathbf{N}$.
3. Prove that (9.49a)–(9.49j) are the pointwise versions of the Lagrangian integral forms of the fundamental laws (9.48a)–(9.48j).
4. Show that conservation of charge (9.49f) is a consequence of Maxwell's equations (9.49g)–(9.49j).

9.3 MODELING OF THE EFFECTIVE ELECTROMAGNETIC FIELDS

Recall that Eulerian modeling of deformable thermo-electro-magneto-mechanical materials involves two sets of spatial (or Eulerian) electromagnetic fields: the **effective fields** $\mathbf{e}^*$, $\mathbf{d}^*$, $\mathbf{p}^*$, $\mathbf{h}^*$, $\mathbf{b}^*$, $\mathbf{m}^*$, σ^*, and $\mathbf{j}^*$, distinguished in our notation by superscript asterisks, and the **standard fields** $\mathbf{e}$, $\mathbf{d}$, $\mathbf{p}$, $\mathbf{h}$, $\mathbf{b}$, $\mathbf{m}$, σ, and $\mathbf{j}$. (Conversely, in Lagrangian modeling, there is no notion of effective or standard fields.) The effective fields are the electromagnetic fields acting on the deforming continuum as seen in its present configuration, measured with respect to a co-moving or rest frame, i.e., a frame affixed to but not deforming with the continuum. The standard fields also act on the deforming continuum as seen in its present configuration, but are instead measured with respect to a fixed or laboratory frame.

Various transformations have been presented in the literature that relate the effective fields to the standard fields.[8] In this section, we present four popular transformation theories, each based on a different set of principles and postulates. As a result, the standard electromagnetic fields generally have different physical connotations from model to model; hence, in what follows, the fields corresponding to a particular model are distinguished with the appropriate subscript notation.

9.3.1 MINKOWSKI MODEL

The Minkowski model [32, 33, 38] is motivated by Einstein's special theory of relativity. In this approximation, the effective fields are related to the standard fields through semirelativistic inverse Lorentz transformations:

$$\begin{gathered} \mathbf{e}^* = \mathbf{e}_\mathrm{M} + \mathbf{v} \times \mathbf{b}_\mathrm{M}, \quad \mathbf{h}^* = \mathbf{h}_\mathrm{M} - \mathbf{v} \times \mathbf{d}_\mathrm{M}, \quad \mathbf{d}^* = \mathbf{d}_\mathrm{M}, \quad \mathbf{b}^* = \mathbf{b}_\mathrm{M}, \\ \mathbf{p}^* = \mathbf{p}_\mathrm{M}, \quad \mathbf{m}^* = \mathbf{m}_\mathrm{M} + \mathbf{v} \times \mathbf{p}_\mathrm{M}, \quad \mathbf{j}^* = \mathbf{j}_\mathrm{M} - \sigma_\mathrm{M}\mathbf{v}, \quad \sigma^* = \sigma_\mathrm{M}, \end{gathered} \tag{9.51}$$

where $(\cdot)_\mathrm{M}$ represents a standard electromagnetic field corresponding to the Minkowski model.

9.3.2 LORENTZ MODEL

The Lorentz model is motivated by Lorentz's theory of electrons [39], which postulates that a body consists of an infinitely large number of rapidly moving charged particles. The motion of these charged particles, in turn, generates rapidly fluctuating microscopic electromagnetic fields. The microscopic fields averaged over a small time interval and an infinitesimal volume are defined as the corresponding macroscopic fields. The aforementioned postulates lead to the following relationships:

$$\begin{gathered} \mathbf{e}^* = \mathbf{e}_\mathrm{L} + \mathbf{v} \times \mathbf{b}_\mathrm{L}, \quad \mathbf{h}^* = \mathbf{h}_\mathrm{L} - \epsilon_0\, \mathbf{v} \times \mathbf{e}_\mathrm{L}, \quad \mathbf{d}^* = \mathbf{d}_\mathrm{L}, \quad \mathbf{b}^* = \mathbf{b}_\mathrm{L}, \\ \mathbf{p}^* = \mathbf{p}_\mathrm{L}, \quad \mathbf{m}^* = \mathbf{m}_\mathrm{L}, \quad \mathbf{j}^* = \mathbf{j}_\mathrm{L} - \sigma_\mathrm{L}\mathbf{v}, \quad \sigma^* = \sigma_\mathrm{L}, \end{gathered} \tag{9.52}$$

where $(\cdot)_\mathrm{L}$ represents a standard electromagnetic field corresponding to the Lorentz model.

9.3.3 STATISTICAL MODEL

The statistical model [40] is a modification of Lorentz's theory, wherein the charged particles are grouped into stable structures (e.g., atoms, molecules, or ions). The field effects of the charged particles within each stable structure are represented by microscopic electric and magnetic multipole moments (e.g., dipole, quadrupole, and octupole moments), and the macroscopic polarization and magnetization fields are defined as statistical averages of these multipole moments over a large number

[8] A constraint on the mathematical forms of these transformations is that the effective fields reduce to the standard fields in the absence of motion. That is, when $\mathbf{v} = \mathbf{0}$, then $\mathbf{e}^*$ should collapse to $\mathbf{e}$, $\mathbf{d}^*$ should collapse to $\mathbf{d}$, and so on.

of stable structures. The transformations presented in this model are identical to Minkowski's, i.e.,

$$\mathbf{e}^* = \mathbf{e}_S + \mathbf{v} \times \mathbf{b}_S, \quad \mathbf{h}^* = \mathbf{h}_S - \mathbf{v} \times \mathbf{d}_S, \quad \mathbf{d}^* = \mathbf{d}_S, \quad \mathbf{b}^* = \mathbf{b}_S,$$
$$\mathbf{p}^* = \mathbf{p}_S, \quad \mathbf{m}^* = \mathbf{m}_S + \mathbf{v} \times \mathbf{p}_S, \quad \mathbf{j}^* = \mathbf{j}_S - \sigma_S \mathbf{v}, \quad \sigma^* = \sigma_S, \tag{9.53}$$

where $(\cdot)_S$ represents a standard electromagnetic field corresponding to the statistical model.

9.3.4 CHU MODEL

The Chu model [41] is based on the postulate that deforming bodies contribute to electromagnetic phenomena by acting, in a macroscopic sense, as electric and magnetic dipole sources for the electromagnetic fields. The transformations for the Chu model are

$$\mathbf{e}^* = \mathbf{e}_C + \mu_0 \mathbf{v} \times \mathbf{h}_C, \quad \mathbf{h}^* = \mathbf{h}_C - \epsilon_0 \mathbf{v} \times \mathbf{e}_C, \quad \mathbf{d}^* = \mathbf{d}_C, \quad \mathbf{b}^* = \mathbf{b}_C,$$
$$\mathbf{p}^* = \mathbf{p}_C + \mathbf{m}_C \times \frac{\mathbf{v}}{c^2}, \quad \mathbf{m}^* = \mathbf{m}_C, \quad \mathbf{j}^* = \mathbf{j}_C - \sigma_C \mathbf{v}, \quad \sigma^* = \sigma_C, \tag{9.54}$$

where $(\cdot)_C$ represents a standard electromagnetic field corresponding to the Chu model.

9.3.5 A COMPARISON OF THE FOUR MODELS

Table 9.2 catalogs Maxwell's equations for each of the four models, deduced by substituting each set of transformations (i.e., (9.51)–(9.54)) into (9.46g)–(9.46j). Recall that since each of the four models is based on a different set of postulates, the standard electromagnetic fields generally have different physical connotations from model to model. We can deduce relationships between the standard fields by comparing the equations in Table 9.2. For instance, the Minkowski and statistical models, although developed from different perspectives, are mathematically equivalent (i.e., $\mathbf{m}_M = \mathbf{m}_S$, $\mathbf{p}_M = \mathbf{p}_S$, etc.).

Table 9.2 Maxwell's Equations for Different Transformation Models

Model	Maxwell's Equations	
Minkowski	$\text{curl}\,\mathbf{e}_M = -\mathbf{b}'_M$ $\text{curl}\,\mathbf{h}_M = \mathbf{d}'_M + \mathbf{j}_M$	$\text{div}\,\mathbf{d}_M = \sigma_M$ $\text{div}\,\mathbf{b}_M = 0$
Lorentz	$\text{curl}\,\mathbf{e}_L = -\mathbf{b}'_L$ $\text{curl}\,\mathbf{h}_L = \mathbf{d}'_L - \text{curl}\,(\mathbf{v} \times \mathbf{p}_L) + \mathbf{j}_L$	$\text{div}\,\mathbf{d}_L = \sigma_L$ $\text{div}\,\mathbf{b}_L = 0$
Statistical	$\text{curl}\,\mathbf{e}_S = -\mathbf{b}'_S$ $\text{curl}\,\mathbf{h}_S = \mathbf{d}'_S + \mathbf{j}_S$	$\text{div}\,\mathbf{d}_S = \sigma_S$ $\text{div}\,\mathbf{b}_S = 0$
Chu	$\text{curl}\,\mathbf{e}_C = -\mathbf{b}'_C + \mu_0\text{curl}\,(\mathbf{v} \times \mathbf{m}_C)$ $\text{curl}\,\mathbf{h}_C = \mathbf{d}'_C - \text{curl}\,(\mathbf{v} \times \mathbf{p}_C) + \mathbf{j}_C$	$\text{div}\,\mathbf{d}_C = \sigma_C$ $\text{div}\,\mathbf{b}_C = 0$

PROBLEM 9.7

Verify the results presented in Table 9.2 for the Minkowski model.

Solution

Straightforward substitution of the Minkowski transformations $(9.51)_3$, $(9.51)_4$, and $(9.51)_8$ into the Eulerian forms of *Gauss's law for magnetism* (9.46g) and *Gauss's law for electricity* (9.46i) gives

$$\operatorname{div} \mathbf{b}_{\mathrm{M}} = 0, \quad \operatorname{div} \mathbf{d}_{\mathrm{M}} = \sigma_{\mathrm{M}}.$$

Then, use of the transformation equations $(9.51)_2$, $(9.51)_3$, $(9.51)_7$, and $(9.51)_8$, i.e.,

$$\mathbf{h}^* = \mathbf{h}_{\mathrm{M}} - \mathbf{v} \times \mathbf{d}_{\mathrm{M}}, \quad \mathbf{d}^* = \mathbf{d}_{\mathrm{M}}, \quad \mathbf{j}^* = \mathbf{j}_{\mathrm{M}} - \sigma_{\mathrm{M}}\mathbf{v}, \quad \sigma^* = \sigma_{\mathrm{M}},$$

in the *Ampère-Maxwell law* (9.46j) leads to

$$\operatorname{curl}\left(\mathbf{h}_{\mathrm{M}} - \mathbf{v} \times \mathbf{d}_{M}\right) = \mathbf{d}'_{\mathrm{M}} + \operatorname{curl}\left(\mathbf{d}_{\mathrm{M}} \times \mathbf{v}\right) + \sigma_{\mathrm{M}}\mathbf{v} + \left(\mathbf{j}_{\mathrm{M}} - \sigma_{\mathrm{M}}\mathbf{v}\right).$$

Anticommutativity of the vector product (2.61) and the distributive property of the curl (2.103a) then imply that the above equation becomes

$$\operatorname{curl} \mathbf{h}_{\mathrm{M}} = \mathbf{d}'_{\mathrm{M}} + \mathbf{j}_{\mathrm{M}}.$$

Similarly, *Faraday's law* (9.46h) is transformed by applying $(9.51)_1$ and $(9.51)_4$:

$$\operatorname{curl}\left(\mathbf{e}_{\mathrm{M}} + \mathbf{v} \times \mathbf{b}_{\mathrm{M}}\right) = -\mathbf{b}'_{\mathrm{M}} - \operatorname{curl}\left(\mathbf{b}_{\mathrm{M}} \times \mathbf{v}\right).$$

Simplification of this result using (2.61) and (2.103a) leads to

$$\operatorname{curl} \mathbf{e}_{\mathrm{M}} = -\mathbf{b}'_{\mathrm{M}}.$$

EXERCISE

1. Verify the results presented in Table 9.2 for the Lorentz, statistical, and Chu models.

9.4 MODELING OF THE ELECTROMAGNETICALLY INDUCED COUPLING TERMS

As discussed in Section 9.1, the thermomechanical equations ((9.46a)–(9.46d) in Eulerian form) are coupled to the electromagnetic equations ((9.46f)–(9.46j) in Eulerian form) through the electromagnetic body force $\mathbf{f}^{\mathrm{em}}$, body couple $\mathbf{c}^{\mathrm{em}}$ (or, equivalently, $\boldsymbol{\Gamma}^{\mathrm{em}}$), and energy supply rate r^{em}. The mathematical forms of these coupling terms are *postulated* on the basis of the interaction theories presented in Section 9.3, which are motivated by either atomic physics or empiricism. As an example, the coupling terms postulated using the Minkowski theory for a polarizable, magnetizable, deformable continuum are, in Eulerian form [32],

$$\rho\mathbf{f}^{\text{em}} = \sigma^*\mathbf{e}^* + \mathbf{j}^* \times \mathbf{b}^* + (\text{grad}\,\mathbf{e}^*)^{\text{T}}\mathbf{p}^* + \mu_o(\text{grad}\,\mathbf{h}^*)^{\text{T}}\mathbf{m}^* + \mathring{\mathbf{d}}^* \times \mathbf{b}^* + \mathbf{d}^* \times \mathring{\mathbf{b}}^*, \quad (9.55a)$$

$$\rho\mathbf{\Gamma}^{\text{em}} = \left(\mathbf{e}^* \otimes \mathbf{p}^* - \mathbf{p}^* \otimes \mathbf{e}^*\right) + \mu_o\left(\mathbf{h}^* \otimes \mathbf{m}^* - \mathbf{m}^* \otimes \mathbf{h}^*\right), \quad (9.55b)$$

$$\rho r^{\text{em}} = \mathbf{j}^* \cdot \mathbf{e}^* + \rho\mathbf{e}^* \cdot \overline{\left(\frac{\mathbf{p}^*}{\rho}\right)}^{\boldsymbol{\cdot}} + \rho\mu_o\mathbf{h}^* \cdot \overline{\left(\frac{\mathbf{m}^*}{\rho}\right)}^{\boldsymbol{\cdot}}, \quad (9.55c)$$

where we recall that () ⊗ () denotes the dyadic product of two vectors, "grad" is the Eulerian gradient, and

$$\mathring{\mathbf{a}} = \mathbf{a}' + \text{curl}\,(\mathbf{a} \times \mathbf{v}) + \mathbf{v}(\text{div}\,\mathbf{a}) \quad (9.56)$$

is a convected rate of an arbitrary vector $\mathbf{a} = \tilde{\mathbf{a}}(\mathbf{x}, t)$. Coupling terms for the Lagrangian forms of the fundamental laws can be found, for instance, in [32].

The Minkowski model (9.55a)–(9.55c) has as special cases interaction theories describing forces exerted by static electric fields in polarizable solids [42] and static magnetic fields in magnetizable solids [43]. Use of (9.55a)–(9.55c) in the Eulerian forms of balance of linear momentum (9.46b), balance of angular momentum (9.46c), and the first law of thermodynamics (9.46d) yields

$$\begin{aligned} \rho\dot{\mathbf{v}} = \rho\mathbf{f}^{\text{m}} &+ \sigma^*\mathbf{e}^* + \mathbf{j}^* \times \mathbf{b}^* + (\text{grad}\,\mathbf{e}^*)^{\text{T}}\mathbf{p}^* + \mu_o(\text{grad}\,\mathbf{h}^*)^{\text{T}}\mathbf{m}^* + \mathring{\mathbf{d}}^* \times \mathbf{b}^* \\ &+ \mathbf{d}^* \times \mathring{\mathbf{b}}^* + \text{div}\,\mathbf{T}, \end{aligned} \quad (9.57a)$$

$$\mathbf{T} - \mathbf{T}^{\text{T}} = \left(\mathbf{p}^* \otimes \mathbf{e}^* - \mathbf{e}^* \otimes \mathbf{p}^*\right) + \mu_o\left(\mathbf{m}^* \otimes \mathbf{h}^* - \mathbf{h}^* \otimes \mathbf{m}^*\right), \quad (9.57b)$$

$$\dot{\varepsilon} = \frac{1}{\rho_{\text{R}}}\mathbf{P} \cdot \dot{\mathbf{F}} + \mathbf{e}^* \cdot \overline{\left(\frac{\mathbf{p}^*}{\rho}\right)}^{\boldsymbol{\cdot}} + \mu_o\mathbf{h}^* \cdot \overline{\left(\frac{\mathbf{m}^*}{\rho}\right)}^{\boldsymbol{\cdot}} + r^{\text{t}} + \frac{1}{\rho}\mathbf{j}^* \cdot \mathbf{e}^* - \frac{1}{\rho}\text{div}\,\mathbf{q}. \quad (9.57c)$$

9.4.1 AN ALTERNATIVE APPROACH

Electromagnetic effects in the thermomechanical balance laws can be modeled as an electromagnetically induced body force, body couple, and energy supply (as discussed previously; see also [32, 33, 44]) or, alternatively, incorporated into the constitutive response (see, for instance, [34–36]). The latter is accomplished by defining a *symmetric* **total stress tensor**

$$\boldsymbol{\tau} = \mathbf{T} + \mathbf{T}^{\text{em}} \quad (9.58)$$

that consists of contributions from the Cauchy stress tensor **T** and the electromagnetic **Maxwell stress tensor** $\mathbf{T}^{\text{em}}$. The Maxwell stress $\mathbf{T}^{\text{em}}$ is defined so that twice its skew part is $\rho\mathbf{\Gamma}^{\text{em}}$ and its divergence is $\rho\mathbf{f}^{\text{em}}$; refer to Problem 9.8. For the Minkowski formulation (9.55a)–(9.55c), it can be shown that

$$\mathbf{T}^{\text{em}} = \mathbf{e}^* \otimes \mathbf{d}^* + \mathbf{h}^* \otimes \mathbf{b}^* - \frac{1}{2}\left(\epsilon_o\mathbf{e}^* \cdot \mathbf{e}^* + \mu_o\mathbf{h}^* \cdot \mathbf{h}^*\right)\mathbf{I}. \quad (9.59)$$

By formulating the first principles and the constitutive equations in terms of the total stress $\boldsymbol{\tau}$ instead of the Cauchy stress $\mathbf{T}$, we eliminate *explicit* coupling between the electromagnetic fields and the thermomechanical fields in the first principles (9.57a)–(9.57c). This coupling is instead accounted for *implicitly* through the constitutive equation for $\boldsymbol{\tau}$.

PROBLEM 9.8

Prove that the divergence of the Maxwell stress tensor $\mathbf{T}^{\mathrm{em}}$ in (9.59) is equal to the electromagnetic body force vector $\rho\mathbf{f}^{\mathrm{em}}$ in (9.55a).

Solution

Taking the divergence of the Maxwell stress tensor (9.59) gives

$$\operatorname{div}\mathbf{T}^{\mathrm{em}} = \operatorname{div}\left[\mathbf{e}^*\otimes\mathbf{d}^* + \mathbf{h}^*\otimes\mathbf{b}^* - \frac{1}{2}(\epsilon_0\mathbf{e}^*\cdot\mathbf{e}^* + \mu_0\mathbf{h}^*\cdot\mathbf{h}^*)\mathbf{I}\right].$$

The divergence is distributive over tensor addition (refer to (2.103b)), which implies that

$$\operatorname{div}\mathbf{T}^{\mathrm{em}} = \operatorname{div}\left(\mathbf{e}^*\otimes\mathbf{d}^*\right) + \operatorname{div}\left(\mathbf{h}^*\otimes\mathbf{b}^*\right) - \frac{1}{2}\left\{\epsilon_0\operatorname{div}\left[(\mathbf{e}^*\cdot\mathbf{e}^*)\mathbf{I}\right] + \mu_0\operatorname{div}\left[(\mathbf{h}^*\cdot\mathbf{h}^*)\mathbf{I}\right]\right\}. \tag{a}$$

The first term in (a) simplifies to

$$\begin{aligned}\operatorname{div}\left(\mathbf{e}^*\otimes\mathbf{d}^*\right) &= (\operatorname{div}\mathbf{d}^*)\mathbf{e}^* + (\operatorname{grad}\mathbf{e}^*)\mathbf{d}^* && \text{(result } (2.99)_7\text{)}\\ &= \sigma^*\mathbf{e}^* + (\operatorname{grad}\mathbf{e}^*)\mathbf{d}^* && \text{(Gauss's law (9.46i)).}\end{aligned} \tag{b}$$

Similarly, the second term reduces to

$$\begin{aligned}\operatorname{div}\left(\mathbf{h}^*\otimes\mathbf{b}^*\right) &= (\operatorname{div}\mathbf{b}^*)\mathbf{h}^* + (\operatorname{grad}\mathbf{h}^*)\mathbf{b}^* && \text{(result } (2.99)_7\text{)}\\ &= (\operatorname{grad}\mathbf{h}^*)\mathbf{b}^* && \text{(Gauss's law (9.46g)).}\end{aligned} \tag{c}$$

The third term in (a) simplifies to

$$\begin{aligned}\frac{1}{2}\epsilon_0\operatorname{div}\left[(\mathbf{e}^*\cdot\mathbf{e}^*)\mathbf{I}\right] &= \frac{1}{2}\epsilon_0\operatorname{grad}(\mathbf{e}^*\cdot\mathbf{e}^*) && \text{(result } (2.99)_6\text{)}\\ &= \epsilon_0(\operatorname{grad}\mathbf{e}^*)^{\mathrm{T}}\mathbf{e}^* && \text{(result } (2.99)_3\text{).}\end{aligned} \tag{d}$$

Similarly, use of results $(2.99)_3$ and $(2.99)_6$ reduces the fourth term to

$$\frac{1}{2}\mu_0\operatorname{div}\left[(\mathbf{h}^*\cdot\mathbf{h}^*)\mathbf{I}\right] = \mu_0(\operatorname{grad}\mathbf{h}^*)^{\mathrm{T}}\mathbf{h}^*. \tag{e}$$

Combining results (b)–(e) in (a), we obtain

$$\operatorname{div}\mathbf{T}^{\mathrm{em}} = \sigma^*\mathbf{e}^* + (\operatorname{grad}\mathbf{e}^*)\mathbf{d}^* + (\operatorname{grad}\mathbf{h}^*)\mathbf{b}^* - \left[\epsilon_0(\operatorname{grad}\mathbf{e}^*)^{\mathrm{T}}\mathbf{e}^* + \mu_0(\operatorname{grad}\mathbf{h}^*)^{\mathrm{T}}\mathbf{h}^*\right]. \tag{f}$$

Relationships $(9.34)_1$ and $(9.40)_1$, i.e.,

$$\mathbf{d}^* = \mathbf{p}^* + \epsilon_0\mathbf{e}^*, \quad \mathbf{b}^* = \mu_0(\mathbf{h}^* + \mathbf{m}^*),$$

allow us to rewrite (f) as

$$\operatorname{div}\mathbf{T}^{em} = \sigma^*\mathbf{e}^* + \left[\operatorname{grad}\mathbf{e}^* - (\operatorname{grad}\mathbf{e}^*)^T\right]\mathbf{d}^* + \left[\operatorname{grad}\mathbf{h}^* - (\operatorname{grad}\mathbf{h}^*)^T\right]\mathbf{b}^*$$
$$+ (\operatorname{grad}\mathbf{e}^*)^T\mathbf{p}^* + \mu_o(\operatorname{grad}\mathbf{h}^*)^T\mathbf{m}^*.$$

Invoking definition (2.100) of the curl implies that

$$\operatorname{div}\mathbf{T}^{em} = \sigma^*\mathbf{e}^* + (\operatorname{curl}\mathbf{e}^*) \times \mathbf{d}^* + (\operatorname{curl}\mathbf{h}^*) \times \mathbf{b}^* + (\operatorname{grad}\mathbf{e}^*)^T\mathbf{p}^* + \mu_o(\operatorname{grad}\mathbf{h}^*)^T\mathbf{m}^*. \tag{g}$$

Faraday's law (9.46h) and the Ampère-Maxwell law (9.46j), i.e.,

$$\operatorname{curl}\mathbf{e}^* = -(\mathbf{b}^*)' - \operatorname{curl}(\mathbf{b}^* \times \mathbf{v}), \quad \operatorname{curl}\mathbf{h}^* = (\mathbf{d}^*)' + \operatorname{curl}(\mathbf{d}^* \times \mathbf{v}) + \sigma^*\mathbf{v} + \mathbf{j}^*,$$

are then used in (g), which, noting anticommutativity of the vector product, leads to

$$\operatorname{div}\mathbf{T}^{em} = \sigma^*\mathbf{e}^* + \mathbf{j}^* \times \mathbf{b}^* + (\operatorname{grad}\mathbf{e}^*)^T\mathbf{p}^* + \mu_o(\operatorname{grad}\mathbf{h}^*)^T\mathbf{m}^*$$
$$+ \left[(\mathbf{d}^*)' + \operatorname{curl}(\mathbf{d}^* \times \mathbf{v}) + \sigma^*\mathbf{v}\right] \times \mathbf{b}^* + \mathbf{d}^* \times \left[(\mathbf{b}^*)' + \operatorname{curl}(\mathbf{b}^* \times \mathbf{v})\right]$$

It then follows that

$$\operatorname{div}\mathbf{T}^{em} = \sigma^*\mathbf{e}^* + \mathbf{j}^* \times \mathbf{b}^* + (\operatorname{grad}\mathbf{e}^*)^T\mathbf{p}^* + \mu_o(\operatorname{grad}\mathbf{h}^*)^T\mathbf{m}^* + \mathring{\mathbf{d}}^* \times \mathbf{b}^* + \mathbf{d}^* \times \mathring{\mathbf{b}}^* = \rho\mathbf{f}^{em},$$

where $\mathring{\mathbf{b}}^*$ and $\mathring{\mathbf{d}}^*$ represent convected rates as defined in (9.56).

EXERCISES

1. Prove that $\mathbf{T} \cdot \mathbf{L} = \frac{1}{J}\mathbf{P} \cdot \dot{\mathbf{F}}$.

2. Verify that use of the electromagnetic energy (9.55c) in the Eulerian form of the first law of thermodynamics (9.46d) leads to (9.57c).

3. Verify that
  (a) twice the skew part of the Maxwell stress tensor $\mathbf{T}^{em}$ in (9.59) is $\rho\mathbf{\Gamma}^{em}$ in (9.55b), and
  (b) the total stress tensor $\boldsymbol{\tau}$ in (9.58) is symmetric.

9.5 THERMO-ELECTRO-MAGNETO-MECHANICAL PROCESS

The pointwise Eulerian field equations (9.46a), (9.46f)–(9.46j), and (9.57a)–(9.57c) constitute the first principles of our model, true for all thermo-electro-magneto-mechanical (TEMM) materials. For the purposes of developing the **constitutive**

equations that supplement these first principles and characterize particular TEMM materials, we conceptually divide the fields appearing in (9.46a), (9.46f)–(9.46j), and (9.57a)–(9.57c) into three groups:

$$\{\mathbf{x}, \eta, \mathbf{p}^*, \mathbf{m}^*\}, \quad \{\mathbf{P}, \Theta, \mathbf{e}^*, \mathbf{h}^*, \varepsilon, \mathbf{q}, \mathbf{j}^*\}, \quad \{\rho, \mathbf{f}^{\mathrm{m}}, r^t, \sigma^*\}, \tag{9.60}$$

denoted the **independent variables**, **dependent variables**, and **balancing terms**, respectively.[9] (Note that the set of independent variables contains slots occupied by one mechanical, one thermal, one electrical, and one magnetic quantity, respectively, from left to right.) Hence, the first Piola-Kirchhoff stress $\mathbf{P}$, temperature Θ, electric field $\mathbf{e}^*$, magnetic field $\mathbf{h}^*$, internal energy ε, heat flux $\mathbf{q}$, and conductive current $\mathbf{j}^*$ are determined from constitutive equations that, in general, depend on the history of the motion $\mathbf{x}$, entropy η, electric polarization $\mathbf{p}^*$, and magnetization $\mathbf{m}^*$, and possibly their rates or gradients. A group of quantities $\mathbf{x}$, η, $\mathbf{p}^*$, $\mathbf{m}^*$, $\mathbf{P}$, Θ, $\mathbf{e}^*$, $\mathbf{h}^*$, ε, $\mathbf{q}$, $\mathbf{j}^*$, ρ, $\mathbf{f}^{\mathrm{m}}$, r^t, and σ^* that satisfy the governing equations (9.46a), (9.46f)–(9.46j), and (9.57a)–(9.57c) for all space and time in the domain of interest describes a **TEMM process**.

9.6 CONSTITUTIVE MODEL DEVELOPMENT FOR THERMO-ELECTRO-MAGNETO-ELASTIC MATERIALS: LARGE-DEFORMATION THEORY

In this section, we specialize to (1) thermo-electro-magneto-*elastic* (TEME) materials that are capable of undergoing large elastic strains before yielding and (2) material response that is reversible (i.e., path independent and rate insensitive). We then illustrate the development of a *general* finite-deformation TEME constitutive framework using the principles of continuum thermodynamics, and show how this framework can be simplified using restrictions imposed by the second law of thermodynamics, invariance, angular momentum, and material symmetry. Similar approaches have been used to develop constitutive models for *particular classes* of smart materials in the finite-deformation regime, e.g., electroactive elastomers and magnetosensitive elastomers; see, for instance, [35–37, 45–60]. The large-deformation constitutive framework presented in this section, on the other hand, is intended to model a more general range of smart material behavior (e.g., magneto-electric coupling).

9.6.1 THE REDUCED CLAUSIUS-DUHEM INEQUALITY, WORK CONJUGATES

We algebraically combine the Eulerian forms of the Clausius-Duhem inequality (9.46e) and the first law of thermodynamics (9.57c) to produce the **reduced Clausius-Duhem inequality**

[9] In this division, the magnetic flux $\mathbf{b}^*$, electric displacement $\mathbf{d}^*$, and Cauchy stress $\mathbf{T}$ are relegated to *secondary dependent variables*, i.e., variables that can be calculated algebraically from the independent and *primary dependent variables* in (9.60). Also note that $\mathbf{v}$, $\mathbf{F}$, and $\mathbf{L}$ can be calculated from $\mathbf{x}$ using tensor calculus.

$$-\dot{\varepsilon}+\frac{1}{\rho_{\mathrm{R}}}\mathbf{P}\cdot\dot{\mathbf{F}}+\Theta\dot{\eta}+\mathbf{e}^{*}\cdot\overline{\left(\frac{\mathbf{p}^{*}}{\rho}\right)}^{\cdot}+\mu_{o}\mathbf{h}^{*}\cdot\overline{\left(\frac{\mathbf{m}^{*}}{\rho}\right)}^{\cdot}+\frac{1}{\rho}\mathbf{j}^{*}\cdot\mathbf{e}^{*}-\frac{1}{\rho\Theta}\mathbf{q}\cdot\operatorname{grad}\Theta\geq 0. \tag{9.61}$$

Analogously to classical thermodynamics [61], this *fundamental statement* of the second law consists of contributions from **conjugate pairs** of thermal, electrical, magnetic, and mechanical quantities. Each of these conjugate pairs (or **work conjugates**) is the product of the rate of an **extensive** quantity in rate form ($\dot{\mathbf{F}}$, $\dot{\eta}$, $\dot{(\mathbf{p}^*/\rho)}$, and $\dot{(\mathbf{m}^*/\rho)}$) and an **intensive** quantity ($\mathbf{P}$, Θ, $\mathbf{e}^*$, and $\mathbf{h}^*$).[10] The utility of the second law, and our identification of the extensive and intensive quantities, will become apparent in the following section.

EXERCISE

1. Verify that the Clausius-Duhem inequality (9.46e) can be algebraically combined with the first law of thermodynamics (9.57c) to produce the reduced Clausius-Duhem inequality (9.61).

9.6.2 THE ALL-EXTENSIVE FORMULATION

Recall that we specialized to reversible TEME material response, which implies that the deformation, although it may be large, is elastic and fully recoverable, and the material only undergoes nondissipative TEMM processes. It follows, then, that these TEMM processes can be completely described through the principles of classical equilibrium thermodynamics[11] [61].

Analogously to classical equilibrium thermodynamics, our fundamental energy potential is the specific internal energy ε, which employs *extensive quantities as its independent variables* [61]. Hence, for the fundamental formulation, the *natural* independent variables are the *extensive* quantities $\mathbf{F}$, η, $\mathbf{p}^*/\rho$, and $\mathbf{m}^*/\rho$ appearing as rates in (9.61), and the *natural* dependent variables are the conjugate *intensive* quantities $\mathbf{P}$, Θ, $\mathbf{e}^*$, and $\mathbf{h}^*$. Thus, for thermodynamic consistency, and to respect the reversible elastic nature of the process, we adjust the division (9.60) so that the dependence of the response on the present position $\mathbf{x}$, polarization $\mathbf{p}^*$, and magnetization $\mathbf{m}^*$ is through the deformation gradient $\mathbf{F} = \operatorname{Grad}\mathbf{x}$, $\mathbf{p}^*/\rho$, and $\mathbf{m}^*/\rho$, respectively. Additionally, we demand that the response depends on these independent variables only through their values at the present time t, not their histories, rates, or gradients, i.e.,

[10] In *classical* thermodynamics, extensive quantities are properties of a thermodynamic system that depend on its size or quantity, whereas intensive quantities are independent of its size or quantity. This intensive-extensive terminology is akin to the notion of generalized thermodynamic force-displacement pairs, with each such pair contributing to the internal energy of the system. In this chapter, we extend these *classical* thermodynamic concepts to the analogous setting of *continuum* thermodynamics [62].

[11] Although we derive our constitutive framework assuming that the system is in thermodynamic equilibrium (i.e., it only undergoes reversible processes), our framework can be used to model irreversible systems that operate in a regime close to equilibrium [61].

$$\mathbf{P} = \breve{\mathbf{P}}\left(\mathbf{F}, \eta, \frac{\mathbf{p}^*}{\rho}, \frac{\mathbf{m}^*}{\rho}\right), \qquad \Theta = \breve{\Theta}\left(\mathbf{F}, \eta, \frac{\mathbf{p}^*}{\rho}, \frac{\mathbf{m}^*}{\rho}\right), \qquad \mathbf{e}^* = \breve{\mathbf{e}}^*\left(\mathbf{F}, \eta, \frac{\mathbf{p}^*}{\rho}, \frac{\mathbf{m}^*}{\rho}\right),$$

$$\mathbf{h}^* = \breve{\mathbf{h}}^*\left(\mathbf{F}, \eta, \frac{\mathbf{p}^*}{\rho}, \frac{\mathbf{m}^*}{\rho}\right), \qquad \varepsilon = \breve{\varepsilon}\left(\mathbf{F}, \eta, \frac{\mathbf{p}^*}{\rho}, \frac{\mathbf{m}^*}{\rho}\right). \tag{9.62}$$

Notation (9.62) indicates that $\mathbf{P}$, Θ, $\mathbf{e}^*$, $\mathbf{h}^*$, and ε are *explicit* functions of $\mathbf{F}$, η, $\mathbf{p}^*/\rho$, and $\mathbf{m}^*/\rho$ evaluated at present position $\mathbf{x}$ and time t, and *implicit* functions of $\mathbf{x}$ and t. Note that a superscript breve is used to distinguish a *function* from its *value* (or output). *As with all materials in the TEMM theory, the response functions (9.62) for a TEME material must satisfy the second law of thermodynamics, invariance requirements, conservation of angular momentum, and material symmetry conditions.*

In what follows, we demonstrate how restrictions imposed by the second law of thermodynamics (9.61) yield a set of constitutive equations that provide the dependent variables $\mathbf{P}$, Θ, $\mathbf{e}^*$, and $\mathbf{h}^*$ (the intensive quantities) as partial derivatives of the specific internal energy ε (the fundamental thermodynamic energy potential) with respect to the independent variables $\mathbf{F}$, η, $\mathbf{p}^*/\rho$, and $\mathbf{m}^*/\rho$ (the extensive quantities), respectively. Use of the chain rule on $\varepsilon = \breve{\varepsilon}\,(\mathbf{F}, \eta, \mathbf{p}^*/\rho, \mathbf{m}^*/\rho)$ gives

$$\dot{\varepsilon} = \frac{\partial \breve{\varepsilon}}{\partial \mathbf{F}} \cdot \dot{\mathbf{F}} + \frac{\partial \breve{\varepsilon}}{\partial \eta}\dot{\eta} + \frac{\partial \breve{\varepsilon}}{\partial \left(\frac{\mathbf{p}^*}{\rho}\right)} \cdot \overline{\left(\frac{\mathbf{p}^*}{\rho}\right)}^{\boldsymbol{\cdot}} + \frac{\partial \breve{\varepsilon}}{\partial \left(\frac{\mathbf{m}^*}{\rho}\right)} \cdot \overline{\left(\frac{\mathbf{m}^*}{\rho}\right)}^{\boldsymbol{\cdot}},$$

and substitution of this result into the second law (9.61) leads to

$$\left(\frac{1}{\rho_{\mathrm{R}}}\mathbf{P} - \frac{\partial \breve{\varepsilon}}{\partial \mathbf{F}}\right) \cdot \dot{\mathbf{F}} + \left(\Theta - \frac{\partial \breve{\varepsilon}}{\partial \eta}\right)\dot{\eta} + \left(\mathbf{e}^* - \frac{\partial \breve{\varepsilon}}{\partial \left(\frac{\mathbf{p}^*}{\rho}\right)}\right) \cdot \overline{\left(\frac{\mathbf{p}^*}{\rho}\right)}^{\boldsymbol{\cdot}}$$

$$+ \left(\mu_0 \mathbf{h}^* - \frac{\partial \breve{\varepsilon}}{\partial \left(\frac{\mathbf{m}^*}{\rho}\right)}\right) \cdot \overline{\left(\frac{\mathbf{m}^*}{\rho}\right)}^{\boldsymbol{\cdot}} + \frac{1}{\rho}\mathbf{j}^* \cdot \mathbf{e}^* - \frac{1}{\rho\Theta}\mathbf{q} \cdot \operatorname{grad}\Theta \geq 0. \tag{9.63}$$

Inequality (9.63) must hold for all processes. Since the coefficients of the rates ($\dot{\mathbf{F}}$, $\dot{\eta}$, etc.) in (9.63) are independent of the rates themselves, and the rates may be varied independently and are arbitrary, it follows that the coefficients vanish, i.e.,

$$\mathbf{P} = \rho_{\mathrm{R}}\frac{\partial \breve{\varepsilon}}{\partial \mathbf{F}}, \qquad \Theta = \frac{\partial \breve{\varepsilon}}{\partial \eta}, \qquad \mathbf{e}^* = \frac{\partial \breve{\varepsilon}}{\partial \left(\frac{\mathbf{p}^*}{\rho}\right)}, \qquad \mathbf{h}^* = \frac{1}{\mu_0}\frac{\partial \breve{\varepsilon}}{\partial \left(\frac{\mathbf{m}^*}{\rho}\right)}. \tag{9.64}$$

What remains of inequality (9.63), i.e.,

$$\mathbf{j}^* \cdot \mathbf{e}^* - \frac{1}{\Theta}\mathbf{q} \cdot \operatorname{grad}\Theta \geq 0, \tag{9.65}$$

is called the **residual dissipation inequality**. The residual dissipation inequality quantifies irreversibilities in a thermodynamic process. In this case, the first term in (9.65) represents Joule heating due to current flow, and the second term represents heat conduction; both are transport processes. Accordingly, unlike the other dependent variables (see (9.64)), the conductive current $\mathbf{j}^*$ and heat flux $\mathbf{q}$ are not derivable from a thermodynamic energy potential. Instead, their constitutive equations are specified empirically (e.g., Ohm's law for $\mathbf{j}^*$ and Fourier's law for $\mathbf{q}$), with the caveat that they must satisfy inequality (9.65).

We collectively coin the set of extensive independent variables $\{\mathbf{F}, \eta, \mathbf{p}^*/\rho, \mathbf{m}^*/\rho\}$, the thermodynamic energy potential $\varepsilon = \breve{\varepsilon}\,(\mathbf{F}, \eta, \mathbf{p}^*/\rho, \mathbf{m}^*/\rho)$, and the constitutive equations (9.64) the *all-extensive formulation*. The all-extensive formulation, along with the other formulations that we present in this section, correlate with a particular experiment, the independent variables being controlled and the dependent variables being the measured responses.

9.6.2.1 Polarization and magnetization as independent variables

From an experimental point of view, it is more practical to control the electric polarization $\mathbf{p}^*$ and magnetization $\mathbf{m}^*$ than the polarization per unit mass $\mathbf{p}^*/\rho$ and magnetization per unit mass $\mathbf{m}^*/\rho$. Hence, we modify the all-extensive formulation presented in Section 9.6.2 to accommodate the use of $\mathbf{p}^*$ and $\mathbf{m}^*$ as independent variables. We proceed by using the chain rule

$$\overline{\left(\frac{\mathbf{p}^*}{\rho}\right)}^{\boldsymbol{\cdot}} = \frac{1}{\rho}\dot{\mathbf{p}}^* + \frac{1}{\rho}\left(\mathbf{F}^{-\mathrm{T}} \cdot \dot{\mathbf{F}}\right)\mathbf{p}^*, \quad \overline{\left(\frac{\mathbf{m}^*}{\rho}\right)}^{\boldsymbol{\cdot}} = \frac{1}{\rho}\dot{\mathbf{m}}^* + \frac{1}{\rho}\left(\mathbf{F}^{-\mathrm{T}} \cdot \dot{\mathbf{F}}\right)\mathbf{m}^* \tag{9.66}$$

to rewrite the fundamental form (9.61) of the Clausius-Duhem inequality:

$$\begin{aligned} &-\dot{\varepsilon} + \left[\frac{1}{\rho_{\mathrm{R}}}\mathbf{P} + \frac{1}{\rho}\left(\mathbf{e}^* \cdot \mathbf{p}^* + \mu_{\mathrm{o}}\mathbf{h}^* \cdot \mathbf{m}^*\right)\mathbf{F}^{-\mathrm{T}}\right] \cdot \dot{\mathbf{F}} + \Theta\dot{\eta} \\ &+ \frac{1}{\rho}\mathbf{e}^* \cdot \dot{\mathbf{p}}^* + \frac{\mu_{\mathrm{o}}}{\rho}\mathbf{h}^* \cdot \dot{\mathbf{m}}^* + \frac{1}{\rho}\mathbf{j}^* \cdot \mathbf{e}^* - \frac{1}{\rho\Theta}\mathbf{q} \cdot \operatorname{grad}\Theta \geq 0, \end{aligned} \tag{9.67}$$

where we have used

$$\overline{\left(\frac{1}{\rho}\right)}^{\boldsymbol{\cdot}} = \frac{1}{\rho}\operatorname{div}\mathbf{v}, \quad \operatorname{div}\mathbf{v} = \operatorname{tr}\mathbf{L} = \operatorname{tr}(\dot{\mathbf{F}}\mathbf{F}^{-1}) = \mathbf{F}^{-\mathrm{T}} \cdot \dot{\mathbf{F}}. \tag{9.68}$$

In the modified form (9.67) of the second law, polarization $\mathbf{p}^*$ and magnetization $\mathbf{m}^*$ appear as rates (i.e., natural independent variables). Accordingly, the thermodynamic energy potential ε in this formulation is a function of $\mathbf{F}$, η, $\mathbf{p}^*$, and $\mathbf{m}^*$, i.e., $\varepsilon = \bar{\varepsilon}\,(\mathbf{F}, \eta, \mathbf{p}^*, \mathbf{m}^*)$; a superscript bar is used instead of a superscript breve to signify a different internal energy *function* with the same *value*. Use of the chain rule gives

$$\dot{\varepsilon} = \frac{\partial\bar{\varepsilon}}{\partial\mathbf{F}} \cdot \dot{\mathbf{F}} + \frac{\partial\bar{\varepsilon}}{\partial\eta}\dot{\eta} + \frac{\partial\bar{\varepsilon}}{\partial\mathbf{p}^*} \cdot \dot{\mathbf{p}}^* + \frac{\partial\bar{\varepsilon}}{\partial\mathbf{m}^*} \cdot \dot{\mathbf{m}}^*,$$

and substitution of this result into (9.67) leads to

$$\left(\frac{1}{\rho_{\mathrm{R}}}\mathbf{P} + \frac{1}{\rho}\left(\mathbf{e}^* \cdot \mathbf{p}^* + \mu_{\mathrm{o}}\mathbf{h}^* \cdot \mathbf{m}^*\right)\mathbf{F}^{-\mathrm{T}} - \frac{\partial\bar{\varepsilon}}{\partial\mathbf{F}}\right) \cdot \dot{\mathbf{F}} + \left(\Theta - \frac{\partial\bar{\varepsilon}}{\partial\eta}\right)\dot{\eta} + \left(\frac{1}{\rho}\mathbf{e}^* - \frac{\partial\bar{\varepsilon}}{\partial\mathbf{p}^*}\right) \cdot \dot{\mathbf{p}}^*$$

$$+\left(\frac{\mu_o}{\rho}\mathbf{h}^* - \frac{\partial \bar{\varepsilon}}{\partial \mathbf{m}^*}\right) \cdot \dot{\mathbf{m}}^* + \frac{1}{\rho}\mathbf{j}^* \cdot \mathbf{e}^* - \frac{1}{\rho \Theta}\mathbf{q} \cdot \operatorname{grad} \Theta \geq 0. \tag{9.69}$$

Since the coefficients of $\dot{\mathbf{F}}$, $\dot{\eta}$, $\dot{\mathbf{p}}^*$, and $\dot{\mathbf{m}}^*$ in inequality (9.69) are independent of the rates, and the rates may be varied independently and are arbitrary, it follows that the coefficients vanish, i.e.,

$$\mathbf{P} = \rho_R \frac{\partial \bar{\varepsilon}}{\partial \mathbf{F}} - J\left(\mathbf{e}^* \cdot \mathbf{p}^* + \mu_o \mathbf{h}^* \cdot \mathbf{m}^*\right) \mathbf{F}^{-T}, \quad \Theta = \frac{\partial \bar{\varepsilon}}{\partial \eta},$$

$$\mathbf{e}^* = \rho \frac{\partial \bar{\varepsilon}}{\partial \mathbf{p}^*}, \quad \mathbf{h}^* = \frac{\rho}{\mu_o} \frac{\partial \bar{\varepsilon}}{\partial \mathbf{m}^*}. \tag{9.70}$$

We collectively coin the set of independent variables $\{\mathbf{F}, \eta, \mathbf{p}^*, \mathbf{m}^*\}$, the thermodynamic energy potential $\varepsilon = \bar{\varepsilon}(\mathbf{F}, \eta, \mathbf{p}^*, \mathbf{m}^*)$, and the constitutive equations (9.70) the *deformation-entropy-polarization-magnetization formulation*.

EXERCISE

1. Verify the relationships in (9.68).

9.6.3 OTHER FORMULATIONS

Knowledge of the internal energy function $\varepsilon = \breve{\varepsilon}(\mathbf{F}, \eta, \mathbf{p}^*/\rho, \mathbf{m}^*/\rho)$ or $\varepsilon = \bar{\varepsilon}(\mathbf{F}, \eta, \mathbf{p}^*, \mathbf{m}^*)$ is sufficient to characterize a reversible TEME material. Said differently, specifying the functional form of the energy potential determines $\mathbf{P}$, Θ, $\mathbf{e}^*$, and $\mathbf{h}^*$ in (9.64) and (9.70). The functional form of the energy potential is ascertained experimentally; this experimental characterization most straightforwardly accomplished when the independent and dependent variables synchronize with those one wishes to control and measure, respectively, in an experiment. From an experimental perspective, it is often more practical to control intensive quantities than extensive quantities; e.g., temperature is easier to control than entropy or internal energy. To change an independent variable from extensive to intensive, a new free energy is defined through a **Legendre transformation** of the internal energy, i.e.,

$$\text{new free energy} = \text{internal energy} - (\text{intensive})(\text{extensive}).$$

Following this blueprint, we present in Tables 9.3–9.6 a catalogue of the 15 possible Legendre transformations of the internal energy. These Legendre transformations are divided into four families, each employing a common set of thermomechanical independent variables: family 1, deformation and entropy (both extensive); family 2, deformation (extensive) and temperature (intensive); family 3, stress (intensive) and entropy (extensive); and family 4, stress and temperature (both intensive). For the sake of brevity, we employ the compact notation E^{abcd} for the Legendre-transformed energy potentials, where the superscript letters a, b, c, and d are placeholders for an appropriate mechanical, thermal, electrical, and magnetic independent variable, respectively. This compact notation denotes that the Legendre-transformed energy potential $E^{F\Theta pm}$, for instance, is a function of deformation $\mathbf{F}$, temperature Θ, electric

Table 9.3 Energy Family 1

Independent Variables	Energy	Legendre Transformation
$\mathbf{F}, \eta, \mathbf{p}^*/\rho, \mathbf{m}^*/\rho$	ε	
$\mathbf{F}, \eta, \mathbf{e}^*, \mathbf{m}^*/\rho$	$E^{F\eta em}$	$E^{F\eta em} = \varepsilon - \mathbf{e}^* \cdot \dfrac{\mathbf{p}^*}{\rho}$
$\mathbf{F}, \eta, \mathbf{p}^*/\rho, \mathbf{h}^*$	$E^{F\eta ph}$	$E^{F\eta ph} = \varepsilon - \mu_0 \mathbf{h}^* \cdot \dfrac{\mathbf{m}^*}{\rho}$
$\mathbf{F}, \eta, \mathbf{e}^*, \mathbf{h}^*$	$E^{F\eta eh}$	$E^{F\eta eh} = \varepsilon - \mathbf{e}^* \cdot \dfrac{\mathbf{p}^*}{\rho} - \mu_0 \mathbf{h}^* \cdot \dfrac{\mathbf{m}^*}{\rho}$

Table 9.4 Energy Family 2

Independent Variables	Energy	Legendre Transformation
$\mathbf{F}, \Theta, \mathbf{p}^*/\rho, \mathbf{m}^*/\rho$	$E^{F\Theta pm}$	$E^{F\Theta pm} = \varepsilon - \Theta\eta$
$\mathbf{F}, \Theta, \mathbf{e}^*, \mathbf{m}^*/\rho$	$E^{F\Theta em}$	$E^{F\Theta em} = \varepsilon - \Theta\eta - \mathbf{e}^* \cdot \dfrac{\mathbf{p}^*}{\rho}$
$\mathbf{F}, \Theta, \mathbf{p}^*/\rho, \mathbf{h}^*$	$E^{F\Theta ph}$	$E^{F\Theta ph} = \varepsilon - \Theta\eta - \mu_0 \mathbf{h}^* \cdot \dfrac{\mathbf{m}^*}{\rho}$
$\mathbf{F}, \Theta, \mathbf{e}^*, \mathbf{h}^*$	$E^{F\Theta eh}$	$E^{F\Theta eh} = \varepsilon - \Theta\eta - \mathbf{e}^* \cdot \dfrac{\mathbf{p}^*}{\rho} - \mu_0 \mathbf{h}^* \cdot \dfrac{\mathbf{m}^*}{\rho}$

Table 9.5 Energy family 3

Independent Variables	Energy	Legendre Transformation
$\mathbf{P}, \eta, \mathbf{p}^*/\rho, \mathbf{m}^*/\rho$	$E^{P\eta pm}$	$E^{P\eta pm} = \varepsilon - \dfrac{1}{\rho_R}\mathbf{P} \cdot \mathbf{F}$
$\mathbf{P}, \eta, \mathbf{e}^*, \mathbf{m}^*/\rho$	$E^{P\eta em}$	$E^{P\eta em} = \varepsilon - \dfrac{1}{\rho_R}\mathbf{P} \cdot \mathbf{F} - \mathbf{e}^* \cdot \dfrac{\mathbf{p}^*}{\rho}$
$\mathbf{P}, \eta, \mathbf{p}^*/\rho, \mathbf{h}^*$	$E^{P\eta ph}$	$E^{P\eta ph} = \varepsilon - \dfrac{1}{\rho_R}\mathbf{P} \cdot \mathbf{F} - \mu_0 \mathbf{h}^* \cdot \dfrac{\mathbf{m}^*}{\rho}$
$\mathbf{P}, \eta, \mathbf{e}^*, \mathbf{h}^*$	$E^{P\eta eh}$	$E^{P\eta eh} = \varepsilon - \dfrac{1}{\rho_R}\mathbf{P} \cdot \mathbf{F} - \mathbf{e}^* \cdot \dfrac{\mathbf{p}^*}{\rho} - \mu_0 \mathbf{h}^* \cdot \dfrac{\mathbf{m}^*}{\rho}$

Table 9.6 Energy Family 4

Independent Variables	Energy	Legendre Transformation
$\mathbf{P}, \Theta, \mathbf{p}^*/\rho, \mathbf{m}^*/\rho$	$E^{P\Theta pm}$	$E^{P\Theta pm} = \varepsilon - \dfrac{1}{\rho_R}\mathbf{P} \cdot \mathbf{F} - \Theta\eta$
$\mathbf{P}, \Theta, \mathbf{e}^*, \mathbf{m}^*/\rho$	$E^{P\Theta em}$	$E^{P\Theta em} = \varepsilon - \dfrac{1}{\rho_R}\mathbf{P} \cdot \mathbf{F} - \Theta\eta - \mathbf{e}^* \cdot \dfrac{\mathbf{p}^*}{\rho}$
$\mathbf{P}, \Theta, \mathbf{p}^*/\rho, \mathbf{h}^*$	$E^{P\Theta ph}$	$E^{P\Theta ph} = \varepsilon - \dfrac{1}{\rho_R}\mathbf{P} \cdot \mathbf{F} - \Theta\eta - \mu_0 \mathbf{h}^* \cdot \dfrac{\mathbf{m}^*}{\rho}$
$\mathbf{P}, \Theta, \mathbf{e}^*, \mathbf{h}^*$	$E^{P\Theta eh}$	$E^{P\Theta eh} = \varepsilon - \dfrac{1}{\rho_R}\mathbf{P} \cdot \mathbf{F} - \Theta\eta - \mathbf{e}^* \cdot \dfrac{\mathbf{p}^*}{\rho} - \mu_0 \mathbf{h}^* \cdot \dfrac{\mathbf{m}^*}{\rho}$

polarization per unit mass $\mathbf{p}^*/\rho$, and magnetization per unit mass $\mathbf{m}^*/\rho$. Note that the Legendre transformations in Tables 9.3–9.6 provide explicit connections between the new free energies E^{abcd} and the primitive internal energy ε of the system, whose evolution is governed by the first law of thermodynamics (9.57c).

The constitutive equations associated with the 15 formulations shown in Tables 9.3–9.6 are obtained via restrictions imposed by the second law of thermodynamics, invariance requirements, conservation of angular momentum, and material symmetry; refer to [62] for a complete catalogue. In the following sections, as a representative example, we investigate the *deformation-temperature-electric field-magnetic field formulation*. However, before we proceed, note the following:

(1) A procedure for introducing free energies that use the electric displacement $\mathbf{d}^*$ and magnetic induction $\mathbf{b}^*$ as independent variables, and the derivation of their associated constitutive equations, is presented in Appendix F.

(2) For formulations that employ either the electric polarization per unit mass $\mathbf{p}^*/\rho$ or the magnetization per unit mass $\mathbf{m}^*/\rho$ as an independent variable, as was the case with the all-extensive formulation in Section 9.6.2, $\mathbf{p}^*$ or $\mathbf{m}^*$ can be introduced as the independent variable ex post facto using the procedure described in Section 9.6.2.1. Refer to Problem 9.9.

9.6.3.1 The deformation-temperature-electric field-magnetic field formulation

In this formulation, $\mathbf{F}$, Θ, $\mathbf{e}^*$, and $\mathbf{h}^*$ are the independent variables. We define the energy potential $E^{F\Theta eh}$ as the Legendre transformation of internal energy $\varepsilon = \breve{\varepsilon}\,(\mathbf{F}, \eta, \mathbf{p}^*/\rho, \mathbf{m}^*/\rho)$ with respect to the thermal, electrical, and magnetic variables, from η to Θ, $\mathbf{p}^*/\rho$ to $\mathbf{e}^*$, and $\mathbf{m}^*/\rho$ to $\mathbf{h}^*$, i.e.,

$$E^{F\Theta eh} = \varepsilon - \Theta\eta - \mathbf{e}^* \cdot \frac{\mathbf{p}^*}{\rho} - \mu_o\mathbf{h}^* \cdot \frac{\mathbf{m}^*}{\rho}. \tag{9.71}$$

Refer to Table 9.4. Taking the rate of (9.71) gives

$$\dot{E}^{F\Theta eh} = \dot{\varepsilon} - \Theta\dot{\eta} - \eta\dot{\Theta} - \mathbf{e}^* \cdot \overline{\left(\frac{\mathbf{p}^*}{\rho}\right)}^{\cdot} - \frac{\mathbf{p}^*}{\rho} \cdot \dot{\mathbf{e}}^* - \mu_o\mathbf{h}^* \cdot \overline{\left(\frac{\mathbf{m}^*}{\rho}\right)}^{\cdot} - \mu_o\frac{\mathbf{m}^*}{\rho} \cdot \dot{\mathbf{h}}^*,$$

and use of this result in (9.61) yields the second law statement

$$-\dot{E}^{F\Theta eh} + \frac{1}{\rho_R}\mathbf{P} \cdot \dot{\mathbf{F}} - \eta\dot{\Theta} - \frac{\mathbf{p}^*}{\rho} \cdot \dot{\mathbf{e}}^* - \mu_o\frac{\mathbf{m}^*}{\rho} \cdot \dot{\mathbf{h}}^* + \frac{1}{\rho}\mathbf{j}^* \cdot \mathbf{e}^* - \frac{1}{\rho\Theta}\mathbf{q} \cdot \text{grad}\,\Theta \geq 0. \tag{9.72}$$

Via the chain rule, we have

$$\dot{E}^{F\Theta eh} = \frac{\partial E^{F\Theta eh}}{\partial \mathbf{F}} \cdot \dot{\mathbf{F}} + \frac{\partial E^{F\Theta eh}}{\partial \Theta}\dot{\Theta} + \frac{\partial E^{F\Theta eh}}{\partial \mathbf{e}^*} \cdot \dot{\mathbf{e}}^* + \frac{\partial E^{F\Theta eh}}{\partial \mathbf{h}^*} \cdot \dot{\mathbf{h}}^*. \tag{9.73}$$

Substitution of (9.73) into (9.72), and use of the standard arguments, leads to

$$\mathbf{P} = \rho_R\frac{\partial E^{F\Theta eh}}{\partial \mathbf{F}}, \quad \eta = -\frac{\partial E^{F\Theta eh}}{\partial \Theta}, \quad \mathbf{p}^* = -\rho\frac{\partial E^{F\Theta eh}}{\partial \mathbf{e}^*}, \quad \mathbf{m}^* = -\frac{\rho}{\mu_o}\frac{\partial E^{F\Theta eh}}{\partial \mathbf{h}^*}. \tag{9.74}$$

PROBLEM 9.9

Using the second law of thermodynamics and the appropriate Legendre transformation, derive the constitutive equations for the *deformation-temperature-polarization-magnetic field* formulation.

Solution

In this formulation, $\mathbf{F}$, Θ, $\mathbf{p}^*/\rho$, and $\mathbf{h}^*$ are the independent variables. We define the energy potential $E^{F\Theta ph}$ as the Legendre transformation of internal energy $\varepsilon = \breve{\varepsilon}\,(\mathbf{F}, \eta, \mathbf{p}^*/\rho, \mathbf{m}^*/\rho)$ with respect to the thermal and magnetic variables, from η to Θ and $\mathbf{m}^*/\rho$ to $\mathbf{h}^*$, i.e.,

$$E^{F\Theta ph} = \varepsilon - \Theta\eta - \mu_0 \mathbf{h}^* \cdot \frac{\mathbf{m}^*}{\rho}.$$

Refer to Table 9.4. The rate of the Legendre transformation is

$$\dot{E}^{F\Theta ph} = \dot{\varepsilon} - \Theta\dot{\eta} - \eta\dot{\Theta} - \mu_0 \mathbf{h}^* \cdot \overline{\left(\frac{\mathbf{m}^*}{\rho}\right)}^{\boldsymbol{\cdot}} - \mu_0 \frac{\mathbf{m}^*}{\rho} \cdot \dot{\mathbf{h}}^*,$$

and subsequent use of this result in the second law of thermodynamics (9.61) leads to

$$-\dot{E}^{F\Theta ph} + \frac{1}{\rho_{\mathrm{R}}}\mathbf{P} \cdot \dot{\mathbf{F}} - \eta\dot{\Theta} + \mathbf{e}^* \cdot \overline{\left(\frac{\mathbf{p}^*}{\rho}\right)}^{\boldsymbol{\cdot}} - \mu_0 \frac{\mathbf{m}^*}{\rho} \cdot \dot{\mathbf{h}}^* + \frac{1}{\rho}\mathbf{j}^* \cdot \mathbf{e}^* - \frac{1}{\rho\Theta}\mathbf{q} \cdot \operatorname{grad}\Theta \geq 0. \tag{a}$$

Use of the chain rule on $\dot{E}^{F\Theta ph}$ gives

$$\dot{E}^{F\Theta ph} = \frac{\partial E^{F\Theta ph}}{\partial \mathbf{F}} \cdot \dot{\mathbf{F}} + \frac{\partial E^{F\Theta ph}}{\partial \Theta}\dot{\Theta} + \frac{\partial E^{F\Theta ph}}{\partial \left(\frac{\mathbf{p}^*}{\rho}\right)} \cdot \overline{\left(\frac{\mathbf{p}^*}{\rho}\right)}^{\boldsymbol{\cdot}} + \frac{\partial E^{F\Theta ph}}{\partial \mathbf{h}^*} \cdot \dot{\mathbf{h}}^*,$$

and substitution of this result into inequality (a) leads to the constitutive equations

$$\mathbf{P} = \rho_{\mathrm{R}} \frac{\partial E^{F\Theta ph}}{\partial \mathbf{F}}, \quad \eta = -\frac{\partial E^{F\Theta ph}}{\partial \Theta}, \quad \mathbf{e}^* = \frac{\partial E^{F\Theta ph}}{\partial \left(\frac{\mathbf{p}^*}{\rho}\right)}, \quad \mathbf{m}^* = -\frac{\rho}{\mu_0}\frac{\partial E^{F\Theta ph}}{\partial \mathbf{h}^*}.$$

We now illustrate how the procedure described in Section 9.6.2.1 can be used to introduce the polarization $\mathbf{p}^*$ (instead of the polarization per unit mass $\mathbf{p}^*/\rho$) as the electrical independent variable. Equation $(9.66)_1$, i.e.,

$$\overline{\left(\frac{\mathbf{p}^*}{\rho}\right)}^{\boldsymbol{\cdot}} = \frac{1}{\rho}\dot{\mathbf{p}}^* + \frac{1}{\rho}\left(\mathbf{F}^{-\mathrm{T}} \cdot \dot{\mathbf{F}}\right)\mathbf{p}^*,$$

is used to rewrite the second law inequality (a) as

$$\begin{aligned}
-\dot{E}^{F\Theta ph} &+ \left[\frac{1}{\rho_{\mathrm{R}}}\mathbf{P} + \frac{1}{\rho}\left(\mathbf{e}^* \cdot \mathbf{p}^*\right)\mathbf{F}^{-\mathrm{T}}\right] \cdot \dot{\mathbf{F}} - \eta\dot{\Theta} + \frac{1}{\rho}\mathbf{e}^* \cdot \dot{\mathbf{p}}^* \\
&- \mu_0 \frac{\mathbf{m}^*}{\rho} \cdot \dot{\mathbf{h}}^* + \frac{1}{\rho}\mathbf{j}^* \cdot \mathbf{e}^* - \frac{1}{\rho\Theta}\mathbf{q} \cdot \operatorname{grad}\Theta \geq 0.
\end{aligned} \tag{b}$$

Note that the energy potential $E^{F\Theta ph}$ in (b) uses the polarization $\mathbf{p}^*$ instead of the polarization per unit mass $\mathbf{p}^*/\rho$ as the electrical independent variable. It then follows that

$$\dot{E}^{F\Theta ph} = \frac{\partial E^{F\Theta ph}}{\partial \mathbf{F}} \cdot \dot{\mathbf{F}} + \frac{\partial E^{F\Theta ph}}{\partial \Theta} \dot{\Theta} + \frac{\partial E^{F\Theta ph}}{\partial \mathbf{p}^*} \cdot \dot{\mathbf{p}}^* + \frac{\partial E^{F\Theta ph}}{\partial \mathbf{h}^*} \cdot \dot{\mathbf{h}}^*.$$

Subsequent use of this result in (b) leads to the constitutive equations

$$\mathbf{P} = \rho_R \frac{\partial E^{F\Theta ph}}{\partial \mathbf{F}} - J\left(\mathbf{e}^* \cdot \mathbf{p}^*\right)\mathbf{F}^{-T}, \quad \eta = -\frac{\partial E^{F\Theta ph}}{\partial \Theta},$$

$$\mathbf{e}^* = \rho \frac{\partial E^{F\Theta ph}}{\partial \mathbf{p}^*}, \quad \mathbf{m}^* = -\frac{\rho}{\mu_o} \frac{\partial E^{F\Theta ph}}{\partial \mathbf{h}^*}.$$

PROBLEM 9.10

Using the second law of thermodynamics and the appropriate Legendre transformation, derive the constitutive equations for the *temperature-stress-electric field-magnetic field* formulation.

Solution

The *temperature-stress-electric field-magnetic field* formulation correlates with a thermodynamic process where $\{\mathbf{P}, \Theta, \mathbf{e}^*, \mathbf{h}^*\}$ are the independent variables and $E^{P\Theta eh}$ is the characterizing thermodynamic energy potential. $E^{P\Theta eh}$ is defined as the Legendre transformation of internal energy $\varepsilon = \breve{\varepsilon}\left(\mathbf{F}, \eta, \mathbf{p}^*/\rho, \mathbf{m}^*/\rho\right)$ with respect to the mechanical, thermal, electrical, and magnetic variables, from $\mathbf{F}$ to $\mathbf{P}$, η to Θ, $\mathbf{p}^*/\rho$ to $\mathbf{e}^*$, and $\mathbf{m}^*/\rho$ to $\mathbf{h}^*$,

$$E^{P\Theta eh} = \varepsilon - \frac{1}{\rho_R}\mathbf{P} \cdot \mathbf{F} - \Theta\eta - \mathbf{e}^* \cdot \frac{\mathbf{p}^*}{\rho} - \mu_o \mathbf{h}^* \cdot \frac{\mathbf{m}^*}{\rho}.$$

Refer to Table 9.6. Taking the rate of the Legendre transformation gives

$$\dot{E}^{P\Theta eh} = \dot{\varepsilon} - \frac{1}{\rho_R}\mathbf{P} \cdot \dot{\mathbf{F}} - \frac{1}{\rho_R}\mathbf{F} \cdot \dot{\mathbf{P}} - \Theta\dot{\eta} - \eta\dot{\Theta} - \mathbf{e}^* \cdot \overline{\left(\frac{\mathbf{p}^*}{\rho}\right)}^{\cdot}$$

$$- \frac{\mathbf{p}^*}{\rho} \cdot \dot{\mathbf{e}}^* - \mu_o \mathbf{h}^* \cdot \overline{\left(\frac{\mathbf{m}^*}{\rho}\right)}^{\cdot} - \mu_o \frac{\mathbf{m}^*}{\rho} \cdot \dot{\mathbf{h}}^*,$$

and substitution of this result into the second law (9.61) yields

$$-\dot{E}^{P\Theta eh} - \frac{1}{\rho_R}\mathbf{F} \cdot \dot{\mathbf{P}} - \eta\dot{\Theta} - \frac{\mathbf{p}^*}{\rho} \cdot \dot{\mathbf{e}}^* - \mu_o \frac{\mathbf{m}^*}{\rho} \cdot \dot{\mathbf{h}}^* + \frac{1}{\rho}\mathbf{j}^* \cdot \mathbf{e}^* - \frac{1}{\rho\Theta}\mathbf{q} \cdot \operatorname{grad}\Theta \geq 0.$$

It can be shown that use of the chain rule on $\dot{E}^{P\Theta eh}$ leads to the constitutive equations

$$\mathbf{F} = -\rho_R \frac{\partial E^{P\Theta eh}}{\partial \mathbf{P}}, \quad \eta = -\frac{\partial E^{P\Theta eh}}{\partial \Theta}, \quad \mathbf{p}^* = -\rho \frac{\partial E^{P\Theta eh}}{\partial \mathbf{e}^*}, \quad \mathbf{m}^* = -\frac{\rho}{\mu_o} \frac{\partial E^{P\Theta eh}}{\partial \mathbf{h}^*},$$

an exercise that we leave to the reader.

9.6.4 RESTRICTIONS IMPOSED BY INVARIANCE UNDER SUPERPOSED RIGID BODY MOTIONS AND CONSERVATION OF ANGULAR MOMENTUM

The constitutive equations (9.74) for the *deformation-temperature-electric field-magnetic field formulation* (or any other formulation for that matter, e.g., (9.64) or (9.70)), obtained from restrictions imposed by the second law, must also satisfy invariance under superposed rigid body motions. In particular, we must have [30]

$$\left(E^{F\Theta eh}\right)^+ = E^{F\Theta eh} \tag{9.75a}$$

when

$$\mathbf{F}^+ = \mathbf{QF}, \quad \Theta^+ = \Theta, \quad \left(\mathbf{e}^*\right)^+ = \mathbf{Qe}^*, \quad \left(\mathbf{h}^*\right)^+ = \mathbf{Qh}^* \tag{9.75b}$$

for all proper orthogonal tensors $\mathbf{Q}(t)$. Equations (9.75a) and (9.75b) are referred to as invariance requirements. It can be shown that these invariance requirements demand that

$$E^{F\Theta eh}\left(\mathbf{F}, \Theta, \mathbf{e}^*, \mathbf{h}^*\right) = \tilde{E}^{F\Theta eh}\left(\mathbf{C}, \Theta, \mathbf{e}_{\mathrm{R}}, \mathbf{h}_{\mathrm{R}}\right), \tag{9.76}$$

where

$$\mathbf{C} = \mathbf{F}^{\mathrm{T}}\mathbf{F}, \quad \mathbf{e}_{\mathrm{R}} = \mathbf{F}^{\mathrm{T}}\mathbf{e}^*, \quad \mathbf{h}_{\mathrm{R}} = \mathbf{F}^{\mathrm{T}}\mathbf{h}^*$$

are the right Cauchy-Green deformation tensor, referential electric field, and referential magnetic field, respectively. We can then verify, through a change of independent variable from $\mathbf{F}$ to $\mathbf{C}$, $\mathbf{e}^*$ to $\mathbf{e}_{\mathrm{R}}$, and $\mathbf{h}^*$ to $\mathbf{h}_{\mathrm{R}}$, that the constitutive equations (9.74) become

$$\mathbf{T} = 2\rho\mathbf{F}\frac{\partial \tilde{E}^{F\Theta eh}}{\partial \mathbf{C}}\mathbf{F}^{\mathrm{T}} - \mathbf{e}^* \otimes \mathbf{p}^* - \mu_0 \mathbf{h}^* \otimes \mathbf{m}^*, \quad \eta = -\frac{\partial \tilde{E}^{F\Theta eh}}{\partial \Theta},$$

$$\mathbf{p}^* = -\rho\mathbf{F}\frac{\partial \tilde{E}^{F\Theta eh}}{\partial \mathbf{e}_{\mathrm{R}}}, \quad \mathbf{m}^* = -\frac{\rho}{\mu_0}\mathbf{F}\frac{\partial \tilde{E}^{F\Theta eh}}{\partial \mathbf{h}_{\mathrm{R}}}. \tag{9.77}$$

These are the invariant forms of the constitutive equations. We emphasize that $\tilde{E}^{F\Theta eh}$ denotes the free energy function with $\mathbf{C}$, Θ, $\mathbf{e}_{\mathrm{R}}$, and $\mathbf{h}_{\mathrm{R}}$ as independent variables. A useful result follows from a change of the mechanical independent variable from the right Cauchy-Green deformation $\mathbf{C}$ to the Lagrangian strain $\mathbf{E}$, where $\mathbf{E} = 1/2$ $(\mathbf{C} - \mathbf{I})$, in which case (9.77) becomes

$$\mathbf{T} = \rho\mathbf{F}\frac{\partial \breve{E}^{F\Theta eh}}{\partial \mathbf{E}}\mathbf{F}^{\mathrm{T}} - \mathbf{e}^* \otimes \mathbf{p}^* - \mu_0 \mathbf{h}^* \otimes \mathbf{m}^*, \quad \eta = -\frac{\partial \breve{E}^{F\Theta eh}}{\partial \Theta},$$

$$\mathbf{p}^* = -\rho\mathbf{F}\frac{\partial \breve{E}^{F\Theta eh}}{\partial \mathbf{e}_{\mathrm{R}}}, \quad \mathbf{m}^* = -\frac{\rho}{\mu_0}\mathbf{F}\frac{\partial \breve{E}^{F\Theta eh}}{\partial \mathbf{h}_{\mathrm{R}}}, \tag{9.78}$$

where $\breve{E}^{F\Theta eh}$ denotes the free energy function with $\mathbf{E}$, Θ, $\mathbf{e}_{\mathrm{R}}$, and $\mathbf{h}_{\mathrm{R}}$ as independent variables. It can be verified that (9.78) satisfies conservation of angular momentum.

In the following section, we use (9.78) as a starting point for developing constitutive equations for materials with *linear*, reversible TEME response.

EXERCISES

1. Prove that the invariance requirements

$$\left(E^{F\Theta eh}\right)^{+} = E^{F\Theta eh}$$

when

$$\mathbf{F}^{+} = \mathbf{QF}, \quad \Theta^{+} = \Theta, \quad \left(\mathbf{e}^{*}\right)^{+} = \mathbf{Qe}^{*}, \quad \left(\mathbf{h}^{*}\right)^{+} = \mathbf{Qh}^{*}$$

for all proper orthogonal tensors $\mathbf{Q}(t)$ demand that

$$E^{F\Theta eh}\left(\mathbf{F}, \Theta, \mathbf{e}^{*}, \mathbf{h}^{*}\right) = \tilde{E}^{F\Theta eh}\left(\mathbf{C}, \Theta, \mathbf{e}_{\mathrm{R}}, \mathbf{h}_{\mathrm{R}}\right).$$

2. Verify that the invariant form of the constitutive equation for the Cauchy stress, i.e.,

$$\mathbf{T} = \rho \mathbf{F} \frac{\partial \breve{E}^{F\Theta eh}}{\partial \mathbf{E}} \mathbf{F}^{\mathrm{T}} - \mathbf{e}^{*} \otimes \mathbf{p}^{*} - \mu_{0} \mathbf{h}^{*} \otimes \mathbf{m}^{*},$$

satisfies conservation of angular momentum (9.57b).

9.7 CONSTITUTIVE MODEL DEVELOPMENT FOR THERMO-ELECTRO-MAGNETO-ELASTIC MATERIALS: SMALL-DEFORMATION THEORY

Materials exhibiting spontaneous polarization or magnetization in the presence of external electric or magnetic fields are broadly classified as ferroic materials. When coupled with mechanical stresses, ferromagnetic materials exhibit *magnetostriction* (i.e., deformation induced by a magnetic field), and ferroelectric materials exhibit *electrostriction* (i.e., deformation induced by an electric field). Materials exhibiting both orders of ferroic behavior are called multiferroics. These ferroic effects are often hysteretic, dissipative, and nonlinear, but for near-equilibrium processes in the small-deformation and small-electromagnetic-field regime, the associated thermo-electro-magneto-elastic (TEME) constitutive equations can often be approximated as *linear* and *reversible*.

9.7.1 SMALL-DEFORMATION KINEMATICS, KINETICS, ELECTROMAGNETIC FIELDS, AND FUNDAMENTAL LAWS

The small-deformation theory (or, equivalently, the small-strain or infinitesimal theory) is customarily obtained by assuming that the displacements **u** are small, and expanding **u** in a power series with respect to a small parameter.[12] As a consequence

[12] The assumption of small displacements implies that both strains and rotations are small.

of this approximation, (3.7), (3.27), (3.36), (3.54), and (3.55) become, to leading order,[13,14]

$$\mathbf{x} = \mathbf{X}, \quad \mathbf{F} = \mathbf{I}, \quad J = 1, \quad \mathbf{E} = \mathbf{e} = \frac{1}{2}\Big[\operatorname{grad}\mathbf{u} + (\operatorname{grad}\mathbf{u})^{\mathrm{T}}\Big]. \tag{9.79}$$

Hence, in the small-deformation theory, the reference and present configurations of the body are indistinguishable, the referential (Lagrangian) and spatial (Eulerian) descriptions of a given quantity are one and the same, and the Eulerian and Lagrangian strains are identical. In the small-deformation theory, we also have

$$\dot{\phi} = \phi', \quad \operatorname{Grad}\phi = \operatorname{grad}\phi, \tag{9.80}$$

where ϕ is a scalar-valued (or vector-valued or tensor-valued) function of position and time; refer to (3.20) and (3.21). It then follows from (9.41) and (9.79) that, to leading order,

$$\begin{aligned} &\mathbf{e}_{\mathrm{R}} = \mathbf{e}^*, \quad &&\mathbf{p}_{\mathrm{R}} = \mathbf{p}^*, \quad &&\mathbf{d}_{\mathrm{R}} = \mathbf{d}^*, \quad &&\mathbf{h}_{\mathrm{R}} = \mathbf{h}^*, \quad &&\mathbf{m}_{\mathrm{R}} = \mathbf{m}^*, \\ &\mathbf{b}_{\mathrm{R}} = \mathbf{b}^*, \quad &&\sigma_{\mathrm{R}} = \sigma^*, \quad &&\mathbf{j}_{\mathrm{R}} = \mathbf{j}^*, \quad &&\mathbf{P} = \mathbf{T}, \quad &&\mathbf{q}_{\mathrm{R}} = \mathbf{q}. \end{aligned} \tag{9.81}$$

We assume that the electromagnetic fields are small, so their contributions to balance of linear momentum and angular momentum emerge as *higher-order* terms. In other words, the electromagnetic body force and body couple vanish at leading order, i.e.,

$$\mathbf{f}^{\mathrm{em}} = \mathbf{0}, \quad \boldsymbol{\Gamma}^{\mathrm{em}} = \mathbf{0}. \tag{9.82}$$

An additional implication of the small-field assumption is that

$$\begin{aligned} &\mathbf{e}^* = \mathbf{e}, \quad &&\mathbf{p}^* = \mathbf{p}, \quad &&\mathbf{d}^* = \mathbf{d}, \quad &&\mathbf{h}^* = \mathbf{h}, \\ &\mathbf{m}^* = \mathbf{m}, \quad &&\mathbf{b}^* = \mathbf{b}, \quad &&\sigma^* = \sigma, \quad &&\mathbf{j}^* = \mathbf{j}. \end{aligned} \tag{9.83}$$

That is, to leading order, the effective fields collapse to the corresponding standard fields, and all standard fields are equivalent (the $\mathbf{v} \to \mathbf{0}$ limit of (9.51)–(9.54)).

As a consequence of (9.79)–(9.83), we hereafter omit adjectives such as "referential," "spatial," "effective," and "standard" that are no longer needed in the small-deformation/small-field theory, and instead refer to $\mathbf{E}$ as the infinitesimal strain, $\mathbf{T}$ the stress, $\mathbf{e}$ the electric field, $\mathbf{p}$ the electric polarization, $\mathbf{d}$ the electric displacement, and so on. Also, (9.79)–(9.83) imply that the fundamental laws (9.49a)–(9.49j), when expressed in Cartesian component notation, become, to leading order,

$$\rho = \rho_{\mathrm{R}}, \tag{9.84a}$$

$$\rho u_{i,tt} = \rho f_i^{\mathrm{m}} + T_{ij,j}, \tag{9.84b}$$

[13]The leading-order terms within a mathematical equation, expression, or model are the terms with the largest magnitude. For instance, in the small-deformation theory, $\mathbf{F} = \mathbf{I} + \mathcal{O}(\epsilon)$, where ϵ is a small parameter and $\mathcal{O}(\epsilon)$ denotes the higher-order (or lower-magnitude) terms.

[14]Note the common notation $\mathbf{e}$ for the Eulerian strain tensor and the electric field. This should not lead to any ambiguity, however, as this is the first and last time the Eulerian strain tensor appears in this chapter.

$$T_{ij} = T_{ji}, \tag{9.84c}$$

$$\rho\varepsilon_{,t} = T_{ij}E_{ij,t} + e_i\, p_{i,t} + \mu_o\, h_i\, m_{i,t} + \rho r^t + j_i\, e_i - q_{i,i}, \tag{9.84d}$$

$$\rho\eta_{,t} \geq \rho\frac{r^t}{\Theta} - \left(\frac{q_i}{\Theta}\right)_{,i}, \tag{9.84e}$$

$$\sigma_{,t} + j_{i,i} = 0, \tag{9.84f}$$

$$b_{i,i} = 0, \tag{9.84g}$$

$$b_{i,t} + \varepsilon_{ijk}e_{k,j} = 0, \tag{9.84h}$$

$$d_{i,i} = \sigma, \tag{9.84i}$$

$$d_{i,t} + j_i = \varepsilon_{ijk}h_{k,j}. \tag{9.84j}$$

The strain-displacement relationship $(9.79)_4$, expressed in Cartesian component notation, becomes

$$E_{ij} = \frac{1}{2}\left(u_{i,j} + u_{j,i}\right). \tag{9.85}$$

Note that $()_{,t}$ denotes partial differentiation with respect to time, e.g.,

$$b_{i,t} \equiv \frac{\partial b_i(x_1, x_2, x_3, t)}{\partial t}.$$

Also note that conservation of angular momentum (9.84c) and definition (9.85) imply that the stress **T** and infinitesimal strain **E** are symmetric.

9.7.2 LINEAR CONSTITUTIVE EQUATIONS

With use of relationships (9.79)–(9.83), the TEME constitutive equations (9.78) become, to leading order in the small-deformation/small-field theory,

$$T_{ij} = \rho\frac{\partial\psi}{\partial E_{ij}}, \quad \eta = -\frac{\partial\psi}{\partial\Theta}, \quad p_i = -\rho\frac{\partial\psi}{\partial e_i}, \quad m_i = -\frac{\rho}{\mu_o}\frac{\partial\psi}{\partial h_i}, \tag{9.86}$$

where we have introduced $\psi = \breve{E}^{F\Theta eh}$ for notational brevity. In what follows, starting from (9.86), we derive linear TEME constitutive equations for ferroic materials within the near-equilibrium regime. To obtain a functional form of the free energy ψ that characterizes this linear, reversible TEME response, we perform a Taylor series expansion of ψ in terms of its independent variables **E**, Θ, **e**, and **h** in the neighborhood of a thermodynamic equilibrium state $\mathbf{x}_o = (\mathbf{E}^o, \Theta^o, \mathbf{e}^o, \mathbf{h}^o)$:

$$\breve{E}^{F\Theta eh} \equiv \psi\left(\mathbf{E}, \Theta, \mathbf{e}, \mathbf{h}\right) = \psi|_{x_o} + \frac{1}{2}\frac{\partial^2\psi}{\partial E_{ij}\partial E_{kl}}\bigg|_{x_o} E_{ij}E_{kl} + \frac{1}{2}\frac{\partial^2\psi}{\partial e_i\partial e_j}\bigg|_{x_o} e_ie_j$$
$$+ \frac{1}{2}\frac{\partial^2\psi}{\partial h_i\partial h_j}\bigg|_{x_o} h_ih_j + \frac{1}{2}\frac{\partial^2\psi}{\partial\Theta^2}\bigg|_{x_o}\Theta^2 + \frac{\partial^2\psi}{\partial E_{ij}\partial e_k}\bigg|_{x_o} E_{ij}e_k$$

$$+ \left.\frac{\partial^2 \psi}{\partial E_{ij}\partial h_k}\right|_{x_o} E_{ij}h_k + \left.\frac{\partial^2 \psi}{\partial e_i \partial h_j}\right|_{x_o} e_i h_j + \left.\frac{\partial^2 \psi}{\partial \Theta \partial E_{ij}}\right|_{x_o} \Theta E_{ij}$$

$$+ \left.\frac{\partial^2 \psi}{\partial \Theta \partial e_i}\right|_{x_o} \Theta e_i + \left.\frac{\partial^2 \psi}{\partial \Theta \partial h_i}\right|_{x_o} \Theta h_i + \text{higher order terms.} \quad (9.87)$$

Note that the TEME quantities E_{ij}, Θ, e_i, and h_i in (9.87) are *perturbed about the equilibrium state* $\mathbf{x}_o$. Disregarding the higher-order terms (i.e., truncating at second order), we substitute the free energy expansion (9.87) into (9.86) to obtain the linear form of the constitutive equations:

$$T_{ij} = \left.\frac{\partial^2 \bar{\psi}}{\partial E_{ij}\partial E_{kl}}\right|_{x_o} E_{kl} + \left.\frac{\partial^2 \bar{\psi}}{\partial E_{ij}\partial e_k}\right|_{x_o} e_k + \left.\frac{\partial^2 \bar{\psi}}{\partial E_{ij}\partial h_k}\right|_{x_o} h_k + \left.\frac{\partial^2 \bar{\psi}}{\partial E_{ij}\partial \Theta}\right|_{x_o} \Theta, \quad (9.88a)$$

$$p_i = -\left.\frac{\partial^2 \bar{\psi}}{\partial E_{jk}\partial e_i}\right|_{x_o} E_{jk} - \left.\frac{\partial^2 \bar{\psi}}{\partial e_i \partial e_j}\right|_{x_o} e_j - \left.\frac{\partial^2 \bar{\psi}}{\partial e_i \partial h_j}\right|_{x_o} h_j - \left.\frac{\partial^2 \bar{\psi}}{\partial e_i \partial \Theta}\right|_{x_o} \Theta, \quad (9.88b)$$

$$\mu_o m_i = -\left.\frac{\partial^2 \bar{\psi}}{\partial E_{jk}\partial h_i}\right|_{x_o} E_{jk} - \left.\frac{\partial^2 \bar{\psi}}{\partial e_j \partial h_i}\right|_{x_o} e_j - \left.\frac{\partial^2 \bar{\psi}}{\partial h_i \partial h_j}\right|_{x_o} h_j - \left.\frac{\partial^2 \bar{\psi}}{\partial h_i \partial \Theta}\right|_{x_o} \Theta, \quad (9.88c)$$

$$\bar{\eta} = -\left.\frac{\partial^2 \bar{\psi}}{\partial E_{ij}\partial \Theta}\right|_{x_o} E_{ij} - \left.\frac{\partial^2 \bar{\psi}}{\partial e_i \partial \Theta}\right|_{x_o} e_i - \left.\frac{\partial^2 \bar{\psi}}{\partial h_i \partial \Theta}\right|_{x_o} h_i - \left.\frac{\partial^2 \bar{\psi}}{\partial \Theta^2}\right|_{x_o} \Theta, \quad (9.88d)$$

where $\bar{\psi} = \rho\psi$ and $\bar{\eta} = \rho\eta$ denote the free energy per unit volume and the free entropy per unit volume, respectively. The constant coefficients arising in the linearized constitutive equations (9.88a)–(9.88d) are characterized using experimental data specific to the material. For instance, $\partial^2\bar{\psi}/(\partial E_{ij}\partial E_{kl})\big|_{x_o}$ represents the component of the elasticity tensor at equilibrium state $\mathbf{x}_o$, which is specific to the material being characterized; refer to Table 9.7. The other coefficients can be described in a similar

Table 9.7 Material Constants and Their Representations for Linear Reversible Processes

Constant	Representation	Constant	Representation
Elasticity constant	$C_{ijkl} = \left.\frac{\partial^2 \bar{\psi}}{\partial E_{ij}\partial E_{kl}}\right\|_{x_o}$	Piezoelectric constant	$d^{e}_{ijk} = -\left.\frac{\partial^2 \bar{\psi}}{\partial E_{ij}\partial e_k}\right\|_{x_o}$
Piezomagnetic constant	$d^{m}_{ijk} = -\left.\frac{\partial^2 \bar{\psi}}{\partial E_{ij}\partial h_k}\right\|_{x_o}$	Coefficient of thermal stress	$\beta_{ij} = -\left.\frac{\partial^2 \bar{\psi}}{\partial \Theta \partial E_{ij}}\right\|_{x_o}$
Electric susceptibility	$\chi^{e}_{ij} = -\left.\frac{\partial^2 \bar{\psi}}{\partial e_i \partial e_j}\right\|_{x_o}$	Magnetoelectric constant	$\chi^{em}_{ij} = -\left.\frac{\partial^2 \bar{\psi}}{\partial e_i \partial h_j}\right\|_{x_o}$
Pyroelectric constant	$L^{e}_{i} = -\left.\frac{\partial^2 \bar{\psi}}{\partial e_i \partial \Theta}\right\|_{x_o}$	Magnetic susceptibility	$\chi^{m}_{ij} = -\left.\frac{\partial^2 \bar{\psi}}{\partial h_i \partial h_j}\right\|_{x_o}$
Pyromagnetic constant	$L^{m}_{i} = -\left.\frac{\partial^2 \bar{\psi}}{\partial h_i \partial \Theta}\right\|_{x_o}$	Specific heat	$c = -\left.\frac{\partial^2 \bar{\psi}}{\partial \Theta^2}\right\|_{x_o}$

manner. With use of the nomenclature shown in Table 9.7, the free energy per unit volume (9.87) becomes

$$\bar{\psi} = \frac{1}{2}C_{ijkl}E_{ij}E_{kl} - \frac{1}{2}\chi_{ij}^{\mathrm{e}}e_ie_j - \frac{1}{2}\chi_{ij}^{\mathrm{m}}h_ih_j - \frac{1}{2}c\Theta^2 - d_{ijk}^{\mathrm{e}}E_{ij}e_k - d_{ijk}^{\mathrm{m}}E_{ij}h_k$$
$$-\beta_{ij}E_{ij}\Theta - \chi_{ij}^{\mathrm{em}}e_ih_j - L_i^{\mathrm{e}}\Theta e_i - L_i^{\mathrm{m}}\Theta h_i, \qquad (9.89)$$

and the linearized set of constitutive equations (9.88a)–(9.88d) simplify to

$$T_{ij} = C_{ijkl}E_{kl} - d_{ijk}^{\mathrm{e}}e_k - d_{ijk}^{\mathrm{m}}h_k - \beta_{ij}\Theta, \qquad (9.90a)$$

$$p_i = d_{jki}^{\mathrm{e}}E_{jk} + \chi_{ij}^{\mathrm{e}}e_j + \chi_{ij}^{\mathrm{em}}h_j + L_i^{\mathrm{e}}\Theta, \qquad (9.90b)$$

$$\mu_o m_i = d_{jki}^{\mathrm{m}}E_{jk} + \chi_{ji}^{\mathrm{em}}e_j + \chi_{ij}^{\mathrm{m}}h_j + L_i^{\mathrm{m}}\Theta, \qquad (9.90c)$$

$$\bar{\eta} = \beta_{ij}E_{ij} + L_i^{\mathrm{e}}e_i + L_i^{\mathrm{m}}h_i + c\Theta. \qquad (9.90d)$$

9.7.3 MATERIAL SYMMETRY

The symmetric infinitesimal strains and stresses, as well as the existence of a free energy function $\bar{\psi}$, lead to the following restrictions on the material constants in the constitutive equations (9.90a)–(9.90d):

$$C_{ijkl} = C_{jikl} = C_{ijlk} = C_{klij}, \qquad d_{ijk}^{\mathrm{e}} = d_{jik}^{\mathrm{e}}, \qquad d_{ijk}^{\mathrm{m}} = d_{jik}^{\mathrm{m}},$$
$$\chi_{ij}^{\mathrm{e}} = \chi_{ji}^{\mathrm{e}}, \qquad \chi_{ij}^{\mathrm{m}} = \chi_{ji}^{\mathrm{m}}, \qquad \beta_{ij} = \beta_{ji}.$$

These restrictions reduce the total number of unknown material constants to 169. The number of unknown material constants can be further reduced using crystal symmetry arguments. Materials that undergo one or more *linear thermo-electro-mechanical processes* (refer to Figure 9.2) can be classified into 32 crystallographic symmetry groups. These groups are based on rotation, reflection, and inversion symmetry of the crystal structure. A detailed description of each of these symmetry groups is provided in [63]. For magnetic materials, the concept of *time inversion symmetry* becomes an additional consideration, which increases the total number of possible symmetry groups from 32 to 122 (90 magnetic and 32 crystallographic symmetry groups).

Symmetry of the stress, strain, and material constant tensors allows further simplification of the constitutive equations (9.90a)–(9.90d) using **Voigt notation**. Voigt notation is a standard mapping, typically used to reduce the order (or rank) of symmetric tensors. The indices are mapped as follows:

$$11 \to 1, \quad 22 \to 2, \quad 33 \to 3, \quad 23 \to 4, \quad 13 \to 5, \quad 12 \to 6.$$

For example, Voigt notation simplifies the customary three-by-three matrix representation of the symmetric second-order stress tensor (refer, for instance, to (2.34)) to a single column matrix with six independent components. Similar reductions are accomplished for matrix representations of the fourth-order elasticity tensor, third-order piezoelectric and piezomagnetic coupling tensors, and second-order strain

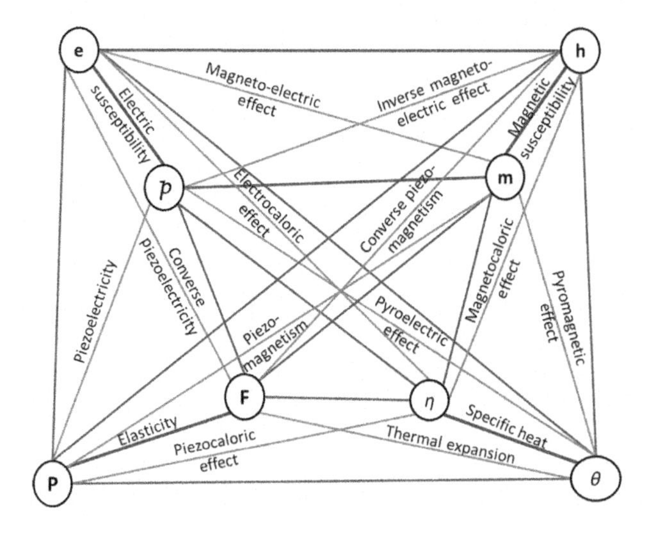

FIGURE 9.2

Multiphysics interaction diagram illustrating various linear TEMM effects.

tensor. With use of this shorthand notation, the constitutive equations (9.90a)–(9.90d), for the special case of a fully coupled TEME material with hexagonal crystal symmetry (i.e., the C_{6v} crystallographic symmetry group), can be presented in matrix form as

$$
\left[\begin{array}{c} T_1 \\ T_2 \\ T_3 \\ T_4 \\ T_5 \\ T_6 \\ \hline p_1 \\ p_2 \\ p_3 \\ \hline \mu_o m_1 \\ \mu_o m_2 \\ \mu_o m_3 \\ \hline \bar{\eta} \end{array}\right]
=
\left[\begin{array}{cccccc|ccc|ccc|c}
C_{11} & C_{12} & C_{13} & 0 & 0 & 0 & 0 & 0 & -d^{e}_{13} & 0 & 0 & -d^{m}_{13} & -\beta_{11} \\
C_{12} & C_{11} & C_{13} & 0 & 0 & 0 & 0 & 0 & -d^{e}_{13} & 0 & 0 & -d^{m}_{13} & -\beta_{11} \\
C_{13} & C_{13} & C_{33} & 0 & 0 & 0 & 0 & 0 & -d^{e}_{33} & 0 & 0 & -d^{m}_{33} & -\beta_{33} \\
0 & 0 & 0 & C_{44} & 0 & 0 & -d^{e}_{41} & -d^{e}_{51} & 0 & 0 & -d^{m}_{51} & 0 & 0 \\
0 & 0 & 0 & 0 & C_{44} & 0 & -d^{e}_{51} & d^{e}_{41} & 0 & -d^{m}_{51} & 0 & 0 & 0 \\
0 & 0 & 0 & 0 & 0 & C_{66} & 0 & 0 & 0 & 0 & 0 & 0 & 0 \\
\hline
0 & 0 & 0 & d^{e}_{41} & d^{e}_{51} & 0 & \chi^{e}_{11} & 0 & 0 & \chi^{em}_{11} & 0 & 0 & 0 \\
0 & 0 & 0 & d^{e}_{51} & -d^{e}_{41} & 0 & 0 & \chi^{e}_{11} & 0 & 0 & \chi^{em}_{11} & 0 & 0 \\
d^{e}_{13} & d^{e}_{13} & d^{e}_{33} & 0 & 0 & 0 & 0 & 0 & \chi^{e}_{33} & 0 & 0 & \chi^{em}_{33} & L^{e}_{3} \\
\hline
0 & 0 & 0 & 0 & d^{m}_{51} & 0 & \chi^{em}_{11} & 0 & 0 & \chi^{m}_{11} & 0 & 0 & 0 \\
0 & 0 & 0 & d^{m}_{51} & 0 & 0 & 0 & \chi^{em}_{11} & 0 & 0 & \chi^{m}_{11} & 0 & 0 \\
d^{m}_{13} & d^{m}_{13} & d^{m}_{33} & 0 & 0 & 0 & 0 & 0 & \chi^{em}_{33} & 0 & 0 & \chi^{m}_{33} & L^{m}_{3} \\
\hline
\beta_{11} & \beta_{11} & \beta_{33} & 0 & 0 & 0 & 0 & 0 & L^{e}_{3} & 0 & 0 & L^{m}_{3} & c
\end{array}\right]
\left[\begin{array}{c} E_1 \\ E_2 \\ E_3 \\ E_4 \\ E_5 \\ E_6 \\ \hline e_1 \\ e_2 \\ e_3 \\ \hline h_1 \\ h_2 \\ h_3 \\ \hline \Theta \end{array}\right]
$$

where $C_{66} = 1/2(C_{11} - C_{12})$. Clearly, crystal symmetry considerations greatly reduce the number of unknown material constants, which in turn reduces the number of experiments needed to completely characterize a material.

9.8 LINEAR, REVERSIBLE, THERMO-ELECTRO-MAGNETO-MECHANICAL PROCESSES

In this section, we discuss the wealth of physical phenomena and material behavior that can be described by the constitutive equations (9.90a)–(9.90d) and characterized

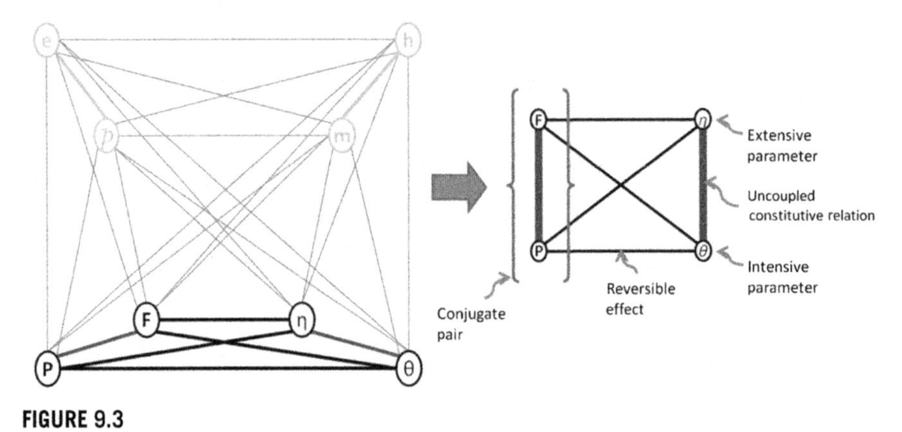

FIGURE 9.3

The thermomechanical panel, a subset of the multiphysics interaction diagram in Figure 9.2.

as linear, reversible, thermo-electro-magneto-mechanical (TEMM) processes. The multiphysics interaction diagram (see Figure 9.2) describes all combinations of linear TEMM processes. Each of the thermal, electrical, magnetic, and mechanical physical effects are defined by their corresponding extensive and intensive variables, marked at the inner and outer quadrilateral corners of the multiphysics interaction diagram, respectively.

The diagonal edges joining the inner and outer quadrilaterals signify the uncoupled processes, i.e., elasticity, polarization, magnetization, and heat capacity. Coupled processes are described through six subset diagrams, each relating two of the four physical effects. Each of the six subset diagrams, the corresponding coupled processes, and the materials that exhibit these properties are highlighted in Table 9.8; also see Figure 9.3.

The coupled processes therein can be categorized as either (1) a **primary process**—a coupled process that relates the intensive parameter of one physical effect to the extensive parameter of the second physical effect—or (2) a **secondary process**—a coupled process that is a superposition of two or more primary processes. In other words, primary processes are direct or one-step processes that describe the coupling between any two physical effects, whereas secondary processes are multistep processes that are a superposition of two or more primary effects.

Owing to the linear nature of the constitutive model under consideration, any coupled TEMM effect can be studied as a superposition of the uncoupled and coupled primary processes highlighted in Table 9.8. For example, a linear thermo-electro-mechanical process can be described as the superposition of a linear thermoelectric (pyroelectric) process and a linear electromechanical (piezoelectric) process.

Depending on the smart material being modeled, appropriate terms can be chosen from equations (9.90a)–(9.90d) to describe its behavior. This will be demonstrated in the next section for the special case of piezoelectric materials.

Table 9.8 Subset Diagrams of the Fully Coupled TEMM Multiphysics Interaction Diagram

Primary Effects	Subset Diagram
Direct and converse piezoelectricity Examples: $BaTiO_3$, lead zirconate titanate	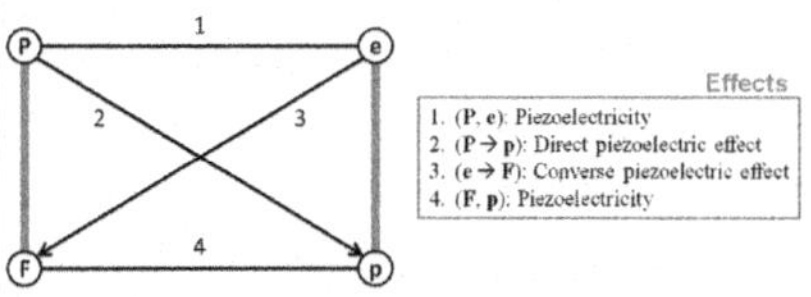 Effects 1. ($\mathbf{P}$, $\mathbf{e}$): Piezoelectricity 2. ($\mathbf{P} \rightarrow \mathbf{p}$): Direct piezoelectric effect 3. ($\mathbf{e} \rightarrow \mathbf{F}$): Converse piezoelectric effect 4. ($\mathbf{F}$, $\mathbf{p}$): Piezoelectricity
Thermal expansion and piezocaloric effect Example: ferroics	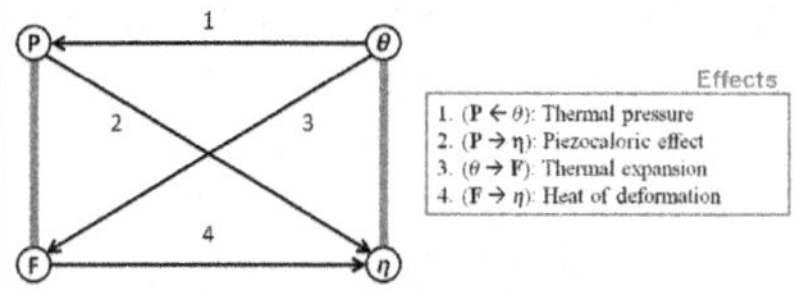 Effects 1. ($\mathbf{P} \leftarrow \theta$): Thermal pressure 2. ($\mathbf{P} \rightarrow \eta$): Piezocaloric effect 3. ($\theta \rightarrow \mathbf{F}$): Thermal expansion 4. ($\mathbf{F} \rightarrow \eta$): Heat of deformation
Pyromagnetic and magnetocaloric effects Examples: gadolinium alloys ($Gd_5Si_2Ge_2$), $PrNi_5$	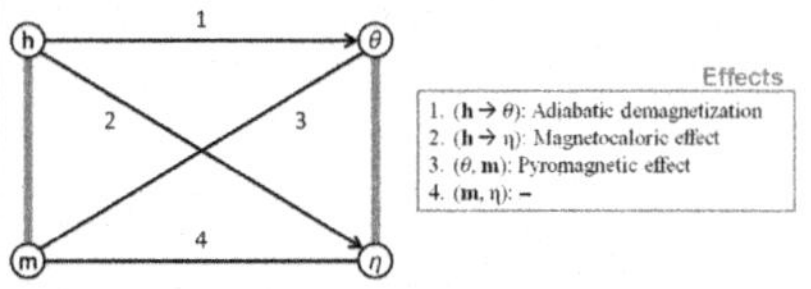 Effects 1. ($\mathbf{h} \rightarrow \theta$): Adiabatic demagnetization 2. ($\mathbf{h} \rightarrow \eta$): Magnetocaloric effect 3. (θ, $\mathbf{m}$): Pyromagnetic effect 4. ($\mathbf{m}$, η): –
Magnetoelectric effects Examples: multiferroics, Cr_2O_3	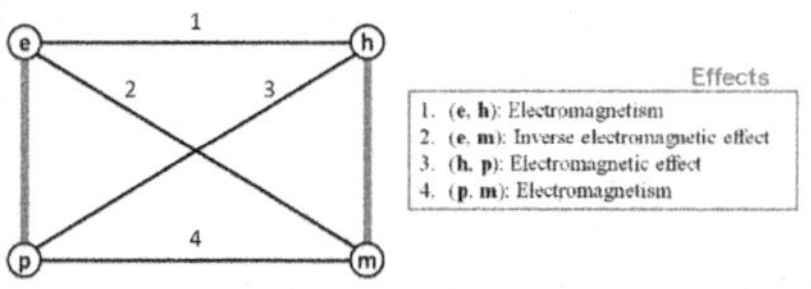 Effects 1. ($\mathbf{e}$, $\mathbf{h}$): Electromagnetism 2. ($\mathbf{e}$, $\mathbf{m}$): Inverse electromagnetic effect 3. ($\mathbf{h}$, $\mathbf{p}$): Electromagnetic effect 4. ($\mathbf{p}$, $\mathbf{m}$): Electromagnetism
Pyroelectric and electrocaloric effects Examples: GaN, $CsNO_3$	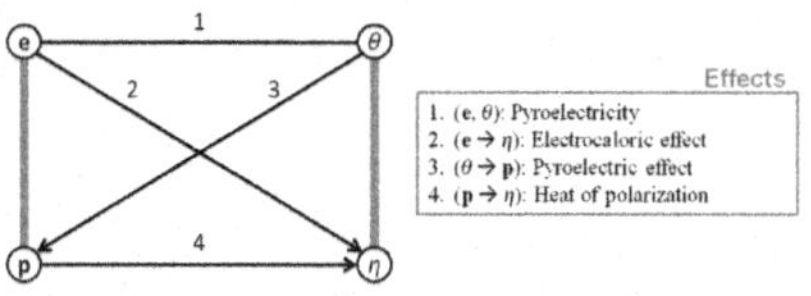 Effects 1. ($\mathbf{e}$, θ): Pyroelectricity 2. ($\mathbf{e} \rightarrow \eta$): Electrocaloric effect 3. ($\theta \rightarrow \mathbf{p}$): Pyroelectric effect 4. ($\mathbf{p} \rightarrow \eta$): Heat of polarization
Piezomagnetic effect Examples: antiferromagnetics like FeMn, NiO	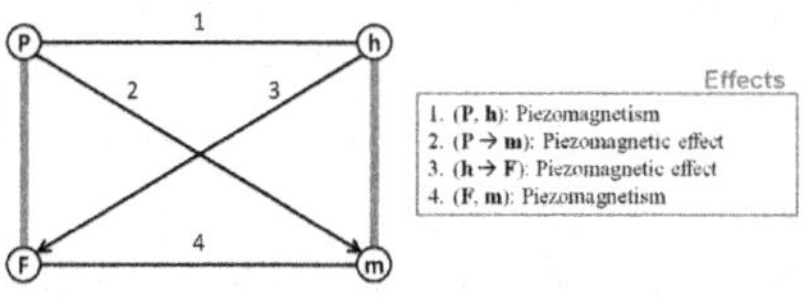 Effects 1. ($\mathbf{P}$, $\mathbf{h}$): Piezomagnetism 2. ($\mathbf{P} \rightarrow \mathbf{m}$): Piezomagnetic effect 3. ($\mathbf{h} \rightarrow \mathbf{F}$): Piezomagnetic effect 4. ($\mathbf{F}$, $\mathbf{m}$): Piezomagnetism

9.9 SPECIALIZATION OF THE SMALL-DEFORMATION THERMO-ELECTRO-MAGNETO-ELASTIC FRAMEWORK TO PIEZOELECTRIC MATERIALS

As discussed in Section 9.7, ferroelectric materials inherently exhibit nonlinear hysteretic behavior. However, for small deformations and small electromagnetic fields, approximately linear responses are observed in materials such as barium titanate ($BaTiO_3$), poly(vinylidene fluoride), and lead zirconate titanate. Ferroelectric materials like these that operate in a predominantly linear regime are known as **piezoelectric materials**.

Piezoelectric materials exhibit spontaneous polarization (at temperatures below the Curie point) in the presence of external electric fields. Below the Curie temperature, these materials exhibit a domain structure that lacks a center of symmetry, i.e., the centers of positive and negative charge are not identical. As a result, each unit cell acts as an electric dipole with a positive end and a negative end. Piezoelectrics change dimension in an electric field because the dipole length can be changed by the field: If a voltage is placed across the material, the dipoles respond to the field and change their dipole length, thereby changing the dimension of the crystal. Alternatively, if the crystal is mechanically stretched or compressed, the length of the dipole is changed, creating a voltage difference if there is no conductive path between the two ends of the dipole.

Since a necessary condition for the occurrence of piezoelectricity is the absence of a center of symmetry, piezoelectric materials are intrinsically *anisotropic*. Since piezoelectricity couples elasticity and polarization, piezoelectric material properties cannot be discussed without reference to the elasticity constant and the electric susceptibility (or electric permittivity); refer to Table 9.7. In what follows, we specialize the linear thermo-electro-magneto-elastic (TEME) framework developed in Section 9.7 to model piezoelectric material behavior. We make the following assumptions for a piezoelectric material operating well below the Curie temperature:

(1) Piezoelectric materials exhibit strains on the order of 10-100 microstrain. Infinitesimal strain theory can thus be used to describe the kinematics.
(2) Piezoelectric materials are used in transducer applications that operate in the low-frequency regime, typically ranging from 10 to 500 Hz. For this range of frequencies, the dynamic behavior of the electromagnetic fields may be ignored. In other words, the electromagnetic fields may be regarded as quasi-static.
(3) Thermal and magnetic effects may be neglected.
(4) The medium is nonconductive. Thus, when an external voltage is applied to the medium, no charge distribution is formed, i.e., $\sigma_i = 0$, and there is no free current, i.e., $j_i = 0$.

These assumptions allow us to simplify the fundamental laws (9.84a)–(9.84j) to

$$T_{ij,j} + \rho f_i^{\mathrm{m}} = \rho u_{i,tt}, \tag{9.91a}$$

$$T_{ij} = T_{ji}, \tag{9.91b}$$

$$d_{i,i} = 0, \tag{9.91c}$$

$$e_i = -\phi_{,i}, \tag{9.91d}$$

where ϕ is the electric potential. Note that (9.91d) is a consequence of Faraday's law: in the absence of time-varying magnetic fields, Faraday's law demands that the electric field is curl free, which, in turn, implies that the electric field is the gradient of a scalar potential.

Similarly, the linear constitutive equations (9.90a)–(9.90d) reduce to

$$T_{ij} = C_{ijkl}E_{kl} - d^{e}_{ijk}e_k, \tag{9.92a}$$

$$p_i = d^{e}_{jki}E_{jk} + \chi^{e}_{ij}e_j, \tag{9.92b}$$

where

$$E_{ij} = \frac{1}{2}(u_{i,j} + u_{i,j}), \quad d_i = \varepsilon_o e_i + p_i.$$

Note that the constitutive equations (9.92a) and (9.92b) relate stress and electric polarization to strain and electric field. Utilizing Voigt notation and imposing hexagonal crystal symmetry (C_{6v} group), we can express (9.92a) and (9.92b) in matrix form as

$$\begin{bmatrix} T_1 \\ T_2 \\ T_3 \\ T_4 \\ T_5 \\ T_6 \\ \hline p_1 \\ p_2 \\ p_3 \end{bmatrix} = \left[\begin{array}{cccccc|ccc} C_{11} & C_{12} & C_{13} & 0 & 0 & 0 & 0 & 0 & -d^{e}_{13} \\ C_{12} & C_{11} & C_{13} & 0 & 0 & 0 & 0 & 0 & -d^{e}_{13} \\ C_{13} & C_{13} & C_{33} & 0 & 0 & 0 & 0 & 0 & -d^{e}_{33} \\ 0 & 0 & 0 & C_{44} & 0 & 0 & -d^{e}_{41} & -d^{e}_{51} & 0 \\ 0 & 0 & 0 & 0 & C_{44} & 0 & -d^{e}_{51} & d^{e}_{41} & 0 \\ 0 & 0 & 0 & 0 & 0 & C_{66} & 0 & 0 & 0 \\ \hline 0 & 0 & 0 & d^{e}_{41} & d^{e}_{51} & 0 & \chi^{e}_{11} & 0 & 0 \\ 0 & 0 & 0 & d^{e}_{51} & -d^{e}_{41} & 0 & 0 & \chi^{e}_{11} & 0 \\ d^{e}_{13} & d^{e}_{13} & d^{e}_{33} & 0 & 0 & 0 & 0 & 0 & \chi^{e}_{33} \end{array}\right] \begin{bmatrix} E_1 \\ E_2 \\ E_3 \\ E_4 \\ E_5 \\ E_6 \\ \hline e_1 \\ e_2 \\ e_3 \end{bmatrix}.$$

EXERCISE

1. Derive a *linear reversible* framework for the constitutive modeling of magnetostrictive materials, i.e., materials that couple magnetic and mechanical fields. Use the magnetic field **h** and stress **T** as the independent variables, and the magnetization **m** and strain **E** as the dependent variables. Disregard thermal and electrical effects.

APPENDIX

Different Notions of Invariance

The subject of invariance requirements is controversial among continuum mechanicists. One group contends that the constitutive equations must be invariant *under all possible superposed rigid body motions of the body*. That is, if two motions of a body composed of the same material differ by only a superposed rigid body motion, then physically the internal response generated in the two motions must be the same, apart from orientation. This is the invariance requirement employed in this textbook; refer, for instance, to Sections 5.2.1, 5.4.1, and 9.6.4. Hooke, Poisson, and Cauchy are historical figures associated with the development of this requirement.

Another group of continuum mechanicists contends that the constitutive equations must be invariant *under an arbitrary motion of the observer*. In other words, the body does not move, but instead the coordinate axes can rotate and translate rigidly, and reflect. To this concept is attached the title *principle of material frame indifference*, *principle of material objectivity*, or *objectivity*, and associated with it are the historical figures Zaremba and Jaumann.

The conceptual bases for these two notions of invariance are very different, but in application the difference between them reduces to the following: the tensor $\mathbf{Q}$ in this book is proper orthogonal ($\mathbf{Q}\mathbf{Q}^{\mathrm{T}} = \mathbf{Q}^{\mathrm{T}}\mathbf{Q} = \mathbf{I}$ and $\det \mathbf{Q} = 1$) for invariance under superposed rigid body motions, whereas $\mathbf{Q}$ is merely orthogonal ($\mathbf{Q}\mathbf{Q}^{\mathrm{T}} = \mathbf{Q}^{\mathrm{T}}\mathbf{Q} = \mathbf{I}$ and $\det \mathbf{Q} = \pm 1$) for use in the principle of material frame indifference. This is because a body cannot turn itself inside out, but a set of coordinate axes can.

Note that a constitutive equation satisfying the principle of material frame indifference is also invariant under superposed rigid body motions: if a statement holds for all *orthogonal* $\mathbf{Q}$, then it holds for all *proper orthogonal* $\mathbf{Q}$ as well (all proper orthogonal tensors are also orthogonal). The converse is not true: a constitutive equation that is invariant under superposed rigid body motions does not necessarily satisfy the principle of material frame indifference. Therefore, the requirement of material frame indifference is more restrictive than the requirement of invariance under superposed rigid body motions.

APPENDIX

B The Physical Basis of Constitutive Assumptions

It is essential to produce a physically meaningful list of arguments for the constitutive functions that characterize a particular material. In the thermomechanical theory (refer to Section 5.3), constitutive functions must be given for the Cauchy stress $\mathbf{T}$, Helmholtz free energy ψ, heat flux vector $\mathbf{q}$, and entropy η, which depend in some manner on the motion $\boldsymbol{\chi}$ and temperature Θ, and possibly their rates, gradients, and histories. Why do we assume that for a thermoelastic material (see Chapter 6), the response functions $\breve{\mathbf{T}}$, $\breve{\psi}$, $\breve{\mathbf{q}}$, and $\breve{\eta}$, evaluated at position $\mathbf{x}$ and time t, depend on $\boldsymbol{\chi}$ and Θ through the list $\mathbf{F} = \text{Grad}\,\boldsymbol{\chi}$, Θ, and $\mathbf{g} = \text{grad}\,\Theta$, also evaluated at position $\mathbf{x}$ and time t? Why, for instance, do we not also include in the list

(a) $\text{Grad}(\text{Grad}\,\boldsymbol{\chi}) = \dfrac{\partial^2 \boldsymbol{\chi}}{\partial \mathbf{X}^2}$,

(b) $\text{grad}(\text{grad}\,\Theta) = \dfrac{\partial^2 \Theta}{\partial \mathbf{x}^2}$,

(c) $\mathbf{L} = \text{grad}\,\mathbf{v}$,

(d) $\mathbf{F}$ evaluated at a position $\mathbf{y} \neq \mathbf{x}$ and time t,

(e) Θ evaluated at time $t + 5$ seconds, or

(f) Θ evaluated at time $t - 5$ seconds?

Some of these terms are excluded from the list of arguments through two postulates on material behavior [64]:

(1) Axiom of determinism: The response of a body is determined by the *history* of motion and temperature for that body. Therefore, the response functions for $\mathbf{T}$, $\mathbf{q}$, ψ, and η at time t can depend only on functions of $\mathbf{x}$ and Θ evaluated at past or present times $\tau \leq t$. Thus, the term "Θ evaluated at time $t + 5$ seconds" cannot be included in the list of arguments.

(2) Axiom of neighborhood: The response of a given particle Y of the body does not depend on the motion and temperature of particles outside a *small neighborhood* of Y in any configuration. Stated differently, the motion and temperature of particles a finite distance from Y in a particular configuration are disregarded in calculating the response of Y. (By the smoothness assumption made on the motion of the body, a particle a finite distance from Y in one configuration is a finite distance from Y in all configurations.) Because of this assumption, the term "$\mathbf{F}$ evaluated at a position $\mathbf{y} \neq \mathbf{x}$ and time t" cannot be included in the list of arguments, unless $\mathbf{y}$ is arbitrarily close to $\mathbf{x}$.

It is also common to assume that the argument list contains gradients of motion and temperature up to and including only first order. Materials satisfying this assumption are called **simple** [64]. In other words, a simple material is one whose constitutive functions are completely determined by the material's response to homogeneous deformations (i.e., deformations for which $\mathbf{F} = \mathbf{F}(t)$ only). Hence, the response functions for a simple material cannot depend on the term "Grad(Grad χ)" or the term "grad(grad Θ)."

Lastly, the response of a thermoelastic material does not depend on the *history* or *rate* of the deformation or temperature. Thus, the terms "$\mathbf{L} = \text{grad}\,\mathbf{v}$" and "$\Theta$ evaluated at time $t - 5$ seconds" cannot be included in the list of arguments. (Recall that the velocity gradient $\mathbf{L}$ is related to the rate of deformation $\dot{\mathbf{F}}$ through $\mathbf{L} = \dot{\mathbf{F}}\mathbf{F}^{-1}$.)

APPENDIX

Isotropic Tensors

Isotropic tensors have the same components in any frame of reference. In this section, we present some useful results for isotropic tensors. Proofs of these results are left as an exercise for the reader.

Let $\mathbf{v}$ be a vector (or first-order tensor). Then $\mathbf{v}$ is isotropic, i.e.,

$$\mathbf{v} = \mathbf{Q}\mathbf{v} \tag{C.1a}$$

for all proper orthogonal tensors $\mathbf{Q}$, if and only if

$$\mathbf{v} = \mathbf{0}. \tag{C.1b}$$

Hence, the only isotropic first-order tensor is the zero vector.

Let $\mathbf{T}$ be a second-order tensor. Then $\mathbf{T}$ is isotropic, i.e.,

$$\mathbf{T} = \mathbf{Q}\mathbf{T}\mathbf{Q}^{\mathrm{T}} \tag{C.2a}$$

for all proper orthogonal tensors $\mathbf{Q}$, if and only if

$$\mathbf{T} = \lambda\mathbf{I}, \tag{C.2b}$$

where λ is a scalar and $\mathbf{I}$ is the identity tensor.

Let $\mathbf{S}^{(3)}$ be a third-order tensor with Cartesian components S_{ijk}. Then $\mathbf{S}^{(3)}$ is isotropic, i.e.,

$$S_{ijk} = Q_{il}Q_{jm}Q_{kn}S_{lmn} \tag{C.3a}$$

for all proper orthogonal Q_{ij}, if and only if

$$S_{ijk} = \lambda\epsilon_{ijk}, \tag{C.3b}$$

where λ is a scalar and ϵ_{ijk} is the permutation symbol.

Let $\mathbf{C}^{(4)}$ be a fourth-order tensor with Cartesian components C_{ijkl}. Then $\mathbf{C}^{(4)}$ is isotropic, i.e.,

$$C_{ijkl} = Q_{im}Q_{jn}Q_{kp}Q_{lq}C_{mnpq} \tag{C.4a}$$

for all proper orthogonal Q_{ij}, if and only if

$$C_{ijkl} = \lambda\delta_{ij}\delta_{kl} + \mu\delta_{ik}\delta_{jl} + \gamma\delta_{il}\delta_{jk}, \tag{C.4b}$$

where λ, μ, and γ are scalars and δ_{ij} is the Kronecker delta. If we further restrict $\mathbf{C}^{(4)}$ such that $C_{ijkl} = C_{jikl}$, then $\mu = \gamma$ in (C.4b), which implies that

$$C_{ijkl} = \lambda\delta_{ij}\delta_{kl} + \mu(\delta_{ik}\delta_{jl} + \delta_{il}\delta_{jk}). \tag{C.5}$$

Note that (C.5) implies the additional symmetry

$$C_{ijkl} = C_{ijlk} = C_{klij}.$$

APPENDIX

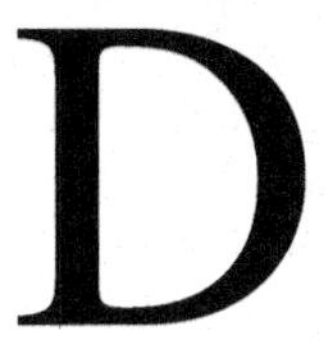

A Family of Thermomechanical Processes

In this appendix, we present a particular family of motions and temperatures, given by the following functions of space and time:

$$\mathbf{x} = \chi(\mathbf{X}, t) = \mathbf{X}_0 + \left[\mathbf{F}_0 + (t - t_0)\mathbf{A}\right](\mathbf{X} - \mathbf{X}_0),$$

$$\Theta = \hat{\Theta}(\mathbf{X}, t) = \Theta_0 + (t - t_0)a + \left[\mathbf{g}_0 + (t - t_0)\mathbf{a}\right] \cdot \mathbf{F}_0(\mathbf{X} - \mathbf{X}_0), \tag{D.1}$$

where t_0, Θ_0, and a are arbitrary constant scalars, $\mathbf{X}_0$, $\mathbf{g}_0$, and $\mathbf{a}$ are arbitrary constant vectors, and $\mathbf{F}_0$ and $\mathbf{A}$ are arbitrary constant tensors. Loosely speaking, (D.1) gives a family of motions and temperatures with an infinite number of members. From (D.1), with the aid of results (2.99) and (3.22), we can explicitly calculate

$$\mathbf{F} = \hat{\mathbf{F}}(\mathbf{X}, t) = \frac{\partial \chi(\mathbf{X}, t)}{\partial \mathbf{X}} = \mathbf{F}_0 + (t - t_0)\mathbf{A},$$

$$\dot{\mathbf{F}} = \frac{\partial \hat{\mathbf{F}}(\mathbf{X}, t)}{\partial t} = \mathbf{A},$$

$$\mathbf{g} = \hat{\mathbf{g}}(\mathbf{X}, t) = \mathbf{F}^{-\mathrm{T}} \frac{\partial \hat{\Theta}(\mathbf{X}, t)}{\partial \mathbf{X}} = \left[\mathbf{F}_0 + (t - t_0)\mathbf{A}\right]^{-\mathrm{T}} \mathbf{F}_0^{\mathrm{T}} \left[\mathbf{g}_0 + (t - t_0)\,\mathbf{a}\right],$$

$$\dot{\mathbf{g}} = \frac{\partial \hat{\mathbf{g}}(\mathbf{X}, t)}{\partial t} = \mathbf{A}^{-\mathrm{T}} \mathbf{F}_0^{\mathrm{T}} \left[\mathbf{g}_0 + (t - t_0)\,\mathbf{a}\right] + \left[\mathbf{F}_0 + (t - t_0)\mathbf{A}\right]^{-\mathrm{T}} \mathbf{F}_0^{\mathrm{T}} \mathbf{a},$$

$$\dot{\Theta} = \frac{\partial \hat{\Theta}(\mathbf{X}, t)}{\partial t} = a + \mathbf{a} \cdot \mathbf{F}_0(\mathbf{X} - \mathbf{X}_0).$$

At the particular place $\mathbf{X} = \mathbf{X}_0$ and time $t = t_0$, the above *functions* take the *values*

$$\mathbf{x} = \chi(\mathbf{X}_0, t_0) = \mathbf{X}_0, \quad \Theta = \hat{\Theta}(\mathbf{X}_0, t_0) = \Theta_0, \quad \mathbf{F} = \mathbf{F}_0,$$

and

$$\dot{\mathbf{F}} = \mathbf{A}, \quad \mathbf{g} = \mathbf{g}_0, \quad \dot{\mathbf{g}} = \mathbf{A}^{-\mathrm{T}} \mathbf{F}_0^{\mathrm{T}} \mathbf{g}_0 + \mathbf{a}, \quad \dot{\Theta} = a.$$

It follows that

$$\mathbf{L} = \dot{\mathbf{F}}\mathbf{F}^{-1} = \mathbf{A}\mathbf{F}_0^{-1}, \quad \mathbf{D} = \frac{1}{2}\left(\mathbf{L} + \mathbf{L}^{\mathrm{T}}\right) = \frac{1}{2}\left(\mathbf{A}\mathbf{F}_0^{-1} + \mathbf{F}_0^{-\mathrm{T}}\mathbf{A}^{\mathrm{T}}\right)$$

at place $\mathbf{x} = \mathbf{X} = \mathbf{X}_0$ and time $t = t_0$.

Recall that t_0, Θ_0, a, $\mathbf{X}_0$, $\mathbf{g}_0$, $\mathbf{a}$, $\mathbf{F}_0$, and $\mathbf{A}$ are *arbitrary* quantities. We have therefore explicitly exhibited a family of thermomechanical processes (D.1) in which the temperature Θ, deformation gradient $\mathbf{F}$, rate of deformation gradient $\dot{\mathbf{F}}$ (or, alternatively, velocity gradient $\mathbf{L}$ or rate of deformation $\mathbf{D}$), temperature gradient $\mathbf{g}$, rate of temperature gradient $\dot{\mathbf{g}}$, and rate of temperature $\dot{\Theta}$ can be chosen *independently* at any place and time.

APPENDIX

Energy Formulations and Stability Conditions for Newtonian Fluids

In this section, we review and augment the thermomechanical theory for **compressible Newtonian fluids** presented in Chapter 7 and discuss the corresponding **stability conditions**.

E.1 GOVERNING EQUATIONS

Recall that the fundamental laws for a continuum in the thermomechanical theory are

$$\dot{\rho} + \rho \operatorname{div} \mathbf{v} = 0, \quad \rho\dot{\mathbf{v}} = \operatorname{div} \mathbf{T} + \rho\mathbf{b}, \quad \rho\dot{\varepsilon} = \mathbf{T}\cdot\mathbf{D} + \rho r - \operatorname{div}\mathbf{q}, \tag{E.1}$$

which correspond to Eulerian statements of conservation of mass, linear momentum, and energy; refer to Section 5.3. In (E.1), ρ is the density, $\mathbf{v}$ is the velocity, $\mathbf{T}$ is the Cauchy stress, $\mathbf{b}$ is the body force per unit mass, ε is the internal energy per unit mass, r is the rate of heat absorbed per unit mass, $\mathbf{q}$ is the heat flux vector, and

$$\mathbf{D} = \frac{1}{2}\left[\operatorname{grad}\mathbf{v} + (\operatorname{grad}\mathbf{v})^{\mathrm{T}}\right]$$

is the rate of deformation. A **compressible Newtonian fluid** (refer to Section 7.2.1.2) is characterized by the constitutive equations

$$\mathbf{T} = -p\mathbf{I} + \lambda(\operatorname{tr}\mathbf{D})\mathbf{I} + 2\mu\mathbf{D}, \quad \mathbf{q} = -k\operatorname{grad}\Theta, \tag{E.2}$$

where p is the pressure, Θ is the absolute temperature, λ and μ are the dilatational and shear viscosities, respectively, and k is the thermal conductivity. The second law of thermodynamics in the form of the Clausius-Duhem inequality is

$$-\rho\dot{\varepsilon} + \mathbf{T}\cdot\mathbf{D} + \rho\Theta\dot{\eta} - \frac{1}{\Theta}\mathbf{q}\cdot\operatorname{grad}\Theta \geq 0, \tag{E.3}$$

where η is the entropy per unit mass. It can be verified that use of the constitutive equations (E.2) in the second law (E.3), and subsequent decomposition of the rate of deformation $\mathbf{D}$ into its spherical (volumetric) and deviatoric (traceless) parts, i.e.,

$$\mathbf{D} = \frac{1}{3}(\operatorname{tr}\mathbf{D})\mathbf{I} + \mathbf{D}_{\mathrm{d}},$$

leads to

$$-\rho\dot{\varepsilon} + \frac{p}{\rho}\dot{\rho} + \rho\Theta\dot{\eta} + \left(\lambda + \frac{2}{3}\mu\right)(\operatorname{tr}\mathbf{D})^2 + 2\mu\mathbf{D}_\mathrm{d}\cdot\mathbf{D}_\mathrm{d} + \frac{k}{\Theta}\left|\operatorname{grad}\Theta\right|^2 \geq 0, \tag{E.4}$$

where we have used

$$\dot{\rho} = -\rho\operatorname{div}\mathbf{v} = -\rho\operatorname{tr}\mathbf{D} = -\rho\mathbf{I}\cdot\mathbf{D}.$$

The mathematical model for a compressible Newtonian fluid is completed by specifying the **equations of state**. In the following sections, we present four equivalent formulations of the equations of state, each employing a different set of independent variables: density ρ or pressure p for the independent mechanical variable, and entropy η or temperature Θ for the independent thermal variable. (Recall that only the density-temperature formulation—with ρ as the independent mechanical variable and Θ as the independent thermal variable—was explored in Section 7.2.1.2.)

E.1.1 DENSITY-ENTROPY FORMULATION

The system formulation (E.1) and (E.2) implicitly uses three choices: density ρ as the independent mechanical variable, entropy η as the independent thermal variable, and **internal energy** $\breve{\varepsilon}(\rho,\eta)$ as the thermodynamic energy potential. Use of the chain rule

$$\dot{\varepsilon} = \frac{\partial\breve{\varepsilon}}{\partial\rho}\dot{\rho} + \frac{\partial\breve{\varepsilon}}{\partial\eta}\dot{\eta}$$

in the second law (E.4) gives

$$\frac{1}{\rho}\left(p - \rho^2\frac{\partial\breve{\varepsilon}}{\partial\rho}\right)\dot{\rho} + \rho\left(\Theta - \frac{\partial\breve{\varepsilon}}{\partial\eta}\right)\dot{\eta} + \left(\lambda + \frac{2}{3}\mu\right)(\operatorname{tr}\mathbf{D})^2$$

$$+ 2\mu\mathbf{D}_\mathrm{d}\cdot\mathbf{D}_\mathrm{d} + \frac{k}{\Theta}\left|\operatorname{grad}\Theta\right|^2 \geq 0. \tag{E.5}$$

Using arguments similar to those employed in Section 7.2.1.2, we can prove that as a consequence of inequality (E.5),

$$p = \rho^2\frac{\partial\breve{\varepsilon}}{\partial\rho}, \qquad \Theta = \frac{\partial\breve{\varepsilon}}{\partial\eta} \tag{E.6a}$$

and

$$\lambda + \frac{2}{3}\mu \geq 0, \quad \mu \geq 0, \quad k \geq 0. \tag{E.6b}$$

With the internal energy $\breve{\varepsilon}(\rho,\eta)$ a *specified function* of ρ and η, the equations of state $(E.6a)_1$ and $(E.6a)_2$ prescribe the *dependent* mechanical and thermal variables p and Θ in terms of the *independent* mechanical and thermal variables ρ and η. Equation (E.6b) demands nonnegativity of the bulk viscosity, shear viscosity, and thermal conductivity. Note that the material parameters λ, μ, and k may depend on the density ρ and entropy η. The conservation laws (E.1), constitutive equations (E.2), and equations of state (E.6a) constitute a closed system for the primitive quantities ρ, η, and $\mathbf{v}$.

E.1.2 DENSITY-TEMPERATURE FORMULATION

In this formulation, density ρ is the independent mechanical variable, temperature Θ is the independent thermal variable, and the **Helmholtz free energy** $\breve{\psi}(\rho, \Theta)$ is the thermodynamic energy potential. The Helmholtz free energy $\breve{\psi}(\rho, \Theta)$ is defined by the **Legendre transformation** of internal energy $\breve{\varepsilon}(\rho, \eta)$ with respect to the thermal variable, from η to Θ,

$$\psi = \varepsilon - \Theta\eta. \tag{E.7}$$

See also (5.33). Taking the material derivative of (E.7), we obtain

$$\dot{\psi} = \dot{\varepsilon} - \Theta\dot{\eta} - \eta\dot{\Theta},$$

so inequality (E.4) becomes

$$-\rho\dot{\psi} + \frac{p}{\rho}\dot{\rho} - \rho\eta\dot{\Theta} + \left(\lambda + \frac{2}{3}\mu\right)(\operatorname{tr}\mathbf{D})^2 + 2\mu\mathbf{D}_\mathrm{d}\cdot\mathbf{D}_\mathrm{d} + \frac{k}{\Theta}\left|\operatorname{grad}\Theta\right|^2 \geq 0. \tag{E.8}$$

Use of the chain rule

$$\dot{\psi} = \frac{\partial\breve{\psi}}{\partial\rho}\dot{\rho} + \frac{\partial\breve{\psi}}{\partial\Theta}\dot{\Theta}$$

in the second law inequality (E.8) leads to

$$\frac{1}{\rho}\left(p - \rho^2\frac{\partial\breve{\psi}}{\partial\rho}\right)\dot{\rho} - \rho\left(\eta + \frac{\partial\breve{\psi}}{\partial\Theta}\right)\dot{\Theta} + \left(\lambda + \frac{2}{3}\mu\right)(\operatorname{tr}\mathbf{D})^2 + 2\mu\mathbf{D}_\mathrm{d}\cdot\mathbf{D}_\mathrm{d} + \frac{k}{\Theta}\left|\operatorname{grad}\Theta\right|^2 \geq 0.$$

Using standard arguments, we obtain

$$p = \rho^2\frac{\partial\breve{\psi}}{\partial\rho}, \quad \eta = -\frac{\partial\breve{\psi}}{\partial\Theta} \tag{E.9a}$$

and

$$\lambda + \frac{2}{3}\mu \geq 0, \quad \mu \geq 0, \quad k \geq 0. \tag{E.9b}$$

See also (7.32). Note that the material parameters λ, μ, and k may depend on the density ρ and temperature Θ. With the Helmholtz free energy $\breve{\psi}(\rho, \Theta)$ a *specified function* of ρ and Θ, (E.1), (E.2), (E.7), and (E.9a) form a closed system for ρ, Θ, and $\mathbf{v}$.

E.1.3 PRESSURE-ENTROPY FORMULATION

In this formulation, pressure p is the independent mechanical variable, entropy η is the independent thermal variable, and **enthalpy** $\breve{\chi}(p, \eta)$ is the thermodynamic energy potential. The enthalpy $\breve{\chi}(p, \eta)$ is defined by the **Legendre transformation** of internal energy $\breve{\varepsilon}(\rho, \eta)$ with respect to the mechanical variable, from specific volume $1/\rho$ to pressure p,

$$\chi = \varepsilon + \frac{1}{\rho} p. \tag{E.10}$$

The rate of the Legendre transformation (E.10) is

$$\dot{\chi} = \dot{\varepsilon} + \frac{1}{\rho}\dot{p} - \frac{p}{\rho^2}\dot{\rho}. \tag{E.11}$$

Substitution of (E.11) into the second law inequality (E.4) gives

$$-\rho\dot{\chi} + \dot{p} + \rho\Theta\dot{\eta} + \left(\lambda + \frac{2}{3}\mu\right)(\operatorname{tr}\mathbf{D})^2 + 2\mu\mathbf{D}_{\mathrm{d}} \cdot \mathbf{D}_{\mathrm{d}} + \frac{k}{\Theta}|\operatorname{grad}\Theta|^2 \geq 0.$$

Subsequent use of the chain rule on $\dot{\chi}$ leads to

$$\left(1 - \rho\frac{\partial\breve{\chi}}{\partial p}\right)\dot{p} + \rho\left(\Theta - \frac{\partial\breve{\chi}}{\partial\eta}\right)\dot{\eta} + \left(\lambda + \frac{2}{3}\mu\right)(\operatorname{tr}\mathbf{D})^2 + 2\mu\mathbf{D}_{\mathrm{d}} \cdot \mathbf{D}_{\mathrm{d}} + \frac{k}{\Theta}|\operatorname{grad}\Theta|^2 \geq 0,$$

from which it follows that

$$\frac{1}{\rho} = \frac{\partial\breve{\chi}}{\partial p}, \qquad \Theta = \frac{\partial\breve{\chi}}{\partial\eta} \tag{E.12a}$$

and

$$\lambda + \frac{2}{3}\mu \geq 0, \quad \mu \geq 0, \quad k \geq 0. \tag{E.12b}$$

Note that the material parameters λ, μ, and k may depend on the pressure p and entropy η. With the enthalpy $\breve{\chi}(p, \eta)$ a *specified function* of p and η, the equations of state (E.12a) together with the conservation laws (E.1), constitutive equations (E.2), and Legendre transformation (E.10) form a closed system for p, η, and $\mathbf{v}$.

E.1.4 PRESSURE-TEMPERATURE FORMULATION

In this formulation, pressure p is the independent mechanical variable, temperature Θ is the independent thermal variable, and the **Gibbs free energy** $\breve{\phi}(p, \Theta)$ is the thermodynamic energy potential. The Gibbs free energy $\breve{\phi}(p, \Theta)$ is defined by the **Legendre transformation** of internal energy $\breve{\varepsilon}(\rho, \eta)$ with respect to the mechanical and thermal variables, from specific volume $1/\rho$ to pressure p and entropy η to temperature Θ,

$$\phi = \varepsilon + \frac{1}{\rho} p - \Theta\eta. \tag{E.13}$$

The material derivative of (E.13) is

$$\dot{\phi} = \dot{\varepsilon} + \frac{1}{\rho}\dot{p} - \frac{p}{\rho^2}\dot{\rho} - \Theta\dot{\eta} - \eta\dot{\Theta}. \tag{E.14}$$

Use of (E.14) in the second law (E.4) yields

$$-\rho\dot{\phi} + \dot{p} - \rho\eta\dot{\Theta} + \left(\lambda + \frac{2}{3}\mu\right)(\operatorname{tr}\mathbf{D})^2 + 2\mu\mathbf{D}_d \cdot \mathbf{D}_d + \frac{k}{\Theta}|\operatorname{grad}\Theta|^2 \geq 0.$$

The chain rule

$$\dot{\phi} = \frac{\partial\breve{\phi}}{\partial p}\dot{p} + \frac{\partial\breve{\phi}}{\partial\Theta}\dot{\Theta}$$

then implies that

$$\left(1 - \rho\frac{\partial\breve{\phi}}{\partial p}\right)\dot{p} - \rho\left(\eta + \frac{\partial\breve{\phi}}{\partial\Theta}\right)\dot{\Theta} + \left(\lambda + \frac{2}{3}\mu\right)(\operatorname{tr}\mathbf{D})^2 + 2\mu\mathbf{D}_d \cdot \mathbf{D}_d + \frac{k}{\Theta}|\operatorname{grad}\Theta|^2 \geq 0.$$

Customary arguments allow us to conclude that

$$\frac{1}{\rho} = \frac{\partial\breve{\phi}}{\partial p}, \quad \eta = -\frac{\partial\breve{\phi}}{\partial\Theta} \tag{E.15a}$$

and

$$\lambda + \frac{2}{3}\mu \geq 0, \quad \mu \geq 0, \quad k \geq 0. \tag{E.15b}$$

Note that the material parameters λ, μ, and k may depend on the pressure p and temperature Θ. With the Gibbs free energy $\breve{\phi}(p, \Theta)$ a *specified function* of p and Θ, (E.1), (E.2), (E.13), and (E.15a) constitute a closed system for p, Θ, and $\mathbf{v}$.

E.2 STABILITY CONDITIONS

Stability of the rest state is required for a theory to be physically valid [22, 65]. For a *compressible* Newtonian fluid, a necessary and sufficient condition for stability of the rest state is that the internal energy ε is a convex function of entropy η and specific volume $\tau = \rho^{-1}$ [66, 67]:

$$\varepsilon_{\tau\tau} \geq 0, \quad \varepsilon_{\eta\eta} \geq 0, \quad \varepsilon_{\tau\tau}\varepsilon_{\eta\eta} - \varepsilon_{\tau\eta}^2 \geq 0, \tag{E.16a}$$

where subscripts denote partial differentiation, e.g.,

$$\varepsilon_{\tau\tau} = \frac{\partial^2\varepsilon(\tau, \eta)}{\partial\tau^2}.$$

With a change of independent variable from specific volume τ to density ρ, condition (E.16a) becomes

$$\frac{2\varepsilon_\rho}{\rho} + \varepsilon_{\rho\rho} \geq 0, \quad \varepsilon_{\eta\eta} \geq 0, \quad \left(\frac{2\varepsilon_\rho}{\rho} + \varepsilon_{\rho\rho}\right)\varepsilon_{\eta\eta} - \varepsilon_{\rho\eta}^2 \geq 0. \tag{E.16b}$$

Note that one need monitor only two of these inequalities, since $(E.16b)_1$ and $(E.16b)_3$ imply $(E.16b)_2$, and $(E.16b)_2$ and $(E.16b)_3$ imply $(E.16b)_1$. It can be verified that the equivalent conditions on the Helmholtz free energy are

$$\psi_{\tau\tau} \geq 0, \quad \psi_{\theta\theta} \leq 0, \quad \psi_{\tau\tau}\psi_{\theta\theta} - \psi_{\tau\theta}^2 \leq 0, \tag{E.17a}$$

or, upon a change of independent variable from specific volume τ to density ρ,

$$\frac{2\psi_\rho}{\rho} + \psi_{\rho\rho} \geq 0, \quad \psi_{\theta\theta} \leq 0, \quad \left(\psi_{\rho\rho} + \frac{\psi_\rho}{\rho}\right)\psi_{\theta\theta} - \psi_{\rho\theta}^2 \leq 0. \tag{E.17b}$$

Similarly, we have

$$\chi_{pp} \leq 0, \quad \chi_{\eta\eta} \geq 0, \quad \chi_{\eta\eta}\chi_{pp} - \chi_{p\eta}^2 \leq 0 \tag{E.18}$$

for the enthalpy and

$$\phi_{pp} \leq 0, \quad \phi_{\theta\theta} \leq 0, \quad \phi_{pp}\phi_{\theta\theta} - \phi_{p\theta}^2 \geq 0 \tag{E.19}$$

for the Gibbs free energy.

It can be shown (see, for instance, [68]) that for a compressible Newtonian fluid, the convexity conditions (E.16b) on the internal energy are equivalent to (1) nonnegativity of the specific heat at constant volume C_V and the isothermal bulk modulus κ, i.e.,

$$C_V \equiv \Theta \frac{\partial \eta(\tau, \Theta)}{\partial \Theta} \geq 0, \quad \kappa \equiv -\tau \frac{\partial p(\tau, \Theta)}{\partial \tau} \geq 0,$$

and (2) an absence of negative decay rates in the linearized stability analysis of the the rest state. Thus, for a compressible Newtonian fluid, there are three *equivalent* conditions for stability of the rest state.

APPENDIX

Additional Energy Formulations for Thermo-Electro-Magneto-Mechanical Materials

Recall from Chapter 9 that the internal energy $\varepsilon = \breve{\varepsilon}\left(\mathbf{F}, \eta, \mathbf{p}^*/\rho, \mathbf{m}^*/\rho\right)$ in large-deformation thermo-electro-magneto mechanics is a function of the *extensive* quantities $\mathbf{F}$, η, $\mathbf{p}^*/\rho$, and $\mathbf{m}^*/\rho$. Free energies that employ one or more of the *intensive* quantities $\mathbf{P}$, Θ, $\mathbf{e}^*$, and $\mathbf{h}^*$ as independent variables were introduced using Legendre transformations of ε; refer to Tables 9.3–9.6. These *intensive-extensive conjugate pairs* ($\mathbf{P}$ and $\mathbf{F}$, Θ and η, $\mathbf{e}^*$ and $\mathbf{p}^*/\rho$, $\mathbf{h}^*$ and $\mathbf{m}^*/\rho$) appearing in the *fundamental statement* of the second law of thermodynamics (9.61) are dictated by the choice (9.55c) of the electromagnetic energy r^{em}. It follows, then, that different choices of r^{em} (of which there are many options available in the literature; see, for instance, [32, 34, 38, 69]) lead to different sets of electromagnetic work conjugates in large-deformation thermo-electro-magneto mechanics. For instance, the electric displacement $\mathbf{d}^*$ and magnetic induction $\mathbf{b}^*$, both extensive quantities, frequently appear as parts of a conjugate pair (see, for instance, [34, 35, 44, 46, 70]). However, since they are not part of the conjugate pairs employed in Chapter 9, $\mathbf{d}^*$ and $\mathbf{b}^*$ cannot be introduced as independent variables using the customary formalism of Legendre transformations. In what follows, we overcome this limitation by presenting an alternative procedure for introducing $\mathbf{d}^*$ and $\mathbf{b}^*$ as independent variables.

F.1 DEFORMATION-TEMPERATURE-ELECTRIC DISPLACEMENT-MAGNETIC INDUCTION FORMULATION

In this section, $\mathbf{F}$, Θ, $\mathbf{d}^*$, and $\mathbf{b}^*$ are selected as the independent variables. However, as discussed above, the electric displacement $\mathbf{d}^*$ and magnetic induction $\mathbf{b}^*$ cannot be introduced as independent variables through a conventional Legendre transformation of the internal energy. We circumvent this by positing a *Legendre-type transformation* of $\varepsilon = \bar{\varepsilon}\left(\mathbf{F}, \eta, \mathbf{p}^*, \mathbf{m}^*\right)$, i.e.,

$$E^{F\Theta db} = \varepsilon - \Theta\eta + \frac{\epsilon_0}{2\rho}\mathbf{e}^* \cdot \mathbf{e}^* + \frac{\mu_0}{2\rho}\mathbf{h}^* \cdot \mathbf{h}^*, \tag{F.1}$$

whose rate form is

$$\dot{E}^{F\Theta db} = \dot{\varepsilon} - \Theta\dot{\eta} - \eta\dot{\Theta} + \frac{1}{2\rho}\left(\epsilon_0\mathbf{e}^* \cdot \mathbf{e}^* + \mu_0\mathbf{h}^* \cdot \mathbf{h}^*\right)\mathbf{F}^{-\mathrm{T}} \cdot \dot{\mathbf{F}} + \frac{1}{\rho}\left(\epsilon_0\mathbf{e}^* \cdot \dot{\mathbf{e}}^* + \mu_0\mathbf{h}^* \cdot \dot{\mathbf{h}}^*\right), \tag{F.2}$$

where we have used (9.68). The last two terms on the right-hand side of (F.1) represent electrical and magnetic energies, respectively, in vacuo. Substitution of (F.2) into (9.67), and subsequent use of the algebraic relationships $(9.34)_1$ and $(9.40)_1$, leads to

$$-\dot{E}^{F\Theta db}+\left[\frac{1}{\rho_R}\mathbf{P}+\frac{1}{\rho}\left(\mathbf{e}^*\cdot\mathbf{d}^*+\mathbf{h}^*\cdot\mathbf{b}^*-\frac{1}{2}\epsilon_o\mathbf{e}^*\cdot\mathbf{e}^*-\frac{1}{2}\mu_o\mathbf{h}^*\cdot\mathbf{h}^*\right)\mathbf{F}^{-T}\right]\cdot\dot{\mathbf{F}}$$

$$-\eta\dot{\Theta}+\frac{1}{\rho}\mathbf{e}^*\cdot\dot{\mathbf{d}}^*+\frac{1}{\rho}\mathbf{h}^*\cdot\dot{\mathbf{b}}^*+\frac{1}{\rho}\mathbf{j}^*\cdot\mathbf{e}^*-\frac{1}{\rho\Theta}\mathbf{q}\cdot\operatorname{grad}\Theta\geq 0, \tag{F.3}$$

the second law statement for this formulation. Note that $\mathbf{d}^*$ and $\mathbf{b}^*$ appear as rates in the second law inequality (F.3), i.e., as natural independent variables. Use of the chain rule on $\dot{E}^{F\Theta db}$ leads to

$$\left[\frac{1}{\rho_R}\mathbf{P}-\frac{\partial E^{F\Theta db}}{\partial\mathbf{F}}+\frac{1}{\rho}\left(\mathbf{e}^*\cdot\mathbf{d}^*+\mathbf{h}^*\cdot\mathbf{b}^*-\frac{1}{2}\epsilon_o\mathbf{e}^*\cdot\mathbf{e}^*-\frac{1}{2}\mu_o\mathbf{h}^*\cdot\mathbf{h}^*\right)\mathbf{F}^{-T}\right]\cdot\dot{\mathbf{F}}$$

$$-\left(\eta+\frac{\partial E^{F\Theta db}}{\partial\Theta}\right)\dot{\Theta}+\left(\frac{1}{\rho}\mathbf{e}^*-\frac{\partial E^{F\Theta db}}{\partial\mathbf{d}^*}\right)\cdot\dot{\mathbf{d}}^*+\left(\frac{1}{\rho}\mathbf{h}^*-\frac{\partial E^{F\Theta db}}{\partial\mathbf{b}^*}\right)\cdot\dot{\mathbf{b}}^*$$

$$+\frac{1}{\rho}\mathbf{j}^*\cdot\mathbf{e}^*-\frac{1}{\rho\Theta}\mathbf{q}\cdot\operatorname{grad}\Theta\geq 0,$$

from which the constitutive equations

$$\mathbf{P}=\rho_R\frac{\partial E^{F\Theta db}}{\partial\mathbf{F}}-J\left(\mathbf{e}^*\cdot\mathbf{d}^*+\mathbf{h}^*\cdot\mathbf{b}^*-\frac{1}{2}\epsilon_o\mathbf{e}^*\cdot\mathbf{e}^*-\frac{1}{2}\mu_o\mathbf{h}^*\cdot\mathbf{h}^*\right)\mathbf{F}^{-T},$$

$$\eta=-\frac{\partial E^{F\Theta db}}{\partial\Theta},\quad \mathbf{e}^*=\rho\frac{\partial E^{F\Theta db}}{\partial\mathbf{d}^*},\quad \mathbf{h}^*=\rho\frac{\partial E^{F\Theta db}}{\partial\mathbf{b}^*} \tag{F.4}$$

and the residual dissipation inequality

$$\mathbf{j}^*\cdot\mathbf{e}^*-\frac{1}{\Theta}\mathbf{q}\cdot\operatorname{grad}\Theta\geq 0$$

follow.

We collectively coin the set of independent variables $\{\mathbf{F},\Theta,\mathbf{d}^*,\mathbf{b}^*\}$ the thermodynamic energy potential $E^{F\Theta db}$, and the constitutive equations (F.4) the *deformation-temperature-electric displacement-magnetic induction formulation.*[1] Legendre-type transformations and constitutive equations for other formulations that employ either $\mathbf{d}^*$ or $\mathbf{b}^*$ as an independent variable can be found in [62].

[1]This formulation makes contact with a formulation presented by Green and Naghdi [44]. This contact is significant as the two formulations were developed from different perspectives. In particular, the terms from the energy supply rate r^{em} that Green and Naghdi [44, p. 184] "transferred to be included in the internal energy," although not explicitly identified by them, are precisely the last two terms on the right-hand side of our Legendre-type transformation (F.1). Green and Naghdi's "augmented" internal energy (although not symbolically differentiated from their original internal energy) is thus equivalent to our transformed energy potential $E^{F\Theta db}$ defined in (F.1).

Bibliography

[1] R. Courant, D. Hilbert, Methods of Mathematical Physics, vol. 1, John Wiley & Sons, New York, 1953.

[2] M.E. Gurtin, An Introduction to Continuum Mechanics, Elsevier Academic Press, San Diego, 2003.

[3] M.E. Gurtin, W.O. Williams, On the Clausius-Duhem inequality, Zeit. Ang. Math. Phys. 17(5) (1966) 626-633.

[4] I.-S. Liu, I. Müller, On the thermodynamics and thermostatics of fluids in electromagnetic fields, Arch. Ration. Mech. Anal. 46(2) (1972) 149-176.

[5] K. Hutter, On thermodynamics and thermostatics of viscous thermoelastic solids in the electromagnetic fields. A Lagrangian formulation, Arch. Ration. Mech. Anal. 58(4) (1975) 339-368.

[6] K. Hutter, A thermodynamic theory of fluids and solids in electromagnetic fields, Arch. Ration. Mech. Anal. 64(3) (1977) 269-298.

[7] A.E. Green, P.M. Naghdi, On thermodynamics and the nature of the second law, Proc. R. Soc. Lond. A 357(1690) (1977) 253-270.

[8] K.R. Rajagopal, A.R. Srinivasa, Mechanics of the inelastic behavior of materials. Part II: Inelastic response, Int. J. Plasticity 14(10-11) (1998) 969-995.

[9] K.R. Rajagopal, A.R. Srinivasa, On thermomechanical restrictions of continua, Proc. R. Soc. Lond. A 460(2042) (2004) 631-651.

[10] P.J. Blatz, W.L. Ko, Application of finite elastic theory to the deformation of rubbery materials, Trans. Soc. Rheol. 6(1) (1962) 223-251.

[11] G.A. Holzapfel, Nonlinear Solid Mechanics: A Continuum Approach for Engineering, John Wiley & Sons, Chichester, 2000.

[12] M. Levinson, I.W. Burgess, A comparison of some simple constitutive relations for slightly compressible rubber-like materials, Int. J. Mech. Sci. 13(6) (1971) 563-572.

[13] R.W. Ogden, Large deformation isotropic elasticity: On the correlation of theory and experiment for compressible rubberlike solids, Proc. R. Soc. Lond. A 328(1575) (1972) 567-583.

[14] L. Anand, A constitutive model for compressible elastomeric solids, Comput. Mech. 18(5) (1996) 339-355.

[15] J.E. Bischoff, E.M. Arruda, K. Grosh, A new constitutive model for the compressibility of elastomers at finite deformations, Rub. Chem. Tech. 74(4) (2001) 541-559.

[16] L.R.G. Treloar, The Physics of Rubber Elasticity, Clarendon Press, Oxford, 1975.

[17] R.W. Ogden, Non-Linear Elastic Deformations, Ellis Horwood, Chichester, 1984.

[18] R.W. Ogden, Elastic deformations of rubberlike solids, in: H.G. Hopkins, M.J. Sewell (Eds.), Mechanics of Solids, The Rodney Hill 60th Anniversary Volume, Pergamon Press, Oxford, 1982, pp. 499-537.

[19] M.C. Boyce, E.M. Arruda, Constitutive models of rubber elasticity: a review, Rub. Chem. Tech. 73(3) (2000) 504-523.

[20] S.E. Bechtel, F.J. Rooney, Q. Wang, A thermodynamic definition of pressure for incompressible viscous fluids, Int. J. Eng. Sci. 42(19-20) (2004) 1987-1994.

[21] S.E. Bechtel, F.J. Rooney, M.G. Forest, Internal constraint theories for the thermal expansion of viscous fluids, Int. J. Eng. Sci. 42(1) (2004) 43-64.
[22] S.E. Bechtel, M.G. Forest, F.J. Rooney, Q. Wang, Thermal expansion models of viscous fluids based on limits of free energy, Phys. Fluids 15(9) (2003) 2681-2693.
[23] M. Mooney, A theory of large elastic deformation, J. Appl. Phys. 11(9) (1940) 582-592.
[24] R.S. Rivlin, Large elastic deformations of isotropic materials. IV. Further developments of the general theory, Philos. Trans. R. Soc. Lond. A 241(835) (1948) 379-397.
[25] R.S. Rivlin, Large elastic deformations of isotropic materials. I. Fundamental concepts, Philos. Trans. R. Soc. Lond. A 240(822) (1948) 459-490.
[26] R.W. Ogden, Large deformation isotropic elasticity—on the correlation of theory and experiment for incompressible rubberlike solids, Proc. R. Soc. Lond. A 326(1567) (1972) 565-584.
[27] E.M. Arruda, M.C. Boyce, A three-dimensional constitutive model for the large stretch behavior of rubber elastic materials, J. Mech. Phys. Solids 41(2) (1993) 389-412.
[28] L.R.G. Treloar, The elasticity of a network of long-chain molecules—II, Trans. Faraday Soc. 39 (1943) 241-246.
[29] L. Anand, Moderate deformations in extension-torsion of incompressible isotropic elastic materials, J. Mech. Phys. Solids 34(3) (1986) 293-304.
[30] A.E. Green, P.M. Naghdi, A unified procedure for construction of theories of deformable media. I. Classical continuum physics, Proc. R. Soc. Lond. A 448(1934) (1995) 335-356.
[31] A.E. Green, P.M. Naghdi, A re-examination of the basic postulates of thermomechanics, Proc. R. Soc. Lond. A 432(1885) (1991) 171-194.
[32] K. Hutter, A.A.F. van de Ven, A. Ursescu, Electromagnetic Field Matter Interactions in Thermoelastic Solids and Viscous Fluids, Springer-Verlag, Berlin, 2006.
[33] Y.-H. Pao, Electromagnetic forces in deformable continua, in: S. Nemat-Nasser (Ed.), Mechanics Today, vol. 4, Pergamon Press, New York, 1978, pp. 209-305.
[34] A. Kovetz, Electromagnetic Theory, Oxford University Press, Oxford, 2000.
[35] S.V. Kankanala, N. Triantafyllidis, On finitely strained magnetorheological elastomers, J. Mech. Phys. Solids 52(12) (2004) 2869-2908.
[36] A. Dorfmann, R.W. Ogden, Some problems in nonlinear magnetoelasticity, Zeit. Ang. Math. Phys. 56(4) (2005) 718-745.
[37] A. Dorfmann, R.W. Ogden, Nonlinear electroelastic deformations, J. Elast. 82(2) (2006) 99-127.
[38] A.C. Eringen, G.A. Maugin, Electrodynamics of Continua I: Foundations and Solid Media, Springer-Verlag, New York, 1990.
[39] H.A. Lorentz, The Theory of Electrons, second ed., Dover, New York, 2004.
[40] S.R. de Groot, L.G. Suttorp, Foundations of Electrodynamics, North-Holland Publishing Company, Amsterdam, 1972.
[41] R.M. Fano, L.J. Chu, R.B. Adler, Electromagnetic Fields, Energy, and Forces, The MIT Press, Cambridge, 1968.
[42] J.C. Maxwell, A Treatise on Electricity and Magnetism, Clarendon Press, Oxford, 1873.
[43] W.F. Brown, Magnetoelastic Interactions, Springer-Verlag, New York, 1966.
[44] A.E. Green, P.M. Naghdi, Aspects of the second law of thermodynamics in the presence of electromagnetic effects, Quart. J. Mech. Appl. Math. 37(2) (1984) 179-193.

[45] A. Dorfmann, R.W. Ogden, Magnetoelastic modelling of elastomers, Euro. J. Mech. A: Solids 22(4) (2003) 497-507.
[46] A. Dorfmann, R.W. Ogden, Nonlinear magnetoelastic deformations, Quart. J. Mech. Appl. Math. 57(4) (2004) 599-622.
[47] A. Dorfmann, R.W. Ogden, Nonlinear magnetoelastic deformations of elastomers, Acta Mech. 167(1-2) (2004) 13-28.
[48] A. Dorfmann, R.W. Ogden, Nonlinear electroelasticity, Acta Mech. 174(3-4) (2005) 167-183.
[49] D.J. Steigmann, Equilibrium theory for magnetic elastomers and magnetoelastic membranes, Int. J. Non-Linear Mech. 39(7) (2004) 1193-1216.
[50] R.M. McMeeking, C.M. Landis, Electrostatic forces and stored energy for deformable dielectric materials, ASME J. Appl. Mech. 72(4) (2005) 581-590.
[51] R.M. McMeeking, C.M. Landis, S.M.A. Jimenez, A principle of virtual work for combined electrostatic and mechanical loading of materials, Int. J. Non-Linear Mech. 42(6) (2007) 831-838.
[52] D.K. Vu, P. Steinmann, G. Possart, Numerical modelling of non-linear electroelasticity, Int. J. Numer. Meth. Eng. 70(6) (2007) 685-704.
[53] P.A. Voltairas, D.I. Fotiadis, C.V. Massalas, A theoretical study of the hyperelasticity of electro-gels, Proc. R. Soc. Lond. A 459(2037) (2003) 2121-2130.
[54] Z. Suo, X. Zhao, W.H. Greene, A nonlinear field theory of deformable dielectrics, J. Mech. Phys. Solids 56(2) (2008) 467-486.
[55] X. Zhao, Z. Suo, Electrostriction in elastic dielectrics undergoing large deformation, J. Appl. Phys. 104(12) (2008) 123530.
[56] J. Zhu, H. Stoyanov, G. Kofod, Z. Suo, Large deformation and electromechanical instability of a dielectric elastomer tube actuator, J. Appl. Phys. 108(7) (2010) 074113.
[57] Z. Qin, L. Librescu, D. Hasanyan, D.R. Ambur, Magnetoelastic modeling of circular cylindrical shells immersed in a magnetic field. Part I: Magnetoelastic loads considering finite dimensional effects, Int. J. Eng. Sci. 41(17) (2003) 2005-2022.
[58] Z. Qin, D. Hasanyan, L. Librescu, D.R. Ambur, Magnetoelastic modeling of circular cylindrical shells immersed in a magnetic field. Part II: Implications of finite dimensional effects on the free vibrations, Int. J. Eng. Sci. 41(17) (2003) 2023-2046.
[59] A.W. Richards, G.M. Odegard, Constitutive modeling of electrostrictive polymers using a hyperelasticity-based approach, ASME J. Appl. Mech. 77(1) (2010) 014502.
[60] W.S. Oates, H. Wang, R.L. Sierakowski, Unusual field-coupled nonlinear continuum mechanics of smart materials, J. Intel. Mater. Syst. Struct. 23(5) (2012) 487-504.
[61] H.B. Callen, Thermodynamics and an Introduction to Thermostatistics, second ed., Wiley, New York, 1985.
[62] S. Santapuri, R.L. Lowe, S.E. Bechtel, M.J. Dapino, Thermodynamic modeling of fully coupled finite-deformation thermo-electro-magneto-mechanical behavior for multifunctional applications, Int. J. Eng. Sci. 72 (2013) 117-139.
[63] R.E. Newnham, Properties of Materials: Anisotropy, Symmetry, Structure, Oxford University Press, New York, 2005.
[64] C. Truesdell, W. Noll, The Non-Linear Field Theories of Mechanics, second ed., Springer-Verlag, Berlin, 1992.
[65] F.J. Rooney, S.E. Bechtel, Constraints, constitutive limits, and instability in finite thermoelasticity, J. Elast. 74(2) (2004) 109-133.

[66] J.W. Gibbs, A method of geometrical representation of the thermodynamic properties of substances by means of surfaces, Trans. Conn. Acad. Arts Sci. 2 (1873) 382-404.

[67] J.W. Gibbs, On the equilibrium of heterogeneous substances, Am. J. Sci. 16(96) (1878) 441-458.

[68] S.E. Bechtel, F.J. Rooney, M.G. Forest, Connections between stability, convexity of internal energy, and the second law for compressible Newtonian fluids, ASME J. Appl. Mech. 72(2) (2005) 299-300.

[69] Y.-H. Pao, K. Hutter, Electrodynamics for moving elastic solids and viscous fluids, Proc. IEEE 63(7) (1975) 1011-1021.

[70] K.R. Rajagopal, M. Růžička, Mathematical modeling of electrorheological materials, Cont. Mech. Thermodyn. 13(1) (2001) 59-78.

Index

Note: Page numbers followed by *f* indicate figures and *t* indicate tables.

L

M

N

O

P

R

Printed in the United States
By Bookmasters